AF380033

Convexity and its Applications in Discrete and Continuous Optimization

Using a pedagogical, unified approach, this book presents both the analytic and combinatorial aspects of convexity and its applications in optimization. On the structural side, this is done via an exposition of classical convex analysis and geometry, along with polyhedral theory and geometry of numbers. On the algorithmic/optimization side, this is done by the first ever exposition of the theory of general mixed-integer convex optimization in a textbook setting. Classical continuous convex optimization and pure integer convex optimization are presented as special cases, without compromising on the depth of either of these areas. For this purpose, several new developments from the past decade are presented for the first time outside technical research articles: discrete Helly numbers, new insights into sublinear functions, and best known bounds on the information and algorithmic complexity of mixed-integer convex optimization. Pedagogical explanations and more than 300 exercises make this book ideal for students and researchers.

AMITABH BASU is Professor of Applied Mathematics and Statistics at Johns Hopkins University. He has received the NSF CAREER award and the Egon Balas Prize from the INFORMS Optimization Society. He serves on the editorial boards of *Mathematics of Operations Research, Mathematical Programming, SIAM Journal on Optimization*, and the *MOS-SIAM Series on Optimization*.

"Written by one of the most brilliant researchers in the field, this book provides an elegant, rigorous, and original presentation of the theory of convexity, describing in a unified way its use in continuous and discrete optimization, and also covering some very recent advancements in these areas."

– **Marco Di Summa**, *University of Padua*

"Convexity is central to most optimization algorithms. This book brings together classical and new developments at the interface between these two vibrant areas of mathematics. It is an essential reference for scholars in optimization. The numerous exercises make it an ideal textbook at the graduate and upper undergraduate levels."

– **Gérard Cornuéjols**, *Carnegie Mellon University*

Convexity and its Applications in Discrete and Continuous Optimization

AMITABH BASU

Johns Hopkins University

CAMBRIDGE
UNIVERSITY PRESS

Shaftesbury Road, Cambridge CB2 8EA, United Kingdom

One Liberty Plaza, 20th Floor, New York, NY 10006, USA

477 Williamstown Road, Port Melbourne, VIC 3207, Australia

314–321, 3rd Floor, Plot 3, Splendor Forum, Jasola District Centre, New Delhi – 110025, India

103 Penang Road, #05–06/07, Visioncrest Commercial, Singapore 238467

Cambridge University Press is part of Cambridge University Press & Assessment,
a department of the University of Cambridge.

We share the University's mission to contribute to society through the pursuit of
education, learning and research at the highest international levels of excellence.

www.cambridge.org
Information on this title: www.cambridge.org/9781108837590

DOI: 10.1017/9781108946650

First published 2025

Printed in the United Kingdom by CPI Group Ltd, Croydon CR0 4YY

A catalogue record for this publication is available from the British Library

A Cataloging-in-Publication data record for this book is available from the Library of Congress

ISBN 978-1-108-83759-0 Hardback

Dedicated to Ma and Baba

Contents

Preface

This book is about convex analysis and geometry, as well as their use in mathematical optimization. Mathematical optimization plays a central role in almost all mathematical and computational disciplines, and its importance in the age of data science, machine learning, and artificial intelligence cannot be overemphasized. Almost every problem in these areas has a mathematical optimization problem at its core. This book focuses primarily on convex optimization (both discrete and continuous) and its mathematical foundations. The reach of convex optimization methods and, more generally, the underlying concepts from convex analysis and geometry in business, engineering, and scientific applications is also significant. The simple idea behind convexity – that the "value" of an average of multiple potential solutions is at least as good as the average "value" of these solutions – has the ability to model an amazing variety of phenomena.

There are enough textbooks on convexity and convex optimization to occupy an entire section of an academic (virtual) library. There are two reasons why I believe this book provides something new and valuable, particularly from the perspective of designing a course that introduces these ideas rigorously. First, I feel that previous textbooks and monographs focus either on the continuous/analytic aspects of convexity [42, 44, 45, 57, 59, 114, 131, 140, 141, 170, 183, 199, 210, 220] or on the discrete/combinatorial aspects of convexity [24, 116, 124, 128, 234], and no one single book surveys in a unified way the main ideas in both of these aspects (theory and applications in optimization).[1] Second, there have been several recent developments, especially in the use of convexity in discrete optimization, that appear only in specialized research papers or surveys. These developments

---

[1] An exception to this is the classic text [218], but we would argue that this book is somewhat limited and outdated in its scope.

have now reached a level of maturity that can be presented in a textbook, with clean and elegant proofs available for the major results. I am convinced that the time is right for a textbook that gives a unified perspective on continuous and discrete optimization via the lens of convexity, and presents some of the recent advances in optimization that have not yet made it to the texts on convexity.

The book is divided into two parts. The first part focuses on the mathematical foundations of convex analysis and geometry, starting with a study of convex sets, followed by an exposition of convex functions, and ending with an introduction to the geometry of numbers. The second part focuses on optimization, presenting discrete and continuous ideas in a common framework. Here are some topics covered in this textbook that, to the best of my knowledge, are not contained in any previous textbook on convex analysis, geometry, or optimization.

1. A careful exposition of the conceptual underpinnings of algorithmic or computational optimization. This topic is approached by continuous and discrete optimizers in related but distinct ways. At the risk of making a sweeping generalization, one could say that computation in continuous optimization has its origins in the traditions of scientific computing and numerical analysis, whereas discrete optimization broadly views computation via the Turing machine model. The different views lead to some friction when trying to cross the boundaries. In the continuous world, one often designs algorithms assuming one can perform exact operations with real numbers (consider, for example, Newton's method), which is impossible in the Turing machine model. In the discrete world, the "input" to a Turing machine becomes a tricky question when dealing with general nonlinear functions and sets. The question of "complexity" of an optimization algorithm is also treated in somewhat different ways in the two communities. For example, it is not clear what the "size" of an optimization problem is when one has nonlinear objective functions and constraints, and "complexity" without a notion of "size" is hard to formulate precisely in the Turing machine model. We believe Section 1.4 and Chapter 5 show how all these issues can be handled in a unified, coherent way, making no distinction whatsoever between "continuous" and "discrete" optimization.

2. Mixed-integer convex optimization, i.e., minimizing a convex function subject to convex constraints where some of the decision variables have to take integer values, is a natural model for presenting continuous and discrete optimization under one umbrella. Continuous convex optimization is the special case where no variable is integer constrained. On the other hand, the use of integer variables to model combinatorial optimization

problems is well known. Chapter 6 of this book presents state-of-the-art results on information and algorithmic complexity of mixed-integer convex optimization.

Information complexity of classical continuous optimization has been well understood since the 1970s due to seminal work by Nemirovski and Yudin. The information complexity in the presence of integer variables was not well developed until research work done in the past decade, and I believe this fundamental topic should become part of the early education in optimization.

On the algorithmic side, decades of work in integer optimization has focused on understanding the best possible dependence of algorithmic complexity on the number n of integer variables. This book presents the best known upper bound of $2^{n \log(n)}$ on the complexity of *deterministic* algorithms for convex integer optimization, which does not appear outside specialized, technical research articles. Moreover, this book gives a general mixed-integer complexity bound allowing for both integer and continuous variables. Such a bound does not explicitly appear anywhere in the literature, to the best of our knowledge; previous work focused on the pure integer case with no continuous variables.

3. Helly's theorem is a classical result in combinatorial convexity that states that if a collection of convex sets in $\mathbb{R}^d$ has empty intersection then there is a subcollection of at most $d + 1$ sets whose intersection is already empty. This fundamental result has many applications, and it has been extended in several directions. One such generalization states that if a collection of convex sets in $\mathbb{R}^n \times \mathbb{R}^d$ has empty intersection with $\mathbb{Z}^n \times \mathbb{R}^d$, then a subcollection of at most $2^n(d + 1)$ sets already has empty intersection with $\mathbb{Z}^n \times \mathbb{R}^d$. This has major consequences in convex optimization in the presence of integer variables, especially in establishing tight bounds on information complexity as discussed in Chapter 6 of this book. The theory of Helly numbers encompassing the above generalization is presented in Section 2.6. This material appeared previously only in specialized surveys and research articles.

4. In classical convex analysis, it is well known that the gauge function of a convex set C containing the origin is the unique *nonnegative* sublinear function whose 1-sublevel set is C. However, when C is not compact, there are other sublinear functions that also represent C as their 1-sublevel set and it was known that the gauge is the largest of all such functions pointwise. In recent work [33, 73], it was shown that there is a unique function that is the pointwise *smallest* such sublinear function and it can be described quite explicitly as the support function of a certain subset of the polar.

In geometric terms, there exists a subset $C^\star \subseteq C^\circ$ of the polar of C such that for all sets D such that $C = D^\circ$, we must have $C^\star \subseteq D \subseteq C^\circ$. The smallest sublinear function representing C is precisely the support function of $C^\star$. A precise description of $C^\star$ is also available in terms of supporting hyperplanes of C. This result, presented in Section 3.3.4, has not appeared in any previous convex analysis textbook. Its use in discrete optimization is discussed in Section 3.6.

5. The algorithmic theory of mixed-integer convex optimization relies on tools from algorithmic geometry of numbers such as lattice basis reduction, Voronoi cell computations, and closest and shortest lattice vector problems (CVP and SVP). The usual presentation of these algorithms, especially lattice basis reduction algorithms, works under the assumption that the lattice is given by rational vectors. This book presents an analysis of these algorithms for nonrational input (working in the real arithmetic model of computation) and the complexity is given in terms of basic properties of the lattice, or equivalently, in terms of condition numbers and norms, in the spirit of numerical analysis (see Section 6.3). These results reduce to the standard results when restricted to rational input.

6. Recent work on the complexity of branch-and-cut methods for mixed-integer convex optimization is presented in Chapter 6. Some recent work on duality for mixed-integer convex optimization is also summarized in Chapter 7; however, the classical duality theory for continuous optimization receives more attention in this chapter.

Suggestions for classroom use. The book grew out of lecture notes for an upper-level undergraduate course on convexity and continuous convex optimization I designed at Johns Hopkins University. Several sections have also been used in a graduate-level discrete optimization course I teach at Hopkins. Based on my experience with these courses, I give some suggestions below for course material at various levels. I hope the textbook can serve as a reference for courses on continuous optimization as well as discrete optimization. In addition, it should be able to serve courses focusing on structural aspects of convex analysis, convex geometry, and discrete geometry. There are over 300 exercises in the textbook, ranging from routine calculations, technical proofs, and tests of conceptual understanding to covering important results that extend the main material. Hints are provided for exercises marked with an asterisk at the end of the book.

1. An upper-level undergraduate or beginning graduate course on convex analysis and continuous convex optimization can be based on Chapters

2, 3, 6, and 7. Chapter 6 can be restricted to simply Section 6.5, except
if a discussion of lower bounds on information complexity is intended
for inclusion. For this, the content of Section 6.2.1 should be presented
for the $n = 0$ setting. Depending on the emphasis of the course, certain
sections of Chapters 2 and 3 can be omitted. For example, in the course I
teach at Hopkins, I skip Section 2.7 on ellipsoids and use relevant results
from this section without proof in my discussion of cutting-plane methods
(ellipsoid or center-of-gravity methods) in Chapter 6. The discussion of
Helly numbers from Section 2.6, if included, should be restricted to the
continuous case. I also skip Section 3.5 on Brunn–Minkowski theory and
appeal to Grunbaum's result on centroids without proof in the discussion
of the center-of-gravity method in Chapter 6. Chapter 7 should also be
restricted to the classical continuous optimization duality theory.

2. A first- or second-year graduate course on convex analysis and geometry,
   as well as geometry of numbers, with no optimization, can be based on
   Chapters 2, 3, and 4. All the material in these chapters can be covered
   in their entirety in one semester. This can serve as a foundation for
   advanced graduate courses on continuous convex optimization or integer
   optimization, discrete geometry, Banach space geometry, or functional
   analytic aspects of convex geometry.

3. A Ph.D.-level course on algorithmic geometry of numbers and mixed-
   integer convex optimization can be based on Chapters 4, 5, and 6. Relevant
   parts of Chapters 2 and 3 should be covered as needed.

Prerequisites. Familiarity with basic linear algebra and real analysis concepts, as reviewed in Sections 1.1, 1.2, and 1.3, will be very helpful. Nevertheless, every year the author has had students in his undergraduate course
at Hopkins who did not have this background at the start of the course, but
it was possible for the motivated ones to pick up this background through
independent reading with Chapter 1 of this book as a guide, as well as
discussions with the instructor and fellow students. The only real prerequisites
are a familiarity with mathematical arguments and proofs.

Acknowledgments. Several colleagues looked at evolving lecture notes and
initial drafts of this book and gave excellent feedback. A special thanks goes
to Marco Di Summa, who meticulously went through an initial draft and gave
immensely useful mathematical suggestions, including corrections in proofs
and exposition, as well as organizational suggestions. I also wish to thank
Michele Conforti, Giacomo Zambelli, Santanu Dey, and Marco Molinaro
for their feedback and encouragement. I am very grateful to Katie Leach at

Cambridge University Press for suggesting the possibility of converting my lecture notes into a textbook, for her immense patience as the vision of the book evolved, and for providing constant encouragement through all the time I worked on the book. I am also very grateful for support from the National Science Foundation (NSF) and the Air Force Office of Scientific Research (AFOSR) provided during the period that I worked on this book. Finally, I would like to express my deepest gratitude to my wife, Deepthi, and my daughter, Maitreyi. This book ultimately owes its existence to time stolen from them.

1

Preliminaries

This chapter is dedicated to establishing the notation and terminology that we will use throughout this book. More specific concepts and definitions will be introduced as and when needed. None of the material in this chapter is intended to be of an expository nature. Rather, it collects together basic notions and facts from linear algebra, real analysis, and fundamentals of computation that are required for our study in later chapters. No proofs of these facts are provided here, and instead we refer the reader to the references listed at the end of this chapter. The handful of exercises in this chapter are included primarily because they will be relevant later. They are in no way meant to be a comprehensive test of familiarity with these subjects.

We will use $\mathbb{N}, \mathbb{Z}, \mathbb{Q}, \mathbb{R}$ to denote the set of natural numbers (starting at 1), the set of integers, the set of rational numbers, and the set of real numbers, respectively. $\mathbb{Z}_+, \mathbb{Q}_+, \mathbb{R}_+$ will denote the nonnegative integers, rationals, and reals, respectively. For any real number $r \in \mathbb{R}$, $|r|$ will denote the absolute value of r. We denote the greatest integer smaller than or equal to r by $\lfloor r \rfloor$, the smallest integer greater than or equal to r by $\lceil r \rceil$, $\{r\}$ will be used to denote $r - \lfloor r \rfloor$, and $\lfloor r \rceil$ will denote the integer closest to r in absolute value. $\ln(r)$ will denote the natural logarithm of r, and $\log(r)$ will denote the logarithm of r in base 2. Logarithms in any other base b will be explicitly denoted by $\log_b(r)$.

δ_{ij} will denote the standard *Kronecker delta function*, i.e., $\delta_{ij} = 1$ if $i = j$ and 0 otherwise (the range of the indices i, j will be clear from context). For any set X, $\#(X)$ will denote the cardinality, i.e., number of elements in X (possibly $+\infty$).[1] 2^X will denote the power set of X. We will also use the following standard asymptotic notations of $O(\cdot)$ and $\Omega(\cdot)$: Given two functions $F(p_1, \ldots, p_k), G(p_1, \ldots, p_k)$ of nonnegative, real-valued

---

[1] We will not worry about different cardinal numbers beyond countable and uncountable sets. When referring to cardinality, $+\infty$ will be used for both countable and uncountable sets.

parameters $p_1, \ldots, p_k$, we say $F = O(G)$ (informally, "F is asymptotically upper bounded by G") and $G = \Omega(F)$ (informally, "G is asymptotically lower bounded by F") if there exist constants $C > 0, M > 0$ such that $|F(p_1, \ldots, p_k)| \leq C|G(p_1, \ldots, p_k)|$ for all $p_1, \ldots, p_k \in \mathbb{R}_+$ with $p_i > M$ for all $i \in \{1, \ldots, k\}$.[2] We mention here that while this version of asymptotic notation for functions with multiple arguments will suffice for our purposes, there are other ways to tackle asymptotics with multiple parameters; see [79] and [144] for a discussion of the subtleties that can arise.

1.1 Euclidean Spaces

All of the action in this book will be in finite-dimensional Euclidean spaces. For any $d \in \mathbb{N}$, we use $\mathbb{R}^d$ to denote the d-dimensional Euclidean space, i.e., $\mathbb{R}^d$ is the set of all d-tuples $\mathbf{x} = (x_1, \ldots, x_d)$ of real numbers; these are also called *vectors* or *points* in $\mathbb{R}^d$. Vectors will be boldface and scalars will be nonbold italics. Subscripts will be used to denote the coordinates, i.e., $\mathbf{x}_i := x_i$. We will use the notation $\mathbb{R}^d_+$ to denote the set of all vectors with nonnegative coordinates. The symbol $\mathbf{e}^i$, $i = 1, \ldots, d$, will denote the ith unit vector, i.e., the vector which has 1 in the ith coordinate and 0 in every other coordinate; to emphasize the dimension we will sometimes use $\mathbf{e}^i_d$. $\mathbf{0}$ and $\mathbf{1}$ will denote the vector of all zeros and all ones, respectively; to emphasize the dimension we will sometimes use $\mathbf{0}_d$ and $\mathbf{1}_d$. $\mathbf{0}$ is also called the *origin*. We will use the coordinate-wise *addition* operation

$$\mathbf{x} + \mathbf{y} := (\mathbf{x}_1 + \mathbf{y}_1, \ldots, \mathbf{x}_d + \mathbf{y}_d) \quad \text{for any } \mathbf{x}, \mathbf{y} \in \mathbb{R}^d,$$

and the operation of *scalar multiplication*

$$\alpha\mathbf{x} := (\alpha\mathbf{x}_1, \ldots, \alpha\mathbf{x}_d) \quad \text{for any } \alpha \in \mathbb{R} \text{ and } \mathbf{x} \in \mathbb{R}^d.$$

These operations impose a vector space structure on $\mathbb{R}^d$. Correspondingly, a subset $L \subseteq \mathbb{R}^d$ is called a *linear subspace* if L is closed under the above operations of addition and scalar multiplication. It is a convention to not consider the empty set as a subspace. Thus, the smallest linear subspace is $\{\mathbf{0}\}$.

Definition 1.1.1 A *norm on* $\mathbb{R}^d$ is a function $N\colon \mathbb{R}^d \to \mathbb{R}_+$ satisfying the following properties:

---

[2] Sometimes, it is more convenient to present asymptotic arguments when some parameters approach 0 as a limit, instead of $+\infty$ as in the definition above. This should be clear from context.

1. $N(\mathbf{x}) = 0$ if and only if $\mathbf{x} = \mathbf{0}$,
2. $N(\alpha\mathbf{x}) = |\alpha| N(\mathbf{x})$ for all $\alpha \in \mathbb{R}$ and $\mathbf{x} \in \mathbb{R}^d$,
3. $N(\mathbf{x} + \mathbf{y}) \le N(\mathbf{x}) + N(\mathbf{y})$ for all $\mathbf{x}, \mathbf{y} \in \mathbb{R}^d$. (Triangle inequality)

Example 1.1.2 For any real number $p \ge 1$, define the ℓ^p norm on $\mathbb{R}^d$ as $\|\mathbf{x}\|_p = (|\mathbf{x}_1|^p + |\mathbf{x}_2|^p + \cdots + |\mathbf{x}_d|^p)^{\frac{1}{p}}$. $p = 2$ is also called the *standard Euclidean norm*; we will drop the subscript 2 to denote the standard norm: $\|\mathbf{x}\| = \sqrt{\mathbf{x}_1^2 + \mathbf{x}_2^2 + \cdots + \mathbf{x}_d^2}$. The ℓ^∞ norm is defined as $\|\mathbf{x}\|_\infty = \max_{i=1}^d |\mathbf{x}_i|$.

The following is an important result that shows that in many situations, the specific choice of a norm does not matter.

Theorem 1.1.3 (Equivalence of norms) *Let N, N' be any two norms on $\mathbb{R}^d$. There exists a constant $C > 0$ (depending on N and N') such that $N'(\mathbf{x}) \le C \cdot N(\mathbf{x})$ for all $\mathbf{x} \in \mathbb{R}^d$.*

Definition 1.1.4 Any norm on $\mathbb{R}^d$ defines a distance between points in $\mathbf{x}, \mathbf{y} \in \mathbb{R}^d$ as $d_N(\mathbf{x}, \mathbf{y}) := N(\mathbf{x} - \mathbf{y})$. This is called the *metric* or *distance induced by the norm*. Such a metric satisfies three important properties:

1. $d_N(\mathbf{x}, \mathbf{y}) = 0$ if and only if $\mathbf{x} = \mathbf{y}$,
2. $d_N(\mathbf{x}, \mathbf{y}) = d_N(\mathbf{y}, \mathbf{x})$ for all $\mathbf{x} \in \mathbb{R}^d$,
3. $d_N(\mathbf{x}, \mathbf{z}) \le d_N(\mathbf{x}, \mathbf{y}) + d_N(\mathbf{y}, \mathbf{z})$ for all $\mathbf{x}, \mathbf{y}, \mathbf{z} \in \mathbb{R}^d$. (Triangle inequality)

We will often identify the set $\mathbb{R}^n \times \mathbb{R}^d$ with the Euclidean space $\mathbb{R}^{n+d}$ by "concatenation," i.e., given $\mathbf{x} \in \mathbb{R}^n$ and $\mathbf{y} \in \mathbb{R}^d$, we will associate the vector $(\mathbf{x}_1, \ldots, \mathbf{x}_n, \mathbf{y}_1, \ldots, \mathbf{y}_d) \in \mathbb{R}^{n+d}$ with the tuple $(\mathbf{x}, \mathbf{y}) \in \mathbb{R}^n \times \mathbb{R}^d$. Given two norms N on $\mathbb{R}^n$ and N' on $\mathbb{R}^d$, $(\mathbf{x}, \mathbf{y}) \mapsto N(\mathbf{x}) + N'(\mathbf{y})$ can be verified to be a new norm on $\mathbb{R}^d \times \mathbb{R}^d \equiv \mathbb{R}^{n+d}$.

Definition 1.1.5 We also utilize the (standard) inner product of $\mathbf{x}, \mathbf{y} \in \mathbb{R}^d$: $\langle \mathbf{x}, \mathbf{y} \rangle = \mathbf{x}_1 \mathbf{y}_1 + \mathbf{x}_2 \mathbf{y}_2 + \cdots + \mathbf{x}_d \mathbf{y}_d$. (Note that $\|\mathbf{x}\|_2^2 = \langle \mathbf{x}, \mathbf{x} \rangle$). We say $\mathbf{x}$ and $\mathbf{y}$ are *orthogonal* if $\langle \mathbf{x}, \mathbf{y} \rangle = 0$. A set of vectors $\mathbf{x}^1, \ldots, \mathbf{x}^k$ is said to be *orthonormal* if $\langle \mathbf{x}^i, \mathbf{x}^j \rangle = \delta_{ij}$ for all $i, j \in \{1, \ldots, k\}$.

Theorem 1.1.6 (Cauchy–Schwarz inequality) *For any $\mathbf{x}, \mathbf{y} \in \mathbb{R}^d$, $|\langle \mathbf{x}, \mathbf{y} \rangle| \le \|\mathbf{x}\| \cdot \|\mathbf{y}\|$.*

Definition 1.1.7 For any norm N and $\mathbf{x} \in \mathbb{R}^d$, $r \in \mathbb{R}_+$, we will call the set $B_N(\mathbf{x}, r) := \{\mathbf{y} \in \mathbb{R}^d : N(\mathbf{y} - \mathbf{x}) \le r\}$ as the *ball around* $\mathbf{x}$ *of radius* r. $B_N(\mathbf{0}, 1)$ will be called the *unit ball for the norm* N. We will drop the subscript N when we speak of the standard Euclidean norm and there is no chance of confusion in the context.

A subset $X \subseteq \mathbb{R}^d$ is said to be *bounded* if there exists $R \geq 0$ such that $X \subseteq B(\mathbf{0}, R)$.

Definition 1.1.8 (Set operations)   Given any set $X \subseteq \mathbb{R}^d$ and a scalar $\alpha \in \mathbb{R}$,

$$\alpha X := \{\alpha \mathbf{x} : \mathbf{x} \in X\}.$$

Given any two sets $X, Y \subseteq \mathbb{R}^d$, we define the *Minkowski sum* of X, Y as

$$X + Y := \{\mathbf{x} + \mathbf{y} : \mathbf{x} \in X, \mathbf{y} \in Y\}.$$

We will write $X + \mathbf{t}$ if $Y = \{\mathbf{t}\}$ is a singleton; this is called the *translate of X by* $\mathbf{t}$.

The *characteristic function*[3] *of X* is defined as $\mathbb{1}_X(\mathbf{x}) := \begin{cases} 1 & \text{if } \mathbf{x} \in X, \\ 0 & \text{otherwise.} \end{cases}$

1.2 Linear Algebra

An important tool in linear algebra is the notion of linear independence.

Definition 1.2.1   Let $\mathbf{x}^1, \ldots, \mathbf{x}^k \in \mathbb{R}^d$. A *linear combination of* $\mathbf{x}^1, \ldots, \mathbf{x}^k$ is any point of the form $\lambda_1 \mathbf{x}_1 + \cdots + \lambda_k \mathbf{x}_k$ where $\lambda_1, \ldots, \lambda_k \in \mathbb{R}$. The set of all linear combinations of $\mathbf{x}^1, \ldots, \mathbf{x}^k$ will be denoted by $\text{span}(\mathbf{x}^1, \ldots, \mathbf{x}^k)$.

A collection of points $\mathbf{x}^1, \ldots, \mathbf{x}^n \in \mathbb{R}^d$ is said to be *linearly independent* if there is no $i \in \{1, \ldots, n\}$ such that $\mathbf{x}^i$ can be expressed as a linear combination of the remaining $\mathbf{x}^j$, $j \neq i$. If there exists such an index i, then $\mathbf{x}^1, \ldots, \mathbf{x}^n$ are said to be *linearly dependent*.

Similarly, a set $X \subseteq \mathbb{R}^d$ (possibly infinite) is said to be linearly independent if no $\mathbf{x} \in X$ can be expressed as a linear combination of (finitely) many points in $X \setminus \{\mathbf{x}\}$; otherwise, X is said to be linearly dependent.

According to the above definition, any set containing the origin and another nonzero vector is linearly dependent. We adopt the standard convention that even the singleton set containing the origin is a linearly dependent set. Then we have the following equivalent formulation of linear independence, which is very useful.

Proposition 1.2.2 *Points* $\mathbf{x}^1, \ldots, \mathbf{x}^k \in \mathbb{R}^d$ *are linearly independent if and only if* $\lambda_1 \mathbf{x}_1 + \cdots + \lambda_k \mathbf{x}_k = \mathbf{0}$ *implies* $\lambda_1 = \lambda_2 = \cdots = \lambda_k = 0$.

A fundamental object of study in linear algebra is the notion of a linear transformation.

---

[3] The alternative terminology "indicator function" will be used for a different but related concept in this book (Example 3.1.11), in line with the practice in convex analysis.

Definition 1.2.3 A *linear map or transformation or function* is a function $T \colon \mathbb{R}^d \to \mathbb{R}^m$ between two Euclidean spaces such that $T(\lambda \mathbf{x} + \gamma \mathbf{y}) = \lambda T(\mathbf{x}) + \gamma T(\mathbf{y})$ for all $\mathbf{x}, \mathbf{y} \in \mathbb{R}^d$ and $\lambda, \gamma \in \mathbb{R}$.

The above definition is "coordinate free," and so it can be extended to maps $T \colon L \to L'$ where $L \subseteq \mathbb{R}^d$ and $L' \subseteq \mathbb{R}^m$ are arbitrary subspaces. This will be useful at a few places in the book (e.g., projections onto linear subspaces).

Any linear transformation between $\mathbb{R}^d$ and $\mathbb{R}^m$ can be represented by an $m \times d$ matrix. In other words, for any linear transformation $T \colon \mathbb{R}^d \to \mathbb{R}^m$, there exists an $m \times d$ matrix A such that $T(\mathbf{x}) = A\mathbf{x}$ for all $\mathbf{x} \in \mathbb{R}^d$, where we view $\mathbf{x}$ as a matrix with a single column and $A\mathbf{x}$ denotes standard matrix multiplication. In particular, $A_{ij} := T(\mathbf{e}^j)_i$. Conversely, any $m \times d$ matrix A gives a linear transformation $\mathbf{x} \mapsto A\mathbf{x}$. This also leads to the observation that the matrix corresponding to a composition of linear transformations is the product of the matrices corresponding to the individual transformations. We will also consider "translated" versions of linear transformations.

Definition 1.2.4 An *affine map* or *transformation* or *function* is a function $T \colon \mathbb{R}^d \to \mathbb{R}^m$ of the form $T(\mathbf{x}) = A\mathbf{x} + \mathbf{b}$, where A is an $m \times d$ matrix and $\mathbf{b} \in \mathbb{R}^m$.

Given a set $X \subseteq \mathbb{R}^d$, we will use the notation $T(X)$ to denote the image of this set under the affine map T; similarly $A(X) := \{A\mathbf{x} : \mathbf{x} \in X\}$. The set of all $m \times d$ matrices will be denoted by $\mathbb{R}^{m \times d}$. We use the notation $\mathbf{0}_{m \times d}$ to denote the $m \times d$ matrix with all zero entries. The *rank of a matrix A* will be denoted by $\mathrm{rk}(A)$ – it is the maximum number of linearly independent rows of A, which is equal to the maximum number of linearly independent columns of A. A matrix is said to have *full row rank* (respectively, *full column rank*) if its rank equals the number of rows (respectively, the number of columns). Any affine transformation $A\mathbf{x} + \mathbf{b}$ is also said to have rank $\mathrm{rk}(A)$. The *transpose A^T* of $A \in \mathbb{R}^{m \times d}$ is defined by $A^T_{ij} = A_{ji}$ for all $i \in \{1, \ldots, d\}$ and $j \in \{1, \ldots, m\}$. The following characterization of the transpose will be important for us.

Proposition 1.2.5 *Let $A \in \mathbb{R}^{m \times d}$ and $B \in \mathbb{R}^{d \times m}$. Then $\langle \mathbf{y}, A\mathbf{x} \rangle = \langle B\mathbf{y}, \mathbf{x} \rangle$ for all $\mathbf{x} \in \mathbb{R}^d$ and $\mathbf{y} \in \mathbb{R}^m$ if and only if $B = A^T$.*

The above also shows that $(AB)^T = B^T A^T$ for any two matrices with the appropriate dimensions.

When $m = d$, we say that the matrix is *square*. For a square matrix, A_{ii} are called the *diagonal entries*. A *diagonal matrix* is a square matrix A such that $A_{ij} = 0$ if $i \neq j$; We use I_d to denote the $d \times d$ *identity matrix*, i.e., the matrix where the diagonal elements are 1 and all other entries are 0. The function

$$\det(A) := \sum_{\substack{\text{Permutations} \\ \sigma \colon \{1,\ldots,d\} \to \{1,\ldots,d\}}} \operatorname{sgn}(\sigma) \prod_{i=1}^{d} A_{i\sigma(i)}$$

defined on the space of all $d \times d$ square matrices is called the *determinant of* A, where $\operatorname{sgn}(\sigma)$ is the signature of the permutation σ.[4] The following formula for computing determinants of products of matrices is very useful.

Theorem 1.2.6 (Cauchy–Binet formula) *Let $m, d \geq 1$ and let $A \in \mathbb{R}^{m \times d}$ and $B \in \mathbb{R}^{d \times m}$. Then*

$$\det(AB) = \sum_{\substack{S \subseteq \{1,\ldots,d\} \\ \#(S) = m}} \det(A_S)\det(B_S),$$

where A_S is the $m \times m$ submatrix of A with columns indexed by S and B_S is the $m \times m$ submatrix of B with rows indexed by S.

Theorem 1.2.7 *Let $T \colon \mathbb{R}^d \to \mathbb{R}^d$ be a linear transformation, and let $A \in \mathbb{R}^{d \times d}$ be the corresponding matrix. The following are equivalent.*

1. *T is injective, i.e., $T(\mathbf{x}) = T(\mathbf{y})$ implies $\mathbf{x} = \mathbf{y}$.*
2. *T is onto, i.e., $T(\mathbb{R}^d) = \mathbb{R}^d$.*
3. *T is bijective and T^{-1} is a linear map as well.*
4. *There exists a matrix $B \in \mathbb{R}^{d \times d}$ such that $AB = BA = I_d$. B is denoted by A^{-1} and is called* the inverse matrix of A.
5. *$\det(A) \neq 0$.*

A matrix satisfying part 4. of Theorem 1.2.7 is called an *invertible matrix*. Theorem 1.2.7 also shows that A^{-1} is the matrix corresponding to the linear map T^{-1}, and $(AB)^{-1} = B^{-1}A^{-1}$. The following formula relating the inverse and transpose is useful.

Proposition 1.2.8 *Let $A \in \mathbb{R}^{d \times d}$. Then A is invertible if and only if A^T is invertible. Moreover, $(A^T)^{-1} = (A^{-1})^T$. We will use the shorthand A^{-T} to denote this matrix.*

The following theorem singles out a special class of matrices that are important in various contexts.

Theorem 1.2.9 *Let $A \in \mathbb{R}^{d \times d}$ be a square matrix. The following are all equivalent.*

---

[4] The signature of a permutation is $+1$ if it has an even number of inversions, and -1 if it has an odd number of inversions.

1. *The columns of A are orthonormal.*
2. *The rows of A are orthonormal.*
3. $A^{-1} = A^T$.
4. $\langle A\mathbf{x}, A\mathbf{y} \rangle = \langle \mathbf{x}, \mathbf{y} \rangle$ *for all* $\mathbf{x}, \mathbf{y} \in \mathbb{R}^d$.
5. $\|A\mathbf{x}\| = \|\mathbf{x}\|$ *for all* $\mathbf{x} \in \mathbb{R}^d$.

Matrices satisfying the conditions in Theorem 1.2.9 are called *orthonormal matrices*.

We now come to a fundamental theorem of linear algebra.

Theorem 1.2.10 *Let $X \subseteq \mathbb{R}^d$. The following are equivalent.*

1. *X is a linear subspace.*
2. *There exists $0 \leq m \leq d$ and linearly independent vectors $\mathbf{v}^1, \ldots, \mathbf{v}^m \in X$ such that, $X = \mathrm{span}(\{\mathbf{v}^1, \ldots, \mathbf{v}^m\})$.*
3. *There exists $0 \leq m \leq d$ and orthonormal vectors $\mathbf{v}^1, \ldots, \mathbf{v}^m \in X$ such that, $X = \mathrm{span}(\{\mathbf{v}^1, \ldots, \mathbf{v}^m\})$.*
4. *There exists $0 \leq m \leq d$ and a matrix $A \in \mathbb{R}^{(d-m) \times d}$ with full row rank such that $X = \{\mathbf{x} \in \mathbb{R}^d : A\mathbf{x} = \mathbf{0}\}$.*

Definition 1.2.11 The number m showing up in items 2., 3., and 4. in Theorem 1.2.10 is called the *dimension* of X. The set of vectors $\{\mathbf{v}^1, \ldots, \mathbf{v}^m\}$ are called a *basis* for the linear subspace.

1.2.1 Singular Value and Eigen Decompositions

Theorem 1.2.12 (Singular Value Decomposition (SVD)) *Let $A \in \mathbb{R}^{m \times d}$ with* $\mathrm{rk}(A) = r$. *There exist scalars $\sigma_1, \ldots, \sigma_r > 0$, and orthonormal sets of vectors $\mathbf{v}^1, \ldots, \mathbf{v}^r \in \mathbb{R}^d$ and $\mathbf{u}^1, \ldots, \mathbf{u}^r \in \mathbb{R}^m$ such that*

$$A\mathbf{x} = \sigma_1 \langle \mathbf{v}^1, \mathbf{x} \rangle \mathbf{u}^1 + \cdots \sigma_r \langle \mathbf{v}^r, \mathbf{x} \rangle \mathbf{u}^r.$$

In other words, $A = U\Sigma V^T$, where Σ is an $r \times r$ diagonal matrix with $\sigma_1, \ldots, \sigma_r$ on the main diagonal, and V and U are the matrices with $\mathbf{v}^1, \ldots, \mathbf{v}^r$ and $\mathbf{u}^1, \ldots, \mathbf{u}^r$ as columns, respectively. Such a factorization of the matrix is known as a singular value decomposition (SVD). $\sigma_1, \ldots, \sigma_r$ are called the singular values, $\mathbf{v}^1, \ldots, \mathbf{v}^r$ are called right singular vectors, and $\mathbf{u}^1, \ldots, \mathbf{u}^r$ are called left singular vectors.

Up to permutations, $\sigma_1, \ldots, \sigma_r$ are unique, i.e., any SVD of A has the same Σ up to permutations of the diagonal entries. However, there may exist two distinct SVDs $U\Sigma V^T$ and $U'\Sigma V'^T$ such that U and U' do not have the same set of columns and/or V and V' do not have the same set of columns.

Definition 1.2.13   A square matrix $A \in \mathbb{R}^{d \times d}$ is called *symmetric* if $A_{ij} = A_{ji}$ for all $i, j \in \{1, \ldots, d\}$, i.e., $A = A^T$.

Definition 1.2.14   Let $A \in \mathbb{R}^{d \times d}$. A vector $\mathbf{v} \in \mathbb{R}^d$ is called an *eigenvector* of A if there exists $\lambda \in \mathbb{R}$ such that $A\mathbf{v} = \lambda\mathbf{v}$. λ is called *the eigenvalue* of A associated with $\mathbf{v}$. $\lambda_{\max}(A)$ and $\lambda_{\min}(A)$ will denote the maximum and minimum eigenvalues of A, respectively.

Theorem 1.2.15 *If $A \in \mathbb{R}^{d \times d}$ is symmetric then it has d orthogonal eigenvectors $\mathbf{v}^1, \ldots, \mathbf{v}^d$ all of unit Euclidean norm, with associated eigenvalues $\lambda_1, \ldots, \lambda_d \in \mathbb{R}$. Moreover, if S is the matrix whose columns are $\mathbf{v}^1, \ldots, \mathbf{v}^d$ and Λ is the diagonal matrix with $\lambda_1, \ldots, \lambda_d$ as the diagonal entries, then $A = S\Lambda S^T$.*

Moreover, $\mathrm{rk}(A)$ equals the number of nonzero eigenvalues.

Theorem 1.2.16 *Let $A \in \mathbb{R}^{d \times d}$ be a symmetric matrix of rank r. The following are equivalent.*

1. *All eigenvalues of A are nonnegative.*
2. *There exists a matrix $B \in \mathbb{R}^{r \times d}$ with linearly independent rows such that $A = B^T B$.*
3. $\mathbf{u}^T A\mathbf{u} \geq 0$ *for all $\mathbf{u} \in \mathbb{R}^d$.*

Definition 1.2.17   A symmetric matrix $A \in \mathbb{R}^{d \times d}$ satisfying any of the three conditions in Theorem 1.2.16 is called a *positive semidefinite* (PSD) matrix. If $\mathrm{rk}(A) = d$, i.e., all its eigenvalues are strictly positive, then A is called *positive definite*. Any matrix B satisfying condition 2. in Theorem 1.2.16 is called a *square root* of A.

The following relationship between the SVD of a matrix A and the eigenvectors and eigenvalues of AA^T and $A^T A$ is useful.

Proposition 1.2.18 *Let $A \in \mathbb{R}^{m \times d}$. The following are true.*

1. *σ is a singular value of A if and only if it is the square root of a positive eigenvalue of AA^T and $A^T A$. In particular, for positive semidefinite matrices, singular values and eigenvalues coincide.*
2. *$\mathbf{u}$ is a left singular vector of A if and only if it is an eigenvector of AA^T.*
3. *$\mathbf{v}$ is a right singular vector of A if and only if it is an eigenvector of $A^T A$.*

1.2.2 Matrix Norms

It is sometimes useful to consider the space $\mathbb{R}^{m \times d}$ matrices as a Euclidean space in its own right. This can be done explicitly by "vectorizing" the matrix, i.e., thinking of $A \in \mathbb{R}^{m \times d}$ as a vector in $\mathbb{R}^{md}$. There are several ways one can do this, depending on how the matrix entries are ordered as coordinates of the corresponding vector. This choice of ordering will not matter at all for what follows. What is important is that the operations of matrix addition and multiplication by a scalar coincide with the standard vector space structure on the corresponding Euclidean space. The notion of a norm and the distance induced by this norm are then well-defined concepts on $\mathbb{R}^{m \times d}$ (or $\mathbb{R}^{md}$). The following particular class of norms will be useful.

Definition 1.2.19 Let N be a norm on $\mathbb{R}^d$ and N' be a norm on $\mathbb{R}^m$. The *induced norm* on $\mathbb{R}^{m \times d}$ is defined as

$$\|A\|_{N,N'} := \sup_{\mathbf{x} \in \mathbb{R}^d \setminus \{\mathbf{0}\}} \frac{N'(A\mathbf{x})}{N(\mathbf{x})}.$$

When $N = \ell_p$ and $N' = \ell_q$ for some $p, q \geq 1$, this will be abbreviated to $\|A\|_{p,q}$, and when $p = q$, to simply $\|A\|_p$. The special case of $p = q = 2$ has the name *spectral norm* of A.

The following is a straightforward consequence of the definitions.

Theorem 1.2.20 *The spectral norm of A is given by the largest singular value of A.*

1.2.3 Exercises

1. Show that any positive definite matrix $A \in \mathbb{R}^{d \times d}$ defines a norm on $\mathbb{R}^d$ via $N_A(\mathbf{x}) = \sqrt{\mathbf{x}^T A \mathbf{x}}$. This norm is called the *norm induced by A*. (When A is the identity matrix I_d, this gives the standard Euclidean norm.)
2. Let $A \in \mathbb{R}^{d \times d}$ be a positive definite matrix. Show the generalized Cauchy–Schwarz inequalities: $|\mathbf{y}^T A \mathbf{x}| \leq N_A(\mathbf{x}) \cdot N_A(\mathbf{y})$ and $|\mathbf{y}^T \mathbf{x}| \leq N_A(\mathbf{x}) \cdot N_{A^{-1}}(\mathbf{y})$.
3. Show that a function $T \colon \mathbb{R}^d \to \mathbb{R}^m$ is an affine transformation if and only if $T(\lambda \mathbf{x} + \gamma \mathbf{y}) = \lambda T(\mathbf{x}) + \gamma T(\mathbf{y})$ for all $\mathbf{x}, \mathbf{y} \in \mathbb{R}^d$ and all $\lambda, \gamma \in \mathbb{R}$ such that $\lambda + \gamma = 1$.
4. Let $A \in \mathbb{R}^{k \times k}$. Show that for any $\mathbf{x} \in \mathbb{R}^k$,

$$\sigma_{\min}(A)\|\mathbf{x}\|_2 \leq \|A\mathbf{x}\|_2 \leq \sigma_{\max}(A)\|\mathbf{x}\|_2,$$

where $\sigma_{\min}(A)$ and $\sigma_{\max}(A)$ are the smallest and largest singular values of the matrix A, respectively, with the convention that if A is not invertible then we take the smallest singular value to be 0.

1.3 Real Analysis

For any subset of real numbers $S \subseteq \mathbb{R}$, we denote the *infimum* by $\inf S$ and the *supremum* by $\sup S$. The following are useful properties of infimums and supremums (we use the set operations from Definition 1.1.8 applied to $\mathbb{R}^d$ with $d = 1$).

Theorem 1.3.1 *Let $A, B \subseteq \mathbb{R}$ and $t \geq 0$. The following are all true.*

1. $\inf(tA) = t\inf(A)$.
2. $\inf(A + B) = \inf(A) + \inf(B)$.
3. *If $A \subseteq B$ then* $\inf(A) \geq \inf(B)$.
4. $\inf(-A) = -\sup(A)$, *where* $-A = \{-x : x \in A\}$.

Definition 1.3.2 Fix a norm N on $\mathbb{R}^d$. A set $X \subseteq \mathbb{R}^d$ is called *open (with respect to N)* if for every $\mathbf{x} \in X$, there exists $r \in \mathbb{R}_+$ such that $B_N(\mathbf{x}, r) \subseteq X$. A set X is *closed (with respect to N)* if its complement $\mathbb{R}^d \setminus X$ is open.

By Theorem 1.1.3, a set $X \subseteq \mathbb{R}^d$ is open with respect to a norm N if and only if X is open with respect to every norm on $\mathbb{R}^d$. Thus, we can drop the qualification "with respect to a norm" for referring to open and closed sets.

Theorem 1.3.3 *The following all hold:*

1. $\emptyset, \mathbb{R}^d$ *are both open and closed.*
2. *An arbitrary union of open sets is open. An arbitrary intersection of closed sets is closed.*
3. *A finite intersection of open sets is open. A finite union of closed sets is closed.*

Definition 1.3.4 A *sequence* in $\mathbb{R}^d$ is a countable ordered set of points, namely $\mathbf{x}^1, \mathbf{x}^2, \mathbf{x}^3, \ldots$, and will often be denoted by $\{\mathbf{x}^i\}_{i \in \mathbb{N}}$. We say that *the sequence converges* or that *the limit of the sequence exists* if there exists a point $\mathbf{x}$ such that for every $\epsilon > 0$, there exists $M \in \mathbb{N}$ such that $N(\mathbf{x} - \mathbf{x}^n) \leq \epsilon$ for all $n \geq M$, for some norm N on $\mathbb{R}^d$. $\mathbf{x}$ is called the *limit point*, or simply the *limit*, of the sequence and will also sometimes be denoted by $\lim_{n \to \infty} \mathbf{x}^n$.

By Theorem 1.1.3, the concept of a limit does not depend on the choice of the norm: $\mathbf{x}$ is the limit of a sequence $\{\mathbf{x}^i\}_{i \in \mathbb{N}}$ under some norm if and only if it is the limit under every norm.

Theorem 1.3.5 *A set $X \subseteq \mathbb{R}^d$ is closed if and only if for every convergent sequence in X, the limit of the sequence is also in X.*

Definition 1.3.6 We introduce three important notions:

1. For any set $X \subseteq \mathbb{R}^d$, the *closure of X* is the smallest (with respect to set inclusion) closed set containing X and will be denoted by $\mathrm{cl}(X)$.
2. For any set $X \subseteq \mathbb{R}^d$, the *interior of X* is the largest (with respect to set inclusion) open set contained inside X and will be denoted by $\mathrm{int}(X)$.
3. For any set $X \subseteq \mathbb{R}^d$, the *boundary of X* is defined as $\mathrm{bd}(X) := \mathrm{cl}(X) \setminus \mathrm{int}(X)$.

Definition 1.3.7 Let $X' \subseteq X \subseteq \mathbb{R}^d$ be two subsets. X' is said to be *dense in X* if $X \subseteq \mathrm{cl}(X')$.

Definition 1.3.8 A set in $\mathbb{R}^d$ that is closed and bounded is called *compact*.

Theorem 1.3.9 (Heine–Borel theorem) *Let $C \subseteq \mathbb{R}^d$ be a compact set, and let $\{U_\lambda : \lambda \in \Lambda\}$ be any (possibly infinite) family of open subsets of $\mathbb{R}^d$ such that $C \subseteq \bigcup_{\lambda \in \Lambda} U_\lambda$. Then there exists a finite subfamily $\Lambda' \subseteq \Lambda$ such that $C \subseteq \bigcup_{\lambda \in \Lambda'} U_\lambda$.*

Theorem 1.3.10 *Let $C \subseteq \mathbb{R}^d$ be a compact set. Then every sequence $\{\mathbf{x}^i\}_{i \in \mathbb{N}}$ contained in C (not necessarily convergent) has a convergent subsequence.*

Theorem 1.3.11 *Let C_λ, $\lambda \in \Lambda$ be any family of closed sets in $\mathbb{R}^d$ such that at least one of them is compact. $\bigcap_{\lambda \in \Lambda} C_\lambda = \emptyset$ if and only if there is a finite set of indices $\lambda_1, \ldots, \lambda_k \in \Lambda$ such that $C_{\lambda_1} \cap \cdots \cap C_{\lambda_k} = \emptyset$.*

Definition 1.3.12 A function $f : \mathbb{R}^d \to \mathbb{R}^n$ is *continuous* at $\mathbf{x} \in \mathbb{R}^d$ if for every convergent sequence $\{\mathbf{x}^i\}_{i \in \mathbb{N}} \subseteq \mathbb{R}^d$ with $\lim_{n \to \infty} \mathbf{x}^i = \mathbf{x}$, we have $\lim_{i \to \infty} f(\mathbf{x}^i) = f(\mathbf{x})$. A function is said to be *continuous* if it is continuous at every $\mathbf{x} \in \mathbb{R}^d$.

Let N_1 be a norm on $\mathbb{R}^d$ and N_2 be a norm on $\mathbb{R}^n$. f is said to be *Lipschitz continuous (with respect to these norms) over a domain* $D \subseteq \mathbb{R}^d$ if there exists a constant L such that $d_{N_2}(f(\mathbf{x}), f(\mathbf{y})) \leq L \cdot d_{N_1}(\mathbf{x}, \mathbf{y})$ for all $\mathbf{x}, \mathbf{y} \in D$. L is called the *Lipschitz constant of f over D with respect to these norms*.

It can be verified that a Lipschitz continuous function is continuous.

Theorem 1.3.13 (Weierstrass' theorem) *Let $f : \mathbb{R}^d \to \mathbb{R}$ be a continuous function. Let $X \subseteq \mathbb{R}^d$ be a nonempty, compact subset. Then $\inf\{f(\mathbf{x}) : \mathbf{x} \in X\}$ is attained, i.e., there exists $\mathbf{x}^{\min} \in X$ such that $f(\mathbf{x}^{\min}) = \inf\{f(\mathbf{x}) : \mathbf{x} \in X\}$. Similarly, there exists $\mathbf{x}^{\max} \in X$ such that $f(\mathbf{x}^{\max}) = \sup\{f(\mathbf{x}) : \mathbf{x} \in X\}$.*

A generalization of the above theorem is the following.

Theorem 1.3.14 *Let $f\colon \mathbb{R}^d \to \mathbb{R}^n$ be a continuous function, and C be a compact set. Then $f(C)$ is compact.*

We will also need to speak of differentiability of functions $f\colon \mathbb{R}^d \to \mathbb{R}^n$.

Definition 1.3.15 We say that $f\colon \mathbb{R}^d \to \mathbb{R}^n$ is differentiable at $\mathbf{x} \in \mathbb{R}^d$ if there exists a linear transformation $A\colon \mathbb{R}^d \to \mathbb{R}^n$ such that

$$\lim_{\mathbf{h}\to 0} \frac{\|f(\mathbf{x}+\mathbf{h}) - f(\mathbf{x}) - A\mathbf{h}\|}{\|\mathbf{h}\|} = 0.$$

If f is differentiable at $\mathbf{x}$, then the linear transformation A is unique. It is commonly called the *differential* or *total derivative of f at $\mathbf{x}$* and is denoted by $f'(\mathbf{x})$. When $n = 1$, $f'(\mathbf{x})$ is a linear functional and can therefore be represented by an inner product, i.e., there exists a vector $\mathbf{v} \in \mathbb{R}^d$ such that $f'(\mathbf{x})(\mathbf{u}) = \langle \mathbf{v}, \mathbf{u}\rangle$ for all $\mathbf{u} \in \mathbb{R}^d$. The vector $\mathbf{v}$ is commonly called the *gradient of f* and is denoted by $\nabla f(\mathbf{x})$.

If $f\colon \mathbb{R}^d \to \mathbb{R}$ is differentiable everywhere, and the gradient function $\nabla f\colon \mathbb{R}^d \to \mathbb{R}^d$ is differentiable at $\mathbf{x}$, then its differential at $\mathbf{x}$ is a linear map from $\mathbb{R}^d$ to $\mathbb{R}^d$. The corresponding matrix (see Section 1.2) is called the *Hessian of f at $\mathbf{x}$*, and it is denoted by $\nabla^2 f(\mathbf{x})$. f is said to be *twice differentiable* if ∇f is differentiable everywhere.

Definition 1.3.16 Let $f\colon \mathbb{R}^d \to \mathbb{R}$ be any function and let $\mathbf{x} \in \mathbb{R}^d$, $\mathbf{r} \in \mathbb{R}^d$. We define the *directional derivative of f at $\mathbf{x}$ in the direction $\mathbf{r}$* as

$$f'(\mathbf{x};\mathbf{r}) := \lim_{t\to 0^+} \frac{f(\mathbf{x}+t\mathbf{r}) - f(\mathbf{x})}{t} \tag{1.3.1}$$

if that limit exists. Note that we consider the limit as t approaches 0 from the right. We will be speaking of $f'(\mathbf{x};\cdot)$ as a function from $\mathbb{R}^d \to \mathbb{R}$.

For any coordinate $i \in \{1, \ldots, d\}$, if $f'(\mathbf{x};\mathbf{e}^i) = -f'(\mathbf{x}, -\mathbf{e}^i)$ then the limit

$$\lim_{h\to 0} \frac{f(\mathbf{x}+h\mathbf{e}^i) - f(\mathbf{x})}{h}$$

exists and is called the *partial derivative of $f\colon \mathbb{R}^d \to \mathbb{R}$ at $\mathbf{x}$ in the ith direction*. It will be denoted by $f_i'(\mathbf{x})$.

Theorem 1.3.17 *If $f\colon \mathbb{R}^d \to \mathbb{R}$ is differentiable at $\mathbf{x}$, then the partial derivatives exist at $\mathbf{x}$ for all $i = 1, \ldots, d$ and the i-coordinate of $\nabla f(\mathbf{x})$ is precisely $f_i'(\mathbf{x})$. Conversely, if the partial derivatives all exist at $\mathbf{x}$ and they are continuous functions at $\mathbf{x}$, then f is differentiable at $\mathbf{x}$.*

Definition 1.3.18 For any subset $X \subseteq \mathbb{R}^d$, $\mathrm{vol}(X)$ will denote the *volume* of X, i.e., $\mathrm{vol}(X) := \int_X d\mathbf{x}$.

The following limit property of volumes will be used.

Theorem 1.3.19 *Let $X_1, X_2, \ldots$ be a sequence of sets in $\mathbb{R}^d$ such that $X_i \subseteq X_j$ for all $i \leq j$. Then* $\mathrm{vol}\left(\bigcup_{i=1}^{\infty} X_i\right) = \lim_{i \to \infty} \mathrm{vol}(X_i)$.

Similarly, if $X_1, X_2, \ldots$ be a sequence of compact sets in $\mathbb{R}^d$ such that $X_i \supseteq X_j$ for all $i \leq j$. Then $\mathrm{vol}\left(\bigcap_{i=1}^{\infty} X_i\right) = \lim_{i \to \infty} \mathrm{vol}(X_i)$.

1.3.1 Exercises

1. Show that if X is compact and Y is closed, then $X + Y$ is closed. Find an example with closed sets X, Y such that $X + Y$ is not closed.
2. Show that if f is differentiable everywhere, it is also continuous everywhere.

1.4 Models of Computation

The second part of this book deals with mathematical optimization, with a heavy emphasis on algorithmic aspects. This necessitates a discussion of a model of computation in which we will carry out all the operations for solving an optimization problem.

We will not enter into a very detailed and formal discussion of computation models. At an intuitive level, an algorithm is simply a piece of computer code that takes as input an instance of an optimization problem and outputs the solution to it. However, there are two related issues that require clarification. First, one has to formalize what it means to "input an instance of an optimization problem." For example, we will discuss at length optimization using functions defined over $\mathbb{R}^d$ and subsets of $\mathbb{R}^d$. We have to define more precisely how we can "input" a function or a subset to a computer. Second, since we will be dealing with functions over $\mathbb{R}^d$ and subsets of $\mathbb{R}^d$, we may have to deal with irrational numbers in $\mathbb{R}$. How are we to represent arbitrary real numbers on a "finite memory" computer?

These two issues bring us face-to-face with two distinct approaches to computation within mathematical optimization. The *continuous* or *numerical* optimization community, with its very close connection to scientific computation and numerical analysis, typically works in the *arithmetic model* of computation. On the other hand, the *discrete* or *combinatorial* optimization community, due to a more significant overlap with computer science, primarily uses the *Turing machine model* of computation. We discuss both of these models in a little more detail below.

1.4.1 Arithmetic Model of Computation

Here one assumes that one can perform addition, subtraction, multiplication, and division of any two real numbers, as well as compare two real numbers to tell whether the first one is lesser than, greater than, or equal to the second number. In some situations, one also allows the possibility of taking square roots. These operations are called *elementary operations*. Thus, this model of computation is over the ordered field of real numbers: arbitrary real numbers can be given as input, be stored in memory, and be outputs. An algorithm in this model of computation is a piece of computer code (with the standard notions of programming variables, iteration loops, branchings, etc.) with the idealization just mentioned: one can store arbitrary real numbers in memory and perform computations on them using the elementary operations. Different algorithms will have different formats for their inputs, but in the end it is simply a list of real numbers that encodes the input.

Given an algorithm in this model, and any input to the algorithm, the *running time* or *time complexity* of the algorithm on this input is the total number of elementary operations performed by the algorithm. Similarly, the *space complexity* of the algorithm is the maximum amount of memory used (i.e., the number of real numbers stored) at any given point during the algorithm's execution.

1.4.2 Turing Machine Model of Computation

Here, one breaks down the notion of computation to even more basic components. For instance, the addition of two numbers is not an elementary operation: intuitively, more computation is needed to add larger numbers compared to smaller numbers. The main difference from the arithmetic model is that everything – inputs, outputs, and intermediate objects in memory – must be stored as finite-length binary strings and one can perform three simple operations: read a bit stored in a particular location in memory (and check if it equals 0 or 1), change the bit stored in a particular location in memory, or write a bit into a new memory location. There are many equivalent ways of formalizing what an algorithm is in this model, which we will not get into here; see the references at the end of the chapter. It suffices for our purposes to think of an algorithm in this model of computation as a piece of computer code (with the standard notions of programming variables, iteration loops, branchings, etc.) with elementary operations restricted to the above operations involving bits.

Given an algorithm in this model, and any input to the algorithm, the *running time* or *time complexity* of the algorithm on this input is again the total number of elementary operations performed by the algorithm. Similarly, the *space complexity* of the algorithm is the maximum amount of memory used (i.e., the number of bits stored) at any given point during the algorithm's execution.

In our opinion, both of these approaches have their advantages and disadvantages and no one approach captures all the subtleties of what it means to solve an optimization problem. Discrete or combinatorial optimization has traditionally been concerned with objects that have a natural encoding as finite binary strings, and so the Turing machine model of computation is a natural way to analyze such algorithms. However, continuous optimization, by its very nature, deals with the ordered set of reals in its computations and the arithmetic model becomes the obvious model. The Turing machine model, with its finitary nature, cannot (exactly) represent the set of all real numbers. From a mathematical perspective, one could argue that the arithmetic model is more general, since it can simulate any computation performed in the Turing machine model, since finite binary strings are simply a list of 0's and 1's, and the elementary operations of a Turing machine can be simulated by the elementary operations of the arithmetic model. In fact, under standard computational complexity assumptions and depending on exactly which programming instructions are allowed, it can be shown rigorously that the arithmetic model is significantly more powerful than the Turing machine model; see Section 1.5 for more on the comparisons between these two models of computation. However, currently, no real (physical) computing machine can implement the arithmetic model. Thus, one could counterargue that the Turing machine model is the one that can actually be implemented and thus more authentic as a mathematical model of computation. We personally think this counterargument is a bit narrow. First, we cannot predict what physical machines may or may not be invented in the future; perhaps, some analog machines will be able to implement the arithmetic model. Second, valuable mathematical insights into numerical optimization can be derived from the arithmetic model of computation, which is impossible to express in the Turing machine model.

We do not mean to enter into a deep philosophical debate on the possibilities and limitations of different computational models. This inevitably leads to foundational questions in mathematics (e.g., the nature of existence of an irrational number) and computation (e.g., designing an algorithm for adding arbitrary real numbers) which is not the focus of this book. Instead, we will adopt a third approach, outlined in Section 1.4.3, that interpolates between these two models and is very convenient for discussing optimization algorithms.

Before we proceed, we point out that Turing machines can perform computations over rational numbers using a standardized encoding which represents any integer using its binary representation and a rational number as a pair of (binary representations of) integers (numerator and denominator). The *encoding size* of any rational number is the size of the binary string used to represent it. We generalize this notion to both models of computation.

Definition 1.4.1   Let X be an arbitrary set. We say that X *has a representation in the arithmetic model of computation* if there is an injective map from X to the set of finite sequences of real numbers. Similarly, we say that X *has a representation in the Turing machine model of computation* if there is an injective map from X to the set of finite binary strings. These maps give "labels" or *encodings* for the elements of X so that they can be stored in memory if needed and an algorithm can process them in its computations. The *encoding size* of any $x \in X$ is the length of the sequence that encodes it.

1.4.3 Oracle-Based Computation

1.4.3.1 Turing Machines Augmented with Real Number Oracles

The tension noted above between the two models led to a hybrid approach that tries to retain the richness of the arithmetic model while staying close to the principle of physical realizability of the Turing machine model. It traces its roots to the constructive philosophy of mathematics [40, 46], and considers the Turing machine model augmented with *oracles*. What this means is that an algorithm can query certain oracles at any point during its execution and use the responses in its computations. The responses from the oracle must be binary strings so that they can be processed by the Turing machine. The most basic kind of an oracle is that representing a set of real numbers.

Definition 1.4.2   Let $\mathcal{I} \subseteq \mathbb{R}$ be a subset of real numbers. A *rational oracle* representing $\mathcal{I}$ is a set of functions $\{q_\epsilon\}_{\epsilon \in \mathbb{Q}_+ \setminus \{0\}}$ indexed by the positive rationals with $q_\epsilon : \mathcal{I} \to \mathbb{Q}$ such that for any $\alpha \in \mathcal{I}$, $|q_\epsilon(\alpha) - \alpha| \leq \epsilon$.

Real number oracles are sometimes equipped with a "size": for every $\alpha \in \mathcal{I}$, there exists a constant $K_\alpha \geq 1$ such that for every $\epsilon > 0$, the encoding size of the rational number q_ϵ is at most K_α times the encoding size of ϵ. Additionally if α is rational, K_α must be at least the encoding size of α. This ensures that any algorithm in the standard Turing machine model has a comparable implementation in the oracle-based model. See [168, Section 1.4] for this important point.

1.4.3.2 General Oracles

In Part II of this book, we will encounter other oracles, e.g., those representing functions on $\mathbb{R}^d$ and subsets of $\mathbb{R}^d$, that generalize the basic real number oracles from Definition 1.4.2. This will also tie into a formalization of the idea of giving an optimization problem as "input" to an algorithm. In this context, it will be useful to extend the notion of oracle-based computation to the arithmetic model as well.

Definition 1.4.3 Let $\mathcal{I}$ be an arbitrary set. An *oracle representing* $\mathcal{I}$ is given by a set $\mathcal{Q}$ of possible *queries* and a set H of possible *answers* or *responses*. Each query $q \in \mathcal{Q}$ is a function $q : \mathcal{I} \to H$. We say that $q(I) \in H$ is the answer (or response) to the query q on the element $I \in \mathcal{I}$. The oracle is said to be *unambiguous* if for every $I, I' \in \mathcal{I}$ with $I \neq I'$, there exists some $q \in \mathcal{Q}$ such that $q(I) \neq q(I')$.

An oracle is *compatible with a model of computation* (arithmetic or Turing machine) if $\mathcal{Q}$ and H both have representations in that model (Definition 1.4.1).

An *oracle-based algorithm for processing* $\mathcal{I}$ in either of these two models of computation that uses an oracle $(\mathcal{Q}, H)$ compatible with that model of computation is an algorithm that has the additional ability to query any $q \in \mathcal{Q}$ and use the response in its computations. Such algorithms do not have an explicit input $I \in \mathcal{I}$, but rather the input is revealed implicitly by the responses $q(I)$ it receives from the queries to the oracle.

In the traditional view of computation the set of possible inputs $\mathcal{I}$ to the algorithm has an encoding itself, i.e., $\mathcal{I}$ has a representation in the model of computation (Definition 1.4.1). One can then associate a natural unambiguous oracle compatible with this model of computation: $\mathcal{Q} = \{q\}$ is a singleton and $q(I)$ is the encoding of $I \in \mathcal{I}$. Algorithms processing $\mathcal{I}$ can then simply query q at the beginning of their computation (equivalent to "receiving the input") and then never make any other oracle queries. Therefore, oracle-based computation strictly expands the scope of computation from the traditional view of "inputs processed to give outputs." In fact, adaptively posing the queries will be a crucial feature of several mathematical optimization algorithms we discuss in Part II. We will see concrete examples of oracles for different kinds of optimization problems (cf. Example 5.2.2 and Section 6.1.1).

The arithmetic model often leads to the most concise and elegant description of algorithmic ideas in optimization and we will mostly use this model in our discussions. Moreover, for some of the algorithms we discuss it is not known if a version in the Turing machine model (with or without oracles) exists. Nevertheless, we will provide references to implementations in the (oracle) Turing machine model for all the algorithms where such implementations are known in the literature.

1.5 Notes and Related Literature

Euclidean spaces and their linear algebraic and real analytic properties are well-studied topics, with many books covering them that fill several shelves in an academic library. We recommend [219] for the linear algebra of Euclidean spaces. Halmos' text [132] is a superb reference for a "coordinate-free" study of finite-dimensional vector spaces and linear algebra. Two classic references for real analysis (both in Euclidean spaces and more generally) are [204, 206]. These three books cover everything that we surveyed in Sections 1.1, 1.2, and 1.3, and much more.

The Turing machine model of computation has been studied for almost a century now. A good introduction is [5, chapter 1], which includes a discussion of related models like *(integer) random access machines (RAMs)*, and [13] dives into the subject in depth.

The arithmetic model has its roots in attempts to formalize numerical algorithms such as Newton's method for finding roots of nonlinear equations. It seems hard to formulate such classical computational procedures in numerical analysis and scientific computation in the Turing machine model, simply because such procedures assume operations over the entire ordered field of real numbers which is impossible to capture in the finitary world of the Turing machine model. Several proposals have been put forward to address this issue; see [1, 53, 56, 118, 123, 129, 155, 158, 168, 189, 191, 224, 230] as a representative list. The arithmetic model, as described in Section 1.4, is essentially the same as the computational models from [56, 191], very closely related to the model in [53], and is also sometimes referred to as the *real RAM* model. As mentioned in Section 1.4.2, it has been shown rigorously that the real RAM model is significantly more powerful, unless one restricts the programming instructions or arithmetic operations [43, 133, 153, 190, 211]. For instance, it can be shown that if something called *indirect indexing* (an operation which is present in every modern programming language) is allowed in the real RAM model, it can solve in polynomial time all problems solvable by Turing machines with polynomial amount of memory. This represents a significant increase in power because if polynomial time is the same as polynomial space for Turing machines, the so-called *polynomial hierarchy* would collapse, and among other things, the class P would equal NP [13]. See [115] for some recent work on the real RAM model that addresses this issue.

The oracle Turing machine model has a long history going back to the original work of Turing [224, 225], and developed further in [129, 158]. We refer to [155, 230] as good textbook expositions. Using this model in the realm of (discrete) optimization seems to have been first explored in [123, 168]. In particular, Definition 1.4.2 is taken directly from [168].

Our definition of a general oracle and oracle-based algorithms in this general setting is not explicitly stated anywhere in the literature. However, these concepts and their use in Part II are very much inspired by the beautiful monograph [222], as well as work in the oracle Turing machine model cited above. Definitions 1.4.1 and 1.4.3 are made with respect to the arithmetic model and the Turing machine model of computation, but they can be easily adapted to any other model of computation, including the ones cited above. This makes the view presented here of algorithmic computation very flexible and general, as well as especially useful for discussing diverse kinds of mathematical optimization algorithms (combinatorial and numerical) under a unifying umbrella in Part II.

PART I

Structural Aspects

2

Convex Sets

2.1 Definitions and Basic Properties

We begin with the most fundamental definition of this book, that of a convex set in a d-dimensional Euclidean space. A subset of $\mathbb{R}^d$ is said to be convex if, for any two points in the set, the line segment connecting them also lies entirely in the set. More formally:

Definition 2.1.1 A set $X \subseteq \mathbb{R}^d$ is called a *convex set* if for all $\mathbf{x}, \mathbf{y} \in X$ and every $\lambda \in [0,1]$, $\lambda \mathbf{x} + (1-\lambda)\mathbf{y} \in X$. We will refer to the set $\{\lambda \mathbf{x} + (1-\lambda)\mathbf{y} : \lambda \in [0,1]\}$ as the *line segment between* $\mathbf{x}$ *and* $\mathbf{y}$ and denote it by $[\mathbf{x}, \mathbf{y}]$.

Example 2.1.2 Some examples of convex sets:

1. In $\mathbb{R}$, the only examples of convex sets are intervals (closed, open, half open): $(a,b), (a,b], [a,b], (-\infty, b]$, etc.
2. Let $\mathbf{a} \in \mathbb{R}^d \setminus \{\mathbf{0}\}$ and $\delta \in \mathbb{R}$. The sets $H^=(\mathbf{a}, \delta) = \{\mathbf{x} \in \mathbb{R}^d : \langle \mathbf{a}, \mathbf{x} \rangle = \delta\}$, $H^\geq(\mathbf{a}, \delta) = \{\mathbf{x} \in \mathbb{R}^d : \langle \mathbf{a}, \mathbf{x} \rangle \geq \delta\}$ and $H^\leq(\mathbf{a}, \delta) = \{\mathbf{x} \in \mathbb{R}^d : \langle \mathbf{a}, \mathbf{x} \rangle \leq \delta\}$ are all convex sets. Sets of the form $H^=(\mathbf{a}, \delta)$ are called *hyperplanes* and sets of the form $H^\geq(\mathbf{a}, \delta), H^\leq(\mathbf{a}, \delta)$ are called *halfspaces*. The vector $\mathbf{a}$ is called the *normal* of the hyperplane or halfspace, and δ is called the *shift*.[1]
3. $\{\mathbf{x} \in \mathbb{R}^d : \|\mathbf{x}\|_p \leq 1\}$ is a convex set for any $p \geq 1$, including $p = \infty$.
4. For $d \geq 1$, the set of polynomials in a single variable of degree $d - 1$ that are nonnegative everywhere may be represented as a subset of $\mathbb{R}^d$:
$\{\mathbf{x} = (x_1, \ldots, x_d) \in \mathbb{R}^d : x_1 + x_2 t + x_3 t^2 + \cdots + x_d t^{d-1} \geq 0 \text{ for all } t \geq 0\}$.
This is a convex set.

---

[1] Since $H^\leq(\mathbf{a}, \delta) = H^\leq(\mathbf{a}', \delta')$ when $\mathbf{a}' = \gamma \mathbf{a}, \delta' = \gamma \delta$ for some $\gamma > 0$, the same halfspace may be represented by different normals and shifts. So technically, it is perhaps better to associate the normal and shift with the inequality $\langle \mathbf{a}, \mathbf{x} \rangle \leq \delta$, as opposed to the geometric object $H^\leq(\mathbf{a}, \delta)$, but this ambiguity should not cause any confusion in this book.

An important idea is to start with some basic convex sets (like halfspaces, unit norm balls, etc.) and use them as building blocks to construct more interesting convex sets. Thus, it is useful to consider operations on sets that preserve convexity.

Theorem 2.1.3 *The following are all true.*

1. *Let X_i, $i \in I$ be an arbitrary family of convex sets. Then $\bigcap_{i \in I} X_i$ is a convex set.*
2. *Let X be a convex set and $\alpha \in \mathbb{R}$, then αX is a convex set.*
3. *Let X, Y be convex sets, then $X + Y$ is convex.*
4. *Let $T \colon \mathbb{R}^d \to \mathbb{R}^m$ be any affine transformation (see Definition 1.2.4). If $X \subseteq \mathbb{R}^d$ is convex, then $T(X)$ is a convex set. If $Y \subseteq \mathbb{R}^m$ is convex, then $T^{-1}(Y)$ is convex.*

Proof   1. Let $\mathbf{x}, \mathbf{y} \in \cap_{i \in I} X_i$. This implies that $\mathbf{x}, \mathbf{y} \in X_i$ for every $i \in I$. Since each X_i is convex, for every $\lambda \in [0, 1]$, $\lambda \mathbf{x} + (1 - \lambda)\mathbf{y} \in X_i$ for all $i \in I$. Therefore, $\lambda \mathbf{x} + (1 - \lambda)\mathbf{y} \in \cap_{i \in I} X_i$.

The proofs of 2., 3., and 4. are very similar and are left for the reader.   □

Remark 2.1.4   Observe that item 4. in Example 2.1.2 can be interpreted as an (uncountable) intersection of halfspaces of the form $H^{\geq}(\mathbf{a}, \delta)$ with $\mathbf{a} = (1, t, t^2, \dots, t^{d-1})$ and $\delta = 0$. Thus, item 2. from that example and Theorem 2.1.3 together give another proof that item 4. describes a convex set.

A special kind of linear combination is central to the study of convex sets.

Definition 2.1.5   Let $\mathbf{y}^1, \dots, \mathbf{y}^n \in \mathbb{R}^d$ be a finite collection of points. A *convex combination* of this collection is a point of the form

$$\lambda_1 \mathbf{y}^1 + \lambda_2 \mathbf{y}^2 + \cdots + \lambda_n \mathbf{y}^n,$$

where $\lambda_1, \dots, \lambda_n \geq 0$, $\lambda_1 + \lambda_2 + \cdots + \lambda_n = 1$.

Proposition 2.1.6 *If X is convex and $\mathbf{y}^1, \dots, \mathbf{y}^n \in X$, then any convex combination of $\mathbf{y}^1, \dots, \mathbf{y}^n$ is in X.*

Proof   We prove it by induction on n. If $n = 1$, then the conclusion is trivial. Else consider any $\lambda_1, \dots, \lambda_n \geq 0$ such that $\lambda_1 + \cdots + \lambda_n = 1$. We may assume $\lambda_i > 0$ for all i; otherwise, we may appeal to the induction hypothesis. Then

$$
\begin{aligned}
&\lambda_1 \mathbf{y}^1 + \lambda_2 \mathbf{y}^2 + \cdots + \lambda_n \mathbf{y}^n \\
&= (\lambda_1 + \cdots + \lambda_{n-1}) \left(\tfrac{\lambda_1}{\lambda_1 + \cdots + \lambda_{n-1}} \mathbf{y}^1 + \tfrac{\lambda_2}{\lambda_1 + \cdots + \lambda_{n-1}} \mathbf{y}^2 + \cdots + \tfrac{\lambda_{n-1}}{\lambda_1 + \cdots + \lambda_{n-1}} \mathbf{y}^{n-1} \right) + \lambda_n \mathbf{y}^n \\
&= (1 - \lambda_n) \tilde{\mathbf{y}} + \lambda_n \mathbf{y}^n,
\end{aligned}
$$

where $\tilde{\mathbf{y}} := \frac{\lambda_1}{\lambda_1+\cdots+\lambda_{n-1}}\mathbf{y}^1 + \frac{\lambda_2}{\lambda_1+\cdots+\lambda_{n-1}}\mathbf{y}^2 + \cdots + \frac{\lambda_{n-1}}{\lambda_1+\cdots+\lambda_{n-1}}\mathbf{y}^{n-1}$ belongs to X by the induction hypothesis. The rest follows from the definition of convexity.

$\square$

Definition 2.1.7 Given any set $X \subseteq \mathbb{R}^d$ (not necessarily convex), the *convex hull of* X, denoted by $\mathrm{conv}(X)$, is a convex set C such that $X \subseteq C$ and for any other convex set C', $X \subseteq C' \Rightarrow C \subseteq C'$, i.e., the convex hull of X is the smallest (with respect to set inclusion) convex set containing X.

Theorem 2.1.8 *For any set $X \subseteq \mathbb{R}^d$ (not necessarily convex),*

$$\mathrm{conv}(X) = \bigcap(C : X \subseteq C, C \text{ convex})$$
$$= \left\{\lambda_1\mathbf{x}^1 + \cdots + \lambda_t\mathbf{x}^t : \mathbf{x}^1,\ldots,\mathbf{x}^t \in X, \lambda_1,\ldots,\lambda_t \geq 0, \sum_{i=1}^{t}\lambda_i = 1\right\}.$$

In other words, the convex hull of X is the union of the set of convex combinations of all possible finite subsets of X.

Proof Let $\hat{C} = \bigcap(C : X \subseteq C, C \text{ convex})$, which is a convex set by Theorem 2.1.3 and by definition $X \subseteq \hat{C}$. Consider any other convex set C' such that $X \subseteq C'$. Then C' appears in the intersection, and thus $\hat{C} \subseteq C'$. Thus, $\hat{C} = \mathrm{conv}(X)$.

Next, let $\tilde{C} = \left\{\lambda_1\mathbf{x}^1 + \cdots + \lambda_t\mathbf{x}^t : \mathbf{x}^1,\ldots,\mathbf{x}^t \in X, \lambda_1,\ldots,\lambda_t \geq 0, \sum_{i=1}^{t}\lambda_i = 1\right\}$. Then,

1. $\tilde{C}$ is convex. Consider two points $\mathbf{z}^1, \mathbf{z}^2 \in \tilde{C}$. Thus there exist two finite index sets I_1, I_2 indexing two finite collections of points from X given by $\mathbf{x}^{1,i} \in X, i \in I_1$ and $\mathbf{x}^{2,i} \in X, i \in I_2$, and two subsets of nonnegative real numbers $\{\lambda_i^1 \geq 0, i \in I_1\}$, $\{\lambda_i^2 \geq 0, i \in I_2\}$ such that $\sum_{i \in I_j} \lambda_i^j = 1$ for $j = 1, 2$, with the following property: $\mathbf{z}_j = \sum_{i \in I_j} \lambda_i^j \mathbf{x}^{j,i}$ for $j = 1, 2$. Then for any $\lambda \in [0, 1]$, $\lambda\mathbf{z}_1 + (1-\lambda)\mathbf{z}_2 = \lambda(\sum_{i \in I_1} \lambda_i^1 \mathbf{x}^{1,i}) + (1-\lambda)(\sum_{i \in I_2} \lambda_i^2 \mathbf{x}^{2,i})$. The right-hand side of this equality is a convex combination of the points $\mathbf{x}^{1,i}, i \in I_1$ and $\mathbf{x}^{2,i}, i \in I_2$ that are all from X. Thus, $\lambda\mathbf{z}_1 + (1 - \lambda)\mathbf{z}_2 \in \tilde{C}$ by definition of $\tilde{C}$.
2. $X \subseteq \tilde{C}$. We simply use $\lambda = 1$ as the multiplier for a point from X.
3. Let C' be any convex set such that $X \subseteq C'$. Since C' is convex, every point of the form $\lambda_1\mathbf{x}_1 + \cdots + \lambda_t\mathbf{x}_t$ where $\mathbf{x}_1,\ldots,\mathbf{x}_t \in X, \lambda_i \geq 0, \sum_{i=1}^{t}\lambda_i = 1$ belongs to C' by Proposition 2.1.6. Thus, $\tilde{C} \subseteq C'$.

From 1., 2., and 3., we get that $\tilde{C} = \mathrm{conv}(X)$.

$\square$

2.1.1 Exercises

1. Show that if $N \colon \mathbb{R}^d \to \mathbb{R}$ is a norm (see Definition 1.1.1), then every ball $B_N(\mathbf{x}, R)$ with respect to N is convex.
2. Prove parts 2., 3., and 4. in Theorem 2.1.3.
3. Show that if $X \subseteq \mathbb{R}^d$ is convex and $\alpha_1, \ldots, \alpha_k \geq 0$, then $\alpha_1 X + \cdots + \alpha_k X = (\alpha_1 + \cdots + \alpha_k)X$.
4. Show that $\mathrm{conv}(\mathrm{conv}(X)) = \mathrm{conv}(X)$ for any set $X \subseteq \mathbb{R}^d$.
5. Show that $\mathrm{conv}(A) \cup \mathrm{conv}(B) \subseteq \mathrm{conv}(A \cup B)$. Give an example where the containment is strict.
6. Show that $\mathrm{conv}(A \cap B) \subseteq \mathrm{conv}(A) \cap \mathrm{conv}(B)$. Give an example where the containment is strict.
7. Let $T \colon \mathbb{R}^d \to \mathbb{R}^m$ be any linear transformation and let $X \subseteq \mathbb{R}^d$ be any set. Is it true that $T(\mathrm{conv}(X)) = \mathrm{conv}(T(X))$?
8. Find a closed set $X \subseteq \mathbb{R}^2$ such that $\mathrm{conv}(X)$ is not closed.
9. Let $X \subseteq \mathbb{R}^d$. Let $\mathbf{u} \notin \mathrm{conv}(X)$. Show that any point $\mathbf{y} \in \mathrm{conv}(X \cup \{\mathbf{u}\})$ can be written as $\mathbf{y} = \lambda \mathbf{u} + (1 - \lambda)\mathbf{x}$ for some $\mathbf{x} \in \mathrm{conv}(X)$ and $\lambda \in [0, 1]$.
10. Let $X \subseteq \mathbb{R}^d$. Let $\mathbf{u}, \mathbf{v} \in \mathbb{R}^d$ be two points such that $\mathbf{u} \notin \mathrm{conv}(X)$ and $\mathbf{v} \notin \mathrm{conv}(X)$. Prove that if $\mathbf{u} \in \mathrm{conv}(X \cup \{\mathbf{v}\})$ and $\mathbf{v} \in \mathrm{conv}(X \cup \{\mathbf{u}\})$, then $\mathbf{u} = \mathbf{v}$.
11. Let $\mathbf{a} \in \mathbb{R}^d, \delta \in \mathbb{R}$ and $\mathbf{x} \in \mathbb{R}^d$. Show that the distance between $\mathbf{x}$ and $H^=(\mathbf{a}, \delta)$ is given by $\frac{|\langle \mathbf{a}, \mathbf{x} \rangle - \delta|}{\|\mathbf{a}\|}$. More precisely,

$$\inf\{\|\mathbf{x} - \mathbf{y}\| : \mathbf{y} \in H^=(\mathbf{a}, \delta)\} = \frac{|\langle \mathbf{a}, \mathbf{x} \rangle - \delta|}{\|\mathbf{a}\|}.$$

12. Let $\mathbf{a} \in \mathbb{R}^d$ and $\delta_1 \leq \delta_2$. Show that the distance between the hyperplanes $H^=(\mathbf{a}, \delta_1)$ and $H^=(\mathbf{a}, \delta_2)$ is given by $\frac{\delta_2 - \delta_1}{\|\mathbf{a}\|}$. More precisely, show that

$$\inf\{\|\mathbf{x} - \mathbf{y}\| : \mathbf{x} \in H^=(\mathbf{a}, \delta_1), \mathbf{y} \in H^=(\mathbf{a}, \delta_2)\} = \frac{\delta_2 - \delta_1}{\|\mathbf{a}\|}.$$

2.2 Convex Cones, Affine Sets, and Dimension

We say X is convex if for all $\mathbf{x}, \mathbf{y} \in X$ and $\lambda, \gamma \geq 0$ such that $\lambda + \gamma = 1$, $\lambda \mathbf{x} + \gamma \mathbf{y} \in X$. What happens if we relax the conditions on λ, γ?

Definition 2.2.1   Let $X \subseteq \mathbb{R}^d$ be a nonempty set. We have three possibilities, depending on which constraints on λ, γ are dropped:

1. We say that $X \subseteq \mathbb{R}^d$ is a *convex cone* if for all $\mathbf{x}, \mathbf{y} \in X$ and $\lambda, \gamma \geq 0$, $\lambda \mathbf{x} + \gamma \mathbf{y} \in X$.

2. We say that $X \subseteq \mathbb{R}^d$ is an *affine set* or an *affine subspace* if for all $\mathbf{x}, \mathbf{y} \in X$ and $\lambda, \gamma \in \mathbb{R}$ such that $\lambda + \gamma = 1$, $\lambda \mathbf{x} + \gamma \mathbf{y} \in X$.
3. We say $X \subseteq \mathbb{R}^d$ is a *linear set* or a *linear subspace* if for all $\mathbf{x}, \mathbf{y} \in X$ and $\lambda, \gamma \in \mathbb{R}$, $\lambda \mathbf{x} + \gamma \mathbf{y} \in X$. This coincides with the usual notion of linear subspace from Chapter 1.

Remark 2.2.2   Since we relaxed the conditions on λ, γ, convex cones, affine sets, and linear sets are all special cases of convex sets. We have the condition that X is nonempty in Definition 2.2.1. Thus, a convex cone and a linear subspace always contain $\{\mathbf{0}\}$. However, we will also consider the empty set as an affine set or affine subspace, just as we consider the empty set to be convex.

Definition 2.2.3   Similar to the definition of the convex hull of an arbitrary subset X, one can define the *conical hull* of X as the set inclusion wise smallest convex cone containing X, denoted by $\mathrm{cone}(X)$. Similarly, the *affine (linear) hull* of X as the set inclusion wise smallest affine (linear) subspace containing X. The affine hull will be denoted by $\mathrm{aff}(X)$, and linear hull will be denoted by $\mathrm{span}(X)$.

One can verify the following analog of Theorem 2.1.8.

Theorem 2.2.4 *Let $X \subseteq \mathbb{R}^d$ be nonempty. The following are all true.*

1. $\mathrm{cone}(X) = \bigcap(C : X \subseteq C, C \text{ is a convex cone}) = \{\lambda_1 \mathbf{x}_1 + \cdots + \lambda_t \mathbf{x}_t : \mathbf{x}_1, \ldots, \mathbf{x}_t \in X, \lambda_1, \ldots, \lambda_t \geq 0\}$.
2. $\mathrm{aff}(X) = \bigcap(C : X \subseteq C, C \text{ is an affine set}) = \{\lambda_1 \mathbf{x}_1 + \cdots + \lambda_t \mathbf{x}_t : \mathbf{x}_1, \ldots, \mathbf{x}_t \in X, \sum_{i=1}^{t} \lambda_i = 1\}$.
3. $\mathrm{span}(X) = \bigcap(C : X \subseteq C, C \text{ is a linear subspace}) = \{\lambda_1 \mathbf{x}_1 + \cdots + \lambda_t \mathbf{x}_t : \mathbf{x}_1, \ldots, \mathbf{x}_t \in X, \lambda_1, \ldots, \lambda_t \in \mathbb{R}\}$.

Proof   Left as an exercise. $\square$

Remark 2.2.5   As per our convention above in Remark 2.2.2, we have $\mathrm{aff}(\emptyset) = \emptyset$. Thus, part 2. of Theorem 2.2.4 holds even for $X = \emptyset$. However, $\mathrm{cone}(\emptyset) = \mathrm{span}(\emptyset) = \{\mathbf{0}\}$ from the definition of conical hull and linear hull, since convex cones and linear subspaces always contain $\{\mathbf{0}\}$.

Theorem 1.2.10 gives different characterizations of a linear subspace, each of which is useful in different contexts. There is an analogous theorem for affine sets, which we now develop. For this, we need the concept of *affine independence* that is analogous to the concept of linear independence.

Definition 2.2.6   A collection of points $\mathbf{x}^1, \ldots, \mathbf{x}^n \in \mathbb{R}^d$ is said to be *affinely independent* if there is no $i \in \{1, \ldots, n\}$ such that $\mathbf{x}^i$ lies in the affine hull of the remaining $\mathbf{x}^j$, $j \neq i$. If there exists such an index i, then $\mathbf{x}^1, \ldots, \mathbf{x}^n$ are said to be *affinely dependent*.

Similarly, a set $X \subseteq \mathbb{R}^d$ (possibly infinite) is said to be affinely independent if no $\mathbf{x} \in X$ lies in $\mathrm{aff}(X \setminus \{\mathbf{x}\})$; otherwise, X is said to be affinely dependent.

We now give several characterizations of affine independence.

Proposition 2.2.7 *Let $X \subseteq \mathbb{R}^d$ be nonempty. The following are equivalent.*

1. *X is an affinely independent set.*
2. *For every $\mathbf{x} \in X$, the set $\{\mathbf{v} - \mathbf{x} : \mathbf{v} \in X \setminus \{\mathbf{x}\}\}$ is linearly independent.*
3. *There exists $\mathbf{x} \in X$ such that the set $\{\mathbf{v} - \mathbf{x} : \mathbf{v} \in X \setminus \{\mathbf{x}\}\}$ is linearly independent.*
4. *The set of vectors $\{(\mathbf{x}, 1) \in \mathbb{R}^{d+1} : \mathbf{x} \in X\}$ is linearly independent.*
5. *X is a finite set of vectors $\mathbf{x}^1, \ldots, \mathbf{x}^m$ such that $\lambda_1 \mathbf{x}^1 + \cdots + \lambda_m \mathbf{x}^m = \mathbf{0}, \lambda_1 + \cdots + \lambda_m = 0$ implies $\lambda_1 = \lambda_2 = \cdots = \lambda_m = 0$.*

The same characterization holds for any finite collection of vectors $\mathbf{x}^1, \ldots, \mathbf{x}^m \in \mathbb{R}^d$.

Proof   1. $\Rightarrow$ 2. Consider an arbitrary $\mathbf{x} \in X$. Suppose to the contrary that $\{\mathbf{v} - \mathbf{x} : \mathbf{v} \in X \setminus \{\mathbf{x}\}\}$ is not linearly independent, i.e., there exist multipliers $\lambda_{\mathbf{v}}$, not all zero, such that $\sum_{\mathbf{v} \in X \setminus \{\mathbf{x}\}} \lambda_{\mathbf{v}}(\mathbf{v} - \mathbf{x}) = \mathbf{0}$. Rearranging terms, we get $\sum_{\mathbf{v} \in X \setminus \{\mathbf{x}\}} \lambda_{\mathbf{v}} \mathbf{v} = (\sum_{\mathbf{v} \in X \setminus \{\mathbf{x}\}} \lambda_{\mathbf{v}})\mathbf{x}$. We now consider two cases:

Case 1: $\sum_{\mathbf{v} \in X \setminus \{\mathbf{x}\}} \lambda_{\mathbf{v}} = 0$. In this case, since not all the $\lambda_{\mathbf{v}}$ are zero, let $\bar{\mathbf{v}} \in X \setminus \{\mathbf{x}\}$ be such that $\lambda_{\bar{\mathbf{v}}} \neq 0$. Since $\sum_{\mathbf{v} \in X \setminus \{\mathbf{x}\}} \lambda_{\mathbf{v}} \mathbf{v} = (\sum_{\mathbf{v} \in X \setminus \{\mathbf{x}\}} \lambda_{\mathbf{v}})\mathbf{x} = \mathbf{0}$, we obtain that $\bar{\mathbf{v}} = \sum_{\mathbf{v} \in X \setminus \{\mathbf{x}, \bar{\mathbf{v}}\}} \frac{\lambda_{\mathbf{v}}}{-\lambda_{\bar{\mathbf{v}}}} \mathbf{v}$. Since $\sum_{\mathbf{v} \in X \setminus \{\mathbf{x}\}} \lambda_{\mathbf{v}} = 0$, we obtain that $\sum_{\mathbf{v} \in X \setminus \{\mathbf{x}, \bar{\mathbf{v}}\}} \frac{\lambda_{\mathbf{v}}}{-\lambda_{\bar{\mathbf{v}}}} = 1$ and thus $\bar{\mathbf{v}} \in \mathrm{aff}(X \setminus \{\mathbf{x}, \bar{\mathbf{v}}\})$, contradicting the assumption that X is affinely independent.

Case 2: $\sum_{\mathbf{v} \in X \setminus \{\mathbf{x}\}} \lambda_{\mathbf{v}} \neq 0$. We can write $\mathbf{x} = \sum_{\mathbf{v} \in X \setminus \{\mathbf{x}\}} \frac{\lambda_{\mathbf{v}}}{\sum_{\mathbf{v} \in X \setminus \{\mathbf{x}\}} \lambda_{\mathbf{v}}} \mathbf{v}$. This implies that $\mathbf{x} \in \mathrm{aff}(X \setminus \{\mathbf{x}\})$ contradicting the assumption that X is affinely independent.

2. $\Rightarrow$ 3. Obvious.

3. $\Rightarrow$ 4. Let $\bar{\mathbf{x}}$ be such that $\{\mathbf{v} - \bar{\mathbf{x}} : \mathbf{v} \in X \setminus \{\bar{\mathbf{x}}\}\}$ is linearly independent. This means that the vectors $\{(\mathbf{v} - \bar{\mathbf{x}}, 0) \in \mathbb{R}^{d+1} : \mathbf{v} \in X \setminus \{\bar{\mathbf{x}}\}\} \cup \{(\bar{\mathbf{x}}, 1)\}$ are also linearly independent. Thus the matrix with these vectors as columns has full column rank. Now if we add the column $(\bar{\mathbf{x}}, 1)$ to the rest of the columns, this does not change the column rank, and thus the columns remain linearly

independent. But the new matrix has precisely $\{(\mathbf{x}, 1) \in \mathbb{R}^{d+1} : \mathbf{x} \in X\}$ as its columns.

4. $\Rightarrow$ 5. If $\{(\mathbf{x}, 1) \in \mathbb{R}^{d+1} : \mathbf{x} \in X\}$ is linearly independent, then the set X must be finite with elements $\mathbf{x}^1, \ldots, \mathbf{x}^m$. Moreover, for any $\lambda_1, \ldots, \lambda_m$ such that $\lambda_1 \mathbf{x}^1 + \cdots + \lambda_m \mathbf{x}^m = \mathbf{0}, \lambda_1 + \cdots + \lambda_m = 0$ we have $\sum_{\mathbf{x} \in X} \lambda_{\mathbf{x}}(\mathbf{x}, 1) = \mathbf{0}$. By linear independence of the set $\{(\mathbf{x}, 1) \in \mathbb{R}^{d+1} : \mathbf{x} \in X\}, \lambda_1 = \cdots = \lambda_m = 0$.

5. $\Rightarrow$ 1. Consider any $\mathbf{x}^i \in X$. If $\mathbf{x}^i \in \mathrm{aff}(X \setminus \{\mathbf{x}^i\})$, by part 2. of Theorem 2.2.4 there exist multipliers $\lambda_j \in \mathbb{R}, j \neq i$ such that $\mathbf{x}^i = \sum_{j \neq i} \lambda_j \mathbf{x}^j$ and $\sum_{j \neq i} \lambda_j = 1$. This implies that $\sum_{j=1}^m \lambda_j \mathbf{x}^j = \mathbf{0}$ where $\lambda_i = -1$, and therefore $\lambda_1 + \cdots + \lambda_m = 0$, contradicting the hypothesis of 5.

The proof for a finite collection of (possibly repeated vectors) $\mathbf{x}^1, \ldots, \mathbf{x}^m$ follows along the same lines. $\qquad\square$

We are now ready to state the affine version of Theorem 1.2.10.

Theorem 2.2.8 *Let $X \subseteq \mathbb{R}^d$ be nonempty. The following are equivalent.*

1. *X is an affine subspace.*
2. *There exists a linear subspace L of dimension $0 \leq m \leq d$ such that $X - \mathbf{x} = L$ for every $\mathbf{x} \in X$.*
3. *There exist affinely independent vectors $\mathbf{v}^1, \ldots, \mathbf{v}^{m+1} \in X$ for $0 \leq m \leq d$ such that $X = \mathrm{aff}(\{\mathbf{v}^1, \ldots, \mathbf{v}^{m+1}\})$.*
4. *There exist affinely independent vectors $\mathbf{v}^1, \ldots, \mathbf{v}^{m+1} \in X$ for $0 \leq m \leq d$ such that $\mathbf{v}^2 - \mathbf{v}^1, \mathbf{v}^3 - \mathbf{v}^1, \ldots, \mathbf{v}^{m+1} - \mathbf{v}^1$ are orthonormal and $X = \mathrm{aff}(\{\mathbf{v}^1, \ldots, \mathbf{v}^{m+1}\})$.*
5. *There exists a matrix $A \in \mathbb{R}^{(d-m) \times d}$ for $0 \leq m \leq d$ with full row rank and a vector $\mathbf{b} \in \mathbb{R}^{d-m}$ for some $0 \leq m \leq d$ such that $X = \{\mathbf{x} \in \mathbb{R}^d : A\mathbf{x} = \mathbf{b}\}$.*

Proof 1. $\Rightarrow$ 2. Fix an arbitrary $\mathbf{x}^\star \in X$. Define $L = X - \mathbf{x}^\star$. We first show that L is a linear subspace: for any $\mathbf{y}^1, \mathbf{y}^2 \in X, \lambda(\mathbf{y}^1 - \mathbf{x}^\star) + \gamma(\mathbf{y}^2 - \mathbf{x}^\star) \in X - \mathbf{x}^\star$ for any $\lambda, \gamma \in \mathbb{R}$. Since $\lambda(\mathbf{y}^1 - \mathbf{x}^\star) + \gamma(\mathbf{y}^2 - \mathbf{x}^\star) + \mathbf{x}^\star = \lambda \mathbf{y}^1 + \gamma \mathbf{y}^2 + (1 - \lambda - \gamma)\mathbf{x}^\star$ and X is an affine subset, we have $\lambda(\mathbf{y}^1 - \mathbf{x}^\star) + \gamma(\mathbf{y}^2 - \mathbf{x}^\star) + \mathbf{x}^\star \in X$. So, $\lambda(\mathbf{y}^1 - \mathbf{x}^\star) + \gamma(\mathbf{y}^2 - \mathbf{x}^\star) \in X - \mathbf{x}^\star = L$. Now, for any other $\bar{\mathbf{x}} \in X$, we need to show that $L = X - \bar{\mathbf{x}}$. Consider any $\mathbf{y} \in L$, i.e., $\mathbf{y} = \mathbf{x} - \mathbf{x}^\star$ for some $\mathbf{x} \in X$. Observe that $\mathbf{y} = (\mathbf{x} + \bar{\mathbf{x}} - \mathbf{x}^\star) - \bar{\mathbf{x}}$ and $\mathbf{x} + \bar{\mathbf{x}} - \mathbf{x}^\star \in X$ (because the coefficients sum to 1 and $\mathbf{x}, \bar{\mathbf{x}}$ and $\mathbf{x}^\star$ are all in X). Therefore, $\mathbf{y} \in X - \bar{\mathbf{x}}$ showing that $L = X - \mathbf{x}^\star \subseteq X - \bar{\mathbf{x}}$. Switching the roles of $\mathbf{x}^\star$ and $\bar{\mathbf{x}}$, one can similarly show that $X - \bar{\mathbf{x}} \subseteq X - \mathbf{x}^\star = L$.

2. $\Rightarrow$ 1. Consider any $\mathbf{y}^1, \mathbf{y}^2 \in X$ and let $\lambda, \gamma \in \mathbb{R}$ such that $\lambda + \gamma = 1$. We need to show that $\lambda \mathbf{y}^1 + \gamma \mathbf{y}^2 \in X$. Since $X - \mathbf{y}^1$ is a linear subspace, $\gamma(\mathbf{y}^2 - \mathbf{y}^1) \in X - \mathbf{y}^1$. Thus, $\gamma(\mathbf{y}^2 - \mathbf{y}^1) + \mathbf{y}^1 = \lambda \mathbf{y}^1 + \gamma \mathbf{y}^2 \in X$.

The equivalence of parts 2., 3., 4., and 5. follows from Theorem 1.2.10 and Proposition 2.2.7. $\qquad\square$

Definition 2.2.9 (Dimension of convex sets) If $X \subseteq \mathbb{R}^d$ is an affine subspace and $\mathbf{x} \in X$, the linear subspace $X - \mathbf{x}$ is called *the linear subspace parallel to X* and the dimension of X is the dimension of the linear subspace $X - \mathbf{x}$. For any nonempty convex set $X \subseteq \mathbb{R}^d$, the dimension of X is defined to be the dimension of $\mathrm{aff}(X)$ and will be denoted by $\dim(X)$. As a matter of convention, we take the dimension of the empty set to be -1. If $\dim(X) = d$, we say X is *full dimensional*.

Lemma 2.2.10 *If X is a set of affinely independent points, then* $\dim(\mathrm{aff}(X)) = (X) - 1$.

Proof If X is empty, then the relation follows from Definition 2.2.9. So we assume X is nonempty and fix any $\mathbf{x} \in X$. By Theorem 2.2.8, $L = \mathrm{aff}(X) - \mathbf{x}$ is a linear subspace. We claim that $(X \setminus \{\mathbf{x}\}) - \mathbf{x}$ is a basis for L. The verification of this claim is left to the reader. $\qquad\square$

Proposition 2.2.11 *Let X be a convex set. $\dim(X)$ equals one less than the maximum number of affinely independent points in X.*

Proof Let $X_0 \subseteq X$ be a maximum-sized set of affinely independent points in X. By Problem 3. from Section 2.2.3, $\mathrm{aff}(X_0) \subseteq \mathrm{aff}(X)$. Since X_0 is a maximum-sized set of affinely independent points in X, any $\mathbf{x} \in X$ must lie in $\mathrm{aff}(X_0)$. Therefore, $X \subseteq \mathrm{aff}(X_0)$. Since $\mathrm{aff}(X_0)$ is an affine set, by definition of affine hull of X, we have $\mathrm{aff}(X) \subseteq \mathrm{aff}(X_0)$. Therefore, $\mathrm{aff}(X) = \mathrm{aff}(X_0)$, implying that $\dim(\mathrm{aff}(X_0)) = \dim(\mathrm{aff}(X))$. By Lemma 2.2.10, we thus obtain $|X_0| - 1 = \dim(\mathrm{aff}(X))$. $\qquad\square$

2.2.1 Coordinatization of the Affine Hull and Volumes

It is often convenient to work in the affine hull of a convex set as the "true ambient space" of the convex set. For this purpose, the following notions are very useful for making such arguments formal.

Proposition 2.2.12 *Let $C \subseteq \mathbb{R}^d$ be a convex set and $\mathbf{x} \in C$. The following are equivalent.*

1. *For all $\mathbf{y} \in \mathrm{aff}(C)$, there exists $\epsilon_{\mathbf{y}} > 0$ such that $\mathbf{x} + \epsilon_{\mathbf{y}}(\mathbf{y} - \mathbf{x}) \in C$.*
2. *There exists $\epsilon > 0$ such that for all $\mathbf{y} \in \mathrm{aff}(C) \setminus \{\mathbf{x}\}$, $\mathbf{x} + \epsilon \left(\frac{\mathbf{y}-\mathbf{x}}{\|\mathbf{y}-\mathbf{x}\|} \right) \in C$.*
3. *There exists $\epsilon > 0$ such that $B(\mathbf{x}, \epsilon) \cap \mathrm{aff}(C) \subseteq C$.*

Proof Left as an exercise. $\qquad\square$

Definition 2.2.13 Let C be a convex set. We define the *relative interior* of C as the set of all $\mathbf{x} \in C$ that satisfy the three (equivalent) conditions in Proposition 2.2.12. We denote it by $\mathrm{relint}(C)$.[2] We define the *relative boundary* of C to be $\mathrm{relbd}(C) := \mathrm{cl}(C) \setminus \mathrm{relint}(C)$.

One can construct a system of coordinates for the affine hull so that one obtains a full-dimensional set in the new space. More formally, suppose $\dim(C) = k$ for some $C \subseteq \mathbb{R}^d$. One selects affinely independent points $\{\mathbf{x}^0, \mathbf{x}^1, \ldots, \mathbf{x}^k\}$ in C or $\mathrm{aff}(C)$ such that $\mathrm{aff}(C) = \mathrm{aff}(\{\mathbf{x}^0, \mathbf{x}^1, \ldots, \mathbf{x}^k\})$. Using part 3. of Theorem 2.2.8, every point $\mathbf{p} \in \mathrm{aff}(C)$ can be written uniquely (see Exercise 6 from Section 2.2.3) as $\mathbf{p} = \lambda_0 \mathbf{x}^0 + \lambda_1 \mathbf{x}^1 + \cdots \lambda_k \mathbf{x}^k$ with $\sum_{i=0}^{k} \lambda_i = 1$. In other words, $\mathbf{p} = \mathbf{x}^0 + \lambda_1 (\mathbf{x}^1 - \mathbf{x}^0) + \cdots + \lambda_k (\mathbf{x}^k - \mathbf{x}^0)$. We define the *coordinatization map with respect to* $\{\mathbf{x}^0, \mathbf{x}^1, \ldots, \mathbf{x}^k\}$ as $T \colon \mathrm{aff}(C) \to \mathbb{R}^k$ as $T(p) = (\lambda_1, \ldots, \lambda_k)$. Observe that $T(\mathbf{x}^0) = \mathbf{0}$. This coordinatization map makes $\mathrm{aff}(C)$ isomorphic to $\mathbb{R}^k$. The convex set C can be viewed as a full-dimensional subset of $\mathbb{R}^k$, i.e., $T(C) \subseteq \mathbb{R}^k$ and is full dimensional in $\mathbb{R}^k$. The relative interior $\mathrm{relint}(C)$ is mapped to $\mathrm{int}(T(C))$ and the relative boundary $\mathrm{relbd}(C)$ is mapped to $\mathrm{bd}(T(C))$; see Exercise 15 from Section 2.2.3. If $\mathbf{x}^0, \mathbf{x}^1, \ldots, \mathbf{x}^k$ are chosen such that $\mathbf{x}^1 - \mathbf{x}^0, \ldots, \mathbf{x}^k - \mathbf{x}^0$ are orthonormal (see part 4. of Theorem 2.2.8), then the coordinatization is called *orthonormal*. We can use the same idea to assign coordinates to any affine subspace with respect to a choice of affinely independent points whose affine hull equals this subspace. Orthonormal coordinates for this affine subspace are defined analogously.

Coordinatization maps are often useful when making arguments that use induction on the dimension of a convex set. Another important use is to define volumes of convex sets that are not full dimensional.

Definition 2.2.14 Let $X \subseteq \mathbb{R}^d$ be any affine subspace with dimension k. Let $T \colon X \to \mathbb{R}^k$ be an orthonormal coordinatization of X as defined above. Then the volume $\mathrm{vol}_X(Y)$ of any set $Y \subseteq X$ with respect to X is defined as the volume of $T(Y)$ when viewed as a set in $\mathbb{R}^k$, i.e.,

$$\mathrm{vol}_X(Y) := \int_{T(Y)} d\mathbf{y},$$

where $\mathbf{y}$ denotes the coordinates under the coordinatization map T. The *intrinsic volume* of a convex set $C \subseteq \mathbb{R}^d$ is defined as $\mathrm{vol}_{\mathrm{aff}(C)}(C)$ (with the convention that for zero-dimensional sets the intrinsic volume is zero).

Exercise 20 from Section 2.2.3 shows that the volume measure defined in Definition 2.2.14 does not depend on the choice of the affinely independent

---

[2] For readers familiar with the concept of a relative topology: the relative interior of C is the interior of C with respect to the relative topology of $\mathrm{aff}(C)$.

points as long as the condition of orthonormality is maintained. Thus, labeling the volume measure with the subscript X is justified and there is no need to parameterize or index with the map T.

One needs to be more careful when comparing the volume measure of two different affine subspaces. For example, the set $\mathrm{conv}\{(0,0,0),(1,0,0)\} \subseteq \mathbb{R}^3$ has volume zero with respect to the two-dimensional subspace $\{\mathbf{x} \in \mathbb{R}^3 : x_3 = 0\}$ but has volume 1 with respect to the one-dimensional subspace $\mathrm{span}(\{(1,0,0)\})$. However, comparing subspaces of the same dimension is easier. Exercise 21 from Section 2.2.3 shows that if $T \colon \mathbb{R}^d \to \mathbb{R}^d$ is an affine transformation given by $T(\mathbf{x}) = A\mathbf{x} + \mathbf{b}$ where $A \in \mathbb{R}^{d \times d}$ is an orthonormal matrix and $\mathbf{b} \in \mathbb{R}^d$, then $\mathrm{vol}_X(Y) = \mathrm{vol}_{T(X)}(T(Y))$, where X is any affine space and $Y \subseteq X$. Since T is invertible, $\dim(X) = \dim(T(X))$ (see Exercise 4 from Section 2.2.3). If a set Y is a subset of two distinct affine subspaces of the same dimension, then it can be shown that the volume of Y with respect to both subspaces is zero. In light of these observations, we will use the shorthand $\mathrm{vol}_k(\cdot)$ in place of $\mathrm{vol}_X(\cdot)$ whenever X is an affine subspace of dimension $k \leq d$. We will use this shorthand in the text wherever convenient. For a convex set C of dimension k, $\mathrm{vol}_k(C)$ coincides with the intrinsic volume. When the ambient dimension d is clear from context, we will drop the subscript and $\mathrm{vol}(\cdot)$ will default to the standard d-dimensional volume as in Definition 1.3.18.

2.2.2 Carathéodory's Theorem

Theorem 2.1.8 tells us that for any $X \subseteq \mathbb{R}^d$, if $\mathbf{x} \in \mathrm{conv}(X)$, there is a finite set of points $X' \subseteq X$ such that $\mathbf{x} \in \mathrm{conv}(X')$. Theorem 2.2.4 is the corresponding statement for the conical, affine, and linear hull of a set. From basic linear algebra facts, one can derive a bound on the number of elements needed in the linear and affine combinations for expressing points in the linear span and the affine hull in Theorems 1.2.10 and 2.2.8, respectively. Carathéodory's theorem below does the same for the conical and convex hulls.

Theorem 2.2.15 (Carathéodory's theorem – cone version) *Let $X \subseteq \mathbb{R}^d$ (not necessarily convex) and let $\mathbf{x} \in \mathrm{cone}(X)$. There exists a subset $X' \subseteq X$ such that X' is linearly independent (and thus, $|X'| \leq d$), and $\mathbf{x} \in \mathrm{cone}(X')$.*

Proof Since $\mathbf{x} \in \mathrm{cone}(X)$, by Theorem 2.2.4, we can find a finite set $\{\mathbf{x}^1, \ldots, \mathbf{x}^k\} \subseteq X$ such that $\mathbf{x} \in \mathrm{cone}(\{\mathbf{x}^1, \ldots, \mathbf{x}^k\})$. Choose a minimal such set, i.e., there is no strict subset of $\{\mathbf{x}^1, \ldots, \mathbf{x}^k\}$ whose conical hull contains $\mathbf{x}$. This implies that $\mathbf{x} = \lambda_1 \mathbf{x}^1 + \cdots + \lambda_k \mathbf{x}^k$ for some $\lambda_i > 0$ for each $i = 1, \ldots, k$. We claim that $\mathbf{x}^1, \ldots, \mathbf{x}^k$ are linearly independent. Suppose to the contrary that there exist multipliers $\gamma_1, \ldots, \gamma_k \in \mathbb{R}$, not all zero, such that $\gamma_1 \mathbf{x}^1 + \cdots + \gamma_k \mathbf{x}^k = \mathbf{0}$. By changing the signs of the γ_i's if necessary, we may assume that there exists $j \in \{1, \ldots, k\}$ such that $\gamma_j > 0$. Define

$$\theta = \min_{j:\gamma_j > 0} \frac{\lambda_j}{\gamma_j}, \qquad \lambda_i' = \lambda_i - \theta \gamma_i \quad \forall i = 1, \ldots, k.$$

Observe that $\lambda_i' \geq 0$ for all $i = 1, \ldots, k$ and

$$\lambda_1' \mathbf{x}^1 + \cdots + \lambda_k' \mathbf{x}^k = \lambda_1 \mathbf{x}^1 + \cdots + \lambda_k \mathbf{x}^k - \theta(\gamma_1 \mathbf{x}^1 + \cdots + \gamma_k \mathbf{x}^k) = \lambda_1 \mathbf{x}^1 + \cdots + \lambda_k \mathbf{x}^k = \mathbf{x}.$$

However, at least one of the λ_i''s is zero (corresponding to an index in $\arg\min_{j:\gamma_j > 0} \frac{\lambda_j}{\gamma_j}$), contradicting the minimal choice of $\{\mathbf{x}^1, \ldots, \mathbf{x}^k\}$. $\qquad\square$

Theorem 2.2.16 (Carathéodory's theorem – convex version) *Let $X \subseteq \mathbb{R}^d$ (not necessarily convex) and let $\mathbf{x} \in \mathrm{conv}(X)$. There exists a subset $X' \subseteq X$ such that X' is affinely independent (and thus, $|X'| \leq d + 1$), and $\mathbf{x} \in \mathrm{conv}(X')$.*

Proof Consider the set $Y \subseteq \mathbb{R}^{d+1}$ defined by $Y := \{(\mathbf{y}, 1) : \mathbf{y} \in X\}$. Now, $\mathbf{x} \in \mathrm{conv}(X)$ is equivalent to saying that $(\mathbf{x}, 1) \in \mathrm{cone}(Y)$. We get the desired result by applying Theorem 2.2.15 and part 4. of Proposition 2.2.7. $\qquad\square$

It is important to keep in mind the following difference between the situation with linear and affine hulls, and the situation with conical and convex hulls. Given $X \subseteq \mathbb{R}^d$, there exists a linearly (affinely) independent subset $X' \subseteq X$ such that every point in the linear (affine) hull of X can be expressed as a linear (affine) combination of points from X'. On the other hand, for conical and convex hulls, for every point in the conical (convex) hull of X, there exists a linearly (affinely) independent subset that expresses this point as a conical (convex) combination; different points may require different linearly (affinely) independent sets. The order of the quantifiers is important.

The following two applications of Carathéodory's theorem prove handy in many situations.

Proposition 2.2.17 *Let $\mathbf{a}^1, \ldots, \mathbf{a}^n \in \mathbb{R}^d$. Then $\mathrm{cone}(\{\mathbf{a}^1, \ldots, \mathbf{a}^n\})$ is closed.*

Proof Consider a convergent sequence $\{\mathbf{x}^i\}_{i \in \mathbb{N}} \subseteq \mathrm{cone}(\{\mathbf{a}^1, \ldots, \mathbf{a}^n\})$ converging to $\mathbf{x} \in \mathbb{R}^d$. By Carathéodory's theorem (Theorem 2.2.15), every $\mathbf{x}^i$ is in the conical hull of some linearly independent subset of $\{\mathbf{a}^1, \ldots, \mathbf{a}^n\}$. Since there are only finitely many linearly independent subsets of $\{\mathbf{a}^1, \ldots, \mathbf{a}^n\}$, the conical hull of one of these subsets contains infinitely many elements of the sequence $\{\mathbf{x}^i\}_{i \in \mathbb{N}}$. Thus, after passing to that subsequence, we may assume that $\{\mathbf{x}^i\}_{i \in \mathbb{N}} \subseteq \mathrm{cone}(\{\bar{\mathbf{a}}^1, \ldots, \bar{\mathbf{a}}^k\})$ where $\bar{\mathbf{a}}^1, \ldots, \bar{\mathbf{a}}^k$ are linearly independent. For each $\mathbf{x}^i$, there exists $\lambda^i \in \mathbb{R}_+^k$ such that $\mathbf{x}^i = \lambda_1^i \bar{\mathbf{a}}^1 + \cdots + \lambda_k^i \bar{\mathbf{a}}^k$. If we denote by $A \in \mathbb{R}^{d \times k}$ the matrix whose columns are $\bar{\mathbf{a}}^1, \ldots, \bar{\mathbf{a}}^k$, then $\mathbf{x}^i = A\lambda^i$. Since the columns of A are linearly independent, there exists a matrix $B \in \mathbb{R}^{k \times d}$ such that BA is the identity matrix. Thus, $B\mathbf{x}^i = BA\lambda^i = \lambda^i$ for every $i \in \mathbb{N}$. Since $\{\mathbf{x}^i\}_{i \in \mathbb{N}}$ is a convergent sequence, it is also a bounded set. This implies that $\{\lambda^i\}_{i \in \mathbb{N}}$ is a bounded set in $\mathbb{R}^k_+$ because it is the image of a bounded set under the linear (and therefore continuous) map given by the matrix B. Thus, by Theorem 1.3.10 there is a convergent subsequence $\lambda^{i_k} \to \lambda \in \mathbb{R}^k_+$. Taking limits,

$$\mathbf{x} = \lim_{k \to \infty} \mathbf{x}^{i_k} = \lim_{k \to \infty} A\lambda^{i_k} = A\lambda.$$

Since $\lambda \in \mathbb{R}^k_+$, we find that $\mathbf{x} \in \mathrm{cone}(\{\bar{\mathbf{a}}^1, \ldots, \bar{\mathbf{a}}^k\}) \subseteq \mathrm{cone}(\{\mathbf{a}^1, \ldots, \mathbf{a}^n\})$. $\qquad\square$

Proposition 2.2.18 *Let $X \subseteq \mathbb{R}^d$ be a compact set (not necessarily convex). Then* $\mathrm{conv}(X)$ *is compact.*

Proof By Theorem 2.2.16, every $\mathbf{x} \in \mathrm{conv}(X)$ is the convex combination of some $d + 1$ points in X. Define the following function $f: \underbrace{\mathbb{R}^d \times \cdots \times \mathbb{R}^d}_{d+1 \text{ times}} \times \mathbb{R}^{d+1} \to \mathbb{R}^d$ as follows:

$$f(\mathbf{y}^1, \ldots, \mathbf{y}^{d+1}, \lambda) = \lambda_1 \mathbf{y}^1 + \cdots + \lambda_{d+1} \mathbf{y}^{d+1}.$$

It is easily verified that f is a continuous function (each coordinate of $f(\cdot)$ is a bilinear quadratic function of the input). We now observe that $\mathrm{conv}(X)$ is the image of $\underbrace{X \times \cdots \times X}_{d+1 \text{ times}} \times \Delta^{d+1}$ under f, where

$$\Delta^{d+1} := \left\{ \lambda \in \mathbb{R}^{d+1}_+ : \lambda_1 + \cdots + \lambda_{d+1} = 1 \right\}.$$

Since X and Δ^{d+1} are compact sets, we obtain the result by applying Theorem 1.3.14. $\qquad\square$

2.2.3 Exercises

1. Prove Theorem 2.2.4.
2. Show that $X \subseteq \mathbb{R}^d$ is a linear subspace if and only if X is both a cone and an affine subset.
3. Show that if $A \subseteq B \subseteq \mathbb{R}^d$ then $\mathrm{conv}(A) \subseteq \mathrm{conv}(B)$. Does the converse hold? Show similarly that $\mathrm{cone}(A) \subseteq \mathrm{cone}(B)$, $\mathrm{aff}(A) \subseteq \mathrm{aff}(B)$, and $\mathrm{span}(A) \subseteq \mathrm{span}(B)$.
4. Let $T: \mathbb{R}^d \to \mathbb{R}^k$ be an affine transformation. Let $\mathbf{x}^1, \ldots, \mathbf{x}^n \in \mathbb{R}^d$. Show that if $T(\mathbf{x}^1), \ldots, T(\mathbf{x}^n)$ are affinely independent, then $\mathbf{x}^1, \ldots, \mathbf{x}^n$

are affinely independent. Does the converse hold? Conclude that for any convex set $C \subseteq \mathbb{R}^d$, $\dim(T(C)) \leq \dim(C)$.

5. Let $C \subseteq \mathbb{R}^d$ be convex. Show that the linear subspace parallel to $\mathrm{aff}(C)$ is given by $\mathrm{span}(C - C)$, where $C - C := C + (-1)C$. $C - C$ is called the *difference body* of C. Thus, $\dim(C) = \dim(\mathrm{span}(C - C))$.

6. Show that in part 3. of Theorem 2.2.8, the decomposition of an arbitrary $\mathbf{x} \in X$ as a linear combination of $\mathbf{v}^1, \ldots, \mathbf{v}^{m+1}$ is unique, i.e., if $\mathbf{x} = \lambda_1 \mathbf{v}^1 + \cdots + \lambda_{m+1} \mathbf{v}^{m+1}$ and $\mathbf{x} = \lambda_1' \mathbf{v}^1 + \cdots + \lambda_{m+1}' \mathbf{v}^{m+1}$, then $\lambda_i = \lambda_i'$ for all $i = 1, \ldots, m + 1$.

7. Suppose $X \subseteq \mathbb{R}^d$ is a set of affinely independent points. Let $X' \subseteq X$. Show that

$$\mathrm{conv}(X') \cap \mathrm{conv}(X \setminus X') = \emptyset.$$

8. Complete the proof of Lemma 2.2.10.

9. Let $X \subseteq \mathbb{R}^d$. Show that the maximum size of an affinely independent set of points in X is equal to the maximum size of an affinely independent set of points in $\mathrm{aff}(X)$.

10. Let $X \subseteq \mathbb{R}^d$. Show that X is a hyperplane if and only if X is an affine set of dimension $d - 1$. [Recall from Example 2.1.2 that a hyperplane is any set of the form $\{\mathbf{x} \in \mathbb{R}^d : \langle \mathbf{a}, \mathbf{x} \rangle = \delta\}$ for some $\mathbf{a} \in \mathbb{R}^d \setminus \{\mathbf{0}\}, \delta \in \mathbb{R}.]$

11. Let C be a convex set and let $\mathbf{a} \in \mathbb{R}^d \setminus \{\mathbf{0}\}$ and $\delta \in \mathbb{R}$ be such that $\langle \mathbf{a}, \mathbf{y} \rangle \leq \delta$ for all $\mathbf{y} \in C$. Prove that for any $\mathbf{x} \in \mathrm{int}(C)$, we must have $\langle \mathbf{a}, \mathbf{x} \rangle < \delta$.

12. Let $X \subseteq \mathbb{R}^d$ be a set of $d + 1$ affinely independent points. Show that $\mathrm{int}(\mathrm{conv}(X)) \neq \emptyset$.

13. Let $C \subseteq \mathbb{R}^d$ be a convex set. Show that the following are equivalent.

    (a) $\mathrm{int}(C) = \emptyset$.
    (b) $\dim(C) < d$.
    (c) There exists a hyperplane that contains C.

14. Prove Proposition 2.2.12.

15. Let $C \subseteq \mathbb{R}^d$ be a convex set with $\dim(C) = k$, and let $T : \mathrm{aff}(C) \to \mathbb{R}^k$ be a coordinatization map (with respect to some set of affinely independent points). Show that $T(C)$ is full dimensional in $\mathbb{R}^k$, $T(\mathrm{relint}(C)) = \mathrm{int}(T(C))$ and $T(\mathrm{relbd}(C)) = \mathrm{bd}(T(C))$.

16. Show that $\mathrm{relint}(C)$ is nonempty for any nonempty convex set $C \subseteq \mathbb{R}^d$.

17. Let $C \subseteq \mathbb{R}^d$ be a convex set. Show that $\mathrm{cl}(C)$ is convex. Show that $\mathrm{relint}(C)$ is convex by formalizing the picture in Figure 2.1.

18. Let $C \subseteq \mathbb{R}^d$ be a convex set. Show that $\mathrm{aff}(C) = \mathrm{aff}(\mathrm{cl}(C))$. Use Figure 2.1 to argue that for any convex set $C \subseteq \mathbb{R}^d$, $\mathrm{relint}(C) = \mathrm{relint}(\mathrm{cl}(C))$. Consequently, $\mathrm{relbd}(C) = \mathrm{relbd}(\mathrm{cl}(C))$.

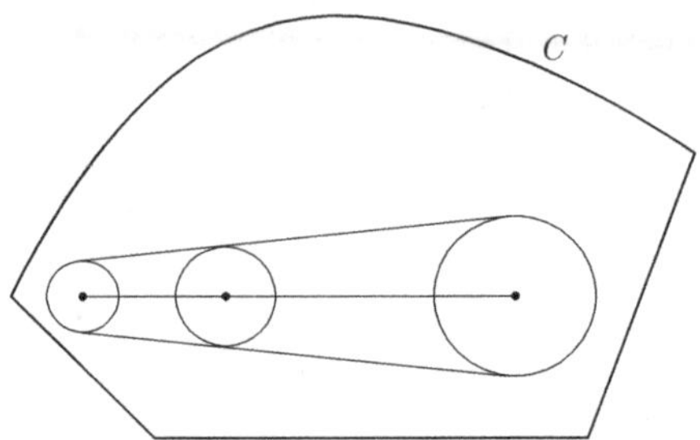

Figure 2.1 The relative interior of a convex set is convex.

19. Let $C \subseteq \mathbb{R}^d$ and let $\mathbf{x} \in \mathrm{relbd}(C)$. Show that there exists a sequence $\{\mathbf{x}^i\}_{i \in \mathbb{N}}$ such that $\mathbf{x}^i \in \mathrm{aff}(C) \setminus C$ for all $i \in \mathbb{N}$ and $\mathbf{x}^i \to \mathbf{x}$ as $i \to \infty$.
20. Let $X \subseteq \mathbb{R}^d$ be any affine subspace with dimension k. Let T and T' be two different orthonormal coordinatizations for X with coordinates $\mathbf{y}$ and $\mathbf{y}'$, respectively. Show that for any $Y \subseteq X$, we have $\int_{T(Y)} d\mathbf{y} = \int_{T'(Y)} d\mathbf{y}'$.
21. Let $T : \mathbb{R}^d \to \mathbb{R}^d$ be an affine transformation given by $T(\mathbf{x}) = A\mathbf{x} + \mathbf{b}$ where $A \in \mathbb{R}^{d \times d}$ is an orthonormal matrix and $\mathbf{b} \in \mathbb{R}^d$. Show that $\mathrm{vol}_X(Y) = \mathrm{vol}_{T(X)}(T(Y))$, where X is any affine space and $Y \subseteq X$.
22. Show that if $S \subseteq \mathbb{R}^d_+$ is closed (not necessarily compact), then $\mathrm{conv}(S) + \mathbb{R}^d_+$ is also closed.

2.3 The Projection Map

A useful tool in linear algebra is the notion of projecting on to a linear subspace. In particular, given a linear subspace $L \subseteq \mathbb{R}^d$ and any $\mathbf{x} \in \mathbb{R}^d$, one can show that there exists a unique $\mathbf{x}^\star \in L$ such that $\|\mathbf{x} - \mathbf{x}^\star\| \leq \|\mathbf{x} - \mathbf{y}\|$ for all $\mathbf{y} \in L$, and $\mathbf{x}^\star$ is called the *projection of* $\mathbf{x}$ *on to* L. In the study of convexity, it is very useful to extend this notion by allowing more general closed, convex sets beyond linear subspaces.

Proposition 2.3.1 *Let $C \subseteq \mathbb{R}^d$ be a nonempty, closed set and let $\mathbf{x} \in \mathbb{R}^d$. Then there is a point $\mathbf{x}^\star \in C$ such that $\|\mathbf{x} - \mathbf{x}^\star\| \leq \|\mathbf{x} - \mathbf{y}\|$ for all $\mathbf{y} \in C$. If C is also convex, then $\mathbf{x}^\star$ is unique and $\langle \mathbf{x} - \mathbf{x}^\star, \mathbf{y} - \mathbf{x}^\star \rangle \leq 0$ for all $\mathbf{y} \in C$.*

Proof If $\mathbf{x} \in C$, then both conclusions are true by setting $\mathbf{x}^\star = \mathbf{x}$. So, we assume $\mathbf{x} \notin C$. Consider any $\bar{\mathbf{x}} \in C$ and let $r = \|\mathbf{x} - \bar{\mathbf{x}}\|$. Let $\bar{C} = C \cap B(\mathbf{x}, r)$. Since C is closed and $B(\mathbf{x}, r)$ is compact, $\bar{C}$ is compact. One can also verify that the function $f(\mathbf{y}) = \|\mathbf{y} - \mathbf{x}\|$ is a continuous function on $\mathbb{R}^d$. Therefore, by Weierstrass' theorem (Theorem 1.3.13), there exists $\mathbf{x}^\star \in \bar{C}$ such that $\|\mathbf{x} - \mathbf{x}^\star\| \leq \|\mathbf{x} - \mathbf{y}\|$ for all $\mathbf{y} \in \bar{C}$, and therefore in fact $\|\mathbf{x} - \mathbf{x}^\star\| \leq \|\mathbf{x} - \mathbf{y}\|$ for all $\mathbf{y} \in C$.

Now suppose C is convex as well. For any $\mathbf{y} \in C$, all the points $\alpha\mathbf{y} + (1 - \alpha)\mathbf{x}^\star$, $\alpha \in (0, 1)$ are in C by convexity. Therefore, by the definition of $\mathbf{x}^\star$, we have

$$\|\mathbf{x} - \mathbf{x}^\star\|^2 \leq \|\mathbf{x} - (\alpha\mathbf{y} + (1 - \alpha)\mathbf{x}^\star)\|^2 \qquad \forall \alpha \in (0, 1),$$
$$\Rightarrow \qquad 0 \leq \alpha^2 \|\mathbf{y} - \mathbf{x}^\star\|^2 - 2\alpha\langle \mathbf{x} - \mathbf{x}^\star, \mathbf{y} - \mathbf{x}^\star \rangle \ \forall \alpha \in (0, 1),$$
$$\Rightarrow 2\langle \mathbf{x} - \mathbf{x}^\star, \mathbf{y} - \mathbf{x}^\star \rangle \leq \alpha\|\mathbf{y} - \mathbf{x}^\star\|^2 \qquad \forall \alpha \in (0, 1).$$

Letting $\alpha \to 0$ in the last inequality yields that $\langle \mathbf{x} - \mathbf{x}^\star, \mathbf{y} - \mathbf{x}^\star \rangle \leq 0$. To show uniqueness of $\mathbf{x}^\star$, consider any $\mathbf{y} \in C \setminus \{\mathbf{x}^\star\}$.

$$\|\mathbf{x} - \mathbf{y}\|^2 = \|(\mathbf{x} - \mathbf{x}^\star) + (\mathbf{x}^\star - \mathbf{y})\|^2$$
$$= \|\mathbf{x} - \mathbf{x}^\star\|^2 + \langle \mathbf{x} - \mathbf{x}^\star, \mathbf{x}^\star - \mathbf{y} \rangle + \|\mathbf{x}^\star - \mathbf{y}\|^2 > \|\mathbf{x} - \mathbf{x}^\star\|^2,$$

where the last inequality follows from the fact that $\mathbf{y} \neq \mathbf{x}^\star$ and $\langle \mathbf{x} - \mathbf{x}^\star, \mathbf{y} - \mathbf{x}^\star \rangle \leq 0$. $\qquad \square$

It turns out that a closed, convex set C is convex if and only if there is a unique closest point in C for every point in $\mathbb{R}^d$; Theorem 2.3.1 covers only one direction of this result. However, we will not need the stronger result and hence skip the proof of the other direction, which can be found in [210, Theorem 1.2.4].

Definition 2.3.2 For any nonempty, closed, convex set C, $\mathrm{Proj}_C(\mathbf{x})$ will denote the unique closest point (under the standard Euclidean norm) in C to $\mathbf{x}$. It is called the *projection map on to C*. As is usual, $\mathrm{Proj}_C(X)$ will denote the image of the set $X \subseteq \mathbb{R}^d$ under this map.

Proposition 2.3.3 *Let C be a nonempty, closed, convex set. $\mathrm{Proj}_C(\mathbf{x})$ is Lipschitz continuous with respect to the standard Euclidean norm with Lipschitz constant 1, i.e.,*

$$\|\mathrm{Proj}_C(\mathbf{x}) - \mathrm{Proj}_C(\mathbf{y})\| \leq \|\mathbf{x} - \mathbf{y}\|$$

for all $\mathbf{x}, \mathbf{y} \in \mathbb{R}^d$.

Proof Left as an exercise. $\qquad \square$

Lemma 2.3.4 *Let C be a nonempty, closed, convex set. For any $\mathbf{x} \in \mathrm{relbd}(C)$, there exists $\mathbf{y} \in \mathrm{aff}(C) \setminus C$ such that $\mathrm{Proj}(\mathbf{y}) = \mathbf{x}$.*

Proof By Exercise 19 from Section 2.2.3, there exists a sequence $\{\mathbf{x}^i\}_{i \in \mathbb{N}}$ such that $\mathbf{x}^i \in \mathrm{aff}(C) \setminus C$ for all $i \in \mathbb{N}$ and $\mathbf{x}^i \to \mathbf{x}$ as $i \to \infty$. Since $\{\mathbf{x}^i\}_{i \in \mathbb{N}}$ is a convergent sequence, it is bounded, and thus there exists $r > 0$ such that the ball $B(\mathbf{x}, r)$ contains the entire sequence. Since $B(\mathbf{x}, r)$ is compact, the half-line $\{\mathrm{Proj}_C(\mathbf{x}^i) + \lambda(\mathbf{x}^i - \mathrm{Proj}_C(\mathbf{x}^i)) : \lambda \geq 0\}$ intersects the boundary of $B(\mathbf{x}, r)$ at some point $\mathbf{y}^i$. Observe that $\mathbf{y}^i \in \mathrm{aff}(C)$ and $\mathrm{Proj}_C(\mathbf{y}^i) = \mathrm{Proj}_C(\mathbf{x}^i)$ (see Exercise 4 in Section 2.3.1).

Since the boundary of $B(\mathbf{x}, r)$ is compact, by Theorem 1.3.10 there exists a convergent subsequence $\mathbf{y}^{i_j}$ converging to a point $\mathbf{y}$ on the boundary of $B(\mathbf{x}, r)$. Since Proj_C is a Lipschitz continuous function,

$$\mathrm{Proj}_C(\mathbf{y}) = \mathrm{Proj}_C\left(\lim_{j\to\infty} \mathbf{y}^{i_j}\right) = \lim_{j\to\infty} \mathrm{Proj}_C\left(\mathbf{y}^{i_j}\right) = \lim_{j\to\infty} \mathrm{Proj}_C\left(\mathbf{x}^{i_j}\right)$$
$$= \mathrm{Proj}_C\left(\lim_{j\to\infty} \mathbf{x}^{i_j}\right) = \mathrm{Proj}_C(\mathbf{x}) = \mathbf{x}.$$

Note that $\mathbf{y} \in \mathrm{aff}(C)$ since $\mathbf{y}^{i_j} \in \mathrm{aff}(C)$ and $\mathrm{aff}(C)$ is a closed set. Moreover, $\mathbf{y} \neq \mathbf{x}$ since $\mathbf{y}$ is on the boundary of $B(\mathbf{x}, r)$ and $r > 0$. This implies that $\mathbf{y} \notin C$ since otherwise $\mathrm{Proj}_C(\mathbf{y}) = \mathbf{y} \neq \mathbf{x}$, contradicting what was derived above. $\qquad\square$

2.3.1 Exercises

1.* Prove Proposition 2.3.3.
2. Let $S \subseteq \mathbb{R}^d$ be an affine subspace, $\mathbf{x} \in \mathbb{R}^d$ and $\mathbf{y} \in S$. Use Proposition 2.3.1 to show that $\mathbf{y} = \mathrm{Proj}_S(\mathbf{x})$ if and only if $\mathbf{x} - \mathbf{y}$ lies in the orthogonal complement of the linear space parallel to S. Conclude that for any linear subspace $L \subseteq \mathbb{R}^d$, $\mathrm{Proj}_L(\cdot)$ is a linear map on $\mathbb{R}^d$.
3. Let $X_1 \subseteq X_2 \subseteq \mathbb{R}^d$ be linear subspaces and let $\mathbf{x} \in \mathbb{R}^d$. Show that $\mathrm{Proj}_{X_1}(\mathbf{x}) = \mathrm{Proj}_{X_1}(\mathrm{Proj}_{X_2}(\mathbf{x}))$. Is this true if $X_1 \subseteq X_2$ are arbitrary convex sets?
4. Let $C \subseteq \mathbb{R}^d$ be a closed, convex set and let $\mathbf{x} \in \mathbb{R}^d$ and $\hat{\mathbf{x}} \in C$. Show that $\hat{\mathbf{x}} = \mathrm{Proj}_C(\mathbf{x})$ if and only if $\langle \mathbf{x} - \hat{\mathbf{x}}, \mathbf{y} - \hat{\mathbf{x}} \rangle \leq 0$ for all $\mathbf{y} \in C$. Conclude that if $\mathbf{z} = \mathrm{Proj}_C(\mathbf{x}) + \lambda(\mathbf{x} - \mathrm{Proj}_C(\mathbf{x}))$ for some $\lambda \geq 0$, then $\mathrm{Proj}_C(\mathbf{z}) = \mathrm{Proj}_C(\mathbf{x})$.
5. Let $X \subseteq \mathbb{R}^d$ be a linear subspace. Is $\mathrm{Proj}_X(C)$ convex for every convex set C? Is $\mathrm{Proj}_X(C)$ closed for every closed, convex set C? Is $\mathrm{Proj}_X(C)$ compact for every compact, convex set C?
6. Let $X \subseteq \mathbb{R}^d$ be a linear subspace of dimension 1, spanned by $\mathbf{r} \in \mathbb{R}^d$. Let $C \subseteq \mathbb{R}^d$ be a convex set. Show that $\mathrm{cl}(\mathrm{Proj}_X(C)) = \{\gamma \hat{\mathbf{r}} : \ell \leq \gamma \leq u\}$, where $\hat{\mathbf{r}} = \frac{\mathbf{r}}{\|\mathbf{r}\|^2}$, $\ell = \inf_{\mathbf{x} \in C} \langle \mathbf{r}, \mathbf{x} \rangle$, and $u = \sup_{\mathbf{x} \in C} \langle \mathbf{r}, \mathbf{x} \rangle$. Moreover, $\mathrm{Proj}_X(C)$ is given by changing the constraint $\ell \leq \gamma$ to a strict inequality if $\inf_{\mathbf{x} \in C} \langle \mathbf{r}, \mathbf{x} \rangle$ is not attained, and retaining the weak inequality if the infimum is attained, and similarly with the constraint $\gamma \leq u$ and $\sup_{\mathbf{x} \in C} \langle \mathbf{r}, \mathbf{x} \rangle$.

2.4 Representations of Convex Sets

A large part of modern convex geometry is concerned with algorithms for computing with or optimizing over convex sets. For algorithmic purposes, we need ways to describe a convex set so that it can be stored in a computer and computations can be performed with it. In this section, we will establish some fundamental results about convex sets motivated by this question.

2.4.1 Extrinsic Description: Separating Hyperplanes

The most primitive convex set in $\mathbb{R}^d$ is the halfspace – see item 2. in Example 2.1.2. Moreover, a halfspace is a *closed* convex set. By Theorem 2.1.3, the intersection of an arbitrary family of halfspaces is a closed convex set. Perhaps the most fundamental theorem of convexity is that the converse is true.

Definition 2.4.1 Let $X \subseteq \mathbb{R}^d$ (not necessarily convex) and let $\mathbf{a} \in \mathbb{R}^d, \delta \in \mathbb{R}$. We say that $\langle \mathbf{a}, \mathbf{x} \rangle \leq \delta$ is a *valid inequality/halfspace* for X if $X \subseteq H^{\leq}(\mathbf{a}, \delta)$.

Theorem 2.4.2 (Separating hyperplane theorem) *Let $C \subseteq \mathbb{R}^d$ be a closed, convex set and let $\mathbf{x} \notin C$. There exists a halfspace that contains C and does not contain $\mathbf{x}$. More precisely, there exists $\mathbf{a} \in \mathbb{R}^d \setminus \{\mathbf{0}\}, \delta \in \mathbb{R}$ such that $\langle \mathbf{a}, \mathbf{y} \rangle \leq \delta$ for all $\mathbf{y} \in C$ and $\langle \mathbf{a}, \mathbf{x} \rangle > \delta$. In other words, $\langle \mathbf{a}, \mathbf{y} \rangle \leq \delta$ is a valid inequality for C that is violated by $\mathbf{x}$. The hyperplane $\{\mathbf{y} \in \mathbb{R}^d : \langle \mathbf{a}, \mathbf{y} \rangle = \delta\}$ is called a* separating hyperplane *for C and $\mathbf{x}$.*

Proof If C is empty, then any halfspace that does not contain $\mathbf{x}$ suffices. Otherwise, let $\mathbf{a} = \mathbf{x} - \mathrm{Proj}_C(\mathbf{x})$ and let $\delta = \langle \mathbf{a}, \mathrm{Proj}_C(\mathbf{x}) \rangle$. Note that $\mathbf{a} \neq \mathbf{0}$ because $\mathbf{x} \notin C$ and $\mathrm{Proj}_C(\mathbf{x}) \in C$. Therefore, $\langle \mathbf{a}, \mathbf{x} \rangle = \langle \mathbf{a}, \mathbf{a} + \mathrm{Proj}_C(\mathbf{x}) \rangle = \|\mathbf{a}\|^2 + \delta > \delta$. Thus, it remains to check that $\langle \mathbf{a}, \mathbf{y} \rangle \leq \delta$ for all $\mathbf{y} \in C$. Proposition 2.3.1 shows that $\langle \mathbf{a}, \mathbf{y} - \mathrm{Proj}_C(\mathbf{x}) \rangle \leq 0$ for all $\mathbf{y} \in C$. In other words, $\langle \mathbf{a}, \mathbf{y} \rangle \leq \langle \mathbf{a}, \mathrm{Proj}_C(\mathbf{x}) \rangle = \delta$. $\square$

Corollary 2.4.3 *Every closed convex set can be written as the intersection of some family of halfspaces, in particular, the family of all halfspaces that contain it. In other words, a subset $X \subseteq \mathbb{R}^d$ is a closed convex set if and only if there exists a family of tuples $(\mathbf{a}^i, \delta^i)$, $i \in I$ (where I may be an uncountable index set) such that $X = \bigcap_{i \in I} H^{\leq}(\mathbf{a}^i, \delta^i)$.*

Definition 2.4.4 A *finite* intersection of halfspaces is called a *polyhedron*. In other words, $P \subseteq \mathbb{R}^d$ is a polyhedron if and only if there exist vectors $\mathbf{a}^1, \ldots, \mathbf{a}^m \in \mathbb{R}^d$ and real numbers $b^1, \ldots, b^m$ such that $P = \{\mathbf{x} \in \mathbb{R}^d : \langle \mathbf{a}^i, \mathbf{x} \rangle \leq b^i \ i = 1, \ldots, m\}$. The shorthand $P = \{\mathbf{x} \in \mathbb{R}^d : A\mathbf{x} \leq \mathbf{b}\}$ is often employed, where A is the $m \times d$ matrix with $\mathbf{a}^1, \ldots, \mathbf{a}^m$ as rows, and

$\mathbf{b} = (b^1, \ldots, b^m) \in \mathbb{R}^m$. Thus, a polyhedron is completely described by specifying a matrix $A \in \mathbb{R}^{m \times d}$ and a vector $\mathbf{b} \in \mathbb{R}^m$.

Another related, and very useful, result is the following.

Theorem 2.4.5 (Supporting hyperplane theorem) *Let $C \subseteq \mathbb{R}^d$ be nonempty, convex and $\mathbf{x} \in \mathrm{relbd}(C)$. There exists $\mathbf{a} \in \mathbb{R}^d \setminus \{\mathbf{0}\}$ and $\delta \in \mathbb{R}$ such that all of the following hold:*

   (i) $\langle \mathbf{a}, \mathbf{y} \rangle \leq \delta$ *for all $\mathbf{y} \in C$, i.e., $\langle \mathbf{a}, \mathbf{y} \rangle \leq \delta$ is a valid inequality for C;*
  (ii) $\langle \mathbf{a}, \mathbf{x} \rangle = \delta$; *and*
 (iii) *there exists $\bar{\mathbf{y}} \in C$ such that $\langle \mathbf{a}, \bar{\mathbf{y}} \rangle < \delta$. This third condition says that C is not completely contained in the hyperplane $\{\mathbf{y} \in \mathbb{R}^d : \langle \mathbf{a}, \mathbf{y} \rangle = \delta\}$.*

Proof Let $\hat{C} = \mathrm{cl}(C)$. By Exercise 18 from Section 2.2.3, $\mathbf{x} \in \mathrm{relbd}(\hat{C})$. By Lemma 2.3.4, there exists $\mathbf{z} \in \mathrm{aff}(\hat{C}) \setminus \hat{C}$ such that $\mathbf{x} = \mathrm{Proj}_{\hat{C}}(\mathbf{z})$. We set $\mathbf{a} = \mathbf{z} - \mathbf{x}$ and $\delta = \langle \mathbf{a}, \mathbf{x} \rangle$, which establishes (ii) by definition, and we follow the proof of Theorem 2.4.2 to obtain (i) (since $C \subseteq \hat{C}$). To establish (iii), suppose to the contrary that C is completely contained in the hyperplane $H^=(\mathbf{a}, \delta)$. By Exercise 18 from Section 2.2.3, $\mathrm{aff}(C) = \mathrm{aff}(\hat{C})$ and thus $\mathbf{z} \in \mathrm{aff}(C)$. Since $H^=(\mathbf{a}, \delta)$ is an affine subspace (see Exercise 10 from Section 2.2.3), $\mathrm{aff}(C) \subseteq H^=(\mathbf{a}, \delta)$, implying that $\mathbf{z} \in H^=(\mathbf{a}, \delta)$, i.e., $\langle \mathbf{a}, \mathbf{z} \rangle = \delta$. However, $\langle \mathbf{a}, \mathbf{z} \rangle = \langle \mathbf{a}, \mathbf{a} + \mathbf{x} \rangle = \|\mathbf{a}\|^2 + \langle \mathbf{a}, \mathbf{x} \rangle > \delta$, since $\mathbf{a} \neq \mathbf{0}$. $\qquad\square$

Definition 2.4.6 For a nonempty, convex set $C \subseteq \mathbb{R}^d$ and $\mathbf{x} \in \mathrm{relbd}(C)$, we say that $\mathbf{a} \in \mathbb{R}^d \setminus \{\mathbf{0}\}$ and $\delta \in \mathbb{R}$ define a *supporting hyperplane for C at $\mathbf{x}$* if the three conditions of Theorem 2.4.5 are satisfied.

2.4.1.1 How to Represent General Convex Sets: Separation Oracles

We have seen that polyhedra can be represented by a matrix A and a right-hand side $\mathbf{b}$. Norm balls can be represented by the center $\mathbf{x}$ and the radius R. Ellipsoids can be represented by positive definite matrices A (see Section 2.7). What about general (closed) convex sets? This problem is gotten around by assuming that one has "black-box" access to the convex set via a *separation oracle*. More formally, we say that a convex set $C \subseteq \mathbb{R}^d$ is equipped with a separation oracle that takes as input any vector $\mathbf{x} \in \mathbb{R}^d$ and gives the following output: If $\mathbf{x} \in C$, the output is "YES," and if $\mathbf{x} \notin C$, then the output is a tuple $(\mathbf{a}, \delta) \in \mathbb{R}^d \times \mathbb{R}$ such that $\{\mathbf{y} \in \mathbb{R}^d : \langle \mathbf{a}, \mathbf{y} \rangle = \delta\}$ is a separating hyperplane for $\mathbf{x}$ and C. We will extensively use this concept in Chapter 6 when we discuss optimization over general closed, convex sets.

2.4.1.2 Duality/Polarity

With every linear space, one can associate a "dual" linear space, which is its orthogonal complement.

Definition 2.4.7   Let $X \subseteq \mathbb{R}^d$ be a linear subspace. We define $X^\perp := \{\mathbf{y} \in \mathbb{R}^d : \langle \mathbf{y}, \mathbf{x} \rangle = 0 \ \forall \mathbf{x} \in X\}$ as the *orthogonal complement* of X.

The following is well known from linear algebra.

Proposition 2.4.8 $X^\perp$ *is a linear subspace. Moreover,* $(X^\perp)^\perp = X$.

There is a way to extend this idea of associating a dual object to general convex sets, beyond linear subspaces. To motivate the construction, let us look at the orthogonal complement of a linear subspace in a slightly different way. Given any linear subspace L, if a halfspace $H^\leq(\mathbf{a}, \delta)$ given by $\mathbf{a} \in \mathbb{R}^d \setminus \{\mathbf{0}\}$ and $\delta \in \mathbb{R}$ contains L, then δ must be nonnegative since $\mathbf{0} \in L$. Thus, after scaling the right-hand side δ, all such halfspaces are of the form $H^\leq(\mathbf{a}, 1)$ or $H^\leq(\mathbf{a}, 0)$. Moreover, for $\mathbf{a} \in \mathbb{R}^d$, $L \subseteq H^\leq(\mathbf{a}, 1)$ is equivalent to $L \subseteq H^\leq(\mathbf{a}, 0)$, and this happens if and only if $\mathbf{a}$ is in the orthogonal complement $L^\perp$. In other words, it is not hard to see that $L^\perp$ is precisely the set of all normal vectors $\mathbf{a} \in \mathbb{R}^d$ such that the corresponding halfspace $H^\leq(\mathbf{a}, 1)$ contains L (allowing for the trivial possibility of $\mathbf{a} = \mathbf{0}$), i.e., $L^\perp = \{\mathbf{a} \in \mathbb{R}^d : \langle \mathbf{a}, \mathbf{x} \rangle \leq 1 \ \forall \mathbf{x} \in L\}$ (see part 4. of Proposition 2.4.10).

Extending this idea to any subset $X \subseteq \mathbb{R}^d$, we want to consider the family of all halfspaces that contain X. Up to scaling, we may assume the right-hand sides defining these halfspaces are -1, 0, or 1. We are led to consider then three classes of normal vectors: $\{\mathbf{a} \in \mathbb{R}^d : \langle \mathbf{a}, \mathbf{x} \rangle \leq -1 \ \forall \mathbf{x} \in X\}$, $\{\mathbf{a} \in \mathbb{R}^d : \langle \mathbf{a}, \mathbf{x} \rangle \leq 0 \ \forall \mathbf{x} \in X\}$, and $\{\mathbf{a} \in \mathbb{R}^d : \langle \mathbf{a}, \mathbf{x} \rangle \leq 1 \ \forall \mathbf{x} \in X\}$. Since the first set is contained in the second set which is contained in the last set, the last set is often all that is needed in the analysis. This leads us to a very important concept.

Definition 2.4.9   Let $X \subseteq \mathbb{R}^d$ be any set. The *polar* of X is defined as

$$X^\circ := \{\mathbf{y} \in \mathbb{R}^d : \langle \mathbf{y}, \mathbf{x} \rangle \leq 1 \ \forall \mathbf{x} \in X\}.$$

Proposition 2.4.10   *The following are all true.*

1. X° *is a closed, convex set for any* $X \subseteq \mathbb{R}^d$ *(not necessarily convex).*
2. $(X^\circ)^\circ = \mathrm{cl}(\mathrm{conv}(X \cup \{\mathbf{0}\}))$. *In particular, if X is a closed, convex set containing the origin, then* $(X^\circ)^\circ = X$.

3. *If X is a convex cone, then $X^\circ = \{\mathbf{y} \in \mathbb{R}^d : \langle \mathbf{y}, \mathbf{x} \rangle \leq 0 \;\; \forall \mathbf{x} \in X\}$.*
4. *If X is a linear subspace, then $X^\circ = X^\perp$.*

Proof   1. The claim follows from the fact that X° can be written as the intersection of closed halfspaces:

$$X^\circ = \bigcap_{\mathbf{x} \in X} \left\{ \mathbf{y} \in \mathbb{R}^d : \langle \mathbf{x}, \mathbf{y} \rangle \leq 1 \right\}.$$

2. Observe that $X \subseteq (X^\circ)^\circ$. Also, $\mathbf{0} \in (X^\circ)^\circ$, because $\mathbf{0}$ is always in the polar of any set. Since $(X^\circ)^\circ$ is a closed convex set by part 1., we must have $\mathrm{cl}(\mathrm{conv}(X \cup \{\mathbf{0}\})) \subseteq (X^\circ)^\circ$.

To show the reverse inclusion, we show that if $\mathbf{y} \notin \mathrm{cl}(\mathrm{conv}(X \cup \{\mathbf{0}\}))$ then $\mathbf{y} \notin (X^\circ)^\circ$. Thus, we need to show that there exists $\mathbf{z} \in X^\circ$ such that $\langle \mathbf{y}, \mathbf{z} \rangle > 1$. Since $\mathbf{y} \notin \mathrm{cl}(\mathrm{conv}(X \cup \{\mathbf{0}\}))$, by Theorem 2.4.2, there exists $\mathbf{a} \in \mathbb{R}^d, \delta \in \mathbb{R}$ such that $\langle \mathbf{a}, \mathbf{y} \rangle > \delta$ and $\langle \mathbf{a}, \mathbf{x} \rangle \leq \delta$ for all $\mathbf{x} \in \mathrm{cl}(\mathrm{conv}(X \cup \{\mathbf{0}\}))$. Since $\mathbf{0} \in \mathrm{cl}(\mathrm{conv}(X \cup \{\mathbf{0}\}))$, we obtain that $0 \leq \delta$. We now consider two cases:

Case 1: $\delta > 0$. Set $\mathbf{z} = \frac{\mathbf{a}}{\delta}$. Now, $\langle \mathbf{z}, \mathbf{x} \rangle \leq 1$ for all $\mathbf{x} \in X$ because $\langle \mathbf{a}, \mathbf{x} \rangle \leq \delta$ for all $\mathbf{x} \in \mathrm{cl}(\mathrm{conv}(X \cup \{\mathbf{0}\})) \supseteq X$. Therefore, $\mathbf{z} \in X^\circ$. Moreover, $\langle \mathbf{z}, \mathbf{y} \rangle > 1$ because $\langle \mathbf{a}, \mathbf{y} \rangle > \delta$. So we are done.

Case 2: $\delta = 0$. Define $\epsilon := \langle \mathbf{a}, \mathbf{y} \rangle > \delta = 0$. Set $\mathbf{z} = \frac{2\mathbf{a}}{\epsilon}$. Then, $\langle \mathbf{z}, \mathbf{y} \rangle = 2 > 1$. Also, for every $\mathbf{x} \in X \subseteq \mathrm{cl}(\mathrm{conv}(X \cup \{\mathbf{0}\}))$, we obtain that $\langle \mathbf{z}, \mathbf{x} \rangle = \frac{2}{\epsilon} \langle \mathbf{a}, \mathbf{x} \rangle \leq \frac{2}{\epsilon} \delta = 0 \leq 1$. Thus, $\mathbf{z} \in X^\circ$. Thus, we are done.

3. and 4. are left to the reader. $\square$

2.4.2 Intrinsic Description: Faces, Recession Cone, Lineality Space

We have seen that given any set X of points in $\mathbb{R}^d$, the convex hull of X – the smallest convex set containing X – can be expressed as the set of all convex combinations of finite subsets of X (Theorem 2.1.8). One possibility to represent a convex set C *intrinsically* is to give a minimal subset $X \subseteq C$ such that all points in C can be expressed as convex combinations of points in X, i.e., $C = \mathrm{conv}(X)$. In particular, if X is a finite set, then we can use X to represent C in a computer implicitly: C is the convex hull of the set X. We are going to get to such a "minimal" intrinsic description.

Definition 2.4.11 (Faces and extreme points)   Let C be a convex set. A convex subset $F \subseteq C$ is called an *extreme subset* or a *face* of C if for any $\mathbf{x} \in F$ the following holds: $\mathbf{x}^1, \mathbf{x}^2 \in C, \frac{\mathbf{x}^1 + \mathbf{x}^2}{2} = \mathbf{x}$ implies that $\mathbf{x}^1, \mathbf{x}^2 \in F$. This is equivalent to saying that there is no point in F that can be expressed as a strict convex combination of points in C, at least one of which is outside F; see Problem 15 from Section 2.4.4.

A face of dimension 0 is called an *extreme point*. In other words, $\mathbf{x}$ is an extreme point of C if the following holds: $\mathbf{x}^1, \mathbf{x}^2 \in C, \frac{\mathbf{x}^1 + \mathbf{x}^2}{2} = \mathbf{x}$ implies that $\mathbf{x}^1 = \mathbf{x}^2 = \mathbf{x}$. We denote the set of extreme points of C by $\mathrm{ext}(C)$. The one-dimensional faces of a convex set are called its *edges*. If $k = \dim(C)$, then the $(k-1)$-dimensional faces are called *facets*. We will see below that the only k-dimensional face of C is C itself. Any face of C that is not C or $\emptyset$ is called a *proper* face of C.

Lemma 2.4.12 *Let C be a convex set of dimension k. The only k-dimensional face of C is C itself.*

Proof If C is empty, then C is the only face of itself. Thus, we assume C is nonempty. Let $F \subsetneq C$ be a proper face of C. Let $\mathbf{x} \in C \setminus F$. Let $X \subseteq F$ be a maximum set of affinely independent points in F. We claim that $X \cup \{\mathbf{x}\}$ is affinely independent. By Proposition 2.2.11, this implies that $\dim(C) > \dim(F)$ and we will be done.

Suppose to the contrary that $\mathbf{x} \in \mathrm{aff}(X) = \mathrm{aff}(F)$. Then consider $\mathbf{x}^\star \in \mathrm{relint}(F)$ (which is nonempty by Exercise 16 from Section 2.2.10). By definition of the relative interior, since $\mathbf{x} \in \mathrm{aff}(F) \setminus F$, there exists $\epsilon > 0$ such that $\mathbf{y} = \mathbf{x}^\star + \epsilon(\mathbf{x} - \mathbf{x}^\star) \in F$ (see part 2. of Proposition 2.2.12). But this means that $\mathbf{y} = (1 - \epsilon)\mathbf{x}^\star + \epsilon\mathbf{x}$. Since $\mathbf{y} \in F$, and $\mathbf{x} \notin F$, this contradicts that F is a face. $\qquad\square$

Lemma 2.4.13 *Let $C \subseteq \mathbb{R}^d$ be convex. Let $\mathbf{a} \in \mathbb{R}^d$ and $\delta \in \mathbb{R}$ be such that $C \subseteq \{\mathbf{x} \in \mathbb{R}^d : \langle \mathbf{a}, \mathbf{x} \rangle \leq \delta\}$. Then, the set $F = C \cap \{\mathbf{x} \in \mathbb{R}^d : \langle \mathbf{a}, \mathbf{x} \rangle = \delta\}$ is a face of C.*

Proof Let $\bar{\mathbf{x}} \in F$ and $\mathbf{x}^1, \mathbf{x}^2 \in C$ such that $\frac{\mathbf{x}^1 + \mathbf{x}^2}{2} = \bar{\mathbf{x}}$. By the hypothesis, $\langle \mathbf{a}, \mathbf{x}^i \rangle \leq \delta$ for $i = 1, 2$. If for either $i = 1, 2$, $\langle \mathbf{a}, \mathbf{x}^i \rangle < \delta$, then

$$\langle \mathbf{a}, \bar{\mathbf{x}} \rangle = \left\langle \mathbf{a}, \frac{\mathbf{x}^1 + \mathbf{x}^2}{2} \right\rangle = \frac{\langle \mathbf{a}, \mathbf{x}^1 \rangle + \langle \mathbf{a}, \mathbf{x}^2 \rangle}{2} < \delta$$

contradicting that $\bar{\mathbf{x}} \in F$. Therefore, we must have $\langle \mathbf{a}, \mathbf{x}^i \rangle = \delta$ for $i = 1, 2$ and thus, $\mathbf{x}^1, \mathbf{x}^2 \in F$. $\qquad\square$

Definition 2.4.14 A face F of a convex set C is called an *exposed face* if there exists $\mathbf{a} \in \mathbb{R}^d$ and $\delta \in \mathbb{R}$ such that $C \subseteq \{\mathbf{x} \in \mathbb{R}^d : \langle \mathbf{a}, \mathbf{x} \rangle \leq \delta\}$ and $F = C \cap \{\mathbf{x} \in \mathbb{R}^d : \langle \mathbf{a}, \mathbf{x} \rangle = \delta\}$. We will sometimes make it explicit and say that F is an *exposed face induced by* $(\mathbf{a}, \delta)$.

An important consequence of the above discussion is the following theorem about the relative boundary of a closed, convex set C.

Theorem 2.4.15 *Let $C \subseteq \mathbb{R}^d$ be a nonempty, closed, convex set and $\mathbf{x} \in C$. $\mathbf{x}$ is contained in a proper face of C if and only if $\mathbf{x} \in \mathrm{relbd}(C)$.*

Proof If $\mathbf{x} \in \mathrm{relbd}(C)$, then by Theorem 2.4.5 there exists $\mathbf{a} \in \mathbb{R}^d$ and $\delta \in \mathbb{R}$ such that the three conditions in Theorem 2.4.5 hold. By Lemma 2.4.13, $F = C \cap \{\mathbf{x} \in \mathbb{R}^d : \langle \mathbf{a}, \mathbf{x} \rangle = \delta\}$ is a face of C, and it is a proper face because of condition (iii) in Theorem 2.4.5.

Now let $\mathbf{x} \in F$ where F is a proper face of C. Since C is closed, it suffices to show that $\mathbf{x} \notin \mathrm{relint}(C)$. Suppose to the contrary that $\mathbf{x} \in \mathrm{relint}(C)$. Let $\bar{\mathbf{x}} \in C \setminus F$. Observe that $2\mathbf{x} - \bar{\mathbf{x}} \in \mathrm{aff}(C)$ and since $\mathbf{x} \neq \bar{\mathbf{x}}$, $2\mathbf{x} - \bar{\mathbf{x}} \neq \mathbf{x}$. Since $\mathbf{x}$ is assumed to be in the relative interior of C, there exists $\epsilon > 0$ such that $\mathbf{y} = \epsilon((2\mathbf{x} - \bar{\mathbf{x}}) - \mathbf{x}) + \mathbf{x} \in C$ (see part 2. of Proposition 2.2.12). Rearranging terms, we obtain that

$$\mathbf{x} = \frac{\epsilon}{\epsilon + 1}\bar{\mathbf{x}} + \frac{1}{\epsilon + 1}\mathbf{y}.$$

Since $\mathbf{x} \in F$ and $\bar{\mathbf{x}} \notin F$, this contradicts the fact that F is a face. Thus, $\mathbf{x} \notin \mathrm{relint}(C)$ and so $\mathbf{x} \in \mathrm{relbd}(C)$. $\qquad\square$

In our search for a subset $X \subseteq C$ such that $C = \mathrm{conv}(X)$, it is clear that X must contain all extreme points. But is it sufficient to include all extreme points? In other words, is it true that $C = \mathrm{conv}(\mathrm{ext}(C))$? No! A simple counterexample is $\mathbb{R}^d_+$. Its only extreme point is $\mathbf{0}$. Another weird example is the set $\{\mathbf{x} \in \mathbb{R}^d : \|\mathbf{x}\| < 1\}$ – this set has NO extreme points! As you might suspect, the problem is that these sets are not compact, i.e., closed and bounded.

Theorem 2.4.16 (Krein–Milman theorem) *If C is a compact convex set, then $C = \mathrm{conv}(\mathrm{ext}(C))$.*

Proof The proof uses induction on the dimension of C. First, if C is the empty set, then the statement is a triviality. So we assume C is nonempty.

For the base case with $\dim(C) = 0$, i.e., $C = \{\mathbf{x}\}$ is a single point, the statement follows because $\{\mathbf{x}\}$ is an extreme point of C, and $C = \mathrm{conv}(\{\mathbf{x}\})$. For the induction step, consider any point $\mathbf{x} \in C$. We consider two cases:

Case 1: $\mathbf{x} \in \mathrm{relbd}(C)$. By Theorem 2.4.15, $\mathbf{x}$ is contained in a proper face F of C; in fact, following the proof of Theorem 2.4.15, F is an exposed face. Thus, F is closed, and since $F \subseteq C$, it is also bounded. In other words, F is a compact, convex set. By Lemma 2.4.12, $\dim(F) < \dim(C)$. By the induction hypothesis applied to F, we can express $\mathbf{x}$ as a convex combination of extreme points of F. Exercise 16 from Section 2.4.4 then shows that $\mathbf{x}$ is a convex combination of extreme points of C.

Case 2: $\mathbf{x} \in \text{relint}(C)$. Let $\ell \subseteq \text{aff}(C)$ be any affine set of dimension one (i.e., a line) going through $\mathbf{x}$. Since C is compact, $\ell \cap C$ is a line segment. The end points $\mathbf{x}^1, \mathbf{x}^2$ of $\ell \cap C$ must be in the relative boundary of C. By the previous case, $\mathbf{x}^1, \mathbf{x}^2$ can be expressed as the convex combination of extreme points in C. Since $\mathbf{x}$ is a convex combination of $\mathbf{x}^1$ and $\mathbf{x}^2$, and a convex combination of convex combinations is a convex combination, we can express $\mathbf{x}$ as the convex combination of extreme points of C. $\qquad\square$

What about noncompact sets? Let us relax the condition of being bounded, i.e., we want to describe closed, convex sets. It turns out that there is a nice way to deal with unboundedness. We introduce the necessary concepts next.

Proposition 2.4.17 *Let C be a nonempty, closed, convex set, and $\mathbf{r} \in \mathbb{R}^d$. The following are equivalent:*

1. *There exists $\mathbf{x} \in C$ such that $\mathbf{x} + \lambda\mathbf{r} \in C$ for all $\lambda \geq 0$.*
2. *For every $\mathbf{x} \in C$, $\mathbf{x} + \lambda\mathbf{r} \in C$ for all $\lambda \geq 0$.*

Proof Since C is nonempty, we only need to show 1. $\Rightarrow$ 2.; the reverse implication is trivial. Let $\bar{\mathbf{x}} \in C$ be such that $\bar{\mathbf{x}} + \lambda\mathbf{r} \in C$ for all $\lambda \geq 0$. Consider any arbitrary $\mathbf{x}^\star \in C$. Suppose to the contrary that there exists $\lambda' \geq 0$ such that $\mathbf{y} = \mathbf{x}^\star + \lambda'\mathbf{r} \notin C$. By Theorem 2.4.2, there exist $\mathbf{a} \in \mathbb{R}^d \setminus \{\mathbf{0}\}, \delta \in \mathbb{R}$ such that $\langle \mathbf{a}, \mathbf{y} \rangle > \delta$ and $\langle \mathbf{a}, \mathbf{x} \rangle \leq \delta$ for all $\mathbf{x} \in C$. This means that $\langle \mathbf{a}, \mathbf{r} \rangle > 0$ because, otherwise, $\langle \mathbf{a}, \mathbf{y} \rangle = \langle \mathbf{a}, \mathbf{x}^\star \rangle + \lambda'\langle \mathbf{a}, \mathbf{r} \rangle \leq \delta + \lambda'\langle \mathbf{a}, \mathbf{r} \rangle \leq \delta$, leading to a contradiction. But then, if we choose $\bar{\lambda} = \frac{|\delta - \langle \mathbf{a}, \bar{\mathbf{x}} \rangle| + 1}{\langle \mathbf{a}, \mathbf{r} \rangle} > 0$, we would obtain that

$$\langle \mathbf{a}, \bar{\mathbf{x}} + \bar{\lambda}\mathbf{r} \rangle = \langle \mathbf{a}, \bar{\mathbf{x}} \rangle + \bar{\lambda}\langle \mathbf{a}, \mathbf{r} \rangle = \langle \mathbf{a}, \bar{\mathbf{x}} \rangle + |\delta - \langle \mathbf{a}, \bar{\mathbf{x}} \rangle| + 1$$
$$\geq \langle \mathbf{a}, \bar{\mathbf{x}} \rangle + \delta - \langle \mathbf{a}, \bar{\mathbf{x}} \rangle + 1 \geq \delta + 1 > \delta,$$

contradicting the assumption that $\bar{\mathbf{x}} + \bar{\lambda}\mathbf{r} \in C$. $\qquad\square$

Definition 2.4.18 Any $\mathbf{r} \in \mathbb{R}^d$ that satisfies the conditions in Proposition 2.4.17 is called a *recession direction* for C.

Proposition 2.4.19 *The set of all recession directions of a nonempty, closed, convex set is a closed, convex cone.*

Proof Consider any nonempty, closed, convex set C. From Definition 2.2.1, we need to verify that the set of recession directions of C is nonempty and, for any two recession directions $\mathbf{r}^1, \mathbf{r}^2$ and $\lambda_1, \lambda_2 \geq 0$, the vector $\lambda_1\mathbf{r}^1 + \lambda_2\mathbf{r}^2$ is also a recession direction. Since $\mathbf{0}$ is always a recession direction, we have nonemptiness. Consider any point $\mathbf{x} \in C$. We need to verify that for any $\lambda \geq 0$,

$\mathbf{x} + \lambda(\lambda_1 \mathbf{r}^1 + \lambda_2 \mathbf{r}^2)$ is also in C. This follows from the fact that $\mathbf{x} + \lambda\lambda_1 \mathbf{r}^1 \in C$ since $\mathbf{r}^1$ is a recession direction, and therefore $(\mathbf{x} + \lambda\lambda_1 \mathbf{r}^1) + \lambda\lambda_2 \mathbf{r}^2$ is also in C since $\mathbf{r}^2$ is a recession direction.

For any sequence of recession directions $\mathbf{r}^i$, $i \in \mathbb{N}$ such that $\mathbf{r} = \lim_{i \to \infty} \mathbf{r}^i$, we verify that $\mathbf{r}$ is also a recession direction. Consider any point $\mathbf{x} \in C$ and $\lambda \geq 0$. Since each $\mathbf{r}^i$ is a recession direction, $\mathbf{x} + \lambda\mathbf{r}^i \in C$ for all $i \in \mathbb{N}$. Since C is closed, we must have that $\mathbf{x} + \lambda\mathbf{r} = \lim_{i \to \infty}(\mathbf{x} + \lambda\mathbf{r}^i)$ is also in C. $\qquad\square$

Definition 2.4.20 Let C be any nonempty, closed, convex set. We call the cone of recession directions the *recession cone* of C and it is denoted by $\mathrm{rec}(C)$. The set $\mathrm{rec}(C) \cap -\mathrm{rec}(C)$ is a linear subspace and is called the *lineality space* of C. It will be denoted by $\mathrm{lin}(C)$. As a matter of convention, we say that $\mathrm{rec}(C) = \mathrm{lin}(C) = \mathbb{R}^d$ when C is empty.

Proposition 2.4.17 immediately gives the following corollary.

Corollary 2.4.21 *Let C be a nonempty, closed convex set and let $F \subseteq C$ be a nonempty, closed, convex subset. Then $\mathrm{rec}(F) \subseteq \mathrm{rec}(C)$.*

Proof Left as an exercise. $\qquad\square$

Here is a characterization of compact convex sets.

Theorem 2.4.22 *A nonempty, closed convex set C is compact if and only if $\mathrm{rec}(C) = \{\mathbf{0}\}$.*

Proof We leave it to the reader to check that if C is compact, then $\mathrm{rec}(C) = \{\mathbf{0}\}$. For the other direction, assume that $\mathrm{rec}(C) = \{\mathbf{0}\}$. Suppose to the contrary that C is not bounded, i.e., there exists a sequence of points $\mathbf{y}^i \in C$ such that $\|\mathbf{y}^i\| \to \infty$. Let $\mathbf{x} \in C$ be any point and consider the set of unit norm vectors $\mathbf{r}^i = \frac{\mathbf{y}^i - \mathbf{x}}{\|\mathbf{y}^i - \mathbf{x}\|}$. Since this is a sequence of unit norm vectors, by Theorem 1.3.10, there is a convergent subsequence $\{\mathbf{r}^{i_k}\}_{k=1}^{\infty}$ converging to $\mathbf{r}$ also with unit norm. We claim that $\mathbf{r}$ is a recession direction, giving a contradiction to $\mathrm{rec}(C) = \{\mathbf{0}\}$. To see this, for any $\lambda \geq 0$, let $N \in \mathbb{N}$ such that $\|\mathbf{y}^{i_k} - \mathbf{x}\| > \lambda$ for all $k \geq N$. We now observe that

$$
\begin{aligned}
\mathbf{x} + \lambda\mathbf{r}^{i_k} &= \frac{(\|\mathbf{y}^{i_k} - \mathbf{x}\| - \lambda)}{\|\mathbf{y}^{i_k} - \mathbf{x}\|}\mathbf{x} + \frac{\lambda}{\|\mathbf{y}^{i_k} - \mathbf{x}\|}(\mathbf{x} + \mathbf{r}^{i_k}\|\mathbf{y}^{i_k} - \mathbf{x}\|) \\
&= \frac{(\|\mathbf{y}^{i_k} - \mathbf{x}\| - \lambda)}{\|\mathbf{y}^{i_k} - \mathbf{x}\|}\mathbf{x} + \frac{\lambda}{\|\mathbf{y}^{i_k} - \mathbf{x}\|}\mathbf{y}^{i_k} \in C
\end{aligned}
$$

for all $k \geq N$. Letting $k \to \infty$, since C is closed, we obtain that $\mathbf{x} + \lambda\mathbf{r} = \lim_{k \to \infty} \mathbf{x} + \lambda\mathbf{r}^{i_k} \in C$. $\qquad\square$

We next consider closed convex sets whose lineality space is $\{\mathbf{0}\}$.

Definition 2.4.23   If $\mathrm{lin}(C) = \{\mathbf{0}\}$ then C is called *pointed*.

The main result about pointed closed convex sets says that you can decompose them into convex combinations of extreme points and recession directions.

Theorem 2.4.24 *If C is a closed, convex set that is pointed, then $C = \mathrm{conv}(\mathrm{ext}(C)) + \mathrm{rec}(C)$.*

Proof   The proof follows the same lines as Theorem 2.4.16. We may assume C is nonempty since otherwise $\mathrm{ext}(C) = \emptyset$ and we are done. We prove by induction on dimension of C. If $\dim(C) = 0$, then C is a single point, and we are done. Consider any $\mathbf{x} \in C$ and then two cases:

Case 1: $\mathbf{x} \in \mathrm{relbd}(C)$. By Theorem 2.4.15, $\mathbf{x}$ is contained in a proper face F of C; in fact, following the proof of Theorem 2.4.15, F is an exposed face. Thus, F is closed, and since $F \subseteq C$, it is also pointed by Corollary 2.4.21. By Lemma 2.4.12, $\dim(F) < \dim(C)$. By the induction hypothesis applied to F, we can express $\mathbf{x} = \mathbf{x}' + \mathbf{d}$, where $\mathbf{x}'$ is a convex combination of extreme points of F and $\mathbf{d}$ is a recession direction for F. By Exercise 16 from Section 2.4.4, $\mathbf{x}'$ is a convex combination of extreme points of C. By Corollary 2.4.21, $\mathbf{d} \in \mathrm{rec}(C)$.

Case 2: $\mathbf{x} \in \mathrm{relint}(C)$. Let $\ell \subseteq \mathrm{aff}(C)$ be any affine set of dimension one (i.e., a line) going through $\mathbf{x}$. Since C contains no lines (C is pointed), $\ell \cap C$ is either a line segment, i.e., $\mathbf{x}$ is the convex combination of $\mathbf{x}^1, \mathbf{x}^2 \in \mathrm{relbd}(C)$, or $\ell \cap C$ is a half-line, i.e., $\mathbf{x} = \mathbf{x}' + \mathbf{d}$, where $\mathbf{x}' \in \mathrm{relbd}(C)$ and $\mathbf{d} \in \mathrm{rec}(C)$.

In the first case, using Case 1, for each $i = 1, 2$, $\mathbf{x}^i$ can be expressed as $\mathbf{x}^i = \mathbf{y}^i + \mathbf{d}^i$, where $\mathbf{y}^i$ is a convex combination of extreme points in C, and $\mathbf{d}^i \in \mathrm{rec}(C)$. Since $\mathbf{x}$ is a convex combination of $\mathbf{x}^1$ and $\mathbf{x}^2$, this shows that $\mathbf{x} \in \mathrm{conv}(\mathrm{ext}(C)) + \mathrm{rec}(C)$, since $\mathrm{rec}(C)$ is a convex cone.

In the second case, applying Case 1 to $\mathbf{x}'$, we express $\mathbf{x}' = \mathbf{y}' + \mathbf{d}'$ where $\mathbf{y}'$ is a convex combination of extreme points in C, and $\mathbf{d}' \in \mathrm{rec}(C)$. Thus, $\mathbf{x} = \mathbf{y}' + \mathbf{d}' + \mathbf{d}$ and we have the desired representation, again using the fact the $\mathrm{rec}(C)$ is a convex cone. $\qquad\square$

Let's make this description even more "minimal." For this we will need to understand the structure of pointed cones.

Proposition 2.4.25 *Let $D \subseteq \mathbb{R}^d$ be a closed, convex cone. The following are equivalent.*

1. D is pointed.
2. D° is full dimensional, i.e., $\dim(D^\circ) = d$.

3. **0** *is an exposed face of D.*
4. *There exists a compact, convex subset $B \subset D \setminus \{\mathbf{0}\}$ such that every $\mathbf{d} \in D \setminus \{\mathbf{0}\}$ can be uniquely written in the form $\mathbf{d} = \lambda\mathbf{b}$, where $\mathbf{b} \in B$ and $\lambda > 0$. In particular, $D = \text{cone}(B)$.*

Proof 1. $\Rightarrow$ 2. If D° is not full dimensional, then $\text{aff}(D^\circ)$ is a linear space of dimension strictly less than d, and so $\text{aff}(D^\circ)^\perp \neq \{\mathbf{0}\}$. Since $D^\circ \subseteq \text{aff}(D^\circ)$, using Exercise 9b from Section 2.4.4, and properties 2. and 4. in Proposition 2.4.10, we obtain that $\text{aff}(D^\circ)^\perp = \text{aff}(D^\circ)^\circ \subseteq (D^\circ)^\circ = D$. Since $\text{aff}(D^\circ)^\perp$ is a linear space, this implies that $\text{aff}(D^\circ)^\perp \subseteq \text{lin}(D)$, contradicting the assumption that D is pointed.

2. $\Rightarrow$ 3. By Exercise 13 from Section 2.2.3, $\text{int}(D^\circ) \neq \emptyset$. Choose any $\bar{\mathbf{y}} \in \text{int}(D^\circ)$. Since $D^\circ = \{\mathbf{y} \in \mathbb{R}^d : \langle \mathbf{x}, \mathbf{y} \rangle \leq 0 \ \forall \mathbf{x} \in D\}$, using Exercise 11 from Section 2.2.3, we obtain that $\langle \bar{\mathbf{y}}, \mathbf{x} \rangle < 0$ for every $\mathbf{x} \in D \setminus \{\mathbf{0}\}$. This shows that the exposed face induced by $(\bar{\mathbf{y}}, 0)$ is exactly $\{\mathbf{0}\}$.

3. $\Rightarrow$ 4. Let **0** be an exposed face induced by $(\mathbf{y}, 0)$. Define $B := D \cap \{\mathbf{x} \in \mathbb{R}^d : \langle \mathbf{y}, \mathbf{x} \rangle = -1\}$. It is clear from the definition that $\mathbf{0} \notin B$. Since it is the intersection of a convex cone and a hyperplane, B is also convex. We now show that B is compact. It is the intersection of closed sets, so it is closed. If B is empty, then B is compact and so we assume B is nonempty. By Theorem 2.4.22, it suffices to show that $\text{rec}(B) = \{\mathbf{0}\}$. Suppose to the contrary that there exists $\mathbf{r} \in \text{rec}(B) \setminus \{\mathbf{0}\}$. Consider any point $\bar{\mathbf{x}} \in B$. Since $\langle \mathbf{y}, \bar{\mathbf{x}} \rangle = -1$ and $\langle \mathbf{y}, \bar{\mathbf{x}} + \mathbf{r} \rangle = -1$, we obtain that $\langle \mathbf{y}, \mathbf{r} \rangle = 0$. Since $B \subseteq D$, $\text{rec}(B) \subseteq \text{rec}(D)$ by Corollary 2.4.21 and so $\mathbf{r} \in \text{rec}(D)$. By Proposition 2.4.17, we obtain that $\mathbf{0} + \mathbf{r} \in D$, i.e., $\mathbf{r} \in D$. But then $\langle \mathbf{y}, \mathbf{r} \rangle = 0$ contradicts the fact that **0** is an exposed face of D induced by $(\mathbf{y}, 0)$.

We next consider any $\mathbf{d} \in D \setminus \{\mathbf{0}\}$. By our assumption, $\langle \mathbf{y}, \mathbf{d} \rangle < 0$. Thus, setting $\mathbf{b} = \frac{\mathbf{d}}{|\langle \mathbf{y}, \mathbf{d} \rangle|}$, we obtain that $\langle \mathbf{y}, \mathbf{b} \rangle = -1$ and thus $\mathbf{b} \in B$. To show uniqueness, consider $\mathbf{b}^1, \mathbf{b}^2 \in B$ both satisfying the condition. This means, $\mathbf{b}^2 = \lambda\mathbf{b}^1$ for some $\lambda > 0$. Therefore,

$$\lambda\langle \mathbf{y}, \mathbf{b}^1 \rangle = \langle \mathbf{y}, \mathbf{b}^2 \rangle = -1 = \langle \mathbf{y}, \mathbf{b}^1 \rangle,$$

showing that $\lambda = 1$. This shows uniqueness of **b**.

4. $\Rightarrow$ 1. If D is not pointed, then there exists $\mathbf{x} \in D \setminus \{\mathbf{0}\}$ such that $-\mathbf{x} \in D$. Moreover, there exists $\lambda_1 > 0$ such that $\mathbf{x}^1 = \lambda_1\mathbf{x} \in B$ and $\lambda_2 > 0$ such that $\mathbf{x}^2 = \lambda_2(-\mathbf{x}) \in B$. Since B is convex, $\frac{\lambda_2}{\lambda_1+\lambda_2}\mathbf{x}^1 + \frac{\lambda_1}{\lambda_1+\lambda_2}\mathbf{x}^2 = \mathbf{0}$ is in B, contradicting the assumption that $B \subset D \setminus \{\mathbf{0}\}$. $\qquad \square$

Definition 2.4.26 For any closed convex cone D, any subset $B \subseteq D$ satisfying condition 4. of Proposition 2.4.25 is called a *base* of D.

The proof of Proposition 2.4.25 also shows the following.

Corollary 2.4.27 *Let D be a closed, convex cone. D is pointed if and only if there exists a hyperplane H such that $H \cap D$ is a base of D.*

Definition 2.4.28 Let D be a closed, convex cone. An edge of D is called an *extreme ray* of D. We say that $\mathbf{r} \in D$ *spans an extreme ray* if $\{\lambda \mathbf{r} : \lambda \geq 0\}$ is an extreme ray. The set of extreme rays of D will be denoted by $\mathrm{extr}(D)$.

Proposition 2.4.29 *Let D be a closed, convex cone and $\mathbf{r} \in D \setminus \{\mathbf{0}\}$. $\mathbf{r}$ spans an extreme ray of D if and only if for all $\mathbf{r}^1, \mathbf{r}^2 \in D$ such that $\mathbf{r} = \frac{\mathbf{r}^1 + \mathbf{r}^2}{2}$, there exist $\lambda_1, \lambda_2 \geq 0$ such that $\mathbf{r}^1 = \lambda_1 \mathbf{r}$ and $\mathbf{r}^2 = \lambda_2 \mathbf{r}$.*

Proof Left as an exercise. $\qquad\square$

Here is an analogue of the Krein–Milman theorem (Theorem 2.4.16) for closed, convex cones.

Theorem 2.4.30 *If D is a pointed, closed, convex cone, then $D = \mathrm{cone}(\mathrm{extr}(D))$.*

Proof By Proposition 2.4.25, there exists a base B for D. Since B is compact, $B = \mathrm{conv}(\mathrm{ext}(B))$ by Theorem 2.4.16. One can verify that the ray spanned by each $\mathbf{r} \in \mathrm{ext}(B)$ is an extreme ray for D, and vice versa, any extreme ray of D is spanned by some $\mathbf{r} \in \mathrm{ext}(B)$. Moreover, using the fact that $B = \mathrm{conv}(\mathrm{ext}(B))$, it follows from Proposition 2.4.25, part 4. that either $D = \{\mathbf{0}\}$ or $D = \mathrm{cone}(\mathrm{extr}(D))$. $\qquad\square$

For a closed convex set C, we will also use $\mathrm{extr}(C)$ to denote $\mathrm{extr}(\mathrm{rec}(C))$. We will also say these are the extreme rays of C. Now we can write a sharper version of Theorem 2.4.24:

Corollary 2.4.31 *If C is a closed, convex set that is pointed, then $C = \mathrm{conv}(\mathrm{ext}(C)) + \mathrm{cone}(\mathrm{extr}(C))$.*

Thus, to describe a pointed closed convex set, we just need to specify its extreme points and its extreme rays. We finally deal with general closed convex sets that are not necessarily pointed. The idea is that the lineality space can be "factored out."

Lemma 2.4.32 *Let C be a closed convex set and define $\hat{C} := C \cap \mathrm{lin}(C)^{\perp}$. Then $C = \hat{C} + \mathrm{lin}(C)$. Moreover, if C is nonempty, then $\hat{C}$ is pointed.*

Proof For any $\mathbf{x} \in C$, we can express $\mathbf{x} = \mathbf{x}' + \mathbf{r}$, where $\mathbf{x}' \in \mathrm{lin}(C)^{\perp}$ and $\mathbf{r} \in \mathrm{lin}(C)$ (since $\mathrm{lin}(C) + \mathrm{lin}(C)^{\perp} = \mathbb{R}^d$). We also know that $\mathbf{x}' = \mathbf{x} - \mathbf{r} \in C$ because $\mathbf{r} \in \mathrm{lin}(C)$. Thus, $\mathbf{x}' \in \hat{C}$ and we obtain $C = \hat{C} + \mathrm{lin}(C)$. This implies that C is nonempty if and only if $\hat{C}$ is nonempty. Moreover, $\hat{C}$ is closed because it is the intersection of two closed sets. By Corollary 2.4.21, $\mathrm{rec}(\hat{C}) \subseteq \mathrm{rec}(C)$. Therefore, $\mathrm{lin}(\hat{C}) = \mathrm{rec}(\hat{C}) \cap -\mathrm{rec}(\hat{C}) \subseteq \mathrm{rec}(C) \cap -\mathrm{rec}(C) = \mathrm{lin}(C)$. By the same reasoning, $\mathrm{lin}(\hat{C}) \subseteq \mathrm{lin}(\mathrm{lin}(C)^{\perp}) = \mathrm{lin}(C)^{\perp}$. Since $\mathrm{lin}(C) \cap \mathrm{lin}(C)^{\perp} = \{\mathbf{0}\}$, we obtain that $\mathrm{lin}(\hat{C}) = \{\mathbf{0}\}$. $\qquad\square$

Theorem 2.4.33 *Let C be a closed convex set and let $\hat{C} = C \cap \mathrm{lin}(C)^{\perp}$. Then*

$$C = \mathrm{conv}(\mathrm{ext}(\hat{C})) + \mathrm{cone}(\mathrm{extr}(\hat{C})) + \mathrm{lin}(C).$$

Proof If C is empty, then $\hat{C}$ is empty, and so is $\mathrm{ext}(\hat{C})$; the equality holds. So we assume C is nonempty. We first observe that $C = \hat{C} + \mathrm{lin}(C)$. By Lemma 2.4.32, $C = \hat{C} + \mathrm{lin}(C)$ and $\hat{C}$ is pointed. Applying Corollary 2.4.31 gives the desired result. $\qquad\square$

Thus, a general closed convex set C can be specified by giving a set of generators for its lineality space $\mathrm{lin}(C)$, the extreme points of the set $C \cap \mathrm{lin}(C)^{\perp}$, and vectors spanning its extreme rays. In Section 2.5, we will see that polyhedra are precisely those convex sets C that have a finite number of extreme points and extreme rays for $C \cap \mathrm{lin}(C)^{\perp}$. So we see that polyhedra are especially easy to describe intrinsically: simply specify the finite list of extreme points, vectors spanning the extreme rays, and a finite list of generators of $\mathrm{lin}(C)$.

2.4.3 A Remark about Extrinsic and Intrinsic Descriptions

The reader may have already observed that although a closed convex set can be represented as the intersection of halfspaces, such a representation is not unique. For example, consider the circle in $\mathbb{R}^2$. One can represent it by intersecting all its tangent halfspaces. On the other hand, if one throws away any finite subset of these halfspaces, one still gets the same set. In fact, there is a representation which uses only countably many halfspaces. Thus, the same convex set can have many different representations as the intersection of halfspaces. Moreover, there is usually no way to choose a "canonical" representation, i.e., there is no set of representing halfspaces such that *any representation* will always include this "canonical" set of halfspaces.

On the other hand, the intrinsic representation for a closed convex set is more "canonical." To begin with, consider the compact case. We express a compact C as $\mathrm{conv}(\mathrm{ext}(C))$. We cannot remove any extreme point, because it cannot be represented as the convex combination of other points. Thus, this representation is minimal/canonical in the sense that for any X such that $C = \mathrm{conv}(X)$, we must have $\mathrm{ext}(C) \subseteq X$. With closed, convex sets that have a nontrivial recession cone, the situation is a bit more subtle. First, there is more flexibility in choosing the representation because one can choose a different set of vectors to span the extreme rays. One might think that this is just a scaling issue and the following result holds: if C is a pointed, closed, convex set, and we consider any "intrinsic" representation

$$C = \mathrm{conv}(E) + \mathrm{cone}(R)$$

for some sets $E, R \subseteq \mathbb{R}^d$, then we must have

(i) $\mathrm{ext}(C) \subseteq E$, and
(ii) for every $\mathbf{r}$ that spans an extreme ray of $\mathrm{rec}(C)$, there must be some nonnegative scaling of $\mathbf{r}$ present in R.

While the above holds for polyhedra and many other closed, convex sets, it is not true in general. We leave it as an exercise to find a closed, convex set that violates the above claim.

2.4.4 Exercises

1.* Let X be a compact, convex set and let Y be a closed, convex set such that $X \cap Y = \emptyset$. Show that there exists a vector $\mathbf{a} \in \mathbb{R}^d \setminus \{\mathbf{0}\}$, $\delta \in \mathbb{R}$ and $\epsilon > 0$ such that $Y \subseteq \{\mathbf{y} : \langle \mathbf{a}, \mathbf{y} \rangle \leq \delta\}$ and $X \subseteq \{\mathbf{y} : \langle \mathbf{a}, \mathbf{y} \rangle \geq \delta + \epsilon\}$.
    Show that this property does not necessarily hold if X is only a closed, convex set, but not compact.
2. Let $C \subseteq \mathbb{R}^d$ be a convex set and two points $\mathbf{z} \notin C, \mathbf{x} \in C$. Show that $\mathbf{x} = \mathrm{Proj}_C(\mathbf{z})$ if and only if $\langle \mathbf{a}, \mathbf{y} \rangle = \langle \mathbf{a}, \mathbf{x} \rangle$ is a supporting hyperplane for C at $\mathbf{x}$, where $\mathbf{a} = \mathbf{z} - \mathbf{x}$.
3. Let $C \subseteq \mathbb{R}^d$ be a convex set (not necessarily closed) and let $\mathbf{x} \in C$. Show that $\mathbf{x} \in \mathrm{bd}(C)$ if and only if there exists $\mathbf{a} \in \mathbb{R}^d \setminus \{\mathbf{0}\}$ and $\delta \in \mathbb{R}$ such that $C \subseteq H^{\leq}(\mathbf{a}, \delta)$ and $\langle \mathbf{a}, \mathbf{x} \rangle = \delta$.
4. What happens when C is a singleton in Lemma 2.3.4 and Theorem 2.4.5?
5. Let $C \subseteq \mathbb{R}^d$ be convex (not necessarily closed) and let $\mathbf{x} \notin C$. Show that there exists $\mathbf{a} \in \mathbb{R}^d \setminus \{\mathbf{0}\}$ and $\delta \in \mathbb{R}$ such that $C \subseteq H^{\leq}(\mathbf{a}, \delta)$ and $\langle \mathbf{a}, \mathbf{x} \rangle \geq \delta$. Conclude that for any disjoint convex sets $X, Y \subseteq \mathbb{R}^d$, there exists $\mathbf{a} \in \mathbb{R}^d \setminus \{\mathbf{0}\}$ and $\delta \in \mathbb{R}$ such that $X \subseteq H^{\leq}(\mathbf{a}, \delta)$ and $Y \subseteq H^{\geq}(\mathbf{a}, \delta)$.
6. Let $S \subseteq \mathbb{R}^d$ (not necessarily convex) and $\mathbf{x} \in \mathbb{R}^d \setminus \mathrm{aff}(S)$. Show that there exists a hyperplane H such that $S \subseteq H$ and $\mathbf{x} \notin H$.
7. Prove parts 3. and 4. in Proposition 2.4.10.
8. What is the polar of $\{\mathbf{0}\}$? What is the polar of $\mathbb{R}^d_+$?
9. Let $X, Y \subseteq \mathbb{R}^d$. Show the following.

(a) Let $A \in \mathbb{R}^{d \times d}$ be an invertible matrix. Show that
$(A(X))^\circ = A^{-T}(X^\circ)$. Conclude that $(\alpha X)^\circ = \frac{1}{\alpha} X^\circ$ for any
$\alpha \in \mathbb{R} \setminus \{0\}$.

(b) If $X \subseteq Y$, then $Y^\circ \subseteq X^\circ$.

(c) $(X \cup Y)^\circ = X^\circ \cap Y^\circ$.

(d) If X is a closed, convex set with $\mathbf{0} \in X$ and Y is a linear subspace,
then $(X \cap Y)^\circ = \mathrm{cl}(X^\circ + Y^\perp)$.

10. Let $r > 0$. Show that $B_{\ell^2}(\mathbf{0}, r)^\circ = B_{\ell^2}(\mathbf{0}, \frac{1}{r})$. Conclude that the unit ball
for the standard Euclidean norm is its own polar. Is that the only set which
is its own polar?

11. Show that $X \subseteq \mathbb{R}^d$ is bounded if and only if X° contains the origin in its
interior.

12. Show that $B_{\ell^1}^\circ(\mathbf{0}, 1) = B_{\ell^\infty}(\mathbf{0}, 1)$.

13. Let $\mathbf{x}^1, \ldots, \mathbf{x}^k \in \mathbb{R}^d$.

(i) Show that if $A = \mathrm{cone}\{\mathbf{x}^1, \ldots, \mathbf{x}^k\}$ then

$$A^\circ = \{\mathbf{y} \in \mathbb{R}^d : \langle \mathbf{x}^i, \mathbf{y} \rangle \le 0 \ \forall i = 1, \ldots, k\}.$$

(ii) Show if $B = \{\mathbf{y} \in \mathbb{R}^d : \langle \mathbf{x}^i, \mathbf{y} \rangle \le 0 \ \forall i = 1, \ldots, k\}$, then
$B^\circ = \mathrm{cone}\{\mathbf{x}^1, \ldots, \mathbf{x}^k\}$.

14. Let $C_1, C_2 \subseteq \mathbb{R}^d$ be closed, convex cones. Show that $(C_1 + C_2)^\circ = C_1^\circ \cap C_2^\circ$.

15. Consider the following definition.

> Let $C \subseteq \mathbb{R}^d$ be a convex set. We say a convex subset $F \subseteq C$ is a *funky face*
> if for all $\mathbf{x} \in F$, the following holds: $\mathbf{x} = \alpha \mathbf{y} + (1 - \alpha)\mathbf{z}$ for some $\mathbf{y}, \mathbf{z} \in C$
> and $0 < \alpha < 1$ implies that $\mathbf{y}, \mathbf{z} \in F$. In other words, a funky face is a
> convex subset F of C such that no point in F is a convex combination of
> points outside F.

Show that F is a funky face of C if and only if F is a face of C.

16. Let $C \subseteq \mathbb{R}^d$ be convex. Let F_1 be a face of C. Let F_2 be a face of F_1.
Show that F_2 is a face of C. Does this remain true if one replaces *face*
with *exposed face* everywhere? In other words, if F_1 is an exposed face of
C, and F_2 is an exposed face of F_1, is it true that F_2 is an exposed face of
C?

17. Let $C \subseteq \mathbb{R}^d$ be convex set and $F_1, F_2 \subseteq C$ be faces of C. Show that
$F_1 \cap F_2$ is a face. Does this remain true if one replaces *face* with *exposed
face* everywhere? In other words, if F_1, F_2 are exposed faces of C, is it
true that $F_1 \cap F_2$ is an exposed face of C?

18. Let $C \subseteq \mathbb{R}^d$ be a closed, convex set. Show that every face of C is closed.

19. Let $A \subseteq \mathbb{R}^d$ (not necessarily convex). Show that if F is a face of $\mathrm{conv}(A)$, then $F = \mathrm{conv}(A \cap F)$.

20. Let $C \subseteq \mathbb{R}^d$ be a convex set and let $F \subseteq C$ be a face of C. Let $F' \subseteq F$. Show that F' is a face of C if and only if F' is a face of F.

21. Let $X \subseteq C \subseteq \mathbb{R}^d$ be two convex sets. Let $\mathbf{x} \in \mathrm{relint}(X)$. Suppose F is a face of C such that $\mathbf{x} \in F$. Show that $X \subseteq F$.

22. Let $X \subseteq C \subseteq \mathbb{R}^d$ be two convex sets. Let $F \subseteq X$ such that F is a face of C. Show that F is a face of X as well.

23. Let $C \subseteq \mathbb{R}^d$ be a convex set that is not a singleton. Show that no point from $\mathrm{relint}(C)$ is an extreme point of C.

24. Let $C \subseteq \mathbb{R}^d$ be a convex set and let $T : \mathbb{R}^d \to \mathbb{R}^k$ be a linear map. Let F be a face of $T(C)$. Show that $T^{-1}(F) \cap C$ is a face of C. Is $T(F)$ always a face of $T(C)$ for a face F of C?

25. Let $C \subseteq \mathbb{R}^d$ be a convex set. Show that $\mathbf{x} \in C$ is an extreme point if and only if $C \setminus \{\mathbf{x}\}$ is convex.

26. Give an example of a compact convex set in $\mathbb{R}^3$ whose set of extreme points is not closed. In other words, there is a sequence of extreme points that converges to something that is not extreme. Can this happen in $\mathbb{R}^2$?

27. Let $C \subseteq \mathbb{R}^d$ be a closed, convex set and let $\mathbf{x} \in C$. Show that $\mathbf{x}$ is an extreme point of C if and only if for every $\epsilon > 0$, there exists a halfspace $H^{\leq}(\mathbf{a}, \delta)$ given by $\mathbf{a} \in \mathbb{R}^d \setminus \{\mathbf{0}\}$ and $\delta \in \mathbb{R}$ such that $\langle \mathbf{a}, \mathbf{x} \rangle < \delta$ and $C \cap H^{\leq}(\mathbf{a}, \delta) \subseteq C \cap B(\mathbf{x}, \epsilon)$.

28.* Construct a convex set that has extreme points that are not exposed. Show that for any compact, convex set, every extreme point is the limit of some sequence of exposed points. (This result is known as *Straszewicz's theorem*.)

29. Show that Proposition 2.4.17 remains true if $\lambda \geq 0$ is replaced by $\lambda \in \mathbb{R}$ in both conditions. Show that $\mathrm{lin}(C)$ is exactly the set of all $\mathbf{r} \in \mathbb{R}^d$ that satisfy these modified conditions.

30. Let $X = \{\mathbf{x} \in \mathbb{R}_+^2 : \mathbf{x}_1 \mathbf{x}_2 \geq 1\}$ (the hyperbola). What is $\mathrm{rec}(X)$?

31. Let $X = \{\mathbf{x} \in \mathbb{R}^2 : \mathbf{x}_1^2 \leq \mathbf{x}_2\}$ (the parabola). What is $\mathrm{rec}(X)$?

32. What is the recession cone of an affine subspace?

33. Prove Corollary 2.4.21.

34. Let C be a closed, convex set. Let $\mathbf{x} \in \mathrm{relint}(C)$ and $\mathbf{r} \in \mathrm{rec}(C)$. Show that $\mathbf{x} + \mathbf{r} \in \mathrm{relint}(C)$.

35. Show that any unbounded convex set in $\mathbb{R}^d$ has either zero d-dimensional volume or infinite d-dimensional volume.

36. Let $C \subseteq \mathbb{R}^d$ be a closed, convex set and let $\mathbf{r} \in \mathrm{rec}(C)$. Show that $\mathbf{r} \in \mathrm{span}(C - C)$, i.e., $\mathbf{r}$ is in the linear space parallel to $\mathrm{aff}(C)$ (see Exercise 5 in Section 2.2.3).

37. Let $C \subseteq \mathbb{R}^d$ be a closed, convex set and let $\mathbf{r} \in \mathrm{span}(C - C)$, i.e., $\mathbf{r}$ is in the linear space parallel to $\mathrm{aff}(C)$ (see Exercise 5 in Section 2.2.3). Show that $\mathrm{relint}(C + \mathrm{span}(\mathbf{r})) = \mathrm{relint}(C) + \mathrm{span}(\mathbf{r})$.

38. Let $C \subseteq \mathbb{R}^d$ be a closed, convex set. Show that $\mathrm{Proj}_{\mathrm{lin}(C)^\perp}(C) = C \cap \mathrm{lin}(C)^\perp$, and $\dim(C) = \dim(\mathrm{Proj}_{\mathrm{lin}(C)^\perp}(C)) + \dim(\mathrm{lin}(C))$. Moreover, $\mathrm{Proj}_{\mathrm{lin}(C)^\perp}(C)$ is compact if and only if $\mathrm{rec}(C) = \mathrm{lin}(C)$.

39. Let $D \subseteq \mathbb{R}^d$ be a closed, convex cone. Show that if $\mathbf{x}$ is an extreme point of D then $\mathbf{x} = \mathbf{0}$. Show that D is pointed if and only if $\{\mathbf{0}\}$ is an extreme point of D.

40. Give an example of a convex cone $D \subseteq \mathbb{R}^d$ (not necessarily closed) such that $\mathbf{0}$ is an extreme point of D, and D contains straight lines.

41. Show that if $D \subseteq \mathbb{R}^d$ is a convex cone, then $T(D)$ is also a convex cone for any linear transformation T. Find an example of a closed, convex cone D and linear transformation T such that $T(D)$ is not closed.

42. Show that if $D_1, D_2 \subseteq \mathbb{R}^d$ are convex cones, then $D_1 + D_2$ is a convex cone. Give an example of closed convex cones D_1, D_2 such that $D_1 + D_2$ is not closed.

43. Let D be a closed, convex cone. Let $\langle \mathbf{a}, \mathbf{x} \rangle \leq \delta$ be a valid inequality for D. Show that $0 \leq \delta$ and that $\langle \mathbf{a}, \mathbf{x} \rangle \leq 0$ is also a valid inequality for D.

44. For any $X \subseteq \mathbb{R}^d$, define the set $X^{\circ,0} := \{\mathbf{y} \in \mathbb{R}^d : \langle \mathbf{y}, \mathbf{x} \rangle \leq 0 \ \forall \mathbf{x} \in X\}$. In other words, $X^{\circ,0}$ is that subset of normal vectors $\mathbf{y}$ from the polar X° such that the halfspace $H^{\leq}(\mathbf{y}, 0)$ contains X. Show that for any set $X \subseteq \mathbb{R}^d$, $(X^{\circ,0})^{\circ,0} = \mathrm{cl}(\mathrm{cone}(X))$.

45. (Optimizing linear functions over convex sets and extreme points) Let $X \subseteq \mathbb{R}^d$ be a compact, convex set. Let $f : \mathbb{R}^d \to \mathbb{R}$ be a linear function given by $f(\mathbf{y}) = \langle \mathbf{a}, \mathbf{y} \rangle$ for some $\mathbf{a} \in \mathbb{R}^d$. Show that there exists an extreme point $\mathbf{v} \in X$ such that $f(\mathbf{v}) \geq f(\mathbf{x})$ for all $\mathbf{x} \in X$. In other words, $\mathbf{v} \in \arg\max_{\mathbf{x} \in X} f(\mathbf{x})$, or equivalently, $f(\mathbf{v}) = \max_{\mathbf{x} \in X} f(\mathbf{x})$.

46. Let $X \subseteq \mathbb{R}^d$ be a compact, convex set. Suppose $\mathbf{v} \in X$ satisfies that $\|\mathbf{v}\| \geq \|\mathbf{x}\|$ for all $\mathbf{x} \in X$. Show that $\mathbf{v}$ is an extreme point of X.

47. Show that if $D \subseteq \mathbb{R}^d$ is a convex cone, then any face of D is a convex cone.

48. Give an example of a closed, convex cone D with a proper face that is not exposed.

49. Let D be a closed, convex cone. Show that if $B \subseteq D$ is a base of D (see Definition 2.4.26) then there exists a hyperplane H not containing $\mathbf{0}$ such that $B = H \cap D$.

50. Let $X \subseteq \mathbb{R}^d$ (not necessarily convex), and let $\mathbf{y} \in \mathrm{conv}(X)$. Suppose H is a halfspace such that $\mathbf{y} \in H$. Show that $H \cap X \neq \emptyset$.

51. Show that a nonempty, closed convex set $C \subseteq \mathbb{R}^d$ has an extreme point if and only if it is line-free, i.e., C contains no lines.

52. Prove Proposition 2.4.29.

53. (Homogenization of a convex set) Let $C \subseteq \mathbb{R}^d$ be a convex set. Define the *homogenization of C* as $D(C) := \mathrm{cl}\left(\mathrm{cone}\left(\{(\mathbf{x}, 1) \in \mathbb{R}^d \times \mathbb{R} : \mathbf{x} \in C\}\right)\right)$. Show the following:

   (a) For any closed, convex set $C \subseteq \mathbb{R}^d$, $C = \{\mathbf{x} \in \mathbb{R}^d : (\mathbf{x}, 1) \in D(C)\}$. In other words, C can be recovered from the "slice" of $D(C)$ obtained from setting the last coordinate to 1.

   (b) For any nonempty, closed, convex set $C \subseteq \mathbb{R}^d$, $\mathrm{rec}(C) = \{\mathbf{x} \in \mathbb{R}^d : (\mathbf{x}, 0) \in D(C)\}$. In other words, $\mathrm{rec}(C)$ can be obtained from the "slice" of $D(C)$ with the last coordinate equal to 0.

   (c) For any convex sets $C_1, C_2 \subseteq \mathbb{R}^d$, $D(\mathrm{conv}(C_1 \cup C_2)) = \mathrm{cl}(D(C_1) + D(C_2))$.

2.5 Polyhedra

Recall that a polyhedron is any convex set that can be obtained by intersecting a finite number of halfspaces (Definition 2.4.4). Polyhedra, in a sense, are the nicest convex sets to work with because of this finiteness property. For example, our first result will be that a polyhedron can have only finitely many extreme points.

Even so, one thing to keep in mind is that the same polyhedron can be described as the intersection of two completely different finite families of halfspaces. This brings into sharp focus the nonuniqueness of extrinsic descriptions discussed in Section 2.4.3. Consider the following systems of halfspace/inequalities.

$$
\begin{array}{rl}
-\mathbf{x}_1 & \leq 0 \\
\mathbf{x}_1 + \mathbf{x}_2 & \leq 0 \\
\mathbf{x}_1 - \mathbf{x}_2 & \leq 0 \\
-\mathbf{x}_1 - \mathbf{x}_2 - \mathbf{x}_3 & \leq 0 \\
\mathbf{x}_2 + \mathbf{x}_3 & \leq 5
\end{array}
\qquad
\begin{array}{rl}
2\mathbf{x}_1 + \mathbf{x}_2 & \leq 0 \\
-\mathbf{x}_1 + \mathbf{x}_2 & \leq 0 \\
\mathbf{x}_1 - 2\mathbf{x}_2 & \leq 0 \\
\mathbf{x}_1 \quad - 2\mathbf{x}_3 & \leq 0 \\
2\mathbf{x}_1 + \mathbf{x}_2 + 2\mathbf{x}_3 & \leq 10
\end{array}
$$

Both these systems describe the same polyhedron $P = \mathrm{conv}\{(0,0,0),(0,0,5)\}$ in $\mathbb{R}^3$. However, if a polyhedron is given by its list of extreme points and extreme rays, this ambiguity disappears. Moreover, having these two alternate extrinsic/intrinsic descriptions is very useful as many properties become easier to see in one description, compared to the other description. Let us, therefore, start by making some important observations about extreme points and extreme rays of a polyhedron.

Definition 2.5.1 Let P be a polyhedron. Let $A \in \mathbb{R}^{m \times d}$ with rows $\mathbf{a}^1, \ldots, \mathbf{a}^m$ and $\mathbf{b} \in \mathbb{R}^m$ such that $P = \{\mathbf{x} \in \mathbb{R}^d : A\mathbf{x} \leq \mathbf{b}\}$. Given any $\mathbf{x} \in P$, define $\text{tight}(\mathbf{x}, A, \mathbf{b}) := \{i : \langle \mathbf{a}^i, \mathbf{x} \rangle = \mathbf{b}_i\}$. For brevity, when A and $\mathbf{b}$ are clear from the context, we will shorten this to $\text{tight}(\mathbf{x})$. We also use the notation $A_{\text{tight}(\mathbf{x})}$ to denote the submatrix formed by taking the rows of A indexed by $\text{tight}(\mathbf{x})$. Similarly, $\mathbf{b}_{\text{tight}(\mathbf{x})}$ will denote the subvector of $\mathbf{b}$ indexed by $\text{tight}(\mathbf{x})$.

Theorem 2.5.2 *Let* $P = \{\mathbf{x} \in \mathbb{R}^d : A\mathbf{x} \leq \mathbf{b}\}$ *be a polyhedron given by* $A \in \mathbb{R}^{m \times d}$ *and* $\mathbf{b} \in \mathbb{R}^m$. *Let* $\mathbf{x} \in P$. *Then,* $\mathbf{x}$ *is an extreme point of* P *if and only if* $A_{\text{tight}(\mathbf{x})}$ *has rank equal to* d, *i.e., the rows of* A *indexed by* $\text{tight}(\mathbf{x})$ *span* $\mathbb{R}^d$.

Proof ($\Leftarrow$) Suppose $A_{\text{tight}(\mathbf{x})}$ has rank equal to d; we want to establish that $\mathbf{x}$ is an extreme point. Consider any $\mathbf{x}^1, \mathbf{x}^2 \in P$ such that $\mathbf{x} = \frac{\mathbf{x}^1 + \mathbf{x}^2}{2}$. For each $i \in \text{tight}(\mathbf{x})$, $\langle \mathbf{a}^i, \mathbf{x}^1 \rangle \leq \mathbf{b}_i$ and similarly, $\langle \mathbf{a}^i, \mathbf{x}^2 \rangle \leq \mathbf{b}_i$. Now, we observe that

$$\mathbf{b}_i = \langle \mathbf{a}^i, \mathbf{x} \rangle = \frac{\langle \mathbf{a}^i, \mathbf{x}^1 \rangle}{2} + \frac{\langle \mathbf{a}^i, \mathbf{x}^2 \rangle}{2} \leq \mathbf{b}_i.$$

Thus, the inequality must be an equality. Therefore, for each $i \in \text{tight}(\mathbf{x})$, $\langle \mathbf{a}^i, \mathbf{x}^1 \rangle = \mathbf{b}_i$ and, similarly, $\langle \mathbf{a}^i, \mathbf{x}^2 \rangle = \mathbf{b}_i$. In other words, we have that $A_{\text{tight}(\mathbf{x})} \mathbf{x} = \mathbf{b}_{\text{tight}(\mathbf{x})}$ and $A_{\text{tight}(\mathbf{x})} \mathbf{x}^j = \mathbf{b}_{\text{tight}(\mathbf{x})}$ for $j = 1, 2$. Since the rank of $A_{\text{tight}(\mathbf{x})}$ is d, the system of equations must have a unique solution. This means $\mathbf{x} = \mathbf{x}^1 = \mathbf{x}^2$. This shows that $\mathbf{x}$ is extreme.

($\Rightarrow$) Suppose to the contrary that $\mathbf{x}$ is extreme and $A_{\text{tight}(\mathbf{x})}$ has rank strictly less than d (note that its rank is less than or equal to d because it has d columns). Thus, there exists a nonzero $\mathbf{r} \in \mathbb{R}^d$ such that $A_{\text{tight}(\mathbf{x})} \mathbf{r} = \mathbf{0}$. Define

$$\epsilon := \min \left\{ \min_j \left\{ \frac{\mathbf{b}_j - \langle \mathbf{a}^j, \mathbf{x} \rangle}{\langle \mathbf{a}^j, \mathbf{r} \rangle} : \langle \mathbf{a}^j, \mathbf{r} \rangle > 0 \right\}, \min_j \left\{ \frac{\mathbf{b}_j - \langle \mathbf{a}^j, \mathbf{x} \rangle}{-\langle \mathbf{a}^j, \mathbf{r} \rangle} : \langle \mathbf{a}^j, \mathbf{r} \rangle < 0 \right\} \right\}.$$

Note that $\epsilon > 0$ because whenever $\langle \mathbf{a}^j, \mathbf{r} \rangle \neq 0$ we have that $j \notin \text{tight}(\mathbf{x})$, and thus all the numerators are strictly positive. We now claim that $\mathbf{x}^1 := \mathbf{x} + \epsilon \mathbf{r} \in P$ and $\mathbf{x}^2 := \mathbf{x} - \epsilon \mathbf{r} \in P$. This would show that $\mathbf{x} = \frac{\mathbf{x}^1 + \mathbf{x}^2}{2}$ with $\mathbf{x}^1 \neq \mathbf{x}^2$ (because $\mathbf{r} \neq \mathbf{0}$ and $\epsilon > 0$), contradicting extremality.

To finish the proof, we need to check that $A\mathbf{x}^1 \leq \mathbf{b}$ and $A\mathbf{x}^2 \leq \mathbf{b}$. We will do the calculations for $\mathbf{x}^1$; the calculations for $\mathbf{x}^2$ are similar. Consider any $j \in \{1, \ldots, m\}$. If $j \in \text{tight}(\mathbf{x})$, then $A_{\text{tight}(\mathbf{x})} \mathbf{r} = \mathbf{0}$; thus, we obtain that $\langle \mathbf{a}^j, \mathbf{x}^1 \rangle = \langle \mathbf{a}^j, \mathbf{x} \rangle + \epsilon \langle \mathbf{a}^j, \mathbf{r} \rangle = \langle \mathbf{a}^j, \mathbf{x} \rangle = \mathbf{b}_j$. If $j \notin \text{tight}(\mathbf{x})$, then we consider two cases:

Case 1: $\langle \mathbf{a}^j, \mathbf{r} \rangle > 0$. Since $\epsilon \leq \frac{\mathbf{b}_j - \langle \mathbf{a}^j, \mathbf{x} \rangle}{\langle \mathbf{a}^j, \mathbf{r} \rangle}$, we obtain that $\langle \mathbf{a}^j, \mathbf{x}^1 \rangle = \langle \mathbf{a}^j, \mathbf{x} \rangle + \epsilon \langle \mathbf{a}^j, \mathbf{r} \rangle \leq \mathbf{b}_j$.

Case 2: $\langle \mathbf{a}^j, \mathbf{r} \rangle < 0$. In this case, $\langle \mathbf{a}^j, \mathbf{x}^1 \rangle = \langle \mathbf{a}^j, \mathbf{x} \rangle + \epsilon \langle \mathbf{a}^j, \mathbf{r} \rangle < \mathbf{b}_j$, simply because $\epsilon > 0$ and $\langle \mathbf{a}^j, \mathbf{r} \rangle < 0$. $\qquad\square$

This immediately gives the following.

Corollary 2.5.3 *Any polyhedron* $P \subseteq \mathbb{R}^d$ *has a finite number of extreme points.*

Proof Let $A \in \mathbb{R}^{m \times d}$ and $\mathbf{b} \in \mathbb{R}^m$ be such that $P = \{\mathbf{x} \in \mathbb{R}^d : A\mathbf{x} \leq \mathbf{b}\}$. From Theorem 2.5.2, for any extreme point, $A_{\text{tight}(\mathbf{x})}$ has rank d. There are only finitely many subsets $I \subseteq \{1, \ldots, m\}$ such that the submatrix A_I is of rank d. Moreover, for any $I \subseteq \{1, \ldots, m\}$ such that A_I has rank d and $A_I\mathbf{x} = \mathbf{b}_I$ has a solution, the set of solutions to $A_I\mathbf{x} = \mathbf{b}_I$ is unique. This shows that there are only finitely many extreme points. $\qquad\square$

What about the extreme rays? First we define *polyhedral cones*.

Definition 2.5.4 A convex cone that is also a polyhedron is called a polyhedral cone.

Proposition 2.5.5 *Let* $D \subseteq \mathbb{R}^d$ *be a convex cone.* D *is a polyhedral cone if and only if there exists a matrix* $A \in \mathbb{R}^{m \times d}$ *for some* $m \in \mathbb{N}$ *such that* $D = \{\mathbf{x} \in \mathbb{R}^d : A\mathbf{x} \leq \mathbf{0}\}$.

Proof We simply have to show the forward direction, the reverse is easy. Assume D is a polyhedral cone. Thus, it is a polyhedron, and so there exists a matrix $A \in \mathbb{R}^{m \times d}$ and $\mathbf{b} \in \mathbb{R}^m$ for some $m \in \mathbb{N}$ such that $D = \{\mathbf{x} : A\mathbf{x} \leq \mathbf{b}\}$. The result now follows from Exercise 43 from Section 2.4.4. $\qquad\square$

Proposition 2.5.5 combined with Exercise 4 from Section 2.5.6 implies the following.

Proposition 2.5.6 *If* P *is a polyhedron, then* $\text{rec}(P)$ *is a polyhedral cone.*

Theorem 2.5.7 *Let* $D = \{\mathbf{x} : A\mathbf{x} \leq \mathbf{0}\}$ *be a polyhedral cone and let* $\mathbf{r} \in D \backslash \{\mathbf{0}\}$. $\mathbf{r}$ *spans an extreme ray if and only if* $A_{\text{tight}(\mathbf{r})}$ *has rank* $d - 1$.

Proof ($\Leftarrow$) Let $A_{\text{tight}(\mathbf{r})}$ have rows $\bar{\mathbf{a}}^1, \ldots, \bar{\mathbf{a}}^k$. Each $F_i := D \cap \{\mathbf{x} : \langle \bar{\mathbf{a}}^i, \mathbf{x} \rangle = 0\}$ for each $i = 1, \ldots, k$ is an exposed face of D. By Problem 17 from Section 2.4.4, $F := \cap_{i=1}^k F_i$ is a face of D. Since $A_{\text{tight}(\mathbf{r})}$ has rank $d - 1$, the set $\{\mathbf{x} : A_{\text{tight}(\mathbf{r})}\mathbf{x} = \mathbf{0}\}$ is a one-dimensional linear subspace. Since $F \subseteq \{\mathbf{x} : A_{\text{tight}(\mathbf{r})}\mathbf{x} = \mathbf{0}\}$, F is a one-dimensional face of D (it cannot be 0 dimensional because it contains $\{\mathbf{0}\}$ and $\mathbf{r} \neq \mathbf{0}$) and hence an extreme ray. Since $\mathbf{r} \in F$, we have that $\mathbf{r}$ spans F.

($\Rightarrow$) Suppose $\mathbf{r}$ spans the one-dimensional face F. Recall that this means that any $\mathbf{x} \in F$ is a scaling of $\mathbf{r}$. Rank of $A_{\text{tight}(\mathbf{r})}$ cannot be d since then $A_{\text{tight}(\mathbf{r})}\mathbf{x} = 0$ has only $\mathbf{0}$ as a solution and $\mathbf{r} \neq \mathbf{0}$. This would contradict that $\mathbf{r}$ spans an extreme ray of D. Thus, rank of $A_{\text{tight}(\mathbf{r})} \leq d - 1$. If it is strictly less, then consider any $\mathbf{r}' \in \{\mathbf{x} : A_{\text{tight}(\mathbf{r})}\mathbf{x} = \mathbf{0}\}$ that is linearly independent to $\mathbf{r}$ (such an $\mathbf{r}'$ exists if rank of $A_{\text{tight}(\mathbf{r})} \leq d - 2$). Define

$$\epsilon := \min \left\{ \min \left\{ \frac{-\langle \mathbf{a}^j, \mathbf{r} \rangle}{\langle \mathbf{a}^j, \mathbf{r}' \rangle} : \langle \mathbf{a}^j, \mathbf{r}' \rangle > 0 \right\}, \min \left\{ \frac{-\langle \mathbf{a}^j, \mathbf{r} \rangle}{-\langle \mathbf{a}^j, \mathbf{r}' \rangle} : \langle \mathbf{a}^j, \mathbf{r}' \rangle < 0 \right\} \right\}.$$

Note that $\epsilon > 0$ since $\langle \mathbf{a}^j, \mathbf{r}' \rangle \neq 0$ implies that $j \notin \text{tight}(\mathbf{r})$. We now claim that $\mathbf{r}^1 := \mathbf{r} + \epsilon \mathbf{r}' \in D$ and $\mathbf{r}^2 := \mathbf{r} - \epsilon \mathbf{r}' \in D$. This would show that $\mathbf{r} = \frac{\mathbf{r}^1 + \mathbf{r}^2}{2}$. Moreover, since $\mathbf{r}'$ and $\mathbf{r}$ are linearly independent, $\mathbf{r}^1, \mathbf{r}^2$ are not scalings of $\mathbf{r}$. This contradicts Proposition 2.4.29.

To finish the proof, we need to check that $A\mathbf{r}^1 \leq \mathbf{0}$ and $A\mathbf{r}^2 \leq \mathbf{0}$. This is the same set of calculations as in the proof of Theorem 2.5.2. $\qquad\square$

Analogous to Corollary 2.5.3, we have:

Corollary 2.5.8 *Any polyhedral cone D has finitely many extreme rays.*

2.5.1 The Minkowski–Weyl Theorem

We can now state the first part of the famous Minkowski–Weyl theorem.

Theorem 2.5.9 (Minkowski–Weyl theorem – Part I) *Let $P \subseteq \mathbb{R}^d$ be a polyhedron. Then there exist finite sets $V, R \subseteq \mathbb{R}^d$ such that $P = \text{conv}(V) + \text{cone}(R)$.*

Proof Let L be a finite set of vectors spanning $\text{lin}(P)$ (L is taken as the empty set if $\text{lin}(P) = \{\mathbf{0}\}$). Note that $\text{lin}(P) = \text{cone}(L \cup -L)$. Define $\hat{P} = P \cap \text{lin}(P)^{\perp}$. By Problem 8 (iii) from Section 2.5.6, $\hat{P}$ is also a polyhedron. By Corollary 2.5.3, we obtain that $V := \text{ext}(\hat{P})$ is a finite set. Moreover, by Proposition 2.5.6, $\text{rec}(\hat{P})$ is a polyhedral cone. By Corollary 2.5.8, $\text{extr}(\text{rec}(\hat{P}))$ is a finite set. Define $R = \text{extr}(\text{rec}(\hat{P})) \cup L \cup -L$. By Theorem 2.4.33, $P = \text{conv}(\text{ext}(\hat{P})) + \text{cone}(\text{extr}(\hat{P})) + \text{lin}(P) = \text{conv}(V) + \text{cone}(R)$. $\qquad\square$

We now make an observation about polars.

Lemma 2.5.10 *Let $V, R \subseteq \mathbb{R}^d$ be finite sets and let $X = \text{conv}(V) + \text{cone}(R)$. Then X is a closed, convex set.*

Proof $\mathrm{conv}(V)$ is compact, by Proposition 2.2.18, and $\mathrm{cone}(R)$ is closed by Proposition 2.2.17. By Problem 1 from Section 1.3.1, we obtain that $X = \mathrm{conv}(V) + \mathrm{conv}(R)$ is closed. Since the Minkowski sum of convex sets is convex (property 3. in Theorem 2.1.3), X is also convex. $\qquad\square$

Theorem 2.5.11 *Let* $V = \{\mathbf{v}^1, \ldots, \mathbf{v}^k\} \subseteq \mathbb{R}^d$, *and* $R = \{\mathbf{r}^1, \ldots, \mathbf{r}^n\} \subseteq \mathbb{R}^d$ *(allowing the possibility that R is empty). Let* $X = \mathrm{conv}(V) + \mathrm{cone}(R)$. *Then*

$$X^\circ = \left\{ \mathbf{y} \in \mathbb{R}^d : \begin{array}{ll} \langle \mathbf{v}^i, \mathbf{y}\rangle \le 1 & i = 1, \ldots, k \\ \langle \mathbf{r}^j, \mathbf{y}\rangle \le 0 & j = 1, \ldots, n \end{array} \right\}.$$

Proof Define $\tilde{X} := \left\{ \mathbf{y} \in \mathbb{R}^d : \begin{array}{ll} \langle \mathbf{v}^i, \mathbf{y}\rangle \le 1 & i = 1, \ldots, k \\ \langle \mathbf{r}^i, \mathbf{y}\rangle \le 0 & i = 1, \ldots, n \end{array} \right\}$. We first verify that $\tilde{X} \subseteq X^\circ$, i.e., $\langle \mathbf{y}, \mathbf{x}\rangle \le 1$ for all $\mathbf{y} \in \tilde{X}$ and $\mathbf{x} \in X$. By definition of X, we can write $\mathbf{x} = \sum_{i=1}^k \lambda_i \mathbf{v}^i + \sum_{j=1}^n \mu_j \mathbf{r}^j$ for some $\lambda_i, \mu_j \ge 0$ such that $\sum_{i=1}^k \lambda_i = 1$. Thus,

$$\langle \mathbf{x}, \mathbf{y}\rangle = \sum_{i=1}^k \lambda_i \langle \mathbf{v}^i, \mathbf{y}\rangle + \sum_{j=1}^n \mu_j \langle \mathbf{r}^j, \mathbf{y}\rangle \le 1$$

since $\langle \mathbf{v}^i, \mathbf{y}\rangle \le 1$ for $i = 1, \ldots, k$, and $\langle \mathbf{r}^j, \mathbf{y}\rangle \le 0$ for $j = 1, \ldots, n$.

To see that $X^\circ \subseteq \tilde{X}$, consider any $\mathbf{y} \in X^\circ$. Since $\langle \mathbf{x}, \mathbf{y}\rangle \le 1$ for all $\mathbf{x} \in X$, we must have $\langle \mathbf{v}^i, \mathbf{y}\rangle \le 1$ for $i = 1, \ldots, k$ since $\mathbf{v}^i \in X$. Suppose to the contrary that $\langle \mathbf{r}^j, \mathbf{y}\rangle > 0$ for some $j \in \{1, \ldots, n\}$. Then there exists $\lambda > 0$ such that $\langle \mathbf{v}^1 + \lambda \mathbf{r}^j, \mathbf{y}\rangle > 1$. But this contradicts the fact that $\langle \mathbf{x}, \mathbf{y}\rangle \le 1$ for all $\mathbf{x} \in X$ because $\mathbf{v}^1 + \lambda \mathbf{r}^j \in X$, by definition of X. Therefore, $\langle \mathbf{r}^j, \mathbf{y}\rangle \le 0$ for $j = 1, \ldots, n$ and thus, $\mathbf{y} \in \tilde{X}$. $\qquad\square$

This has the following corollary.

Corollary 2.5.12 *Let P be a nonempty polyhedron. Then P° is a polyhedron.*

Proof By Theorem 2.5.9, there exist finite sets $V, R \subseteq \mathbb{R}^d$ such that $P = \mathrm{conv}(V) + \mathrm{cone}(R)$. By Theorem 2.5.11, P° is the intersection of finitely many halfspaces and is thus a polyhedron. $\qquad\square$

We now prove the converse of Theorem 2.5.9.

Theorem 2.5.13 (Minkowski–Weyl theorem – Part II) *Let $V, R \subseteq \mathbb{R}^d$ be finite sets and let $X = \mathrm{conv}(V) + \mathrm{cone}(R)$. Then X is a polyhedron.*

Proof The case when X is empty is trivial. So we consider X to be nonempty. Take any $\mathbf{t} \in X$ and define $X' = X - \mathbf{t}$. Now, it is easy to see X is a polyhedron if and only if X' is a polyhedron (Verify!!). So it suffices to show that X' is a polyhedron. Note that $X' = \text{conv}(V') + \text{cone}(R)$ where $V' = V - \mathbf{t}$, which is a nonempty set because V is nonempty (since X is assumed to be nonempty). By Theorem 2.5.11, $(X')^\circ$ is a polyhedron. By Lemma 2.5.10, X' is a closed, convex set, and also $\mathbf{0} \in X'$. Therefore, $X' = ((X')^\circ)^\circ$ by condition 2. in Theorem 2.4.10. Applying Corollary 2.5.12 with $P = (X')^\circ$, we obtain that $((X')^\circ)^\circ = X'$ is a polyhedron. $\qquad\square$

Collecting Theorems 2.5.9 and 2.5.13 together, we have the full-blown Minkowski–Weyl theorem.

Theorem 2.5.14 (Minkowski–Weyl theorem – full version) *Let $X \subseteq \mathbb{R}^d$. Then the following are equivalent.*

(i) ($\mathcal{H}$-description) There exists $m \in \mathbb{N}$, a matrix $A \in \mathbb{R}^{m \times d}$ and a vector $\mathbf{b} \in \mathbb{R}^m$ such that $X = \{\mathbf{x} \in \mathbb{R}^d : A\mathbf{x} \leq \mathbf{b}\}$.
(ii) ($\mathcal{V}$-description) There exist finite sets $V, R \subseteq \mathbb{R}^d$ such that $X = \text{conv}(V) + \text{cone}(R)$.

A compact version is often useful.

Definition 2.5.15 A bounded polyhedron is called a *polytope*.

Theorem 2.5.16 (Minkowski–Weyl theorem – compact version) *Let $X \subseteq \mathbb{R}^d$. Then X is a polytope if and only if X is the convex hull of a finite set of points.*

Proof Left as an exercise. $\qquad\square$

2.5.2 Farkas' Lemma

In linear algebra, Gaussian elimination gives a nice characterization of solutions to systems of linear equations, which can be viewed as the most basic type of "theorem of the alternative."

Theorem 2.5.17 *Let $A \in \mathbb{R}^{d \times n}$ and $\mathbf{b} \in \mathbb{R}^d$. Exactly one of the following is true.*

1. $A\mathbf{x} = \mathbf{b}$ has a solution.
2. There exists $\mathbf{u} \in \mathbb{R}^d$ such that $\mathbf{u}^T A = \mathbf{0}$ and $\mathbf{u}^T \mathbf{b} \neq 0$.

What if we are interested in *nonnegative solutions* to linear equations? Farkas' lemma is a characterization of such solutions. As we will see, this result has profound implications in the study of polyhedra and mathematical optimization.

Theorem 2.5.18 (Farkas' Lemma) *Let $A \in \mathbb{R}^{d \times n}$ and $\mathbf{b} \in \mathbb{R}^d$. Exactly one of the following is true.*

1. *$A\mathbf{x} = \mathbf{b}, \ \mathbf{x} \geq \mathbf{0}$ has a solution.*
2. *There exists $\mathbf{u} \in \mathbb{R}^d$ such that $\mathbf{u}^T A \leq \mathbf{0}$ and $\mathbf{u}^T \mathbf{b} > 0$.*

Proof Let $\mathbf{a}^1, \ldots, \mathbf{a}^n \in \mathbb{R}^d$ be the columns of the matrix A. By Proposition 2.2.17, the cone $C := \{A\mathbf{x} : \mathbf{x} \geq \mathbf{0}\}$ is closed. We now have two cases, either $\mathbf{b} \in C$ or $\mathbf{b} \notin C$. In the first case, we end up in Case 1 of the statement of the theorem. In the second case, by Theorem 2.4.2, there exists $\mathbf{u} \in \mathbb{R}^d$ and $\delta \in \mathbb{R}$ such that $\langle \mathbf{u}, \mathbf{y} \rangle \leq \delta$ for all $\mathbf{y} \in C$ and $\langle \mathbf{u}, \mathbf{b} \rangle > \delta$. Since $\mathbf{0} \in C$, we must have $\delta \geq \langle \mathbf{u}, \mathbf{0} \rangle = 0$. This already shows that $\langle \mathbf{u}, \mathbf{b} \rangle > 0$.

Now suppose to the contrary that for some $\mathbf{a}^i$, $\langle \mathbf{u}, \mathbf{a}^i \rangle > 0$. Thus, there exists $\bar{\lambda} \geq 0$ such that $\bar{\lambda} \langle \mathbf{u}, \mathbf{a}^i \rangle > \delta$ (for example, take $\bar{\lambda} = \frac{\delta + 1}{\langle \mathbf{u}, \mathbf{a}^i \rangle}$). Since $\mathbf{y} := \bar{\lambda} \mathbf{a}^i \in C$, this implies that $\langle \mathbf{u}, \mathbf{y} \rangle > \delta$, contradicting that $\langle \mathbf{u}, \mathbf{y} \rangle \leq \delta$ for all $\mathbf{y} \in C$. $\qquad\square$

2.5.3 Valid Inequalities and Feasibility

Consider a polyhedron $P = \{\mathbf{x} \in \mathbb{R}^d : A\mathbf{x} \leq \mathbf{b}\}$ with $A \in \mathbb{R}^{m \times d}, \mathbf{b} \in \mathbb{R}^m$. For any vector $\mathbf{y} \in \mathbb{R}^m_+$, the inequality $\langle \mathbf{y}^T A, \mathbf{x} \rangle \leq \mathbf{y}^T \mathbf{b}$ is clearly a valid inequality for P, since this inequality is simply a nonnegative combination of the defining inequalities in $A\mathbf{x} \leq \mathbf{b}$. The next theorem says that all valid inequalities are of this form, up to a shift.

Theorem 2.5.19 *Let $P = \{\mathbf{x} \in \mathbb{R}^d : A\mathbf{x} \leq \mathbf{b}\}$ with $A \in \mathbb{R}^{m \times d}, \mathbf{b} \in \mathbb{R}^m$ be a nonempty polyhedron. Let $\mathbf{c} \in \mathbb{R}^d, \delta \in \mathbb{R}$. Then $\langle \mathbf{c}, \mathbf{x} \rangle \leq \delta$ is a valid inequality for P if and only if there exists $\mathbf{y} \in \mathbb{R}^m_+$ such that $\mathbf{c}^T = \mathbf{y}^T A$ and $\mathbf{y}^T \mathbf{b} \leq \delta$.*

Proof ($\Leftarrow$) Suppose there exists $\mathbf{y} \in \mathbb{R}^m_+$ such that $\mathbf{c}^T = \mathbf{y}^T A$ and $\mathbf{y}^T \mathbf{b} \leq \delta$. The validity of $\langle \mathbf{c}, \mathbf{x} \rangle \leq \delta$ is clear from the following relations for any $\mathbf{x} \in P$:

$$\langle \mathbf{c}, \mathbf{x} \rangle = \mathbf{y}^T A \mathbf{x} = \mathbf{y}^T (A\mathbf{x}) \leq \mathbf{y}^T \mathbf{b} \leq \delta,$$

where the first inequality follows from the fact that $\mathbf{x} \in P$ implies $A\mathbf{x} \leq \mathbf{b}$ and $\mathbf{y}$ is nonnegative.

($\Rightarrow$) Let $\langle \mathbf{c}, \mathbf{x} \rangle \leq \delta$ be a valid inequality for P. Suppose to the contrary that there is no nonnegative solution to $\mathbf{c}^T = \mathbf{y}^T A$ and $\mathbf{y}^T \mathbf{b} \leq \delta$. This is equivalent to saying that the following system has no solution in $\mathbf{y}, \lambda$:

$$A^T \mathbf{y} = \mathbf{c}, \quad \mathbf{b}^T \mathbf{y} + \lambda = \delta, \quad \mathbf{y} \geq 0, \lambda \geq 0.$$

Setting this up in matrix notation, we have no nonnegative solutions to

$$\begin{bmatrix} A^T & \mathbf{0} \\ \mathbf{b}^T & 1 \end{bmatrix} \begin{bmatrix} \mathbf{y} \\ \lambda \end{bmatrix} = \begin{bmatrix} \mathbf{c} \\ \delta \end{bmatrix}.$$

By Farkas' Lemma (Theorem 2.5.18), there exists $\mathbf{u} = (\bar{\mathbf{u}}, \mathbf{u}_{d+1}) \in \mathbb{R}^{d+1}$ such that

$$\bar{\mathbf{u}}^T A^T + \mathbf{u}_{d+1} \mathbf{b}^T \leq \mathbf{0}, \quad \mathbf{u}_{d+1} \leq 0, \quad \text{and} \quad \bar{\mathbf{u}}^T \mathbf{c} + \mathbf{u}_{d+1}\delta > 0. \tag{2.5.1}$$

We now consider two cases:

Case 1: $\mathbf{u}_{d+1} = 0$. Plugging into (2.5.1), we obtain $\bar{\mathbf{u}}^T A^T \leq \mathbf{0}$, i.e., $A\bar{\mathbf{u}} \leq \mathbf{0}$, and $\langle \mathbf{c}, \bar{\mathbf{u}} \rangle > 0$. By Problem 4 from Section 2.5.6, $\bar{\mathbf{u}} \in \mathrm{rec}(P)$. Consider any $\mathbf{x} \in P$ (we assume P is nonempty). Let $\mu = \frac{1+(\delta-\langle \mathbf{c}, \mathbf{x}\rangle)}{\langle \mathbf{c}, \bar{\mathbf{u}}\rangle} > 0$. Now $\mathbf{x} + \mu\bar{\mathbf{u}} \in P$ since $\bar{\mathbf{u}} \in \mathrm{rec}(P)$. However, $\langle \mathbf{c}, \mathbf{x} + \mu\bar{\mathbf{u}} \rangle = \delta + 1 > \delta$, contradicting that $\langle \mathbf{c}, \mathbf{x} \rangle \leq \delta$ is a valid inequality for P.

Case 2: $\mathbf{u}_{d+1} < 0$. By rearranging (2.5.1), we have $A\bar{\mathbf{u}} \leq (-\mathbf{u}_{d+1})\mathbf{b}$ and $\langle \mathbf{c}, \bar{\mathbf{u}} \rangle > (-\mathbf{u}_{d+1})\delta$. By setting $\mathbf{x} = \frac{\bar{\mathbf{u}}}{-\mathbf{u}_{d+1}}$, we obtain that $A\mathbf{x} \leq \mathbf{b}$ and $\langle \mathbf{c}, \mathbf{x} \rangle > \delta$, contradicting that $\langle \mathbf{c}, \mathbf{x} \rangle \leq \delta$ is a valid inequality for P. $\qquad\square$

Definition 2.5.20 Let $\mathbf{c} \in \mathbb{R}^d$ and $\delta_1, \delta_2 \in \mathbb{R}$. If $\delta_1 \leq \delta_2$, then the inequality (halfspace) $\langle \mathbf{c}, \mathbf{x} \rangle \leq \delta_1$ is said to *dominate* the inequality (halfspace) $\langle \mathbf{c}, \mathbf{x} \rangle \leq \delta_2$.

Remark 2.5.21 Let $P = \{\mathbf{x} \in \mathbb{R}^d : A\mathbf{x} \leq \mathbf{b}\}$ with $A \in \mathbb{R}^{m \times d}, \mathbf{b} \in \mathbb{R}^m$ be a polyhedron. Then $\langle \mathbf{c}, \mathbf{x} \rangle \leq \delta$ is called a *consequence of* $A\mathbf{x} \leq \mathbf{b}$ if there exists $\mathbf{y} \in \mathbb{R}^m_+$ such that $\mathbf{c}^T = \mathbf{y}^T A$ and $\delta = \mathbf{y}^T \mathbf{b}$. Another way to think of Theorem 2.5.19 is that it says the geometric property of being a valid inequality is the same as the algebraic property of being a consequence:

> [Alternate version of Theorem 2.5.19] Let $P = \{\mathbf{x} \in \mathbb{R}^d : A\mathbf{x} \leq \mathbf{b}\}$ be a nonempty polyhedron. Then $\langle \mathbf{c}, \mathbf{x} \rangle \leq \delta$ is a valid inequality for P if and only if $\langle \mathbf{c}, \mathbf{x} \rangle \leq \delta$ is dominated by a consequence of $A\mathbf{x} \leq \mathbf{b}$.

A version of Theorem 2.5.19 for empty polyhedra is also useful, which we state next. It can be interpreted as the existence of a short certificate of infeasibility of polyhedra. As one can see from the proof, it can also be viewed as an alternate statement of Farkas' Lemma (Theorem 2.5.18), adapted to general inequality systems.

Theorem 2.5.22 *Let $P = \{\mathbf{x} \in \mathbb{R}^d : A\mathbf{x} \leq \mathbf{b}\}$ with $A \in \mathbb{R}^{m \times d}, \mathbf{b} \in \mathbb{R}^m$ be a polyhedron. Then $P = \emptyset$ if and only if $\langle \mathbf{0}, \mathbf{x} \rangle \leq -1$ is a consequence of $A\mathbf{x} \leq \mathbf{b}$.*

Proof It is easy to see that if $\langle \mathbf{0}, \mathbf{x} \rangle \leq -1$ is a consequence of $A\mathbf{x} \leq \mathbf{b}$ then $P = \emptyset$, because any point that satisfies $A\mathbf{x} \leq \mathbf{b}$ must satisfy every consequence of it, and no point satisfies $\langle \mathbf{0}, \mathbf{x} \rangle \leq -1$.

So now assume $P = \emptyset$. This means that there is no solution to $A\mathbf{x} \leq \mathbf{b}$. This is equivalent to saying that there is no solution to $A\mathbf{x}^1 - A\mathbf{x}^2 + \mathbf{s} = \mathbf{b}$

with $\mathbf{x}^1, \mathbf{x}^2, \mathbf{s} \geq \mathbf{0}$.[3] In matrix notation, this means that there are no nonnegative solutions to

$$\begin{bmatrix} A & -A & I \end{bmatrix} \begin{bmatrix} \mathbf{x}^1 \\ \mathbf{x}^2 \\ \mathbf{s} \end{bmatrix} = \mathbf{b}.$$

By Farkas' Lemma (Theorem 2.5.18), there exists $\mathbf{u} \in \mathbb{R}^m$ such that

$$\mathbf{u}^T A \leq \mathbf{0}, \quad \mathbf{u}^T(-A) \leq \mathbf{0}, \quad \mathbf{u} \leq \mathbf{0}, \quad \text{and} \quad \mathbf{u}^T \mathbf{b} > 0.$$

Define $\mathbf{y} = \frac{-\mathbf{u}}{\mathbf{u}^T \mathbf{b}} \geq \mathbf{0}$. Then $\mathbf{y}^T A = \mathbf{0}$ and $\mathbf{y}^T \mathbf{b} = -1$, showing that $\langle \mathbf{0}, \mathbf{x} \rangle \leq -1$ is a consequence of $A\mathbf{x} \leq \mathbf{b}$. $\qquad\qquad\square$

2.5.4 Faces of Polyhedra

Faces for polyhedra are very structured. First, every face is an exposed face – something that is not true for general closed, convex sets. Second, there is an algebraic characterization of faces in terms of the describing inequalities of a polyhedron.

Theorem 2.5.23 *Let $P = \{\mathbf{x} \in \mathbb{R}^d : A\mathbf{x} \leq \mathbf{b}\}$ with $A \in \mathbb{R}^{m \times d}, \mathbf{b} \in \mathbb{R}^m$. Let $F \subseteq P$ such that $F \neq \emptyset, P$. The following are equivalent.*

(i) F is a face of P.
(ii) F is an exposed face of P.
(iii) There exists a subset $I \subseteq \{1, \ldots, m\}$ such that $F = \{\mathbf{x} \in P : A_I \mathbf{x} = \mathbf{b}_I\}$.

Proof (i) $\Rightarrow$ (ii). Consider $\bar{\mathbf{x}} \in \mathrm{relint}(F)$ (which exists by Exercise 16 from Section 2.2.3). Since F is a proper face, by Theorem 2.4.15, $\bar{\mathbf{x}} \in \mathrm{relbd}(P)$. By Theorem 2.4.5, there exists a supporting hyperplane at $\bar{\mathbf{x}}$ given by $\langle \mathbf{a}, \mathbf{x} \rangle \leq \delta$. Let $\{\mathbf{y} \in P : \langle \mathbf{a}, \mathbf{y} \rangle = \delta\}$ be the corresponding exposed face. Since $\bar{\mathbf{x}} \in \mathrm{relint}(F)$, one can show that $F \subseteq \{\mathbf{y} \in P : \langle \mathbf{a}, \mathbf{y} \rangle = \delta\}$ (Verify!). Thus, there exists an exposed face containing F. Let F' be the minimal (with respect to set inclusion) exposed face of P that contains F, i.e., for any other exposed face $F'' \supseteq F$, we have $F' \subseteq F''$. Note that such a minimal exposed face exists because a proper face of any convex set has strictly lower dimension by Lemma 2.4.12. Let this exposed face F' be defined by the valid inequality $\langle \mathbf{c}^1, \mathbf{x} \rangle \leq \delta_1$ for P.

If $F = F'$, then we are done because F' is an exposed face. Otherwise, $F \subsetneq F'$, and so F is a face of F'. Therefore, $\bar{\mathbf{x}} \in \mathrm{relbd}(F')$ by Theorem 2.4.15. Applying Theorem 2.4.5 to F' and $\bar{\mathbf{x}}$, we obtain $\mathbf{c}^2 \in \mathbb{R}^d, \delta_2 \in \mathbb{R}$

---

[3] This is easily seen by the transformation $\mathbf{x} = \mathbf{x}^1 - \mathbf{x}^2$.

such that $F \subseteq F' \cap \{\mathbf{y} \in \mathbb{R}^d : \langle \mathbf{c}^2, \mathbf{y} \rangle = \delta_2\}$, and there exists $\bar{\mathbf{y}} \in F'$ such that $\langle \mathbf{c}^2, \bar{\mathbf{y}} \rangle < \delta_2$. Using Theorem 2.5.14, we find finite sets V, R such that $P = \text{conv}(V) + \text{cone}(R)$. Notice that since $P \subseteq H^{\leq}(\mathbf{c}^1, \delta_1)$, we must have $\langle \mathbf{c}^1, \mathbf{v} \rangle \leq \delta_1$ for all $\mathbf{v} \in V$ and $\langle \mathbf{c}^1, \mathbf{r} \rangle \leq 0$ for all $\mathbf{r} \in R$ (see Exercise 22 from Section 2.5.6).

Claim 1 One can always choose $\lambda \geq 0$ such that $\lambda \mathbf{c}^1 + \mathbf{c}^2, \lambda\delta_1 + \delta_2$ satisfy

$$\langle \lambda \mathbf{c}^1 + \mathbf{c}^2, \mathbf{v} \rangle \leq \lambda\delta_1 + \delta_2 \text{ for all } \mathbf{v} \in V, \qquad \langle \lambda \mathbf{c}^1 + \mathbf{c}^2, \mathbf{r} \rangle \leq 0 \text{ for all } \mathbf{r} \in R,$$

and the inequalities hold strictly for $\mathbf{v} \in V \setminus F'$ and any $\mathbf{r} \in R \setminus \text{rec}(F')$.

Proof of Claim The relations can be rearranged to say

$$\langle \mathbf{c}^2, \mathbf{v} \rangle - \delta_2 \leq \lambda(\delta_1 - \langle \mathbf{c}^1, \mathbf{v} \rangle) \text{ for all } \mathbf{v} \in V \qquad \langle \mathbf{c}^2, \mathbf{r} \rangle \leq \lambda(-\langle \mathbf{c}^1, \mathbf{r} \rangle) \text{ for all } \mathbf{r} \in R.$$
$$(2.5.2)$$

First, recall that $0 \leq \delta_1 - \langle \mathbf{c}^1, \mathbf{v} \rangle$ for all $\mathbf{v} \in V$ and $0 \leq -\langle \mathbf{c}^1, \mathbf{r} \rangle$ for all $\mathbf{r} \in R$. Notice that since $F' \subseteq H^{\leq}(\mathbf{c}^2, \delta_2)$, if $\langle \mathbf{c}^1, \mathbf{v} \rangle = \delta_1$ for some $\mathbf{v} \in V$, this means that $\mathbf{v} \in F'$ and therefore $\langle \mathbf{c}^2, \mathbf{v} \rangle \leq \delta_2$. Similarly, if $\langle \mathbf{c}^1, \mathbf{r} \rangle = 0$ for some $\mathbf{r} \in R$, this means that $\mathbf{r} \in \text{rec}(F')$ and therefore $\langle \mathbf{c}^2, \mathbf{r} \rangle \leq 0$. Thus, the following choice of

$$\lambda := \max \left\{ 0, \max_{\mathbf{v} \in V : \delta_1 - \langle \mathbf{c}^1, \mathbf{v} \rangle > 0} \frac{\langle \mathbf{c}^2, \mathbf{v} \rangle - \delta_2}{\delta_1 - \langle \mathbf{c}^1, \mathbf{v} \rangle}, \max_{\mathbf{r} \in R : -\langle \mathbf{c}^1, \mathbf{r} \rangle > 0} \frac{\langle \mathbf{c}^2, \mathbf{r} \rangle}{-\langle \mathbf{c}^1, \mathbf{r} \rangle} \right\} + 1$$

satisfies (2.5.2). The "+1" ensures that for any $\mathbf{v} \in V \setminus F'$ and any $\mathbf{r} \in R \setminus \text{rec}(F')$, (2.5.2) is satisfied strictly. $\square$

Using the λ from the above claim, $X = P \cap \{\mathbf{y} \in \mathbb{R}^d : \langle \lambda \mathbf{c}^1 + \mathbf{c}^2, \mathbf{y} \rangle = \lambda\delta_1 + \delta_2\}$ is an exposed face of P containing F. Moreover, $\langle \lambda \mathbf{c}^1 + \mathbf{c}^2, \mathbf{y} \rangle \leq \lambda\delta_1 + \delta_2$ is valid for F' because the inequality is a nonnegative combination of the two valid inequalities $\langle \mathbf{c}^1, \mathbf{y} \rangle \leq \delta_1$, $\langle \mathbf{c}^2, \mathbf{y} \rangle \leq \delta_2$ for F'. Furthermore, any $\mathbf{x} \in P \setminus F'$ can be expressed as $\mathbf{x} = \sum_{\mathbf{v} \in V} \mu_{\mathbf{v}} \mathbf{v} + \sum_{\mathbf{r} \in R} \gamma_{\mathbf{r}} \mathbf{r}$ with a strictly positive coefficient on either some $\mathbf{v} \in V \setminus F'$ or $\mathbf{r} \in R \setminus \text{rec}(F')$. By the second part of the claim above, $\langle \lambda \mathbf{c}^1 + \mathbf{c}^2, \mathbf{x} \rangle < \lambda\delta_1 + \delta_2$. Therefore, $X \subseteq F'$. But $\bar{\mathbf{y}} \in F'$ also satisfies this inequality strictly, because it satisfies $\langle \mathbf{c}^2, \bar{\mathbf{y}} \rangle < \delta_2$, so $X \subsetneq F'$. This contradicts the minimality of F'.

(ii) $\Rightarrow$ (iii). Let $\mathbf{c} \in \mathbb{R}^d, \delta \in \mathbb{R}$ be such that $P \subseteq H^{\leq}(\mathbf{c}, \delta)$ and $F = P \cap \{\mathbf{x} : \langle \mathbf{c}, \mathbf{x} \rangle = \delta\}$. By Theorem 2.5.19, there exists $\mathbf{y} \in \mathbb{R}^m_+$ such that $\mathbf{c}^T = \mathbf{y}^T A$ and $\delta \geq \mathbf{y}^T \mathbf{b}$. Consider any $\mathbf{x} \in F$ (recall that F is assumed to be nonempty). Then

$$\delta = \langle \mathbf{c}, \mathbf{x} \rangle = \mathbf{y}^T A\mathbf{x} \leq \mathbf{y}^T \mathbf{b} \leq \delta. \tag{2.5.3}$$

Thus, equality must hold everywhere and $\mathbf{y}^T \mathbf{b} = \delta$. Moreover, $\mathbf{y}^T A\mathbf{x} = \mathbf{y}^T \mathbf{b}$ for all $\mathbf{x} \in F$, which implies that $\mathbf{y}^T(A\mathbf{x} - \mathbf{b}) = \mathbf{0}$ for all $\mathbf{x} \in F$. This last relation says that for any $i \in \{1, \ldots, m\}$, if $\mathbf{y}_i > 0$ then $\langle \mathbf{a}^i, \mathbf{x} \rangle = \mathbf{b}_i$ for every $\mathbf{x} \in F$. Thus, setting $I = \{i : \mathbf{y}_i > 0\}$, we immediately obtain that $A_I \mathbf{x} = \mathbf{b}_I$ for all $\mathbf{x} \in F$. Consider any $\bar{\mathbf{x}} \in P$ satisfying $A_I \bar{\mathbf{x}} = \mathbf{b}_I$. Therefore, $\mathbf{y}^T A \bar{\mathbf{x}} = \mathbf{y}^T \mathbf{b}$ since $\mathbf{y}_i = 0$ for $i \notin I$. Therefore, $\mathbf{c}^T \bar{\mathbf{x}} = \mathbf{y}^T A \bar{\mathbf{x}} = \mathbf{y}^T \mathbf{b} = \delta$, and thus, $\bar{\mathbf{x}} \in P \cap \{\mathbf{x} : \langle \mathbf{c}, \mathbf{x} \rangle = \delta\} = F$.

(iii) $\Rightarrow$ (i). By definition, $F = \cap_{i \in I} F_i$, where $F_i = \{\mathbf{x} \in P : \langle \mathbf{a}^i, \mathbf{x} \rangle = \mathbf{b}_i\}$. By definition, each F_i is an exposed face, and thus a face. By Problem 17 from Section 2.4.4, the intersection of faces is a face and thus, F is a face. $\qquad \square$

Here are some nice consequences of Theorem 2.5.23.

Theorem 2.5.24 *The following are both true.*

1. *Every polyhedron has finitely many faces.*
2. *Every face of a polyhedron is a polyhedron.*

2.5.5 Implicit Equalities, Dimension of Polyhedra, and Facets

Given a polyhedron $P = \{\mathbf{x} : A\mathbf{x} \leq \mathbf{b}\}$, can we express the dimension of P in terms of A and $\mathbf{b}$? The concept of implicit equalities is important for this.

Definition 2.5.25 Let $A \in \mathbb{R}^{m \times d}$ and $\mathbf{b} \in \mathbb{R}^m$. We say that the inequality $\langle \mathbf{a}^i, \mathbf{x} \rangle \leq \mathbf{b}_i$ for some $i \in \{1, \ldots, m\}$ is an *implicit equality for the polyhedron* $P = \{\mathbf{x} : A\mathbf{x} \leq \mathbf{b}\}$ if $P \subseteq \{\mathbf{x} : \langle \mathbf{a}^i, \mathbf{x} \rangle = \mathbf{b}_i\}$, i.e., $P \subseteq H^=(\mathbf{a}^i, \mathbf{b}_i)$. We denote the subsystem of implicit equalities of $A\mathbf{x} \leq \mathbf{b}$ by $A^=\mathbf{x} \leq \mathbf{b}^=$. We will also use $A^+\mathbf{x} \leq \mathbf{b}^+$ to denote the inequalities in $A\mathbf{x} \leq \mathbf{b}$ that are *not* implicit equalities.

Note that for each i such that $\langle \mathbf{a}^i, \mathbf{x} \rangle \leq \mathbf{b}$ is not an implicit equality, there exists $\mathbf{x} \in P$ such that $\langle \mathbf{a}^i, \mathbf{x} \rangle < \mathbf{b}_i$.

We can completely characterize the affine hull of a polyhedron, and consequently its dimension, in terms of the implicit equalities.

Proposition 2.5.26 *Let $A \in \mathbb{R}^{m \times d}$ and $\mathbf{b} \in \mathbb{R}^m$ and $P = \{\mathbf{x} : A\mathbf{x} \leq \mathbf{b}\}$. Then*

$$\mathrm{aff}(P) = \{\mathbf{x} \in \mathbb{R}^d : A^=\mathbf{x} = \mathbf{b}^=\} = \{\mathbf{x} \in \mathbb{R}^d : A^=\mathbf{x} \leq \mathbf{b}^=\}.$$

Proof Since $\{\mathbf{x} \in \mathbb{R}^d : A^=\mathbf{x} = \mathbf{b}^=\}$ is an affine subset, we have that $\mathrm{aff}(P) \subseteq \{\mathbf{x} \in \mathbb{R}^d : A^=\mathbf{x} = \mathbf{b}^=\} \subseteq \{\mathbf{x} \in \mathbb{R}^d : A^=\mathbf{x} \leq \mathbf{b}^=\}$. We show that $\{\mathbf{x} \in \mathbb{R}^d : A^=\mathbf{x} \leq \mathbf{b}^=\} \subseteq \mathrm{aff}(P)$. Consider any $\mathbf{y}$ satisfying $A^=\mathbf{y} \leq \mathbf{b}^=$. If $A^+\mathbf{y} \leq \mathbf{b}^+$, then $\mathbf{y} \in P \subseteq \mathrm{aff}(P)$ and we are done. Otherwise, using Exercise 18 from Section 2.5.6, choose any $\bar{\mathbf{x}} \in P$ such that $A^=\bar{\mathbf{x}} = \mathbf{b}^=$ and $A^+\bar{\mathbf{x}} < \mathbf{b}^+$. Set

$$\mu := \min_{i:\langle \mathbf{a}^i, \mathbf{y} \rangle > \mathbf{b}_i} \left\{ \frac{\mathbf{b}_i - \langle \mathbf{a}^i, \bar{\mathbf{x}} \rangle}{\langle \mathbf{a}^i, \mathbf{y} \rangle - \langle \mathbf{a}^i, \bar{\mathbf{x}} \rangle} \right\}.$$

Observe that since $\langle \mathbf{a}^i, \mathbf{y} \rangle > \mathbf{b}_i > \langle \mathbf{a}^i, \bar{\mathbf{x}} \rangle$ for each i considered in the minimum, we have $0 < \mu < 1$. One can check that $(1 - \mu)\bar{\mathbf{x}} + \mu \mathbf{y} \in P$. This shows that $\mathbf{y} \in \mathrm{aff}(P)$, because $\mathbf{y}$ is on the line joining two points in P, namely $\bar{\mathbf{x}}$ and $(1 - \mu)\bar{\mathbf{x}} + \mu \mathbf{y}$. $\qquad\square$

Combined with part 4. of Theorem 2.2.8, Theorem 2.5.26 gives the following corollary.

Corollary 2.5.27 *Let $A \in \mathbb{R}^{m \times d}$ and $\mathbf{b} \in \mathbb{R}^m$ and $P = \{\mathbf{x} : A\mathbf{x} \leq \mathbf{b}\}$. Then*

$$\dim(P) = d - rank(A^=).$$

As we have seen before, a given description $P = \{\mathbf{x} : A\mathbf{x} \leq \mathbf{b}\}$ for a polyhedron may be redundant, in the sense that we can remove some of the inequalities and still have the same set P. This motivates the following definition.

Definition 2.5.28 Let $A \in \mathbb{R}^{m \times d}$ and $\mathbf{b} \in \mathbb{R}^m$. We say that the inequality $\langle \mathbf{a}^i, \mathbf{x} \rangle \leq \mathbf{b}_i$ for some $i \in \{1, \ldots, m\}$ is *redundant for the polyhedron* $P = \{\mathbf{x} : A\mathbf{x} \leq \mathbf{b}\}$ if $P = \{\mathbf{x} : A_{-i}\mathbf{x} \leq \mathbf{b}_{-i}\}$, where A_{-i} denotes the matrix A without row i and $\mathbf{b}_{-i}$ is the vector $\mathbf{b}$ with the ith coordinate removed. Otherwise, if $P \subsetneq \{\mathbf{x} : A_{-i}\mathbf{x} \leq \mathbf{b}_{-i}\}$, then $\langle \mathbf{a}^i, \mathbf{x} \rangle \leq \mathbf{b}_i$ is said to be *irredundant for P*. The system $A\mathbf{x} \leq \mathbf{b}$ is said to be an irredundant system if every inequality is irredundant for $P = \{\mathbf{x} : A\mathbf{x} \leq \mathbf{b}\}$.

The following characterization of facets of a polyhedron is quite useful, specially in combinatorial optimization and polyhedral combinatorics.

Theorem 2.5.29 *Let $P = \{\mathbf{x} \in \mathbb{R}^d : A\mathbf{x} \leq \mathbf{b}\}$ be nonempty with $A \in \mathbb{R}^{m \times d}, \mathbf{b} \in \mathbb{R}^m$ giving an irredundant system. Let $F \subseteq P$. The following are equivalent.*

(i) *F is a facet of P, i.e., F is a face with $\dim(F) = \dim(P) - 1$.*
(ii) *F is a maximal, proper face of P; i.e., for any proper face $F' \supseteq F$, we must have $F' = F$.*
(iii) *There exists a unique $i \in \{1, \ldots, m\}$ such that*
$$F = \{\mathbf{x} \in P : \langle \mathbf{a}^i, \mathbf{x} \rangle = \mathbf{b}_i\} \text{ and } \langle \mathbf{a}^i, \mathbf{x} \rangle \leq \mathbf{b}_i \text{ is not an implicit equality.}$$

Proof (i) $\Rightarrow$ (ii). Suppose to the contrary that there exists a proper face $F' \supsetneq F$. Observe that F is a face of F' by Problem 20 from Section 2.4.10, and so F is a proper face of F'. By Lemma 2.4.12, $\dim(F') > \dim(F) = \dim(P) - 1$. So, $\dim(F') = \dim(P)$. This contradicts the fact that F' is a proper face, by Lemma 2.4.12.

(ii) $\Rightarrow$ (iii). By Theorem 2.5.23, there exists a subset of indices $I \subseteq \{1, \ldots, m\}$ such that $F = \{\mathbf{x} \in \mathbb{R}^d : A\mathbf{x} \leq \mathbf{b}, \quad A_I\mathbf{x} = \mathbf{b}_I\}$. If all the inequalities indexed by I are implicit equalities for P, then $F = P$, contradicting the assumption that F is a proper face. So there exists $i \in I$ such that $\langle \mathbf{a}^i, \mathbf{x} \rangle \leq \mathbf{b}_i$ is not an implicit equality. Let $F' = \{\mathbf{x} \in P : \langle \mathbf{a}^i, \mathbf{x} \rangle = \mathbf{b}_i\}$ be the face defined by this inequality; since $\langle \mathbf{a}^i, \mathbf{x} \rangle \leq \mathbf{b}_i$ is not an implicit equality, F' is a proper face of P. Also observe that $F \subseteq F'$. Hence $F = F' = \{\mathbf{x} \in P : \langle \mathbf{a}^i, \mathbf{x} \rangle = \mathbf{b}_i\}$. To show uniqueness of i, we exhibit $\mathbf{x}^0 \in F$ with the following property: for any $j \neq i$ such that $\langle \mathbf{a}^j, \mathbf{x} \rangle \leq \mathbf{b}_j$ is not an implicit equality, we have $\langle \mathbf{a}^j, \mathbf{x}^0 \rangle < \mathbf{b}_j$. To see this, let $\mathbf{x}^1 \in P$ such that $A^=\mathbf{x}^1 = \mathbf{b}^=$ and $A^+\mathbf{x}^1 < \mathbf{b}^+$ (such an $\mathbf{x}^1$ exists by Exercise 18 from Section 2.5.6). Since $A\mathbf{x} \leq \mathbf{b}$ is an irredundant system, if we remove the inequality indexed by i, then we get some new points that satisfy the rest of the inequalities, but which violate $\langle \mathbf{a}^i, \mathbf{x} \rangle \leq \mathbf{b}_i$. More precisely, there exists $\mathbf{x}^2 \in \mathbb{R}^d$ such that $A^=\mathbf{x}^2 = \mathbf{b}^=$, $A^+_{-i}\mathbf{x}^2 \leq \mathbf{b}^+_{-i}$, and $\langle \mathbf{a}^i, \mathbf{x}^2 \rangle > \mathbf{b}_i$, where $A^+_{-i}\mathbf{x} \leq \mathbf{b}^+_{-i}$ denotes the system $A^+\mathbf{x} \leq \mathbf{b}^+$ without the inequality indexed by i. Since $\langle \mathbf{a}^i, \mathbf{x}^1 \rangle < \mathbf{b}_i$ and $\langle \mathbf{a}^i, \mathbf{x}^2 \rangle > \mathbf{b}_i$, there exists a convex combination of $\mathbf{x}^1, \mathbf{x}^2$ such that this convex combination $\mathbf{x}^0$ satisfies $\langle \mathbf{a}^i, \mathbf{x}^0 \rangle = \mathbf{b}_i$. Since $A^=\mathbf{x}^1 = \mathbf{b}^=$ and $A^=\mathbf{x}^2 = \mathbf{b}^=$, we must have $A^=\mathbf{x}^0 = \mathbf{b}^=$. Moreover, since $A^+\mathbf{x}^1 < \mathbf{b}^+$ and $A^+_{-i}\mathbf{x}^2 \leq \mathbf{b}^+_{-i}$, we must have that for any $j \neq i$ indexing an inequality in $A^+\mathbf{x} \leq \mathbf{b}^+$, $\mathbf{x}^0$ must satisfy $\langle \mathbf{a}^j, \mathbf{x}^0 \rangle < \mathbf{b}_j$. Thus, we are done.

(iii) $\Rightarrow$ (i). By Theorem 2.5.23, F is a face. We now establish that $\dim(F) = \dim(P) - 1$. Let $\mathcal{J}$ denote the set of indices that index inequalities in $A\mathbf{x} \leq \mathbf{b}$ that are not implicit equalities. Since there exists a unique $i \in \mathcal{J}$ such that $F = \{\mathbf{x} \in P : \langle \mathbf{a}^i, \mathbf{x} \rangle = \mathbf{b}_i\}$, this means that for any $j \in \mathcal{J} \setminus i$, there exists $\mathbf{x}^j \in F$ such that $\langle \mathbf{a}^j, \mathbf{x}^j \rangle < \mathbf{b}_j$. Now let $\mathbf{x}^0 = \frac{1}{|\mathcal{J}|-1} \sum_{j \in \mathcal{J} \setminus \{i\}} \mathbf{x}^j$, and observe that $\mathbf{x}^0 \in F$ (since F is convex) and for any $j \in \mathcal{J} \setminus i$, we have $\langle \mathbf{a}^j, \mathbf{x}^0 \rangle < \mathbf{b}_j$. Let us describe the polyhedron F by the system $\tilde{A}\mathbf{x} \leq \tilde{\mathbf{b}}$ that appends the inequality $\langle -\mathbf{a}^i, \mathbf{x} \rangle \leq -\mathbf{b}_i$ to the system $A\mathbf{x} \leq \mathbf{b}$.

Claim 2 $\operatorname{rank}(\tilde{A}^=) = \operatorname{rank}(A^=) + 1$.

Proof The properties of $\mathbf{x}^0$ show that the matrix $\tilde{A}^=$ is simply the matrix $A^=$ appended with $\mathbf{a}^i$ and $-\mathbf{a}^i$. So it suffices to show that $\mathbf{a}^i$ is not a linear combination of the rows of $A^=$. Suppose to the contrary that $\mathbf{a}^i = \mathbf{y}^T A^=$ for some $\mathbf{y} \in \mathbb{R}^k$, where k is the number of rows of $A^=$. If $\mathbf{b}_i < \mathbf{y}^T\mathbf{b}^=$, then P is empty because any $\mathbf{x} \in P$ satisfies $A^=\mathbf{x} = \mathbf{b}^=$ and therefore must satisfy $\mathbf{y}^T A^=\mathbf{x} = \mathbf{y}^T\mathbf{b}^=$, and this contradicts $\mathbf{y}^T A^=\mathbf{x} = \langle \mathbf{a}^i, \mathbf{x} \rangle \leq \mathbf{b}_i$. If $\mathbf{b}_i \geq \mathbf{y}^T\mathbf{b}^=$, then $\langle \mathbf{a}^i, \mathbf{x} \rangle \leq \mathbf{b}_i$ is redundant for P, as every $\mathbf{x}$ satisfying $A^=\mathbf{x} = \mathbf{b}^=$ satisfies $\langle \mathbf{a}^i, \mathbf{x} \rangle \leq \mathbf{b}_i$. $\qquad\square$

Using Corollary 2.5.27, we obtain that $\dim(F) = d - \mathrm{rank}(\tilde{A}^=) = d - \mathrm{rank}(A^=) - 1 = \dim(P) - 1$. $\qquad\square$

A consequence of this characterization of facets is that full-dimensional polyhedra have a unique system describing them, up to scaling.

Definition 2.5.30 We say that the inequality $\langle \mathbf{a}, \mathbf{x} \rangle \leq \delta$ is *equivalent* to the inequality $\langle \mathbf{a}', \mathbf{x} \rangle \leq \delta'$ if there exists $\lambda \geq 0$ such that $\mathbf{a}' = \lambda \mathbf{a}$ and $\delta' = \lambda \delta$. Equivalent inequalities define the same halfspace, i.e., $H^\leq(\mathbf{a}, \delta) = H^\leq(\mathbf{a}', \delta')$.

Theorem 2.5.31 *Let P be a full-dimensional polyhedron. Let $A \in \mathbb{R}^{m \times d}$, $A' \in \mathbb{R}^{p \times d}$, $\mathbf{b} \in \mathbb{R}^m$, and $\mathbf{b}' \in \mathbb{R}^p$ be such that $A\mathbf{x} \leq \mathbf{b}$ and $A'\mathbf{x} \leq \mathbf{b}'$ are both irredundant systems describing P, i.e.,*

$$\{\mathbf{x} \in \mathbb{R}^d : A\mathbf{x} \leq \mathbf{b}\} = \{\mathbf{x} \in \mathbb{R}^d : A'\mathbf{x} \leq \mathbf{b}'\} = P.$$

Then both systems are the same up to permutation and scaling. More precisely, the following holds:

1. *$m = p$.*
2. *There exists a permutation $\sigma \colon \{1, \ldots, m\} \to \{1, \ldots, m\}$ such that for each $i \in \{1, \ldots, m\}$, $\langle \mathbf{a}^i, \mathbf{x} \rangle \leq \mathbf{b}_i$ is equivalent to $\langle \mathbf{a}'^{\sigma(i)}, \mathbf{x} \rangle \leq \mathbf{b}'_{\sigma(i)}$.*

Proof Left as an exercise. $\qquad\square$

2.5.6 Exercises

1. Let $A \in \mathbb{R}^{m \times d}$ and $\mathbf{b} \in \mathbb{R}^m$. Consider the polyhedron $P = \{\mathbf{x} \in \mathbb{R}^d : A\mathbf{x} \leq \mathbf{b}\}$. Suppose there exists $\bar{\mathbf{x}} \in \mathbb{R}^d$ such that $A\bar{\mathbf{x}} < \mathbf{b}$, i.e., all inequalities are satisfied *strictly*. Show that P is full dimensional.
2. Let $A \in \mathbb{R}^{m \times d}$ and $\mathbf{b} \in \mathbb{R}^m$. Consider the polyhedron $P = \{\mathbf{x} \in \mathbb{R}^d : A\mathbf{x} \leq \mathbf{b}\}$. Suppose that for each $j = 1, \ldots, m$, there exists $\mathbf{x}^j \in P$ such that $\langle \mathbf{a}^j, \mathbf{x}^j \rangle < \mathbf{b}_j$, i.e., for every inequality, there is a point in P that satisfies this inequality *strictly*. Show that P is full dimensional.
3. Prove Theorem 2.5.16.
4. Let $A \in \mathbb{R}^{m \times d}$ and $\mathbf{b} \in \mathbb{R}^m$. Consider the polyhedron $P = \{\mathbf{x} \in \mathbb{R}^d : A\mathbf{x} \leq \mathbf{b}\}$ and suppose that it is nonempty. Show that

$$\mathrm{rec}(P) = \{\mathbf{x} \in \mathbb{R}^d : A\mathbf{x} \leq \mathbf{0}\}$$

and

$$\mathrm{lin}(P) = \{\mathbf{x} \in \mathbb{R}^d : A\mathbf{x} = \mathbf{0}\}.$$

5. Let P be a nonempty polyhedron such that $\mathrm{rec}(P)$ is full dimensional. Show that for any $\mathbf{r} \in \mathrm{int}(\mathrm{rec}(P))$ and any $\mathbf{x} \in \mathbb{R}^d$ (not necessarily in P), there exists $\lambda \geq 0$ such that $\mathbf{x} + \lambda \mathbf{r} \in P$.

6. Let $P = \{\mathbf{x} \in \mathbb{R}^d : A\mathbf{x} \leq \mathbf{b}\}$ be a nonempty polyhedron. Show that P is bounded if and only if the conical hull of the rows of A is $\mathbb{R}^d$.

7. Let $C = \{\mathbf{x} \in \mathbb{R}^d : A\mathbf{x} \leq \mathbf{0}\}$ be a polyhedral cone for some $A \in \mathbb{R}^{m \times d}$ such that all rows of A are nonzero. Show that C is full dimensional if and only if $\mathbf{0}$ is not in the convex hull of the rows of A.

8. (Operations that preserve polyhedrality). Show that the following are all true.

    (a) Let $X \subseteq \mathbb{R}^d$ be a polyhedron and $\mathbf{t} \in \mathbb{R}^d$. Show that $X + \mathbf{t}$ is a polyhedron.
    (b) Let $X_1, X_2 \subseteq \mathbb{R}^d$ be polyhedra. Then $X_1 \cap X_2$ is a polyhedron.
    (c) Let $X \subseteq \mathbb{R}^d$ be a polyhedron and H be an affine subspace. Show that $X \cap H$ is a polyhedron.
    (d) Let $X_1, X_2 \subseteq \mathbb{R}^d$ be polyhedra. Then $X_1 + X_2$ is a polyhedron.
    (e) Let $X \subseteq \mathbb{R}^d$ be a polyhedron and let $T : \mathbb{R}^d \to \mathbb{R}^n$ be a linear transformation. Then $T(X)$ is a polyhedron.

Note that this exercise shows the power of having both the extrinsic and intrinsic descriptions of polyhedra (the Minkowski–Weyl theorem – Theorem 2.5.14). With the exception of part 8a, each one is easy to show in one description and nontrivial to show in the other.

9. Let $P \subseteq \mathbb{R}^d$ be a polyhedron such that $P \subseteq \mathbb{R}^d_+$. Show that P is pointed, i.e., $\mathrm{lin}(P) = \{\mathbf{0}\}$.

10. Let $P \subseteq \mathbb{R}^d$ be a pointed polyhedron. Then Corollary 2.4.31 shows that $P = \mathrm{conv}(\mathrm{ext}(P)) + \mathrm{cone}(\mathrm{extr}(P))$. Show that this description is minimal in the sense that if P is expressed as $\mathrm{conv}(V) + \mathrm{conv}(R)$ as per the Minkowski–Weyl theorem (Theorem 2.5.14), then $\mathrm{ext}(P) \subseteq V$ and $\mathrm{extr}(P) \subseteq R$.

11. Let $P \subseteq \mathbb{R}^d$ be a nonempty polyhedron (not necessarily bounded) and let $\mathbf{c} \in \mathbb{R}^d$. Show that either $\sup\{\langle \mathbf{c}, \mathbf{x} \rangle : \mathbf{x} \in P\} = +\infty$ or there exists $\mathbf{x}^\star \in P$ such that $\langle \mathbf{c}, \mathbf{x}^\star \rangle = \sup\{\langle \mathbf{c}, \mathbf{x} \rangle : \mathbf{x} \in P\}$. In other words, either the supremum is $+\infty$ or attained at a point in P.

12. Let $P = \{\mathbf{x} \in \mathbb{R}^d : A\mathbf{x} \leq \mathbf{b}\}$ such that $A \in \mathbb{Z}^{m \times d}$ and $\mathbf{b} \in \mathbb{Z}^m$, i.e., both A and $\mathbf{b}$ have only integer entries. Moreover, assume $m \geq d$ and that the determinant of every $d \times d$ submatrix of A is either $0, 1, -1$. Show that if $\mathbf{v} \in P$ is an extreme point, then $\mathbf{v} \in \mathbb{Z}^d$, i.e., $\mathbf{v}$ has all integer coordinates.

13. Prove the following Farkas' type results. Let $A \in \mathbb{R}^{m \times d}$ and $\mathbf{b} \in \mathbb{R}^m$.

   (i) Exactly one of the following is true: either there exists $\mathbf{x} \geq \mathbf{0}$ satisfying $A\mathbf{x} \leq \mathbf{b}$, or there exists $\mathbf{y} \geq \mathbf{0}$ such that $\mathbf{y}^T A \geq 0$ and $\mathbf{y}^T \mathbf{b} < 0$.

  (ii) Exactly one of the following is true: either there exists $\mathbf{x} > \mathbf{0}$ satisfying $A\mathbf{x} = \mathbf{0}$, or there exists $\mathbf{y} \in \mathbb{R}^m$ such $\mathbf{y}^T A$ is a nonnegative vector with at least one nonzero coordinate.

 (iii) Exactly one of the following is true: either there exists $\mathbf{x} \neq \mathbf{0}$ satisfying $\mathbf{x} \geq \mathbf{0}$ and $A\mathbf{x} = \mathbf{0}$, or there exists $\mathbf{y} \in \mathbb{R}^m$ satisfying $\mathbf{y}^T A > \mathbf{0}$.

14. Let P be an affine subspace, and let $\langle \mathbf{c}, \mathbf{x} \rangle \leq \delta$ be a valid inequality for P. Show that there exists $\Delta \in \mathbb{R}$ such that $\langle \mathbf{c}, \mathbf{x} \rangle = \Delta$ for all $\mathbf{x} \in P$, i.e., the function $f(\mathbf{x}) := \langle \mathbf{c}, \mathbf{x} \rangle$ is constant on P.

15. Let $V \subseteq \mathbb{R}^d$ be a finite subset. We say that V is in *convex position* if for every $\mathbf{v} \in V$, $\mathbf{v} \notin \mathrm{conv}(V \setminus \{\mathbf{v}\})$. Show that V is in convex position if and only if $\mathrm{ext}(\mathrm{conv}(V)) = V$. Thus conclude that if $V \subseteq \{0, 1\}^d$, then $\mathrm{ext}(\mathrm{conv}(V)) = V$.

16. Let P be a polyhedron. Let $\mathbf{v}, \mathbf{w} \in \mathrm{ext}(P)$ be distinct extreme points of P, and let $E = \mathrm{conv}(\{\mathbf{v}, \mathbf{w}\})$. Show that E is an edge of P if and only if the midpoint $\frac{\mathbf{v}+\mathbf{w}}{2}$ is not the convex combination of two points in $P \setminus E$.

17. Let P be a polyhedron. A nonempty face $F \subseteq P$ is said to be *minimal* if no nonempty face of P is strictly contained in F. Show that a subset of P is a minimal face if and only if it is of the form $\mathbf{w} + \mathrm{lin}(P)$ where $\mathbf{w}$ is an extreme point of $P \cap \mathrm{lin}(P)^{\perp}$.

18. Let $P = \{\mathbf{x} : A\mathbf{x} \leq \mathbf{b}\}$. Show that there exists $\bar{\mathbf{x}} \in P$ such that $A^=\bar{\mathbf{x}} = \mathbf{b}^=$ and $A^+\bar{\mathbf{x}} < \mathbf{b}^+$. Show the stronger statement that $\mathrm{relint}(P) = \{\mathbf{x} \in \mathbb{R}^d : A^=\mathbf{x} = \mathbf{b}^=, \ A^+\mathbf{x} < \mathbf{b}^+\}$ (see Exercise 16 from Section 2.2.3).

19. Let P be a polyhedron, and let H, G be faces of P such that $H \subseteq G \subseteq P$. Show that there exists a sequence of faces $F_0, \ldots, F_k$ of P such that $H = F_0 \subseteq F_1 \subseteq \cdots \subseteq F_k = G$ and F_i is a facet of F_{i+1} for each $i = 0, \ldots, k-1$.

20. Let $P = \{\mathbf{x} \in \mathbb{R}^d : A\mathbf{x} \leq \mathbf{b}\}$ be a polyhedron, with $A \in \mathbb{R}^{m \times d}$ and $\mathbf{b} \in \mathbb{R}^m$. Show that an inequality $\langle \mathbf{a}^i, \mathbf{x} \rangle \leq \mathbf{b}_i$, $i = 1, \ldots, m$ in the description is irredundant for P if and only if there exists $\bar{\mathbf{x}} \in P$ such that $\langle \mathbf{a}^i, \bar{\mathbf{x}} \rangle = \mathbf{b}_i$ and $\langle \mathbf{a}^j, \bar{\mathbf{x}} \rangle < \mathbf{b}_j$ for all $j \in \{1, \ldots, m\}$ such that $\langle \mathbf{a}^j, \mathbf{x} \rangle \leq \mathbf{b}_j$ is not an implicit equality for P.

21. Let $P = \{\mathbf{x} \in \mathbb{R}^d : A\mathbf{x} \leq \mathbf{b}\}$ be a polyhedron and let $F \subseteq P$ be a face exposed by the valid inequality $\langle \mathbf{c}, \mathbf{x} \rangle \leq \delta$. Show that $\langle \mathbf{c}, \mathbf{x} \rangle \leq \delta$ is a consequence of $d - \dim(F)$ inequalities coming from $A\mathbf{x} \leq \mathbf{b}$. In other words, $\langle \mathbf{c}, \mathbf{x} \rangle \leq \delta$ can be written as a nonnegative combination of $d - \dim(F)$ inequalities from the description of P.

22. Let P be a polyhedron given as $P = \text{conv}(V) + \text{cone}(R)$ for some finite sets $V, R \subseteq \mathbb{R}^d$. Show the following analogues of Theorems 2.5.19 and 2.5.22, and Proposition 2.5.26.

    (a) $\langle \mathbf{c}, \mathbf{x} \rangle \leq \delta$ is a valid inequality for P if and only if $\langle \mathbf{c}, \mathbf{v} \rangle \leq \delta$ for all $\mathbf{v} \in V$ and $\langle \mathbf{c}, \mathbf{r} \rangle \leq 0$ for all $\mathbf{r} \in R$.
    (b) $P = \emptyset$ if and only if $V = \emptyset$.
    (c) $\text{aff}(P) = \text{aff}(V + (\{0\} \cup R))$.

23. Consider the polyhedron in $\mathbb{R}^2$ described by

$$
\begin{aligned}
\mathbf{x}_1 - \mathbf{x}_2 &\leq 0 \\
-\mathbf{x}_1 + \mathbf{x}_2 &\leq 1 \\
- 2\mathbf{x}_2 &\leq -5 \\
8\mathbf{x}_1 - \mathbf{x}_2 &\leq 16 \\
-\mathbf{x}_1 - \mathbf{x}_2 &\leq -4
\end{aligned}
$$

    (i) Find the dimension of P.
    (ii) Find an interior point of P (if it exists).
    (iii) Describe all the faces of P.
    (iv) Consider each of the faces $F^i = P \cap \{x \in \mathbb{R}^2 : \langle \mathbf{a}^i, \mathbf{x} \rangle = \mathbf{b}_i\}$ for $i = 1, \dots, 5$. What is the dimension of each F_i?
    (v) Are there any redundant inequalities in the system?
    (vi) Which inequalities describe facets?

24. Show that a closed, convex set $C \subseteq \mathbb{R}^d$ is a polyhedron if and only if it has a finite number of faces.

25. Prove Theorem 2.5.31.

26. Let $P = \{\mathbf{x} \in \mathbb{R}^d : A\mathbf{x} \leq \mathbf{b}\}$ be a polytope and let $S \subseteq \mathbb{R}^d$ be a closed subset such that $P \cap S = \emptyset$. Show that there exists $\epsilon > 0$ such that $P_\epsilon \cap S = \emptyset$ where $P_\epsilon := \{\mathbf{x} \in \mathbb{R}^d : A\mathbf{x} \leq \mathbf{b} + \epsilon \mathbf{1}\}$.

2.6 Helly Numbers

In Part II, we will devote ourselves to the study of optimization with emphasis on those optimization problems that involve convexity. Optimization problems come with constraints and it becomes an important question to decide if a given collection of constraints can be satisfied or not. The results in this subsection have an important bearing on this question.

Theorem 2.6.1 (Radon's theorem) *Let $X \subseteq \mathbb{R}^d$ be a set of size at least $d + 2$. Then X can be partitioned as $X = X_1 \uplus X_2$ into disjoint nonempty sets X_1, X_2, such that* $\text{conv}(X_1) \cap \text{conv}(X_2) \neq \emptyset$.

Proof Since we can have at most $d + 1$ affinely independent points in $\mathbb{R}^d$ (see Proposition 2.2.7), and X has at least $d + 2$ points, there exists a subset $\{\mathbf{x}^1, \ldots, \mathbf{x}^k\} \subseteq X$ such that $\{\mathbf{x}^1, \ldots, \mathbf{x}^k\}$ is affinely dependent. By using characterization 5. in Proposition 2.2.7, there exist multipliers $\lambda_1, \ldots, \lambda_k \in \mathbb{R}$, not all zero, such that $\lambda_1 + \cdots + \lambda_k = 0$ and $\lambda_1 \mathbf{x}^1 + \cdots + \lambda_k \mathbf{x}^k = 0$. Define $P := \{i : \lambda_i \geq 0\}$ and $N := \{i : \lambda_i < 0\}$. Since the λ_i's are not all zero and $\lambda_1 + \cdots + \lambda_k = 0$, P and N both contain indices such that corresponding multiplier is nonzero. Moreover, $\sum_{i \in P} \lambda_i = \sum_{i \in N}(-\lambda_i)$ since $\lambda_1 + \cdots + \lambda_k = 0$, and $\sum_{j \in P} \lambda_j \mathbf{x}^j = \sum_{j \in N}(-\lambda_j)\mathbf{x}^j$ since $\lambda_1 \mathbf{x}^1 + \cdots + \lambda_k \mathbf{x}^k = \mathbf{0}$. Thus, we obtain that

$$\mathbf{y} := \sum_{j \in P} \frac{\lambda_j}{\sum_{i \in P} \lambda_i} \mathbf{x}^j = \sum_{j \in N} \frac{(-\lambda_j)}{\sum_{i \in N}(-\lambda_i)} \mathbf{x}^j,$$

showing that $\mathbf{y} \in \mathrm{conv}(X_P) \cap \mathrm{conv}(X_N)$ where $X_P = \{\mathbf{x}^i : i \in P\}$ and $X_N = \{\mathbf{x}^i : i \in N\}$. One can now simply define $X_1 = X_P$ and $X_2 = X \setminus X_P$. $\qquad \square$

An important corollary of Radon's theorem (Theorem 2.6.1) is known as Helly's theorem, which is about the intersection of a family of convex sets.

Theorem 2.6.2 (Helly's theorem) *Let $X_1, \ldots, X_k \subseteq \mathbb{R}^d$ be a family of convex sets. If $X_1 \cap \cdots \cap X_k = \emptyset$, then there is a subfamily $X_{i_1}, \ldots, X_{i_m}$ for some $m \leq d+1$, with $i_h \in \{1, \ldots, k\}$ for each $h = 1, \ldots, m$ such that $X_{i_1} \cap \cdots \cap X_{i_m} = \emptyset$. In other words, there is a subfamily of size at most $d + 1$ that already certifies the empty intersection.*

Proof Let $X_{i_1}, \ldots, X_{i_m}$ be a minimal subfamily of $X_1, \ldots, X_k$ with the property that $X_{i_1} \cap \cdots \cap X_{i_m} = \emptyset$. This implies that for any $j \in \{i_1, \ldots, i_m\}$, there exists $\mathbf{p}^j \in \bigcap_{i \in \{i_1, \ldots, i_m\} \setminus \{j\}} X_i$. If $m \geq d + 2$, then by Radon's theorem (Theorem 2.6.1), $\{i_1, \ldots, i_m\}$ can be partitioned into two nonempty sets P, N such that $\mathrm{conv}(\{\mathbf{p}^i\}_{i \in P}) \cap \mathrm{conv}(\{\mathbf{p}^i\}_{i \in N})$ is nonempty. Note that for any $j \in \{i_1, \ldots, i_m\}$, X_j contains all points $\mathbf{p}^i$, $i \neq j$. Therefore, X_j contains either $\mathrm{conv}(\{\mathbf{p}^i\}_{i \in P})$ (if $j \in N$) or contains $\mathrm{conv}(\{\mathbf{p}^i\}_{i \in N})$ (if $j \in P$). Therefore, $\mathrm{conv}(\{\mathbf{p}^j\}_{j \in P}) \cap \mathrm{conv}(\{\mathbf{p}^j\}_{j \in N}) \subseteq \bigcap_{j \in \{i_1, \ldots, i_m\}} X_j$, contradicting the fact that $X_{i_1} \cap \cdots \cap X_{i_m} = \emptyset$. Thus, $m \leq d + 1$. $\qquad \square$

A corollary for infinite families is often useful.

Corollary 2.6.3 *Let $\mathcal{X}$ be a (possibly infinite) family of closed, convex sets such that at least one of the sets is compact. If $\bigcap_{X \in \mathcal{X}} X = \emptyset$, then there is a subfamily $X_{i_1}, \ldots, X_{i_m}$ for some $m \leq d + 1$, with $i_h \in \{1, \ldots, k\}$ for each $h = 1, \ldots, m$ such that $X_{i_1} \cap \cdots \cap X_{i_m} = \emptyset$. In other words, there is a subfamily of size at most $d + 1$ that already certifies the empty intersection.*

Proof By Theorem 1.3.11, there is a finite subfamily whose intersection is also empty. One can now apply Theorem 2.6.2 to this finite subfamily and obtain a subfamily of size at most $d + 1$. $\qquad\square$

An application to learning theory: VC dimension of halfspaces. An important concept in learning theory is the *Vapnik–Çervonenkis (VC) dimension* of a family of subsets [228]. Let $\mathcal{F}$ be a family of subsets of $\mathbb{R}^d$ (possibly infinite).

Definition 2.6.4 A set $X \subseteq \mathbb{R}^d$ is said to be *shattered* by $\mathcal{F}$ if for every subset $X' \subseteq X$, there exists a set $F \in \mathcal{F}$ such that $X' = F \cap X$. The VC-dimension of $\mathcal{F}$ is defined as

$$\sup\{m \in \mathbb{N}: \text{ there exists a set } X \subseteq \mathbb{R}^d \text{ of size } m \text{ that can be shattered by} \mathcal{F}\}.$$

Proposition 2.6.5 *Let $\mathcal{F}$ be the family of halfspaces in $\mathbb{R}^d$. The VC dimension of $\mathcal{F}$ is $d + 1$.*

Proof For any $m \le d + 1$, let X be a set of m affinely independent points. Now, for any subset $X' \subseteq X$, we claim that $\mathrm{conv}(X') \cap \mathrm{conv}(X \setminus X') = \emptyset$ (Exercise 7 from Section 2.2.3). The Minkowski–Weyl theorem (Theorem 2.5.16) implies that $\mathrm{conv}(X')$ and $\mathrm{conv}(X \setminus X')$ are compact convex sets. By Exercise 1 from Section 2.4.4, there exists a separating hyperplane for these two sets, giving a halfspace H such that $X' = H \cap X$.
Let $m \ge d + 2$. Consider any set X with m points. By Radon's theorem (Theorem 2.6.1), one can partition $X = X_1 \uplus X_2$ such that there exists $\mathbf{y} \in \mathrm{conv}(X_1) \cap \mathrm{conv}(X_2)$. Let $X' = X_1$. Consider any halfspace H such that $X' \subseteq H$. Since H is convex, $\mathbf{y} \in H$. By Exercise 50 from Section 2.4.4, we obtain that $H \cap X_2 \ne \emptyset$. Thus, X cannot be shattered by the family of halfspaces in $\mathbb{R}^d$. $\qquad\square$

See Chapters 12 and 13 of [98] for more on VC-dimension.

The ideas behind the proof of Helly's theorem (Theorem 2.6.2) have been generalized and applied in a variety of contexts, including mathematical optimization. We begin with an extension of the notion of convex sets.

Definition 2.6.6 Let $S \subseteq \mathbb{R}^d$. A set is said to be S *convex* if and only if it is of the form $S \cap C$, where $C \subseteq \mathbb{R}^d$ is convex (possibly empty).

Inspired by Helly's theorem (Theorem 2.6.2), we associate the following number with the family of S-convex sets for any $S \subseteq \mathbb{R}^d$.

Definition 2.6.7 Let $S \subseteq \mathbb{R}^d$ be any nonempty subset. A family $X_1, \ldots, X_m$ of S-convex sets is said to be an *S-critical family* if $X_1 \cap \cdots \cap X_m = \emptyset$ and every proper subfamily has nonempty intersection. The *Helly number $h(S)$ of S* is the size of the largest S-critical family (defined to be $+\infty$ if there are S-critical families of unbounded size).

The following alternative characterization of the Helly number ties in with Theorem 2.6.2.

Proposition 2.6.8 *Let $S \subseteq \mathbb{R}^d$ be any nonempty subset. Then $h(S)$ is the smallest natural number $m \in \mathbb{N}$ such that for any family $X_1, \ldots, X_k$ of S-convex sets such that $X_1 \cap \cdots \cap X_k = \emptyset$, there exists a subfamily $X_{i_1}, \ldots, X_{i_m}$ such that $X_{i_1} \cap \cdots \cap X_{i_m} = \emptyset$.*

In this language, one can restate Theorem 2.6.2 as saying $h(\mathbb{R}^d) \leq d + 1$. Exercise 1 from Section 2.6.1 shows that, in fact, $h(\mathbb{R}^d) = d+1$. The following characterization of the Helly number generalizes the idea behind the proof of Theorem 2.6.2.

Proposition 2.6.9 *Let $S \subseteq \mathbb{R}^d$ be any nonempty subset. Then,*

$$h(S) = \sup \left\{ |V| : V \subseteq S \text{ finite}, \bigcap_{\mathbf{v} \in V} (\mathrm{conv}(V \setminus \{\mathbf{v}\}) \cap S) = \emptyset \right\}.$$

Proof We first show that if $X_1, \ldots, X_m$ is an S-critical family, then there exists $V = \{\mathbf{v}_1, \ldots, \mathbf{v}_m\} \subseteq S$ such that $\bigcap_{i=1}^{m} \mathrm{conv}((V \setminus \{\mathbf{v}_i\}) \cap S) = \emptyset$. Indeed, for every $j \in \{1, \ldots, m\}$, there exists $\mathbf{v}^j \in \bigcap_{i \neq j} X_i$ since $X_1, \ldots, X_m$ is an S-critical family; let $V = \{\mathbf{v}_1, \ldots, \mathbf{v}_m\}$. We now check that $\bigcap_{i=1}^{m} (\mathrm{conv}(V \setminus \{\mathbf{v}_i\}) \cap S) = \emptyset$. Observe that for every $i = 1, \ldots, m$, $\mathbf{v}^i \in X_j$ for all $j \neq i$. Thus, $V \setminus \{\mathbf{v}^i\} \subseteq X_i$ for every $i = 1, \ldots, m$. Moreover, since X_i is an S-convex set, we also have $\mathrm{conv}(V \setminus \{\mathbf{v}^i\}) \cap S \subseteq X_i$. Thus, $\bigcap_{i=1}^{m} (\mathrm{conv}(V \setminus \{\mathbf{v}_i\}) \cap S) \subseteq \bigcap_{i=1}^{m} X_i = \emptyset$. This implies that $h(S) \leq \sup \{|V| : V \subseteq S, \bigcap_{\mathbf{v} \in V} (\mathrm{conv}(V \setminus \{\mathbf{v}\}) \cap S) = \emptyset\}$.

In the other direction, let $V \subseteq S$ be a finite set such that $\bigcap_{\mathbf{v} \in V} (\mathrm{conv}(V \setminus \{\mathbf{v}\}) \cap S) = \emptyset$. Observe that $X_{\mathbf{v}} := (\mathrm{conv}(V \setminus \{\mathbf{v}\}) \cap S)$ is an S-convex set for every $\mathbf{v} \in V$. Moreover, for any $\mathbf{v} \in V$, $\bigcap_{\mathbf{v}' \neq \mathbf{v}} X_{\mathbf{v}'}$ contains $\mathbf{v} \in S$. Thus, $X_{\mathbf{v}}, \mathbf{v} \in V$ is an S-critical family. Therefore, $|V| \leq h(S)$ and $\sup \{|V| : V \subseteq S, \bigcap_{\mathbf{v} \in V} (\mathrm{conv}(V \setminus \{\mathbf{v}\}) \cap S) = \emptyset\} \leq h(S)$. $\square$

An important observation that was made in the proof of the above proposition is that for any finite $V \subseteq S$ such that $\bigcap_{\mathbf{v} \in V} (\mathrm{conv}(V \setminus \{\mathbf{v}\}) \cap S) = \emptyset$, the family $X_{\mathbf{v}} := (\mathrm{conv}(V \setminus \{\mathbf{v}\}) \cap S)$ is an S-critical family. Therefore, for a set S with finite Helly number, one may restrict attention to critical families consisting of S-convex sets obtained by intersecting S with polytopes since $\mathrm{conv}(V \setminus \{\mathbf{v}\})$ is a polytope when V is a finite set (Theorem 2.5.16). This further implies that one may restrict attention to critical families consisting of S-convex sets obtained by intersecting S with halfspaces.

Corollary 2.6.10 *Let $S \subseteq \mathbb{R}^d$ be any nonempty subset. Then $h(S)$ is the supremum over all natural numbers m such that there exists a critical family $X_1, \ldots, X_m$ where $X_i = S \cap H_i$ for some halfpsace $H_i \subseteq \mathbb{R}^d$ for every $i = 1, \ldots, m$.*

Proof Left as an exercise. $\qquad\qquad\square$

Another useful characterization of the Helly number is the following.

Definition 2.6.11 Let $S \subseteq \mathbb{R}^d$. A subset $V \subseteq S$ is said to be in *S-convex position* if $\mathrm{conv}(V) \cap S = \mathrm{ext}(\mathrm{conv}(V))$.

Proposition 2.6.12 *Let $S \subseteq \mathbb{R}^d$ be any nonempty subset. Then*

$$h(S) \geq \sup\{|V| : V \subseteq S \text{ finite, } V \text{ is in S-convex position}\}.$$

Moreover, if S is discrete, *i.e., any bounded subset of $\mathbb{R}^d$ contains finitely many elements from S, then*

$$h(S) = \sup\{|V| : V \subseteq S \text{ finite, } V \text{ is in S-convex position}\}.$$

Proof Note that if V is in S-convex position, then $\bigcap_{\mathbf{v} \in V}(\mathrm{conv}(V \setminus \{\mathbf{v}\}) \cap S) = \emptyset$. Thus, the inequality in the statement of the proposition follows from Proposition 2.6.9.

Now suppose S is discrete. We will show that given any finite $V \subseteq S$ such that $\bigcap_{\mathbf{v} \in V}(\mathrm{conv}(V \setminus \{\mathbf{v}\}) \cap S) = \emptyset$, there exists $V^\star$ such that $|V^\star| = |V|$ and $V^\star$ is in S-convex position. Using Proposition 2.6.9, this will establish $h(S) \leq \sup\{|V| : V \subseteq S \text{ finite, } V \text{ is in S-convex position}\}$ and we will have equality.

Define the family $\mathcal{F} := \{V \subseteq S : V \text{ is finite, } \bigcap_{\mathbf{v} \in V}(\mathrm{conv}(V \setminus \{\mathbf{v}\}) \cap S) = \emptyset\}$ and define $P_V := \mathrm{conv}(V)$ for all $V \in \mathcal{F}$. We introduce the following partial order on $\mathcal{F}$: $V \preceq V'$ if $P_V \subseteq P_{V'}$. Fix an arbitrary $k \in \mathbb{N}$, let $V^\star \in \mathcal{F}$ be a minimal element among all $V \in \mathcal{F}$ with $|V| = k$. The existence of a minimal element is guaranteed because for any $V \in \mathcal{F}$, $P_V \cap S$ is finite since P_V is a polytope (Theorem 2.5.16) and S is discrete.

It suffices to show that $V^\star$ is in S-convex position. Observe that no $\bar{\mathbf{v}} \in V^\star$ is in the convex hull of the rest of the elements of $V^\star$; otherwise, $\bar{\mathbf{v}} \in \bigcap_{\mathbf{v} \in V^\star}(\mathrm{conv}(V^\star \setminus \{\mathbf{v}\}) \cap S)$ contradicting that $V^\star \in \mathcal{F}$. Thus, it suffices to show that $(P_{V^\star} \cap S) \setminus V^\star = \emptyset$. Suppose to the contrary. For any $\mathbf{p} \in (P_{V^\star} \cap S) \setminus V^\star$, define $V_{\mathbf{p}}^{\star} := \{\mathbf{v} \in V^{\star} : \mathbf{p} \in \text{conv}(V^{\star} \setminus \{\mathbf{v}\})\}$ and let $\mathbf{v}^{\star} \in \arg\max\{|V_{\mathbf{p}}^{\star}| : \mathbf{p} \in (P_{V^{\star}} \cap S) \setminus V^{\star}\}$. Since $V^{\star} \in \mathcal{F}$, there exists $\hat{\mathbf{v}} \in V^{\star}$ such that $\mathbf{v}^{\star} \notin \text{conv}(V^{\star} \setminus \{\hat{\mathbf{v}}\})$. Consider $\widehat{V} = (V^{\star} \setminus \{\hat{\mathbf{v}}\}) \cup \{\mathbf{v}^{\star}\}$. Since $\mathbf{v}^{\star} \in (P_{V^{\star}} \cap S) \setminus V^{\star}$ and $\hat{\mathbf{v}} \in V^{\star}$, $|\widehat{V}| = |V^{\star}| = k$ and $P_{\widehat{V}} \subsetneq P_{V^{\star}}$. By minimality of $V^{\star}$, $\widehat{V} \notin \mathcal{F}$. In other words, there exists $\mathbf{v}_{\text{new}} \in \text{conv}(\widehat{V} \setminus \{\mathbf{v}\}) \cap S$ for all $\mathbf{v} \in \widehat{V}$. This implies that $\mathbf{v}_{\text{new}} \in (P_{V^{\star}} \cap S) \setminus V^{\star}$.

We first argue that $V_{\mathbf{v}^{\star}}^{\star} \subseteq V_{\mathbf{v}_{\text{new}}}^{\star}$. If $V_{\mathbf{v}^{\star}}^{\star} = \emptyset$, then this is trivial. Else, consider any $\mathbf{z} \in V^{\star}$ such that $\mathbf{v}^{\star} \in \text{conv}(V^{\star} \setminus \{\mathbf{z}\})$. Then $\mathbf{z} \neq \hat{\mathbf{v}}$ and therefore $\mathbf{z} \in \widehat{V}$. Moreover, $\text{conv}(\widehat{V} \setminus \{\mathbf{z}\}) \subseteq \text{conv}(V^{\star} \setminus \{\mathbf{z}\})$ since $\widehat{V} \setminus \{\mathbf{z}\} = V^{\star} \setminus \{\hat{\mathbf{v}}, \mathbf{z}\} \cup \{\mathbf{v}^{\star}\} \subseteq V^{\star} \setminus \{\mathbf{z}\} \cup \{\mathbf{v}^{\star}\}$ and $\mathbf{v}^{\star} \in \text{conv}(V^{\star} \setminus \{\mathbf{z}\})$. Since $\mathbf{v}_{\text{new}} \in \text{conv}(\widehat{V} \setminus \{\mathbf{z}\})$, this implies that $V_{\mathbf{v}^{\star}}^{\star} \subseteq V_{\mathbf{v}_{\text{new}}}^{\star}$. Additionally, $\mathbf{v}_{\text{new}} \in \text{conv}(\widehat{V} \setminus \{\mathbf{v}^{\star}\}) = \text{conv}(V^{\star} \setminus \{\hat{\mathbf{v}}\})$. Thus, $\hat{\mathbf{v}} \in V_{\mathbf{v}_{\text{new}}}^{\star}$, but by definition $\hat{\mathbf{v}} \notin V_{\mathbf{v}^{\star}}^{\star}$. Therefore, $V_{\mathbf{v}^{\star}}^{\star} \subsetneq V_{\mathbf{v}_{\text{new}}}^{\star}$ contradicting the choice of $\mathbf{v}^{\star}$ as the maximizing element in $(P_{V^{\star}} \cap S) \setminus V^{\star}$. $\qquad\square$

The following theorem has been obtained independently by different authors in different contexts and will be important in our study of optimization with integer constrained decision variables.

Corollary 2.6.13 $h(\mathbb{Z}^d) = 2^d$.

Proof By Proposition 2.6.12, it suffices to show that the maximum size of a subset of $\mathbb{Z}^d$ in $\mathbb{Z}^d$-convex position is 2^d. The subset $\{0, 1\}^d \subseteq \mathbb{Z}^d$ has size 2^d and is in $\mathbb{Z}^d$-convex position. Consider any $V \subseteq \mathbb{Z}^d$ such that $|V| > 2^d$. Then, there must exist two points $\mathbf{v}_1, \mathbf{v}_2 \in V$ such that the coordinates of $\mathbf{v}_1$ and $\mathbf{v}_2$ are identical modulo 2. This implies that $\bar{\mathbf{v}} := \frac{\mathbf{v}_1 + \mathbf{v}_2}{2} \in \mathbb{Z}^d$ and V is not in $\mathbb{Z}^d$-convex position. $\qquad\square$

Finally, we prove a generalization of both Theorem 2.6.2 and Corollary 2.6.13.

Theorem 2.6.14 *Let n, d be nonnegative integers such that $n + d \geq 1$. Consider $S = \mathbb{Z}^n \times \mathbb{R}^d$ as a subset of $\mathbb{R}^n \times \mathbb{R}^d$. Then, $h(\mathbb{Z}^n \times \mathbb{R}^d) = (d + 1)2^n$.*

Proof We prove $h(\mathbb{Z}^n \times \mathbb{R}^d) \leq (d + 1)2^n$ and leave the reverse inequality as an exercise. Using Proposition 2.6.9, it suffices to show that for any finite subset $V \subseteq \mathbb{Z}^n \times \mathbb{R}^d$ with $\bigcap_{\mathbf{v} \in V}(\text{conv}(V \setminus \{\mathbf{v}\}) \cap (\mathbb{Z}^n \times \mathbb{R}^d)) = \emptyset$, we must have $|V| \leq (d + 1)2^n$. By Theorem 2.5.16, $P_{\mathbf{v}} := \text{conv}(V \setminus \{\mathbf{v}\})$ is a polytope for every $\mathbf{v} \in V$. Let $A^{\mathbf{v}}\mathbf{x} \leq \mathbf{b}^{\mathbf{v}}$ be a halfspace description of $P_{\mathbf{v}}$.

Case 1: $\bigcap_{\mathbf{v} \in V} P_{\mathbf{v}} = \emptyset$: By Theorem 2.6.2, there exist halfspaces $H_1, \ldots, H_m$ coming from the descriptions $A^{\mathbf{v}}\mathbf{x} \leq \mathbf{b}^{\mathbf{v}}$, $\mathbf{v} \in V$ such that $H_1 \cap \cdots \cap H_m = \emptyset$ and $m \leq n + d + 1 \leq (d + 1)2^n$. Note that for each $\mathbf{v} \in V$, at least one of the halfspaces from $A^{\mathbf{v}}\mathbf{x} \leq \mathbf{b}^{\mathbf{v}}$ must be used since otherwise $\mathbf{v} \in H_1 \cap \cdots \cap H_m$. Therefore, $|V| \leq m \leq (d + 1)2^n$.

Case 2: $\bigcap_{\mathbf{v}\in V} P_{\mathbf{v}} \neq \emptyset$: Since $P_{\mathbf{v}}$ is compact for each $\mathbf{v} \in V$, $\bigcap_{\mathbf{v}\in V} P_{\mathbf{v}}$ is compact as well; i.e., it is a polytope. Since $\mathbb{Z}^n \times \mathbb{R}^d$ is closed, by Exercise 26 from Section 2.5.6 there exists $\epsilon > 0$ such that $\bigcap_{\mathbf{v}\in V}\{\mathbf{x} \in \mathbb{R}^d : A^{\mathbf{v}}\mathbf{x} \leq \mathbf{b}^{\mathbf{v}} + \epsilon\mathbf{1}\} \cap (\mathbb{Z}^n \times \mathbb{R}^d) = \emptyset$. Observe that $P := \bigcap_{\mathbf{v}\in V}\{\mathbf{x} \in \mathbb{R}^d : A^{\mathbf{v}}\mathbf{x} \leq \mathbf{b}^{\mathbf{v}} + \epsilon\mathbf{1}\}$ is full dimensional (see Exercise 1 from Section 2.4.4).

We may assume $n \geq 1$ since, otherwise, the result follows from Theorem 2.6.2. Consider the projection Q of P on to the linear subspace $\mathbb{R}^n \times \{\mathbf{0}\}$. Since projections are linear maps (Exercise 2 from Section 2.3.1) and linear transformations of polyhedra are polyhedra (Exercise 8 from Section 2.5.6), Q is a polyhedron. Moreover, since P is full dimensional, it is not hard to verify that Q has dimension n. Finally, $Q \cap (\mathbb{Z}^n \times \{\mathbf{0}\}) = \emptyset$ since $P \cap (\mathbb{Z}^n \times \mathbb{R}^d) = \emptyset$. Viewing both Q and $\mathbb{Z}^n \times \{\mathbf{0}\}$ as subsets of $\mathbb{R}^n$, we can apply Corollary 2.6.13 and obtain a subset of $\ell \leq 2^n$ facets $F_1, \ldots, F_\ell$ of Q such that the intersection of the corresponding halfspaces contains no point from $\mathbb{Z}^n \times \{\mathbf{0}\}$. Let these facet defining halfspaces for Q be given by $(\mathbf{a}^i, \mathbf{c}^i) \in \mathbb{R}^n \times \mathbb{R}^d$ and $\delta_i \in \mathbb{R}$ for $i = 1, \ldots, \ell$; in other words, $F_i = Q \cap \{(\mathbf{x},\mathbf{y}) \in \mathbb{R}^n \times \mathbb{R}^d : \langle \mathbf{a}^i, \mathbf{x}\rangle + \langle \mathbf{c}^i, \mathbf{y}\rangle = \delta_i\}$ and $\langle \mathbf{a}^i, \mathbf{x}\rangle + \langle \mathbf{c}^i, \mathbf{y}\rangle \leq \delta_i$ is a valid inequality for Q. Observe that the halfspace $\{(\mathbf{x},\mathbf{y}) \in \mathbb{R}^n \times \mathbb{R}^d : \langle \mathbf{a}^i, \mathbf{x}\rangle \leq \delta_i\}$ exposes the same facet F_i of Q, and $\langle \mathbf{a}^i, \mathbf{x}\rangle \leq \delta_i$ is a valid inequality for P. Also, $\bigcap_{i=1}^{\ell}\{(\mathbf{x},\mathbf{y}) \in \mathbb{R}^n \times \mathbb{R}^d : \langle \mathbf{a}^i, \mathbf{x}\rangle \leq \delta_i\}$ contains no point from $\mathbb{Z}^n \times \mathbb{R}^d$, since Q contains no point from $\mathbb{Z}^n \times \{\mathbf{0}\}$.

The facet F_i is exactly the projection of the face $G_i := P \cap \{(\mathbf{x},\mathbf{y}) \in \mathbb{R}^n \times \mathbb{R}^d : \langle \mathbf{a}^i, \mathbf{x}\rangle = \delta_i\}$ of P. Since projections are linear maps, by Exercise 4 from Section 2.2.3, $\dim(G_i) \geq \dim(F_i) = n - 1$, using the fact that Q has dimension n and F_i is a facet of Q. Using Exercise 21 from Section 2.5.6, for every $i = 1, \ldots, \ell$, there exist at most $n + d - (n - 1) = d + 1$ halfspaces in the description $\bigcap_{\mathbf{v}\in V}\{\mathbf{x} \in \mathbb{R}^d : A^{\mathbf{v}}\mathbf{x} \leq \mathbf{b}^{\mathbf{v}} + \epsilon\mathbf{1}\}$ of P such that their intersection is contained in the halfspace $\{(\mathbf{x},\mathbf{y}) \in \mathbb{R}^n \times \mathbb{R}^d : \langle \mathbf{a}^i, \mathbf{x}\rangle \leq \delta_i\}$. Thus, we obtain at most $(d + 1)\ell \leq (d + 1)2^n$ halfspaces from the description $\bigcap_{\mathbf{v}\in V}\{\mathbf{x} \in \mathbb{R}^d : A^{\mathbf{v}}\mathbf{x} \leq \mathbf{b}^{\mathbf{v}} + \epsilon\mathbf{1}\}$ of P whose intersection contains no point from $\mathbb{Z}^n \times \mathbb{R}^d$. As observed in Case 1. above, we must include at least one inequality from $A^{\mathbf{v}}\mathbf{x} \leq \mathbf{b}^{\mathbf{v}} + \epsilon\mathbf{1}$ for each $\mathbf{v} \in V$; otherwise, $\mathbf{v}$ ends up in the intersection. Therefore, $|V| \leq (d + 1)2^n$. $\qquad\square$

2.6.1 Exercises

1. Show that $h(\mathbb{R}^d) = d + 1$.
2. Use Proposition 2.6.9 to prove Corollary 2.6.10.
3. Use Corollary 2.6.10, Theorem 2.5.22, and Carathéodory's theorem (Theorem 2.2.15) to give an alternate proof of the fact that $h(\mathbb{R}^d) \leq d + 1$.

4. Prove the following generalization of Theorem 2.6.2: Let $C_1, \ldots, C_m \subseteq \mathbb{R}^d$ be convex sets and let $k \leq d + 1$. If every family of k subsets from $C_1, \ldots, C_m$ has nonempty intersection, then for every linear subspace L of dimension $d - k + 1$, there exists $\mathbf{t} \in \mathbb{R}^d$ such that the affine subspace $L + \mathbf{t}$ intersects all $C_1, \ldots, C_m$.

5.* Prove the following generalization of Theorem 2.6.2: Let $C_1, \ldots, C_m$ and X be convex sets in $\mathbb{R}^d$. Suppose that for every $d + 1$ subfamily $C_{i_1}, \ldots, C_{i_k}$, there exists $\mathbf{t} \in \mathbb{R}^d$ such that $X + \mathbf{t}$ intersects all of $C_{i_1}, \ldots, C_{i_k}$. Show that there exists $\mathbf{t} \in \mathbb{R}^d$ such that $X + \mathbf{t}$ intersects *all* of $C_1, \ldots, C_m$.

   Why is this a generalization of Theorem 2.6.2?

6. Show that the statement in Exercise 5 above holds when "*intersects*" is replaced by "*is contained in*" or by "*contains.*"

7.* (Approximation of functions). Let $f_i : \mathbb{R}^d \to \mathbb{R}$, $i = 1, \ldots, k$ be functions and let $g : \mathbb{R}^d \to \mathbb{R}$ be a function. Given a set $T \subseteq \mathbb{R}^d$ and $\epsilon > 0$, we say that $f_1, \ldots, f_k$ uniformly ϵ-approximate g on T if there exist $\mu_1, \ldots, \mu_k \in \mathbb{R}$ such that $|g(\mathbf{x}) - \sum_{i=1}^{k} \mu_i f_i(\mathbf{x})| \leq \epsilon$ for all $\mathbf{x} \in T$. Fix $\epsilon > 0$ and some finite set T. Show that $f_1, \ldots, f_k$ uniformly ϵ-approximate g on T if and only if $f_1, \ldots, f_k$ uniformly ϵ-approximate g on every subset $T' \subseteq T$ with $|T'| = k + 1$.

   [Remark: The theorem can be extended to the setting when the set T is infinite, by assuming further regularity conditions on $f_1, \ldots, f_k, g$.]

8. Prove that if a compact, convex set $C \subseteq \mathbb{R}^d$ is contained in the union of some (possibly infinite) family of halfspaces in $\mathbb{R}^d$, then C is contained in the union of some $d + 1$ (or fewer) halfspaces from the family.

9. Let $S \subseteq \mathbb{R}^d$. Show that if $V \subseteq S$ is in S-convex position, then $\mathrm{conv}(V) \cap S = V = \mathrm{ext}(\mathrm{conv}(V))$ (see also Exercise 15 from Section 2.5.6).

10. For any subset $S \subseteq \{1, \ldots, d\}$, define the halfspace $H_S := \{\mathbf{x} \in \mathbb{R}^d : \sum_{i \in S} \mathbf{x}_i + \sum_{i \notin S}(1 - \mathbf{x}_i) \leq d - 1\}$. Show that $X_S := \mathbb{Z}^d \cap H_S$, $S \subseteq \{1, \ldots, d\}$ is a $\mathbb{Z}^d$-critical family. This gives an alternate proof that $h(\mathbb{Z}^d) \geq 2^d$.

11. Complete the proof of Theorem 2.6.14 by showing that $h(\mathbb{Z}^n \times \mathbb{R}^d) \geq (d + 1)2^n$.

2.7 Ellipsoids

An important class of convex sets derives from norms induced by positive definite matrices (see Exercise 1 in Section 1.2.3), which form a generalization of the unit ball associated with the standard Euclidean norm.

Theorem 2.7.1 *Let $E \subseteq \mathbb{R}^d$. The following are equivalent.*

1. *There exists a positive definite matrix $A \in \mathbb{R}^{d \times d}$ and $\mathbf{c} \in \mathbb{R}^d$ such that $E = \{\mathbf{x} \in \mathbb{R}^d : N_A(\mathbf{x} - \mathbf{c}) \leq 1\}$; i.e., E is a translation of the unit ball with respect to the norm N_A.*
2. *There exists an invertible affine transformation $T : \mathbb{R}^d \to \mathbb{R}^d$ such that $E = T(B(\mathbf{0}_d, 1))$; i.e., E is the affine image of the unit ball with respect to the standard Euclidean norm.*
3. *There exists $k \in \mathbb{N}$ and an affine transformation $T : \mathbb{R}^k \to \mathbb{R}^d$ with rank d such that $E = T(B(\mathbf{0}_k, 1))$; i.e., E is the affine image of the unit ball (in $\mathbb{R}^k$) with respect to the standard Euclidean norm.*
4. *There exist orthonormal vectors $\mathbf{p}^1, \ldots, \mathbf{p}^d \in \mathbb{R}^d$, $\sigma_1, \ldots, \sigma_d > 0$ and $\mathbf{v} \in \mathbb{R}^d$ such that*

$$E = \left\{ \mathbf{v} + \lambda_1 \mathbf{p}^1 + \cdots + \lambda_d \mathbf{p}^d : \frac{\lambda_1^2}{\sigma_1^2} + \cdots + \frac{\lambda_d^2}{\sigma_d^2} \leq 1 \right\}. \tag{2.7.1}$$

Proof (1. $\Rightarrow$ 2.) By Theorem 1.2.16, there exists an invertible matrix $R \in \mathbb{R}^{d \times d}$ such that $A = R^T R$. Thus,

$$\begin{aligned}
E &= \{\mathbf{x} \in \mathbb{R}^d : \sqrt{(\mathbf{x} - \mathbf{c}^T)A(\mathbf{x} - \mathbf{c})} \leq 1\} \\
&= \{\mathbf{x} \in \mathbb{R}^d : \sqrt{(\mathbf{x} - \mathbf{c}^T)R^T R(\mathbf{x} - \mathbf{c})} \leq 1\} \\
&= \{\mathbf{x} \in \mathbb{R}^d : \|R(\mathbf{x} - \mathbf{c})\| \leq 1\} \\
&= \{R^{-1}\mathbf{y} + \mathbf{c} : \|\mathbf{y}\| \leq 1\}.
\end{aligned}$$

Thus, the affine transformation $T(\mathbf{y}) = R^{-1}(\mathbf{y}) + \mathbf{c}$ works.

(2. $\Rightarrow$ 3.) This needs no explanation: one simply takes $k = d$.

(3. $\Rightarrow$ 4.) Let the affine transformation be given by $T(\mathbf{x}) = R\mathbf{x} + \mathbf{p}$ for some matrix $R \in \mathbb{R}^{d \times k}$ with rank d and $\mathbf{p} \in \mathbb{R}^d$. Let $\mathbf{u}^1, \ldots, \mathbf{u}^d$ be left singular vectors of R (see Theorem 1.2.12 and the discussion below it), let $\sigma_1, \ldots, \sigma_d > 0$ be the corresponding singular values, and let $\mathbf{v}^1, \ldots, \mathbf{v}^d$ be the corresponding right singular vectors.

Extend $\mathbf{v}^1, \ldots, \mathbf{v}^d$ to an orthonormal set $\{\mathbf{v}^1, \ldots, \mathbf{v}^k\}$ in $\mathbb{R}^k$. Using the SVD $R = U\Sigma V^T$, $R\mathbf{v}^j = \mathbf{0}_d$ for all $j = d+1, \ldots, k$. We can express the unit sphere in $\mathbb{R}^k$ as $B(\mathbf{0}_k, 1) = \{\gamma_1 \mathbf{v}^1 + \cdots + \gamma_k \mathbf{v}^k : \gamma_1^2 + \cdots + \gamma_k^2 \leq 1\}$. Then

$$\begin{aligned}
E &= \{R\mathbf{x} + \mathbf{p} : \mathbf{x} \in B(\mathbf{0}_k, 1)\} \\
&= \{R(\gamma_1 \mathbf{v}^1 + \cdots + \gamma_k \mathbf{v}^k) + \mathbf{p} : \gamma_1^2 + \cdots + \gamma_k^2 \leq 1\} \\
&= \left\{ \sigma_1 \gamma_1 \mathbf{u}^1 + \cdots + \sigma_d \gamma_d \mathbf{u}^d + \mathbf{p} : \begin{array}{l} \gamma_1, \ldots, \gamma_d \in \mathbb{R} \text{ such that there exist} \\ \gamma_{d+1}, \ldots, \gamma_k \in \mathbb{R} \text{ with } \gamma_1^2 + \cdots + \gamma_k^2 \leq 1 \end{array} \right\} \\
&= \{\sigma_1 \gamma_1 \mathbf{u}^1 + \cdots + \sigma_d \gamma_d \mathbf{u}^d + \mathbf{p} : \gamma_1^2 + \cdots + \gamma_d^2 \leq 1\} \\
&= \left\{ \lambda_1 \mathbf{u}^1 + \cdots + \lambda_d \mathbf{u}^d + \mathbf{p} : \frac{\lambda_1^2}{\sigma_1^2} + \cdots + \frac{\lambda_d^2}{\sigma_d^2} \leq 1 \right\}.
\end{aligned}$$

Setting $\mathbf{p}^i := \mathbf{u}^i$ and $\mathbf{v} := \mathbf{p}$ finishes the proof.

$(4. \Rightarrow 1.)$ Define $A = R^{-T} R^{-1}$ where R is the matrix with $\sigma_1 \mathbf{p}^1, \ldots, \sigma_d \mathbf{p}^d$ as columns, and $\mathbf{c} = \mathbf{v}$. One can check that $\left\{ \mathbf{v} + \lambda_1 \mathbf{p}^1 + \cdots + \lambda_d \mathbf{p}^d : \frac{\lambda_1^2}{\sigma_1^2} + \cdots + \frac{\lambda_d^2}{\sigma_d^2} \leq 1 \right\} = \{\mathbf{x} \in \mathbb{R}^d : N_A(\mathbf{x} - \mathbf{c}) \leq 1\}.$ $\qquad\square$

While all four descriptions in Theorem 2.7.1 are useful, we will use the first one to introduce formal notation and terminology. We draw attention to the inverse operation used in the next definition, in comparison to part 1. of Theorem 2.7.1. This makes calculations much easier (as can perhaps be inferred from the proof of Theorem 2.7.1).

Definition 2.7.2   Let $A \in \mathbb{R}^{d \times d}$ be a positive definite matrix and $\mathbf{c} \in \mathbb{R}^d$. The set

$$E(A, \mathbf{c}) := \{\mathbf{x} \in \mathbb{R}^d : (\mathbf{x} - \mathbf{c})^T A^{-1} (\mathbf{x} - \mathbf{c}) \leq 1\} = \{\mathbf{x} \in \mathbb{R}^d : N_{A^{-1}}(\mathbf{x} - \mathbf{c}) \leq 1\}$$

is called an *ellipsoid centered at* $\mathbf{c}$.

Given an ellipsoid, consider any $\mathbf{v}, \mathbf{p}^1, \ldots, \mathbf{p}^d \in \mathbb{R}^d$ and $\sigma_1, \ldots, \sigma_d > 0$ satisfying part 4. of Theorem 2.7.1. The line segments $[\mathbf{v} - \sigma_i \mathbf{p}^i, \mathbf{v} + \sigma_i \mathbf{p}^i] = \mathrm{conv}(\{\mathbf{v} - \sigma_i \mathbf{p}^i, \mathbf{v} + \sigma_i \mathbf{p}^i\})$, $i = 1, \ldots, d$ are called *a set of principal axes of the ellipsoid*.

Note that the set of principal axes of an ellipsoid is not unique since there are potentially many different $\mathbf{p}^1, \ldots, \mathbf{p}^d \in \mathbb{R}^d$ satisfying part 4. of Theorem 2.7.1 ($\sigma_1, \ldots, \sigma_d > 0$ are unique up to permutations – see Section 1.2.1).

The following formulas for the volume of an ellipsoid will prove to be handy.

Theorem 2.7.3 *The following are all true.*

1. *Let $A \in \mathbb{R}^{d \times d}$ be a positive definite matrix and $\mathbf{c} \in \mathbb{R}^d$. Then $\mathrm{vol}(E(A, \mathbf{c})) = \sqrt{\det(A)}\, \mathrm{vol}(B(\mathbf{0}_d, 1))$.*
2. *Let $T(\mathbf{x}) = R\mathbf{x} + \mathbf{p}$ be an affine transformation with $R \in \mathbb{R}^{d \times k}$ and $\mathbf{p} \in \mathbb{R}^d$. Then $\mathrm{vol}(T(B(\mathbf{0}_k, 1))) = \sqrt{\det(R R^T)} \cdot \mathrm{vol}(B(\mathbf{0}_k, 1))$.*
3. *Let $\mathbf{p}^1, \ldots, \mathbf{p}^d \in \mathbb{R}^d$ be orthonormal vectors, $\sigma_1, \ldots, \sigma_d > 0$ and $\mathbf{v} \in \mathbb{R}^d$. Then,*

$$\mathrm{vol}(E) = \left(\prod_{i=1}^{d} \sigma_i \right) \mathrm{vol}(B(\mathbf{0}_d, 1)),$$

where $E = \left\{ \mathbf{v} + \lambda_1 \mathbf{p}^1 + \cdots + \lambda_d \mathbf{p}^d : \frac{\lambda_1^2}{\sigma_1^2} + \cdots + \frac{\lambda_d^2}{\sigma_d^2} \leq 1 \right\}.$

Proof   Left as an exercise. $\qquad\qquad\qquad\qquad\qquad\qquad\qquad\qquad\qquad\square$

The following estimates of the volume of $B(\mathbf{0}_d, 1)$ are useful.

Theorem 2.7.4 $\operatorname{vol}(B(\mathbf{0}_d, 1)) = \dfrac{\pi^{d/2}}{\Gamma\left(\frac{d}{2}+1\right)}$, *where* $\Gamma(x) := \int_0^\infty e^{-t} t^{x-1} dt$ *for* $x > 0$ *is the so-called* gamma function.[4] *Moreover,*

$$\left(\frac{2}{\sqrt{d}}\right)^d \leq \operatorname{vol}(B(\mathbf{0}, 1)) \leq 2^d.$$

Proof The bounds follow from the fact that $\left[-\frac{1}{\sqrt{d}}, \frac{1}{\sqrt{d}}\right]^d \subseteq B(\mathbf{0}, 1) \subseteq [-1, 1]^d$. $\qquad\square$

The following is a useful formula for coordinate projections of ellipsoids.

Proposition 2.7.5 *Let* $I \subseteq \{1, \ldots, d\}$. *Define the linear transformation* $T_I \colon \mathbb{R}^d \to \mathbb{R}^{|I|}$ *as* $T_I(\mathbf{x}) = (\mathbf{x}_i)_{i \in I}$. *Then* $T_I(E(A, \mathbf{c})) = E(A_I, T_I(\mathbf{c}))$ *where* A_I *is the* $|I| \times |I|$ *submatrix of* A *obtained by taking the rows and columns corresponding to indices in* I.

Proof Let P denote the matrix corresponding to the linear transformation T_I, i.e., P is the matrix with the unit vectors $\mathbf{e}^i \in \mathbb{R}^d$, $i \in I$ as rows. Following the proof of the first implication in Theorem 2.7.1, $E(A, \mathbf{c}) = T(B(\mathbf{0}_d, 1))$ where $T \colon \mathbb{R}^d \to \mathbb{R}^d$ can be taken as $T(\mathbf{x}) = R\mathbf{x} + \mathbf{c}$ where $A = RR^T$ for an invertible $R \in \mathbb{R}^{d \times d}$. Thus, $T_I(E(A, \mathbf{c})) = \{PR\mathbf{x} + P\mathbf{c} : \mathbf{x} \in B(\mathbf{0}_d, 1)\}$. Following the proof of the third implication in Theorem 2.7.1, $T_I(E(A, \mathbf{c})) = \left\{\mathbf{v} + \lambda_1 \mathbf{p}^1 + \cdots + \lambda_k \mathbf{p}^k : \frac{\lambda_1^2}{\sigma_1^2} + \cdots + \frac{\lambda_k^2}{\sigma_k^2} \leq 1\right\}$, where $\mathbf{p}^1, \ldots, \mathbf{p}^k$ are left singular vectors and $\sigma_1, \ldots, \sigma_k$ are the singular values of the matrix PR with $k = |I|$, and $\mathbf{v} = P\mathbf{c}$. By Theorem 1.2.18, $\mathbf{p}^1, \ldots, \mathbf{p}^k$ are eigenvectors of $PR(PR)^T = PRR^T P^T = PAP^T = A_I$ and $\sigma_1^2, \ldots, \sigma_k^2$ are eigenvalues of A_I. Thus, $A_I = MM^T$ where M is the matrix with $\sigma_1 \mathbf{p}^1, \ldots, \sigma_k \mathbf{p}^k$ as columns. Following the proof of the fourth implication in Theorem 2.7.1 finishes the proof. $\qquad\square$

2.7.1 Ellipsoidal Approximations

We have all the tools in place to prove two fundamental results. The first one has proved to be a fundamental tool in mathematical optimization. The second one will follow as a consequence, but it is immensely important in its own right.

---

[4] Stirling's approximation can be used to show that $\Gamma(x + 1) = \Theta\left(\sqrt{2\pi x} \left(\frac{x}{e}\right)^x\right)$.

The main motivation behind the first result is that it is sometimes easier to design algorithms that work with ellipsoids because of their special structure, as opposed to general convex sets. Thus, we would like to maintain ellipsoidal approximations of the convex set in question. Consider the special case of approximating a polytope P with an ellipsoid, which is a bounded intersection of finitely many halfspaces. One strategy is to start with a large sphere $B(\mathbf{0}, R)$ (with large enough radius R) that is guaranteed to contain P. This forms our initial approximation. Now, iteratively we add each of the defining halfspaces for P and update our ellipsoid. For this, we need to find a new ellipsoid that is "smaller" but contains the intersection of the previous ellipsoid and the newly introduced halfspace. The following result gives a guarantee on how small we can make the updated ellipsoid, assuming the halfspace is "deep enough." Thus, this iterative procedure will stop when the remaining halfspaces are not "deep." This idea will be fleshed out in Section 6.3.2.

Theorem 2.7.6 (Circumscribing ellipsoid for ellipsoidal section) *Let $d \geq 2$ and let $E \subseteq \mathbb{R}^d$ be an ellipsoid with center $\mathbf{c} \in \mathbb{R}^d$. Let $H \subseteq \mathbb{R}^d$ be a halfspace and let $0 \leq \beta < \frac{1}{d}$ such that H does not contain $\mathbf{c} + \beta(E - \mathbf{c})$; i.e., H does not contain the β scaling of E about its center. Then, there exists another ellipsoid E' such that $E \cap H \subseteq E'$ and*

$$\operatorname{vol}(E') \leq e^{-\frac{(1-\beta d)^2}{2(d+1)}} \operatorname{vol}(E).$$

Proof We may assume that $E = E(I_d, \mathbf{0})$, i.e., E is the unit ball $B(\mathbf{0}, 1)$, since affine transformations affect the volume on both sides of the inequality by the same factor (see Theorem 2.7.3). Since we assume that H does not contain $\beta B(\mathbf{0}, 1)$, we may assume H is of the form $x_1 \leq \beta' < \beta$. It then suffices to show that there exists an ellipsoid E' such that $B(\mathbf{0}, 1) \cap H \subseteq E'$ and $\operatorname{vol}(E') \leq e^{-\frac{(1-\beta d)^2}{2(d+1)}} \operatorname{vol}(B(\mathbf{0}, 1))$. Set

$$\mathbf{v} = -\left(\frac{1 - \beta d}{d + 1}\right) \mathbf{e}^1, \quad \sigma := \frac{(1 + \beta)d}{d + 1} < 1, \quad \sigma' := \sqrt{\frac{(1 - \beta^2)d^2}{d^2 - 1}} > 1. \tag{2.7.2}$$

We claim that the ellipsoid

$$E' := \left\{ \mathbf{v} + \lambda_1 \mathbf{e}^1 + \cdots + \lambda_d \mathbf{e}^d : \frac{\lambda_1^2}{\sigma^2} + \sum_{i=2}^{d} \frac{\lambda_i^2}{\sigma'^2} \leq 1 \right\} \tag{2.7.3}$$

satisfies these two conditions, even with $\beta' = \beta$. These numbers become a bit transparent by looking at Figure 2.2. The figure depicts the two-dimensional "slice" obtained by intersecting E, H, and E' with the linear subspace $\{\mathbf{x} \in \mathbb{R}^d : \mathbf{x}_3 = \cdots = \mathbf{x}_d = 0\}$. E' is obtained by "rotating" this picture about

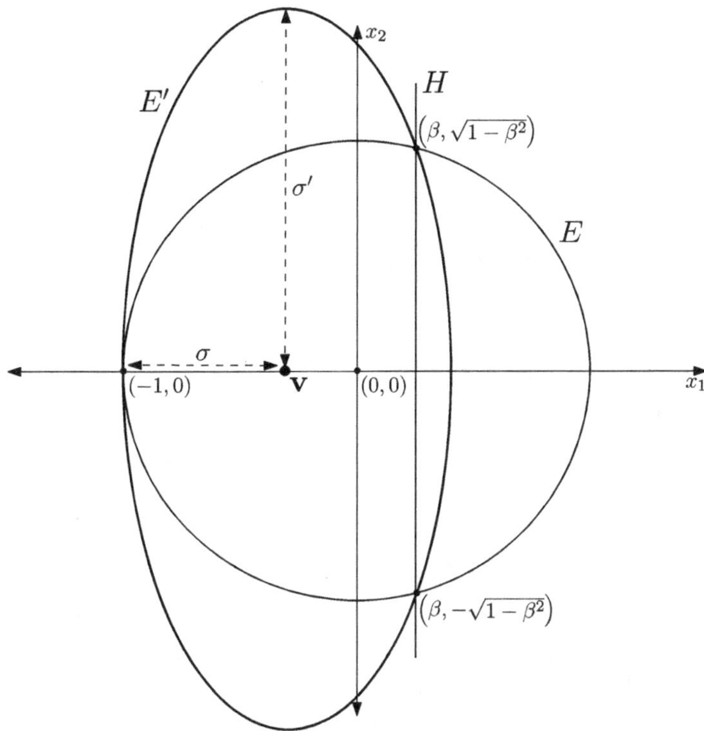

Figure 2.2 A picture illustrating the relationship between E, H, and E'.

the x_1-axis. The parameters σ, σ' are found by computing the two-dimensional ellipse centered at $\mathbf{v}$ passing through $(-1,0)$ and the intersection of $B(\mathbf{0},1)$ and H (restricted to $\{\mathbf{x} \in \mathbb{R}^d : \mathbf{x}_3 = \cdots = \mathbf{x}_d = 0\}$).[5] By the symmetry about the x_1-axis, it should be intuitively clear that $E \cap H \subseteq E'$. One can verify this algebraically as well (see Exercise 10 from Section 2.7.2). We now show that $\mathrm{vol}(E') \le e^{-\frac{(1-\beta d)^2}{2(d+1)}} \mathrm{vol}(B(\mathbf{0},1))$. Using the formula in part 3. of Theorem 2.7.3, $\mathrm{vol}(E') = \sigma(\sigma')^{d-1} \mathrm{vol}(B(\mathbf{0},1))$. We now verify that

[5] There is flexibility in choosing the center $\mathbf{v}$. It turns out that the choice made here makes E' the ellipsoid with minimum volume that contains $E \cap H$. By symmetry, one can argue that the center of this ellipsoid must be on the x_1-axis, i.e., of the form $(t,0,0,\ldots,0)$ and be symmetric about the x_1-axis, i.e., $\sigma_2 = \sigma_3 = \cdots = \sigma_d$. By imposing the condition that $(-1,0)$ and $(\beta, \pm\sqrt{1-\beta^2})$ must lie on the two-dimensional slice of this ellipsoid (depicted in Figure 2.2), one can express σ and σ' in terms of t. One then solves the one-dimensional minimization problem in t of minimizing $\sigma \sigma'^{d-1}$.

$$
\begin{aligned}
\sigma(\sigma')^{d-1} &= \frac{(1+\beta)d}{d+1}\left(\frac{(1-\beta^2)d^2}{d^2-1}\right)^{(d-1)/2} \\
&= \frac{d+1+\beta d-1}{d+1}\left(\frac{d^2-1-\beta^2 d^2+1}{d^2-1}\right)^{(d-1)/2} \\
&= \left(1-\frac{1-\beta d}{d+1}\right)\left(1+\frac{1-\beta^2 d^2}{d^2-1}\right)^{(d-1)2} \\
&\le e^{-\frac{1-\beta d}{d+1}}\, e^{\left(\frac{1-\beta^2 d^2}{d^2-1}\right)\left(\frac{d-1}{2}\right)} \\
&= e^{-\frac{1-\beta d}{d+1}+\frac{1-\beta^2 d^2}{2(d+1)}} \\
&= e^{-\frac{(1-\beta d)^2}{2(d+1)}},
\end{aligned}
$$

where the inequality follows from the fact that $(1+x) \le e^x$ for all $x \in \mathbb{R}$ (see Exercise 7 in Section 3.4.1). $\qquad\square$

We now derive a powerful consequence of Theorem 2.7.6: any full-dimensional, compact convex set can be approximated by an ellipsoid, "up to a factor of the dimension." Here is the formal statement.

Theorem 2.7.7 (Löwner–John ellipsoidal approximation) *Let $C \subseteq \mathbb{R}^d$ be a full-dimensional, compact, convex set. There exists an ellipsoid E containing C such that it has minimum volume among all ellipsoids containing C. Further, $\mathbf{c}+\frac{1}{d}(E-\mathbf{c}) \subseteq C \subseteq E$, where $\mathbf{c}$ is the center of E. In other words, C is contained in E and contains the $1/d$ scaling of E about its center.*

Proof We first show the existence of the minimum volume ellipsoid. Since C is full dimensional, it contains $B(\mathbf{z},\epsilon)$ for some $\mathbf{z} \in C$ and $\epsilon > 0$. Since C is compact, it is bounded and there exists $R > 0$ such that $C \subseteq B(\mathbf{0}, R)$. Consider the (nonempty) set of ellipsoids that contain C and have volume at most $\mathrm{vol}(B(\mathbf{0}, R))$. Consider any such ellipsoid given by (2.7.1) for some $\mathbf{v} \in \mathbb{R}^d$, orthonormal set $\{\mathbf{p}^1,\ldots,\mathbf{p}^d\}$ and $\sigma_1,\ldots,\sigma_d > 0$. By Exercise 7 from Section 2.7.2, the shortest principal axis has length at least 2ϵ, which means the smallest σ_d is at least ϵ. Since the volume of $E(A,\mathbf{c})$ is bounded by $\mathrm{vol}(B(\mathbf{0}, R))$, the largest σ_d is upper bounded by $\frac{\mathrm{vol}(B(\mathbf{0},R))}{\epsilon^{d-1}}$ (Theorem 2.7.3). This implies that the standard Euclidean distance between $\mathbf{v}$ and any point in E and therefore in C is bounded by $\frac{\mathrm{vol}(B(\mathbf{0},R))}{\epsilon^{d-1}}$ and, consequently, $\|\mathbf{v}\|_2 \le R + \frac{\mathrm{vol}(B(\mathbf{0},R))}{\epsilon^{d-1}}$. Consider the set

$$
X := \left\{ (\mathbf{v},\mathbf{p}^1,\ldots,\mathbf{p}^d,\sigma_1,\ldots,\sigma_d) \in \mathbb{R}^d \times \underbrace{\mathbb{R}^d \times \cdots \times \mathbb{R}^d}_{d \text{ times}} \times \underbrace{\mathbb{R} \times \cdots \times \mathbb{R}}_{d \text{ times}} :
\begin{array}{l}
\mathbf{p}^1,\ldots,\mathbf{p}^d \text{ orthonormal,} \\
\sigma_1,\ldots,\sigma_d > 0, \\
\text{such that } C \subseteq E, \\
E \text{ given by (2.7.1),} \\
\mathrm{vol}(E) \le \mathrm{vol}(B(\mathbf{0}, R))
\end{array}
\right\}.
$$

$$\tag{2.7.4}$$

By the arguments above, X is bounded under the norm $(\mathbf{v}, \mathbf{p}^1, \ldots, \mathbf{p}^d, \sigma_1, \ldots, \sigma_d) \mapsto \|\mathbf{v}\|_2 + \sum_{i=1}^d \|\mathbf{p}^i\|_2 + \sum_{i=1}^d |\sigma_i|$. It can also be verified that X is closed (see Exercise 11 from Section 2.7.2). The function $f(\mathbf{v}, \mathbf{p}^1, \ldots, \mathbf{p}^d, \sigma_1, \ldots, \sigma_d) = (\prod_{i=1}^d \sigma_i) \operatorname{vol}(B(\mathbf{0}_d, 1))$, giving the volume of the corresponding ellipsoid, is a continuous function. Thus, by Theorem 1.3.13, there is a minimizer which gives the minimum volume ellipsoid E. Let its center be $\mathbf{c}$.

To show that $\mathbf{c} + \frac{1}{d}(E - \mathbf{c}) \subseteq C$, consider any $0 \le \beta < \frac{1}{d}$. If $\mathbf{c} + \beta(E - \mathbf{c}) \not\subseteq C$, then there exists $\mathbf{y} \in (\mathbf{c} + \beta(E - \mathbf{c})) \setminus C$. Consider a separating hyperplane that gives a halfspace H with $C \subseteq H$ and $\mathbf{y} \notin H$. In other words, H does not contain $\mathbf{c} + \beta(E - \mathbf{c})$. By Theorem 2.7.6, there exists an ellipsoid E' such that $E' \supseteq E \cap H \supseteq C$ and the volume of E' is strictly smaller than E. This contradicts the fact that E is a minimum volume ellipsoid containing C. Therefore, we must have $\mathbf{c} + \beta(E - \mathbf{c}) \subseteq C$ for all $0 \le \beta < \frac{1}{d}$, i.e., $\left(\bigcup_{0 \le \beta < \frac{1}{d}} \mathbf{c} + \beta(E - \mathbf{c}) \right) \subseteq C$. Since C is closed, the closure of $\left(\bigcup_{0 \le \beta < \frac{1}{d}} \mathbf{c} + \beta(E - \mathbf{c}) \right)$, which is $\mathbf{c} + \frac{1}{d}(E - \mathbf{c})$, is contained in C. $\qquad\square$

2.7.2 Exercises

1. Check the equality in the final line of the proof of Theorem 2.7.1.
2. Show that an ellipsoid is convex.
3. If $E = T(B(\mathbf{0}, 1))$ for some affine transformation $T(\mathbf{x}) = R\mathbf{x} + \mathbf{p}$, show that $\mathbf{p}$ is the center of E.
4. If E is of the form (2.7.1), show that $\mathbf{v}$ is the center of the ellipsoid.
5. Let $A \in \mathbb{R}^{d \times d}$ be a positive definite matrix with eigenvectors $\mathbf{v}^1, \ldots, \mathbf{v}^d$ and eigenvalues $\lambda_1, \ldots, \lambda_d$. Then the line segments $[\mathbf{c} - \sqrt{\lambda_i}\,\mathbf{v}^i, \mathbf{c} + \sqrt{\lambda_i}\,\mathbf{v}^i]$, $i = 1, \ldots, d$ form a set of principal axes for $E(A, \mathbf{c})$ for any $\mathbf{c} \in \mathbb{R}^d$.
6. Let $T(\mathbf{x}) = R\mathbf{x} + \mathbf{p}$ be an affine transformation where $R \in \mathbb{R}^{k \times d}$ has rank d, and $\mathbf{u}^1, \ldots, \mathbf{u}^d \in \mathbb{R}^d$ are left singular vectors of R and $\sigma_1, \ldots, \sigma_d > 0$ are the singular values, then $[\mathbf{p} - \sigma_i \mathbf{u}^i, \mathbf{p} + \sigma_i \mathbf{u}^i]$, $i = 1, \ldots, d$ form a set of principal axes for the ellipsoid $T(B(\mathbf{0}, 1))$.
7. Let E be an ellipsoid and suppose $B(\mathbf{z}, \epsilon) \subseteq E$ for some $\mathbf{z} \in \mathbb{R}^d$ and $\epsilon > 0$. Show that the length of every principal axis is at least 2ϵ.
8.* Prove Theorem 2.7.3.
9. Let $A \in \mathbb{R}^{d \times d}$ be a positive definite matrix. Then $E(A, \mathbf{0})^\circ = E(A^{-1}, \mathbf{0})$.
10. Show that $B(\mathbf{0}, 1) \cap \{\mathbf{x} \in \mathbb{R}^d : \mathbf{x}_1 \le \beta\} \subseteq E'$ where E' is as defined in (2.7.3).
11.* Show that the set X defined in (2.7.4) is closed.
12.* Show that for any full-dimensional, compact convex set, there exists an ellipsoid contained in it with maximum volume.

13. Suppose $E \subseteq \mathbb{R}^d$ is a *minimal* ellipsoid containing a convex set $C \subseteq \mathbb{R}^d$; i.e., there is no ellipsoid containing C that is strictly contained in E. Is it true that E is a minimum volume ellipsoid containing C?

14. Let $A \in \mathbb{R}^{d \times d}$ be a positive definite matrix and $\mathbf{c} \in \mathbb{R}^d$. Let $\mathbf{a} \in \mathbb{R}^d \setminus \{\mathbf{0}\}$ and $\delta \in \mathbb{R}$. Show that when $E = E(A, \mathbf{c})$ and $H = H^=(\mathbf{a}, \delta)$ in Theorem 2.7.6, the construction of $E' = E(A', \mathbf{c}')$ in the proof can be expressed using the formulas $A' = \frac{d^2(1-\beta^2)}{d^2-1} \left(A - \left(\frac{2(1-d\beta)}{(d+1)(1-\beta)} \right) \frac{A\mathbf{a}\mathbf{a}^T A}{\mathbf{a}^T A \mathbf{a}} \right)$ and $\mathbf{c}' = \mathbf{c} - \left(\frac{1-d\beta}{d+1} \right) \frac{A\mathbf{a}}{\sqrt{\mathbf{a}^T A \mathbf{a}}}$. Further, show that $\lambda_{\max}(A') \le \sigma' \lambda_{\max}(A)$ and $\lambda_{\min}(A') \ge \sigma \lambda_{\min}(A)$, where σ, σ' are defined in (2.7.2).

15. Prove the following symmetric versions of Theorems 2.7.6 and 2.7.7.

    (a) Let $d \ge 2$ and let $E \subseteq \mathbb{R}^d$ be an ellipsoid with center $\mathbf{c} \in \mathbb{R}^d$. Let $H = \{\mathbf{x} \in \mathbb{R}^d : \langle \mathbf{a}, \mathbf{x} \rangle \le \langle \mathbf{a}, \mathbf{c} \rangle + \delta\}$ be a halfspace, with $\mathbf{a} \in \mathbb{R}^d$ and $\delta > 0$. Let $0 \le \beta < \frac{1}{\sqrt{d}}$ be such that $H^{\le}(\mathbf{a}, \delta)$ does not contain $\mathbf{c} + \beta(E - \mathbf{c})$. Then, there exists another ellipsoid E' with the same center $\mathbf{c}$ such that

$$E \cap \{\mathbf{x} \in \mathbb{R}^d : \langle \mathbf{a}, \mathbf{c} \rangle - \delta \le \langle \mathbf{a}, \mathbf{x} \rangle \le \langle \mathbf{a}, \mathbf{c} \rangle + \delta\} \subseteq E'$$

    and $\operatorname{vol}(E') < \operatorname{vol}(E)$. See Figure 2.3.

    (b) Let $C \subseteq \mathbb{R}^d$ be a full-dimensional, $\mathbf{0}$-symmetric, compact, convex set. There exists an ellipsoid E centered at $\mathbf{0}$ containing C such that it has minimum volume among all ellipsoids containing C. Further, $\frac{1}{\sqrt{d}} E \subseteq C \subseteq E$. Conclude that for any norm N on $\mathbb{R}^d$, there exists a positive definite matrix A such that $\frac{1}{\sqrt{d}} N_A \le N \le N_A$.

2.8 Notes and Related Literature

The material presented in Sections 2.1– 2.4 is classical convex analysis and geometry. We have only covered aspects that we considered most relevant for Part II of the book. Interested readers can find much more in [124, 140, 199, 210]. We have used proofs, as well as ideas for exercises, from these texts in a few places in this chapter. We thank Marco Di Summa [104] for pointing out the counterexample mentioned at the end of Section 2.4.3.

There is an intimate connection between functional analysis and convex geometry that we do not touch upon in this chapter. For example, the separating hyperplane theorem (Theorem 2.4.2) can be obtained as a special case of the celebrated *Hahn–Banach theorem* from functional analysis. In fact, many of the notions and results of this chapter, e.g., the Krein–Milman theorem

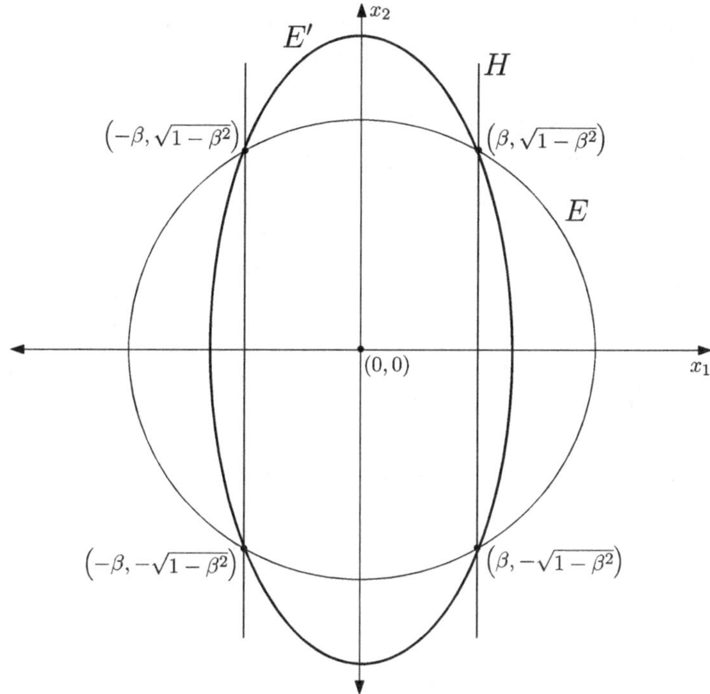

Figure 2.3 A picture illustrating the relationship between E, H, and E'.

(Theorem 2.4.16), have extensions to more general vector spaces beyond finite-dimensional Euclidean spaces. See the textbooks [24, 143] for expositions of some of these aspects. The textbook [24] by Barvinok is also an excellent resource for many modern applications of convexity. Going back to functional analysis, connections with Banach space geometry form the basis of much research in modern convex geometry. See [20, 187] and the recent lecture notes by Rothvoss [200] for an introduction to this deep and fascinating area.

Section 2.5 covers only the very basics of polyhedral theory. The reader should consult [116, 124, 128, 234] for more on the discrete aspects of convex geometry. Some of the exercises in Section 2.5.6 were taken from these books as well as [74, Chapter 3]. The material from Section 2.6 traces its origins to work by Helly [135, 136] and Radon [194, 195]. In fact, Carathéodory's theorem (Theorems 2.2.15 and 2.2.16), proved in [65], is also intimately connected to the theorems of Helly and Radon (Theorems 2.6.1 and 2.6.2) – any one of these results, combined with more elementary arguments, can be used to derive the other two (see Exercise 3 from Section 2.6.1). See [82] for an extensive discussion of these interconnections and also generalizations of these theorems. Many of the exercises from Section 2.6.1 have been taken from sections 2 and 4 of this survey paper, as well as from Barvinok's book [24]. In particular, Exercise 7 from Section 2.6.1 is based on [24, Section I.6]. The different characterizations of the Helly number presented in Section 2.6 were first noted explicitly by Hoffman [142], who also stated Theorem 2.6.14 but without proof. The first complete proof of this important result appears in [17]. The special case of Corollary 2.6.13 has its own history. It was independently discovered by Bell [41], Doignon [106], and Scarf [208], with completely different motivations. Bell and Scarf came at it from an optimization perspective and Doignon was looking at lattice structures arising in crystallography. Their arguments foreshadow the characterizations of the Helly number from Hoffman's paper. In recent years, there has been quite a bit of research activity in this area; see [9, 10, 14, 16, 75, 91–95] for a small sample of recent research articles, and the monograph [23].

Minimum volume ellipsoids and Löwner–John ellipsoids appear in the work of Karel Löwner and Fritz John [149]. The existence and uniqueness of the minimum volume circumscribing ellipsoid were known to Löwner, and John showed that it approximated the convex body up to a factor of the dimension (Theorem 2.7.7) and up to a square root of the dimension for symmetric convex bodies (Exercise 15 from Section 2.7.2). See [137] for a history and short survey of these important objects in convex geometry and their various applications. Their use in optimization via ellipsoidal sections (Theorem 2.7.6) can be traced back to the work of Shor [213–215] and Yudin–Nemirosvkii [232, 233], with Yudin and Nemirovskii making some fundamental insights using Theorem 2.7.6. We will see some of this in action in Chapter 6. The minimum volume ellipsoid containing the intersection of an ellipsoid with a halfspace has been studied from various perspectives and it is a completely solved problem; see [123, Chapter 3] and [49] for the full story. Our computations in Theorem 2.7.6 are based on the excellent lecture notes of Lap Chi Lau [162, Lecture 15]. Our proof of the existence of the minimum volume ellipsoid in Theorem 2.7.7 is adapted from Barvinok [24, Section V.2].

3
Convex Functions

We now turn our attention to convex functions, as a step toward optimization. In this context, we will need to sometimes talk about the extended real numbers $\mathbb{R} \cup \{-\infty, +\infty\}$. One reason is that in optimization problems, many times, a supremum may be $+\infty$ or an infimum may be $-\infty$, and using them on the same footing as the reals makes certain statements nicer, without having to exclude annoying special cases. For this, one needs to set up some convenient rules for arithmetic over $\mathbb{R} \cup \{-\infty, +\infty\}$:

- $x + \infty = \infty$ for any $x \in \mathbb{R} \cup \{+\infty\}$. Similarly, $x + (-\infty) = -\infty$ for any $x \in \mathbb{R} \cup \{-\infty\}$.
- $x \cdot (+\infty) = +\infty$ and $x \cdot (-\infty) = -\infty$ for all $x > 0$. We will avoid situations where we need to consider $0 \cdot (+\infty)$ or $0 \cdot (-\infty)$.
- $-\infty < x < \infty$ for all $x \in \mathbb{R}$.

3.1 General Properties, Epigraphs, Subgradients

Definition 3.1.1   A function $f : \mathbb{R}^d \to \mathbb{R} \cup \{+\infty\}$ is called *convex* if

$$f(\lambda \mathbf{x} + (1 - \lambda)\mathbf{y}) \leq \lambda f(\mathbf{x}) + (1 - \lambda)f(\mathbf{y})$$

for all $\mathbf{x}, \mathbf{y} \in \mathbb{R}^d$ and $\lambda \in (0, 1)$. If the inequality is strict for all $\mathbf{x} \neq \mathbf{y}$, then the function is called *strictly convex*. The *domain* (sometimes also called *effective domain*) of f is defined as

$$\mathrm{dom}(f) := \{\mathbf{x} \in \mathbb{R}^d : f(\mathbf{x}) < +\infty\}.$$

A function g is said to be *(strictly) concave* if $-g$ is (strictly) convex.

The domain of a convex function is easily seen to be convex.

89

Proposition 3.1.2 *Let* $f\colon \mathbb{R}^d \to \mathbb{R} \cup \{+\infty\}$ *be a convex function. Then* $\mathrm{dom}(f)$ *is a convex set.*

The following subfamily of convex functions is nicer to deal with from an algorithmic perspective.

Definition 3.1.3 A function $f\colon \mathbb{R}^d \to \mathbb{R} \cup \{+\infty\}$ is called *strongly convex* with *modulus of strong convexity* $c > 0$ if

$$f(\lambda \mathbf{x} + (1-\lambda)\mathbf{y}) \le \lambda f(\mathbf{x}) + (1-\lambda)f(\mathbf{y}) - \frac{1}{2}c\lambda(1-\lambda)\|\mathbf{x}-\mathbf{y}\|^2$$

for all $\mathbf{x}, \mathbf{y} \in \mathbb{R}^d$ and $\lambda \in (0,1)$.

Since $c > 0$, the term $\frac{1}{2}c\lambda(1-\lambda)\|\mathbf{x}-\mathbf{y}\|^2$ in the definition of strongly convex functions is strictly positive for $\mathbf{x} \ne \mathbf{y}$ and so strongly convex functions are strictly convex. However, the above definition actually endows quite a bit more structure on the function. The following proposition gives some more intuition about strongly convex functions.

Proposition 3.1.4 *A function* $f\colon \mathbb{R}^d \to \mathbb{R} \cup \{+\infty\}$ *is strongly convex with modulus of strong convexity* $c > 0$ *if and only if the function* $g(\mathbf{x}) := f(\mathbf{x}) - \frac{1}{2}c\|\mathbf{x}\|^2$ *is convex (or equivalently,* $f = g + \frac{1}{2}c\|\cdot\|^2$ *for some convex function* g*).*

Proof Left as an exercise. $\square$

After we introduce subgradients and subdifferentials below, strongly convex functions can be understood as those convex functions that have a quadratic underestimator everywhere (see Theorems 3.1.20 and 3.2.8). We will see more on this in Part II of the book, when we start looking at optimization algorithms.

Definition 3.1.5 It will be sometimes convenient to extend the definition of convex functions in the following way. Let $f\colon \mathbb{R}^d \to \mathbb{R} \cup \{+\infty\}$ and let $C \subseteq \mathbb{R}^d$ be a convex set. f is said to be *convex over* C if $f(\lambda \mathbf{x} + (1-\lambda)\mathbf{y}) \le \lambda f(\mathbf{x}) + (1-\lambda)f(\mathbf{y})$ for all $\mathbf{x}, \mathbf{y} \in C$ and $\lambda \in (0,1)$. Similarly for strictly and strongly convex functions over C.

Since $\mathrm{dom}(f)$ is a convex set, it follows from the definitions that $f\colon \mathbb{R}^d \to \mathbb{R} \cup \{+\infty\}$ is convex if and only if f is convex over $\mathrm{dom}(f)$.

Convex functions have a natural convex set associated with them, called the *epigraph*. Many properties of convex functions can be obtained by just analyzing the corresponding epigraph and using all the technology built in Chapter 2. We give the formal definition for general functions below; very informally, it is "the region above the graph of a function."

Definition 3.1.6 Let $f : \mathbb{R}^d \to \mathbb{R} \cup \{+\infty\}$ be any function (not necessarily convex). The *epigraph* of f is defined as

$$\mathrm{epi}(f) := \{(\mathbf{x}, t) \in \mathbb{R}^n \times \mathbb{R} : f(\mathbf{x}) \leq t\}.$$

Note that $\mathrm{epi}(f) \subseteq \mathbb{R}^d \times \mathbb{R}$, so it lives in a space whose dimension is one more than the space over which the function is defined, just like the graph of the function. Note also that the epigraph is nonempty if and only if the function is not identically equal to $+\infty$. Convex functions are precisely those functions whose epigraphs are convex.

Proposition 3.1.7 *Let $f : \mathbb{R}^d \to \mathbb{R} \cup \{+\infty\}$ be any function. f is convex if and only if $\mathrm{epi}(f)$ is a convex set.*

Proof ($\Rightarrow$) Consider any $(\mathbf{x}^1, t_1), (\mathbf{x}^2, t_2) \in \mathrm{epi}(f)$, and any $\lambda \in (0, 1)$.

The result is a consequence of the following sequence of implications:

$$(\mathbf{x}^1, t_1) \in \mathrm{epi}(f), \quad (\mathbf{x}^2, t_2) \in \mathrm{epi}(f), \quad f \text{ is convex}$$
$$\Rightarrow f(\mathbf{x}^1) \leq t_1, \quad f(\mathbf{x}^2) \leq t_2, \quad f(\lambda \mathbf{x}^1 + (1 - \lambda)\mathbf{x}^2) \leq \lambda f(\mathbf{x}^1) + (1 - \lambda) f(\mathbf{x}^2)$$
$$\Rightarrow f(\lambda \mathbf{x}^1 + (1 - \lambda)\mathbf{x}^2) \leq \lambda t_1 + (1 - \lambda) t_2$$
$$\Rightarrow (\lambda \mathbf{x}^1 + (1 - \lambda)\mathbf{x}^2, \lambda t_1 + (1 - \lambda) t_2) \in \mathrm{epi}(f).$$

($\Leftarrow$) Consider any $\mathbf{x}^1, \mathbf{x}^2 \in \mathbb{R}^d$ and $\lambda \in (0, 1)$. We wish to show that $f(\lambda \mathbf{x}^1 + (1 - \lambda)\mathbf{x}^2) \leq \lambda f(\mathbf{x}^1) + (1 - \lambda) f(\mathbf{x}^2)$. If $f(\mathbf{x}^1) = +\infty$ or $f(\mathbf{x}^2) = +\infty$, then the relation holds trivially. So we assume $f(\mathbf{x}^1), f(\mathbf{x}^2) < +\infty$. The points $(\mathbf{x}^1, f(\mathbf{x}^1)), (\mathbf{x}^2, f(\mathbf{x}^2))$ both lie in $\mathrm{epi}(f)$. By convexity of $\mathrm{epi}(f)$, we have that $(\lambda \mathbf{x}^1 + (1 - \lambda)\mathbf{x}^2, \lambda f(\mathbf{x}^1) + (1 - \lambda) f(\mathbf{x}^2)) \in \mathrm{epi}(f)$. This implies that $f(\lambda \mathbf{x}^1 + (1 - \lambda)\mathbf{x}^2) \leq \lambda f(\mathbf{x}^1) + (1 - \lambda) f(\mathbf{x}^2)$, showing that f is convex. $\square$

Just like the class of *closed*, convex sets are nicer to deal with compared to sets that are simply convex but not closed (mainly because of the separating/supporting hyperplane theorem), it will be convenient to isolate a similar class of "nicer" convex functions.

Definition 3.1.8 A function is said to be a *closed, convex function* if its epigraph is a closed, convex set.

One can associate another family of convex sets with a convex function.

Definition 3.1.9 Let $f : \mathbb{R}^d \to \mathbb{R} \cup \{+\infty\}$ be any function. Given $\alpha \in \mathbb{R}$, the *α-sublevel set* of f is the set

$$f_\alpha := \{\mathbf{x} \in \mathbb{R}^d : f(\mathbf{x}) \leq \alpha\}.$$

The following can be verified by the reader.

Proposition 3.1.10 *All sublevel sets of a convex function are convex sets.*

The converse of Proposition 3.1.10 is *not true*. Functions whose sublevel sets are all convex are called *quasiconvex*.

Example 3.1.11 1. *Indicator function.* For any subset $X \subseteq \mathbb{R}^d$, define

$$I_X(\mathbf{x}) := \begin{cases} 0 & \text{if } \mathbf{x} \in X, \\ +\infty & \text{if } \mathbf{x} \notin X. \end{cases}$$

Then I_X is convex if and only if X is convex.

2. *Linear/affine function.* Let $\mathbf{a} \in \mathbb{R}^d$ and $\delta \in \mathbb{R}$. Then the function $\mathbf{x} \mapsto \langle \mathbf{a}, \mathbf{x} \rangle + \delta$ is called an *affine function* (if $\delta = 0$, this is a *linear function*). It is easily verified that affine functions are convex. Epigraphs of affine functions are halfspaces.

3. *Norms and distances.* Let $N : \mathbb{R}^d \to \mathbb{R}$ be a norm (see Definition 1.1.1). Then N is convex. Let C be a nonempty convex set. Then the distance function associated with the norm N, defined as

$$d_C^N(\mathbf{x}) := \inf_{\mathbf{y} \in C} N(\mathbf{y} - \mathbf{x}),$$

is a convex function.

4. *Maximum of affine functions/piecewise linear function/polyhedral function.* Let $\mathbf{a}^1, \ldots, \mathbf{a}^m \in \mathbb{R}^d$ and $\delta_1, \ldots, \delta_m \in \mathbb{R}$. The function

$$f(\mathbf{x}) := \max_{i=1,\ldots,m} (\langle \mathbf{a}^i, \mathbf{x} \rangle + \delta_i)$$

is a convex function. Let us verify this. Consider any $\mathbf{x}^1, \mathbf{x}^2 \in \mathbb{R}^d$ and $\lambda \in (0, 1)$. Then,

$$\begin{aligned}
f(\lambda \mathbf{x}^1 + (1 - \lambda)\mathbf{x}^2) &= \max_{i=1,\ldots,m} (\langle \mathbf{a}^i, \lambda \mathbf{x}^1 + (1 - \lambda)\mathbf{x}^2 \rangle + \delta_i) \\
&= \max_{i=1,\ldots,m} \left(\lambda(\langle \mathbf{a}^i, \mathbf{x}^1 \rangle + \delta_i) + (1 - \lambda)(\langle \mathbf{a}^i, \mathbf{x}^2 \rangle + \delta_i) \right) \\
&\leq \max_{i=1,\ldots,m} \left(\lambda(\langle \mathbf{a}^i, \mathbf{x}^1 \rangle + \delta_i) \right) + \max_{i=1,\ldots,m} \left((1 - \lambda)(\langle \mathbf{a}^i, \mathbf{x}^2 \rangle + \delta_i) \right) \\
&= \lambda \max_{i=1,\ldots,m} (\langle \mathbf{a}^i, \mathbf{x}^1 \rangle + \delta_i) + (1 - \lambda) \max_{i=1,\ldots,m} (\langle \mathbf{a}^i, \mathbf{x}^2 \rangle + \delta_i) \\
&= \lambda f(\mathbf{x}^1) + (1 - \lambda) f(\mathbf{x}^2).
\end{aligned}$$

The inequality follows from the fact that if $\ell_1, \ldots, \ell_m$ and $u_1, \ldots, u_m$ are two sets of m real numbers for some $m \in \mathbb{N}$, then $\max_{i=1,\ldots,m}(\ell_i + u_i) \leq \max_{i=1,\ldots,m} \ell_i + \max_{i=1,\ldots,m} u_i$ (see also parts 2. and 4. of Theorem 1.3.1).

An important consequence of the definition of convexity for functions is Jensen's inequality, which sees its uses in diverse areas of science and engineering.

Theorem 3.1.12 (Jensen's Inequality) *Let $f : \mathbb{R}^d \to \mathbb{R} \cup \{+\infty\}$ be any function. Then f is convex if and only if for any finite set of points $\mathbf{x}^1, \ldots, \mathbf{x}^n \in \mathbb{R}^d$ and $\lambda_1, \ldots, \lambda_n > 0$ such that $\lambda_1 + \cdots + \lambda_n = 1$, the following holds:*

$$f(\lambda_1 \mathbf{x}^1 + \cdots + \lambda_n \mathbf{x}^n) \leq \lambda_1 f(\mathbf{x}^1) + \cdots + \lambda_n f(\mathbf{x}^n).$$

Proof ($\Leftarrow$) Use the hypothesis with $n = 2$.

($\Rightarrow$) If any $f(\mathbf{x}^i)$ is $+\infty$, then the inequality holds trivially. So we assume that each $f(\mathbf{x}^i) < +\infty$. By Proposition 3.1.7, epi(f) is a convex set. For each $i = 1, \ldots, m$, the point $(\mathbf{x}^i, f(\mathbf{x}^i)) \in$ epi(f) by definition of epi(f). Since epi(f) is convex, Proposition 2.1.6 implies that $\sum_{i=1}^m \lambda_i (\mathbf{x}^i, f(\mathbf{x}^i)) \in$ epi(f), i.e., $(\lambda_1 \mathbf{x}^1 + \cdots + \lambda_n \mathbf{x}^n, \lambda_1 f(\mathbf{x}^1) + \cdots + \lambda_n f(\mathbf{x}^n)) \in$ epi(f). Therefore, $f(\lambda_1 \mathbf{x}^1 + \cdots + \lambda_n \mathbf{x}^n) \leq \lambda_1 f(\mathbf{x}^1) + \cdots + \lambda_n f(\mathbf{x}^n)$. $\square$

Recall Theorem 2.1.3, which shows the convexity of a set is preserved under certain operations. We would like to develop a similar result for convex functions.

Theorem 3.1.13 (Operations that preserve the property of being a (closed) convex function) *Let $f_i \colon \mathbb{R}^d \to \mathbb{R} \cup \{+\infty\}$, $i \in I$ be a family of (closed) convex functions where the index set I is potentially infinite. The following are all true.*

1. *(Nonnegative combinations). If I is a finite set, and $\alpha_i \geq 0$, $i \in I$ is a corresponding set of nonnegative reals, then $\sum_{i \in I} \alpha_i f_i$ is a (closed) convex function.*
2. *(Taking supremums). The function defined as $g(\mathbf{x}) := \sup_{i \in I} f_i(\mathbf{x})$ is a (closed) convex function (even when I is uncountably infinite).*
3. *(Precomposition with an affine function). Let $A \in \mathbb{R}^{m \times d}$ and $\mathbf{b} \in \mathbb{R}^m$ and let $f \colon \mathbb{R}^m \to \mathbb{R}$ be any (closed) convex function on $\mathbb{R}^m$. Then $g(\mathbf{x}) := f(A\mathbf{x} + \mathbf{b})$ as a function from $\mathbb{R}^d \to \mathbb{R}$ is a (closed) convex function.*
4. *(Postcomposition with an increasing convex function). Let $h \colon \mathbb{R} \to \mathbb{R} \cup \{+\infty\}$ be a (closed) convex function that is also increasing, i.e., $h(x) \geq h(y)$ when $x \geq y$. Let $f \colon \mathbb{R}^d \to \mathbb{R} \cup \{+\infty\}$ be a (closed) convex function. We adopt the convention that $h(+\infty) = +\infty$. Then $h(f(\mathbf{x}))$ as a function from $\mathbb{R}^d \to \mathbb{R} \cup \{+\infty\}$ is a (closed) convex function.*

Proof 1. Let $F = \sum_{i \in I} \alpha_i f_i$. Consider any $\mathbf{x}, \mathbf{y} \in \mathbb{R}^d$ and $\lambda \in (0, 1)$. Then

$$
\begin{aligned}
F(\lambda \mathbf{x} + (1 - \lambda)\mathbf{y}) &= \sum_{i \in I} \alpha_i f_i(\lambda \mathbf{x} + (1 - \lambda)\mathbf{y}) \\
&\leq \sum_{i \in I} \alpha_i (\lambda f_i(\mathbf{x}) + (1 - \lambda) f_i(\mathbf{y})) \\
&= \lambda \sum_{i \in I} \alpha_i f_i(\mathbf{x}) + (1 - \lambda) \sum_{i \in I} \alpha_i f_i(\mathbf{y}) \\
&= \lambda F(\mathbf{x}) + (1 - \lambda) F(\mathbf{y}).
\end{aligned}
$$

We use the nonnegativity of α_i in the inequality on the second displayed line above. The proof of closedness is left as an exercise.

2. The main observation is that epi(g) $= \cap_{i \in I}$ epi(f_i) because $g(\mathbf{x}) \leq t$ if and only if $f_i(\mathbf{x}) \leq t$ for all $i \in I$. Since the intersection of (closed) convex sets is a (closed) convex set (part 1. of Theorem 2.1.3), we have the result.

3. The main observation is that for any $\mathbf{x} \in \mathbb{R}^d$ and $t \in \mathbb{R}$, $(\mathbf{x}, t) \in \mathrm{epi}(g)$ if and only if $(A\mathbf{x} + \mathbf{b}, t) \in \mathrm{epi}(f)$. Define the affine map $T \colon \mathbb{R}^d \times \mathbb{R} \to \mathbb{R}^m \times \mathbb{R}$ as follows: $T(\mathbf{x}, t) = (A\mathbf{x} + b, t)$. Then $\mathrm{epi}(g) = T^{-1}(\mathrm{epi}(f))$. Since the preimage of a (closed) convex set with respect to an affine transformation is (closed) convex (part 4. of Theorem 2.1.3), we obtain that $\mathrm{epi}(g)$ is (closed) convex.

4. Left as an exercise.

$\square$

We can now see some more interesting examples of convex functions.

Example 3.1.14   1. Let $\mathbf{a}^i \in \mathbb{R}^d$ and $\delta_i \in \mathbb{R}$, $i \in I$ for some index set I. Then the function

$$f(\mathbf{x}) := \sup_{i \in I} (\langle \mathbf{a}^i, \mathbf{x} \rangle + \delta_i)$$

is closed convex. This is an alternate proof of the convexity of the maximum of finitely many affine functions – part 4. of Example 3.1.11.

2. Consider the vector space V of symmetric $n \times n$ matrices, with entrywise addition and scaling. One can view V as $\mathbb{R}^{\frac{n(n+1)}{2}}$. Let $k \le n$. Consider the function $f_k \colon V \to \mathbb{R}$ which takes a matrix X and maps it to $f_k(X)$, which is the sum of the k largest eigenvalues of X. Then f_k is a convex function. This is seen by the following argument. Given any $Z \in V$ define the linear function A_Z on V as follows: $A_Z(X) = \sum_{i,j} X_{ij} Z_{ij}$. Then

$$f_k(X) = \sup_{Y \in \Omega} A_{YY^T}(X),$$

where Ω is the set of $n \times k$ matrices with k orthonormal columns in $\mathbb{R}^n$. This shows that f_k is the supremum of linear functions, and by Theorem 3.1.13, it is closed convex.

We see in part 1. of Example 3.1.14 that the supremum of affine functions is convex. We will show below that, in fact, every convex function is the supremum of some family of affine functions. This is analogous to the fact that all closed convex sets are the intersection of some family of halfspaces (Corollary 2.4.3). We build up to this with an important definition.

Definition 3.1.15   Let $f \colon \mathbb{R}^d \to \mathbb{R} \cup \{+\infty\}$ be any function. Let $\mathbf{x} \in \mathrm{dom}(f)$. Then $\mathbf{a} \in \mathbb{R}^d$ is said to define a *subgradient*[1] *of* f *at* $\mathbf{x}$ if $f(\mathbf{y}) \ge f(\mathbf{x}) + \langle \mathbf{a}, \mathbf{y} - \mathbf{x} \rangle$ for all $\mathbf{y} \in \mathbb{R}^d$. The set of all subgradients at $\mathbf{x}$ is denoted by $\partial f(\mathbf{x})$ and is called the *subdifferential* of f at $\mathbf{x}$.

---

[1] A subgradient at a point is also called an *affine support* at that point by some authors.

A useful picture to keep in mind is the following fact: $\mathbf{a} \in \mathbb{R}^d$ is a subgradient of f at $\mathbf{x}$ if and only if the hyperplane $H = ((\mathbf{a}, -1), \langle \mathbf{a}, \mathbf{x} \rangle - f(\mathbf{x}))$ is a supporting hyperplane (Definition 2.4.6) for the epigraph of f at $(\mathbf{x}, f(\mathbf{x}))$. In other words, the hyperplane in $\mathbb{R}^d \times \mathbb{R}$ with $(\mathbf{a}, -1)$ as normal vector passing through $(\mathbf{x}, f(\mathbf{x}))$ is a supporting hyperplane for the epigraph of f at $(\mathbf{x}, f(\mathbf{x}))$.

Theorem 3.1.16 *Let* $f : \mathbb{R}^d \to \mathbb{R}$ *be any function. Then* f *is closed convex if and only if there exists a subgradient of* f *at every* $\mathbf{x} \in \mathbb{R}^d$.

Proof ($\Rightarrow$) Consider any $\mathbf{x} \in \mathbb{R}^d$. By definition of closed convex, $\mathrm{epi}(f)$ is a closed convex set. Moreover, $(\mathbf{x}, f(\mathbf{x})) \in \mathrm{bd}(\mathrm{epi}(f))$. By Theorem 2.4.5, there exists $(\bar{\mathbf{a}}, r) \in \mathbb{R}^d \times \mathbb{R}$ and $\delta \in \mathbb{R}$ such that $\bar{\mathbf{a}}$ and r are not both zero, $\langle \bar{\mathbf{a}}, \mathbf{y} \rangle + rt \leq \delta$ for all $(\mathbf{y}, t) \in \mathrm{epi}(f)$, and $\langle \bar{\mathbf{a}}, \mathbf{x} \rangle + rf(\mathbf{x}) = \delta$.

We claim that $r < 0$. Suppose to the contrary that $r \geq 0$. First consider the case that $\bar{\mathbf{a}} = \mathbf{0}$, then $r > 0$. $(\mathbf{x}, t) \in \mathrm{epi}(f)$ for all $t \geq f(\mathbf{x})$. But this contradicts that $rt = \langle \bar{\mathbf{a}}, \mathbf{y} \rangle + rt \leq \delta$ for all $t \geq f(\mathbf{x})$ because we can choose $t \geq f(\mathbf{x})$ large enough to violate $rt \leq \delta$. Next consider the case that $\bar{\mathbf{a}} \neq \mathbf{0}$. Consider any $\mathbf{y} \in \mathbb{R}^d$ satisfying $\langle \bar{\mathbf{a}}, \mathbf{y} \rangle > \delta$. Since f is real valued, there exists $(\mathbf{y}, t) \in \mathrm{epi}(f)$ for some $t \geq 0$. Since $r \geq 0$, this contradicts that $\langle \bar{\mathbf{a}}, \mathbf{y} \rangle + rt \leq \delta$.

Now set $\mathbf{a} = \frac{\bar{\mathbf{a}}}{-r}$. $\langle \bar{\mathbf{a}}, \mathbf{x} \rangle + rf(\mathbf{x}) = \delta$ and $\langle \bar{\mathbf{a}}, \mathbf{y} \rangle + rf(\mathbf{y}) \leq \delta$ for all $\mathbf{y} \in \mathbb{R}^d$ together imply that $\langle \bar{\mathbf{a}}, \mathbf{y} \rangle \leq (-r)f(\mathbf{y}) + \langle \bar{\mathbf{a}}, \mathbf{x} \rangle + rf(\mathbf{x})$. Rearranging, we obtain that $f(\mathbf{y}) \geq f(\mathbf{x}) + \langle \mathbf{a}, \mathbf{y} - \mathbf{x} \rangle$ for all $\mathbf{y} \in \mathbb{R}^d$.

($\Leftarrow$) By definition of subgradient, for every $\mathbf{x} \in \mathbb{R}^d$, there exists $\mathbf{a}_\mathbf{x} \in \mathbb{R}^d$ such that $f(\mathbf{y}) \geq f(\mathbf{x}) + \langle \mathbf{a}_\mathbf{x}, \mathbf{y} - \mathbf{x} \rangle$ for all $\mathbf{y} \in \mathbb{R}^d$. This implies that, in fact,

$$f(\mathbf{y}) = \sup_{\mathbf{x} \in \mathbb{R}^d} \left(f(\mathbf{x}) + \langle \mathbf{a}_\mathbf{x}, \mathbf{y} - \mathbf{x} \rangle \right)$$

because setting $\mathbf{x} = \mathbf{y}$ on the right-hand side gives $f(\mathbf{y})$. Thus, f is the supremum of a family of affine functions, which, by part 1. of Example 3.1.14, shows that f is closed convex. $\square$

Corollary 3.1.17 *Let* $f : \mathbb{R}^d \to \mathbb{R} \cup \{+\infty\}$ *be convex. For any* $\mathbf{x} \in \mathrm{relint}(\mathrm{dom}(f))$, *there exists a subgradient of* f *at* $\mathbf{x}$.

Proof Note that $\mathbf{x} \in \mathrm{relbd}(\mathrm{epi}(f))$. By Theorem 2.4.5, there exists $(\bar{\mathbf{a}}, r) \in \mathbb{R}^d \times \mathbb{R}$ and $\delta \in \mathbb{R}$ such that $\bar{\mathbf{a}}$ and r are not both zero, $\langle \bar{\mathbf{a}}, \mathbf{y} \rangle + rt \leq \delta$ for all $(\mathbf{y}, t) \in \mathrm{epi}(f)$, and $\langle \bar{\mathbf{a}}, \mathbf{x} \rangle + rf(\mathbf{x}) = \delta$. By Exercise 10 in Section 3.1.1, we have $r \leq 0$. Let L be the linear space parallel to $\mathrm{aff}(\mathrm{dom}(f))$ (Definition 2.2.9). If $\bar{\mathbf{a}} \in L^\perp$ and $r = 0$ then we obtain that $\langle \bar{\mathbf{a}}, \mathbf{y} - \mathbf{x} \rangle + r(t - f(\mathbf{x})) = 0$ for all $(\mathbf{y}, t) \in \mathrm{epi}(f)$ since $\mathbf{y} - \mathbf{x} \in L$. Thus, $\langle \bar{\mathbf{a}}, \mathbf{y} \rangle + rt = \langle \bar{\mathbf{a}}, \mathbf{x} \rangle + rf(\mathbf{x}) = \delta$ for all $(\mathbf{y}, t) \in \mathrm{epi}(f)$. This contradicts condition (iii) of Theorem 2.4.5. Therefore, either $r < 0$ or $\bar{\mathbf{a}}$ has a nonzero projection $\hat{\mathbf{a}}$ on to L. In the first case, we obtain the subgradient $\mathbf{a} = \frac{\bar{\mathbf{a}}}{-r}$ following the same calculations as in the proof of Theorem 3.1.16. In the second case, we first observe that $\langle \hat{\mathbf{a}}, \mathbf{y} - \mathbf{x} \rangle = \langle \bar{\mathbf{a}}, \mathbf{y} - \mathbf{x} \rangle$ for all $\mathbf{y} \in \mathrm{aff}(\mathrm{dom}(f))$ since $\bar{\mathbf{a}} - \hat{\mathbf{a}} \in L^\perp$ and $\mathbf{y} - \mathbf{x} \in L$. This implies that $\langle \hat{\mathbf{a}}, \mathbf{y} - \mathbf{x} \rangle + r(t - f(\mathbf{x})) = \langle \bar{\mathbf{a}}, \mathbf{y} - \mathbf{x} \rangle + r(t - f(\mathbf{x})) \leq 0$ because $\langle \bar{\mathbf{a}}, \mathbf{y} \rangle + rt \leq \delta$ for all $(\mathbf{y}, t) \in \mathrm{epi}(f)$, and $\langle \bar{\mathbf{a}}, \mathbf{x} \rangle + rf(\mathbf{x}) = \delta$. In particular, we have $\langle \hat{\mathbf{a}}, \mathbf{y} \rangle + rt \leq \hat{\delta}$ for all $(\mathbf{y}, t) \in \mathrm{epi}(f)$, where $\langle \hat{\mathbf{a}}, \mathbf{x} \rangle + rf(\mathbf{x}) = \hat{\delta}$. Since $\hat{\mathbf{a}} \in L$, we can now follow the proof of Theorem 3.1.16 to obtain the subgradient. $\qquad\square$

Remark 3.1.18 1. Any convex function that is finite valued everywhere is closed convex. This follows from a continuity result we will prove in Section 3.2.1 (Theorem 3.2.3). Thus, in the forward direction of Theorem 3.1.16, one may weaken the hypothesis to just convex, as opposed to closed convex.

2. In the reverse direction of Theorem 3.1.16, one may weaken the hypothesis to having a *local* subgradient everywhere. A function $f : \mathbb{R}^d \to \mathbb{R}$ is said to have a local subgradient at $\mathbf{x}$ if there exists $\epsilon > 0$ (depending on $\mathbf{x}$) such that $f(\mathbf{y}) \geq f(\mathbf{x}) + \langle \mathbf{a}, \mathbf{y} - \mathbf{x} \rangle$ for all $\mathbf{y} \in B(\mathbf{x}, \epsilon)$. We will omit the proof of this extension of Theorem 3.1.16 here. See the chapter on "Convex Functions" in [124] for a complete proof.

Theorem 3.1.19 *Let* $f : \mathbb{R}^d \to \mathbb{R} \cup \{+\infty\}$ *be a convex function. For any* $\mathbf{x} \in \mathrm{dom}(f)$, *the subdifferential* $\partial f(\mathbf{x})$ *at* $\mathbf{x}$ *is a closed, convex set.*

Proof Note that

$$\partial f(\mathbf{x}) := \{\mathbf{a} \in \mathbb{R}^d : \langle \mathbf{y} - \mathbf{x}, \mathbf{a} \rangle \leq f(\mathbf{y}) - f(\mathbf{x}) \ \forall \mathbf{y} \in \mathbb{R}^d\}.$$

The above set is the intersection of a family of halfspaces and so $\partial f(\mathbf{x})$ is a closed, convex set. $\qquad\square$

The following version of Theorem 3.1.16 for strictly and strongly convex functions can be established after we develop the calculus of subdifferentials in Section 3.4 (Theorem 3.4.5); see Exercise 2 from Section 3.4.1. Subdifferential calculus is needed for the strongly convex part of the theorem; the reader should be able to prove the strictly convex statement using just the definitions.

Theorem 3.1.20 *Let* $f : \mathbb{R}^d \to \mathbb{R}$ *be convex. Then the following are true.*

1. f *is strictly convex if and only if for every* $\mathbf{x} \in \mathbb{R}^d$ *and* $\mathbf{s} \in \partial f(\mathbf{x})$, *we have* $f(\mathbf{y}) > f(\mathbf{x}) + \langle \mathbf{s}, \mathbf{y} - \mathbf{x} \rangle$ *for all* $\mathbf{y} \in \mathbb{R}^d \setminus \{\mathbf{x}\}$.

2. f *is strongly convex with modulus of strong convexity* $c > 0$ *if and only if for every* $\mathbf{x} \in \mathbb{R}^d$ *and* $\mathbf{s} \in \partial f(\mathbf{x})$, *we have* $f(\mathbf{y}) \geq f(\mathbf{x}) + \langle \mathbf{s}, \mathbf{y} - \mathbf{x} \rangle + \frac{1}{2}c\|\mathbf{y} - \mathbf{x}\|^2$ *for all* $\mathbf{y} \in \mathbb{R}^d$.

3.1.1 Exercises

1. Prove Proposition 3.1.4.

2. Give an example of a function that is not convex, but all of whose sublevel sets are convex.

3. Prove the assertions made in parts 1., 2., and 3. of Example 3.1.11.

4. Fill in the missing arguments in the proof of Theorem 3.1.13.

5. Let $f: \mathbb{R}^d \to \mathbb{R} \cup \{+\infty\}$ be a convex function. Let $\mathbf{x}^1, \ldots, \mathbf{x}^k \in \operatorname{dom}(f)$. Show that $f(\mathbf{x}) \leq \max_{i=1,\ldots,k} f(\mathbf{x}^i)$ for all $\mathbf{x} \in \operatorname{conv}(\{\mathbf{x}^1, \ldots, \mathbf{x}^k\})$. In other words, f attains its maximum value on $\operatorname{conv}(\{\mathbf{x}^1, \ldots, \mathbf{x}^k\})$ at one of the points $\mathbf{x}^1, \ldots, \mathbf{x}^k$ (there could be other maximizers).

6. Let $f: \mathbb{R}^d \to \mathbb{R}$ be a convex function. Let $C \subseteq \mathbb{R}^d$ be a compact, convex set. Show that there exists an extreme point $\mathbf{x}^\star$ of C such that $f(\mathbf{x}^\star) = \sup_{\mathbf{x} \in C} f(\mathbf{x})$. In other words, there is always an extreme point solution to the problem of maximizing a convex function over a compact convex set.

7. Prove the following more general version of Jensen's inequality (Theorem 3.1.12). Let X be any random variable that takes values in $\mathbb{R}^d$ (the multivariate Gaussian distribution is an example). Let $f: \mathbb{R}^d \to \mathbb{R}$ be any convex function. Show that $f(\mathbb{E}[X]) \leq \mathbb{E}[f(X)]$, where the expectation on both sides is taken over the random variable X.

    [Note: The inequality proved in Theorem 3.1.12 is a special case of the above, where the random variable X takes only finitely many values $\{\mathbf{x}^1, \ldots, \mathbf{x}^n\} \subseteq \mathbb{R}^d$.]

8. Is the domain of a closed, convex function always closed?

9. Let $E \subseteq \mathbb{R}^d \times \mathbb{R}$ be a closed, convex set. Show that E is the epigraph of a closed, convex function if and only if $\{(\mathbf{0}, \lambda) : \lambda \geq 0\} \subseteq \operatorname{rec}(E)$ and E does not contain any line of the form $\{(\mathbf{x}, \lambda) : \lambda \in \mathbb{R}\}$ for some $\mathbf{x} \in \mathbb{R}^d$.

10. Let $f: \mathbb{R}^d \to \mathbb{R} \cup \{+\infty\}$ be a convex function that is not identically $+\infty$; i.e., $\operatorname{epi}(f)$ is a nonempty convex set. Show that if $(\mathbf{y}, \eta) \in \mathbb{R}^d \times \mathbb{R}$ and $\alpha \in \mathbb{R}$ are such that $(\mathbf{y}, \eta) \neq (\mathbf{0}, 0)$ and $\operatorname{epi}(f) \subseteq H^{\leq}((\mathbf{y}, \eta), \alpha)$, then $\eta \leq 0$.

11. Let $f: \mathbb{R}^d \to \mathbb{R}$ be a convex function (so $\operatorname{dom}(f) = \mathbb{R}^d$). Show that if $(\mathbf{y}, \eta) \in \mathbb{R}^d \times \mathbb{R}$ and $\alpha \in \mathbb{R}$ are such that $(\mathbf{y}, \eta) \neq (\mathbf{0}, 0)$ and $\operatorname{epi}(f) \subseteq H^{\leq}((\mathbf{y}, \eta), \alpha)$, then $\eta < 0$.

12. Complete the details in the proof of Corollary 3.1.17.

13. Suppose $f: \mathbb{R}^d \to \mathbb{R} \cup \{+\infty\}$ is a closed, convex function such that $\operatorname{epi}(f)$ is a polyhedron. Show that for any $\mathbf{x} \in \operatorname{dom}(f)$, there exist $(\mathbf{y}, \eta) \in \mathbb{R}^d \times \mathbb{R}$ and $\alpha \in \mathbb{R}$ such that $H^{\leq}((\mathbf{y}, \eta), \alpha)$ is a supporting halfspace for $\operatorname{epi}(f)$ at $(\mathbf{x}, f(\mathbf{x}))$ with $\eta < 0$.

14. Let $f: \mathbb{R}^d \to \mathbb{R} \cup \{+\infty\}$ be a closed, convex function. Let $\mathcal{M}$ be the set of all affine linear functions $\ell: \mathbb{R}^d \to \mathbb{R}$ such that $\ell \leq f$ everywhere. Show

that $f(\mathbf{y}) = \sup_{\ell \in \mathcal{M}} \ell(\mathbf{y})$. In other words, every closed, convex function is the supremum of all affine linear functions that underestimate it.

15. Let $f \colon \mathbb{R}^d \to \mathbb{R} \cup \{+\infty\}$ be a convex function, and let $\alpha \in \mathbb{R}$. Show that

$$\{\mathbf{y} \in \mathbb{R}^d : f(\mathbf{y}) \le \alpha\} = \{\mathbf{y} \in \mathbb{R}^d : (\mathbf{y}, \alpha) \in \mathrm{epi}(f)\}.$$

In other words, the α-sublevel set for f can be obtained from the "slice" of the epigraph of f at the level α.

16. Suppose $f \colon \mathbb{R}^d \to \mathbb{R} \cup \{+\infty\}$ is a closed, convex function, and let $\mathbf{x} \in \mathrm{int}(\mathrm{dom}(f))$. Show that $\partial f(\mathbf{x})$ is a nonempty, compact convex set.

3.2 Analytic Properties

3.2.1 Continuity

Convex functions enjoy strong continuity properties in the relative interior of their domains. This fact is very useful in many contexts, especially in optimization, because this is useful in showing that minimizers and maximizers exist when optimizing convex functions that show up in practice, via Weierstrass' theorem (Theorem 1.3.13).

The proof of continuity proceeds in two steps. We first show that for any point in the relative interior of the domain of a convex function, there is a neighborhood of that point where the function is bounded above and below (Proposition 3.2.1). We then show that any convex function that is bounded in a neighborhood of a point is Lipschitz continuous over a small neighborhood of that point (Proposition 3.2.2). These two results together imply the main continuity result, which is formally stated in Theorem 3.2.3.

Proposition 3.2.1 *Let* $f \colon \mathbb{R}^d \to \mathbb{R} \cup \{+\infty\}$ *be a convex function and let* $\mathbf{x}^\star \in \mathrm{relint}(\mathrm{dom}(f))$. *Then there is an* $\epsilon_{\mathbf{x}^\star} > 0$ *and values* $m_{\mathbf{x}^\star}, M_{\mathbf{x}^\star} \in \mathbb{R}$ *so that* $B(\mathbf{x}^\star, \epsilon_{\mathbf{x}^\star}) \cap \mathrm{dom}(f) = B(\mathbf{x}^\star, \epsilon_{\mathbf{x}^\star}) \cap \mathrm{aff}(\mathrm{dom}(f))$, *and*

$$m_{\mathbf{x}^\star} \le f(\mathbf{x}) \le M_{\mathbf{x}^\star}$$

for all $\mathbf{x} \in B(\mathbf{x}^\star, \epsilon_{\mathbf{x}^\star}) \cap \mathrm{dom}(f)$.

Proof Let $\mathbf{v}^1, \ldots, \mathbf{v}^\ell$ be vectors that span the linear space parallel to $\mathrm{aff}(\mathrm{dom}(f))$ (see Theorem 2.2.8). Since $\mathbf{x}^\star \in \mathrm{relint}(\mathrm{dom}(f))$, there exists $\epsilon > 0$ such that $\mathbf{x}^\star + \epsilon \mathbf{v}^j$ and $\mathbf{x}^\star - \epsilon \mathbf{v}^j$ are both in $\mathrm{dom}(f)$ for $j = 1, \ldots, \ell$ (see Theorem 2.2.12). Denote the set of points $\mathbf{x}^\star \pm \epsilon \mathbf{v}^j$ as $\mathbf{x}_1, \ldots, \mathbf{x}_k \in \mathrm{dom}(f)$ ($k = 2\ell$), and define $S := \mathrm{conv}\{\mathbf{x}_1, \ldots, \mathbf{x}_k\}$. Observe that $\mathbf{x}^\star \in \mathrm{relint}(S)$ and $\mathrm{aff}(S) = \mathrm{aff}(\mathrm{dom}(f))$. Therefore, there is some $\epsilon_{\mathbf{x}^\star} > 0$

so that $B(\mathbf{x}^\star, \epsilon_{\mathbf{x}^\star}) \cap \mathrm{aff}(\mathrm{dom}(f)) \subseteq S$. Therefore, $B(\mathbf{x}^\star, \epsilon_{\mathbf{x}^\star}) \cap \mathrm{dom}(f) \supseteq B(\mathbf{x}^\star, \epsilon_{\mathbf{x}^\star}) \cap S \supseteq B(\mathbf{x}^\star, \epsilon_{\mathbf{x}^\star}) \cap \mathrm{aff}(\mathrm{dom}(f)) \supseteq B(\mathbf{x}^\star, \epsilon_{\mathbf{x}^\star}) \cap \mathrm{dom}(f)$. Thus, these three sets are all equal to each other.

Set $M_{\mathbf{x}^\star} = \max\{f(\mathbf{x}_i) : i = 1, \ldots, k\}$. Using Exercise 5 from Section 3.1.1, it follows that $f(\mathbf{x}) \leq M_{\mathbf{x}^\star}$ for all $\mathbf{x} \in S$. Since f is convex, by Corollary 3.1.17, there is a subgradient $\mathbf{a}$ of f at $\mathbf{x}^\star$ giving the affine function $\ell(\mathbf{x}) = f(\mathbf{x}^\star) + \langle \mathbf{a}, (\mathbf{x} - \mathbf{x}^\star)\rangle$. Define $m_{\mathbf{x}^\star} = \min\{\ell(\mathbf{x}_i) : i = 1, \ldots, k\}$. Consider any point $\mathbf{x} = \sum_{i=1}^{k} \lambda_i \mathbf{x}_i \in S$, where $\lambda_1, \ldots, \lambda_k$ are convex coefficients and observe that

$$\ell(\mathbf{x}) = \left\langle \mathbf{a}, \left(\sum_{i=1}^{k} \lambda_i \mathbf{x}_i\right) - \mathbf{x}^\star \right\rangle + f(\mathbf{x}^\star) = \sum_{i=1}^{k} \lambda_i \left(\langle \mathbf{a}, \mathbf{x}_i - \mathbf{x}^\star\rangle + f(\mathbf{x}^\star)\right)$$

$$= \sum_{i=1}^{k} \lambda_i \ell(\mathbf{x}_i) \geq m_{\mathbf{x}^\star}.$$

Since ℓ is an affine function coming from a subgradient, it follows that $f(\mathbf{x}) \geq \ell(\mathbf{x}) \geq m_{\mathbf{x}^\star}$ for all $\mathbf{x} \in S$. $\qquad\square$

Proposition 3.2.2 *Let $f: \mathbb{R}^d \to \mathbb{R} \cup \{+\infty\}$ be a convex function. Take $\mathbf{x}^\star \in \mathrm{dom}(f)$ and suppose that for some $\epsilon > 0$ and $m, M \in \mathbb{R}$, we have $B(\mathbf{x}^\star, 2\epsilon) \cap \mathrm{dom}(f) = B(\mathbf{x}^\star, 2\epsilon) \cap \mathrm{aff}(\mathrm{dom}(f))$ and the inequalities*

$$m \leq f(\mathbf{x}) \leq M$$

hold for all $\mathbf{x} \in B(\mathbf{x}^\star, 2\epsilon) \cap \mathrm{dom}(f)$. Then for all $\mathbf{x}, \mathbf{y} \in B(\mathbf{x}^\star, \epsilon) \cap \mathrm{dom}(f)$, it holds that

$$|f(\mathbf{x}) - f(\mathbf{y})| \leq \left(\frac{M - m}{\epsilon}\right) \|\mathbf{x} - \mathbf{y}\|. \tag{3.2.1}$$

In particular, f is Lipschitz continuous over $B(\mathbf{x}^\star, \epsilon) \cap \mathrm{dom}(f)$.

Proof Take $\mathbf{x}, \mathbf{y} \in B(\mathbf{x}^\star, \epsilon) \cap \mathrm{dom}(f)$ with $\mathbf{x} \neq \mathbf{y}$. Define $\mathbf{z} = \mathbf{y} + \epsilon\left(\frac{\mathbf{y}-\mathbf{x}}{\|\mathbf{y}-\mathbf{x}\|}\right)$. Note that

$$\|\mathbf{z} - \mathbf{x}^\star\| = \left\|\mathbf{y} + \epsilon\left(\frac{\mathbf{y} - \mathbf{x}}{\|\mathbf{y} - \mathbf{x}\|}\right) - \mathbf{x}^\star\right\| \leq \|\mathbf{y} - \mathbf{x}^\star\| + \left\|\epsilon\left(\frac{\mathbf{y} - \mathbf{x}}{\|\mathbf{y} - \mathbf{x}\|}\right)\right\| \leq \epsilon + \epsilon = 2\epsilon.$$

Thus $\mathbf{z} \in B(\mathbf{x}^\star, 2\epsilon)$. Moreover, since $\mathbf{x}, \mathbf{y} \in \mathrm{dom}(f)$, $\mathbf{z} \in \mathrm{aff}(\mathrm{dom}(f))$. Thus, $\mathbf{z} \in B(\mathbf{x}^\star, 2\epsilon) \cap \mathrm{aff}(\mathrm{dom}(f)) = B(\mathbf{x}^\star, 2\epsilon) \cap \mathrm{dom}(f)$. Also,

$$\mathbf{y} = \left(\frac{\|\mathbf{y} - \mathbf{x}\|}{\epsilon + \|\mathbf{y} - \mathbf{x}\|}\right)\mathbf{z} + \left(1 - \frac{\|\mathbf{y} - \mathbf{x}\|}{\epsilon + \|\mathbf{y} - \mathbf{x}\|}\right)\mathbf{x},$$

showing that $\mathbf{y}$ is a convex combination of $\mathbf{x}$ and $\mathbf{z}$. Therefore we may apply the convexity of f to obtain

$$f(\mathbf{y}) \leq \left(\frac{\|\mathbf{y}-\mathbf{x}\|}{\epsilon + \|\mathbf{y}-\mathbf{x}\|}\right) f(\mathbf{z}) + \left(1 - \frac{\|\mathbf{y}-\mathbf{x}\|}{\epsilon + \|\mathbf{y}-\mathbf{x}\|}\right) f(\mathbf{x})$$

$$= f(\mathbf{x}) + \left(\frac{\|\mathbf{y}-\mathbf{x}\|}{\epsilon + \|\mathbf{y}-\mathbf{x}\|}\right)(f(\mathbf{z}) - f(\mathbf{x}))$$

$$\leq f(\mathbf{x}) + \left(\frac{\|\mathbf{y}-\mathbf{x}\|}{\epsilon}\right)(M - m)$$

using the bounds on f in $B(\mathbf{x}^\star, 2\epsilon) \cap \mathrm{dom}(f)$.

Hence $f(\mathbf{y}) - f(\mathbf{x}) \leq \left(\frac{\|\mathbf{y}-\mathbf{x}\|}{\epsilon}\right)(M - m)$. Repeating this argument by swapping the roles of $\mathbf{x}$ and $\mathbf{y}$, we get $f(\mathbf{x}) - f(\mathbf{y}) \leq \left(\frac{\|\mathbf{y}-\mathbf{x}\|}{\epsilon}\right)(M - m)$, and the result follows. $\square$

Theorem 3.2.3 *Let* $f: \mathbb{R}^d \to \mathbb{R} \cup \{+\infty\}$ *be a convex function. Let* $D \subseteq$ *relint*$(\mathrm{dom}(f))$ *be a convex, compact subset. Then there is a constant* $L \geq 0$ *(depending on D) so that*

$$|f(\mathbf{x}) - f(\mathbf{y})| \leq L\|\mathbf{x} - \mathbf{y}\| \tag{3.2.2}$$

for all $\mathbf{x}, \mathbf{y} \in D$. *In particular,* f *is locally Lipschitz continuous over the relative interior of its domain.*

Proof From Proposition 3.2.1, for every $\mathbf{x} \in D$, there exists $\epsilon_{\mathbf{x}} > 0$, and $m_{\mathbf{x}}, M_{\mathbf{x}} \in \mathbb{R}$ such that $m_{\mathbf{x}} \leq f(\mathbf{y}) \leq M_{\mathbf{x}}$ for all $\mathbf{y} \in B(\mathbf{x}, 2\epsilon_{\mathbf{x}}) \cap \mathrm{dom}(f) = B(\mathbf{x}, 2\epsilon_{\mathbf{x}}) \cap \mathrm{aff}(\mathrm{dom}(f))$. Proposition 3.2.2 then implies that there is some $L_{\mathbf{x}} \geq 0$ so that $|f(\mathbf{y}) - f(\mathbf{z})| \leq L_{\mathbf{x}}\|\mathbf{z} - \mathbf{y}\|$ for all $\mathbf{z}, \mathbf{y} \in B(\mathbf{x}, \epsilon_{\mathbf{x}}) \cap \mathrm{dom}(f)$. Since D is compact, using the Heine–Borel theorem (Theorem 1.3.9) on a coordinatization of $\mathrm{aff}(\mathrm{dom}(f))$ (see Section 2.2.1), there exists a finite set $\{\mathbf{x}_1, ..., \mathbf{x}_k\} \subset D$ so that $D \subseteq \bigcup_{i=1}^{k} B(\mathbf{x}_i, \epsilon_{\mathbf{x}_i}) \cap \mathrm{dom}(f)$. Set $L = \max\{L_{\mathbf{x}_i} : i \in \{1, \ldots, k\}\}$.

Now take $\mathbf{y}, \mathbf{z} \in D$. The line segment $[\mathbf{y}, \mathbf{z}]$ can be divided into finitely many segments $[\mathbf{y}, \mathbf{z}] = [\mathbf{y}_1, \mathbf{y}_2] \cup [\mathbf{y}_2, \mathbf{y}_3] \cup \cdots \cup [\mathbf{y}_{q-1}, \mathbf{y}_q]$, where $\mathbf{y}_1 = \mathbf{y}$, $\mathbf{y}_q = \mathbf{z}$, and each interval $[\mathbf{y}_i, \mathbf{y}_{i+1}]$ is contained in some ball $B(\mathbf{x}_j, \epsilon_{\mathbf{x}_j})$ for $j \in \{1, \ldots, k\}$. Without loss of generality, we may assume that $q - 1 \leq k$ and $[\mathbf{y}_i, \mathbf{y}_{i+1}] \subseteq B(\mathbf{x}_i, \epsilon_{\mathbf{x}_i})$ for each $i \in \{1, \ldots, q - 1\}$. It follows that

$$|f(\mathbf{y}) - f(\mathbf{z})| = \left| f(\mathbf{y}_1) + \left(\sum_{i=2}^{q-1} f(\mathbf{y}_i)\right) - \left(\sum_{i=2}^{q-1} f(\mathbf{y}_i)\right) - f(\mathbf{y}_q) \right|$$

$$= \left| \sum_{i=1}^{q-1} f(\mathbf{y}_i) - f(\mathbf{y}_{i+1}) \right|$$

$$\leq \sum_{i=1}^{q-1} |f(\mathbf{y}_i) - f(\mathbf{y}_{i+1})|$$

$$\leq \sum_{i=1}^{q-1} L_{\mathbf{x}_i} \|\mathbf{y}_i - \mathbf{y}_{i+1}\|$$

$$\leq \sum_{i=1}^{q-1} L \|\mathbf{y}_i - \mathbf{y}_{i+1}\|$$

$$= L \|\mathbf{y}_1 - \mathbf{y}_q\| = L \|\mathbf{y} - \mathbf{z}\|.$$

Hence f is Lipschitz continuous over D with constant L. $\qquad\square$

3.2.2 First-Order Derivative Properties

A convex function need not be differentiable, e.g., the ℓ_1 norm $\|\mathbf{x}\|_1$, which is not differentiable at $\mathbf{0}$. Nevertheless, there is a result in real analysis called *Rademacher's theorem* [229, Theorem 10.8(ii)], which states that any Lipschitz continuous function over an open subset of $\mathbb{R}^d$ is differentiable almost everywhere in that open subset. Combined with Theorem 3.2.3, we obtain the following.

Theorem 3.2.4 (Rademacher's theorem) *Let $f : \mathbb{R}^d \to \mathbb{R} \cup \{+\infty\}$ be a convex function. Then f is differentiable almost everywhere in* $\mathrm{int}(\mathrm{dom}(f))$, *i.e., the subset of* $\mathrm{int}(\mathrm{dom}(f))$ *where f is not differentiable has zero Lebesgue measure.*

In this section, however, we will study convex functions that are differentiable everywhere in the interior of the domain. The key insight is the following fundamental property of one-dimensional convex functions.

Proposition 3.2.5 *Let $f : \mathbb{R} \to \mathbb{R} \cup \{+\infty\}$ be a convex function. Then for any real numbers $x < y < z$ in the domain of f, we must have (Figure 3.1)*

$$\frac{f(y) - f(x)}{y - x} \leq \frac{f(z) - f(x)}{z - x} \leq \frac{f(z) - f(y)}{z - y}.$$

Moreover, if f is strictly convex, then these inequalities are strict.

Proof Since $y \in (x, z)$, there exists $\alpha \in (0, 1)$ such that $y = \alpha x + (1 - \alpha)z$. Now we follow the inequalities:

$$\begin{aligned}
\frac{f(y) - f(x)}{y - x} &= \frac{f(\alpha x + (1 - \alpha)z) - f(x)}{\alpha x + (1 - \alpha)z - x} \\
&\leq \frac{\alpha f(x) + (1 - \alpha) f(z) - f(x)}{\alpha x + (1 - \alpha)z - x} \\
&= \frac{f(z) - f(x)}{z - x}.
\end{aligned}$$

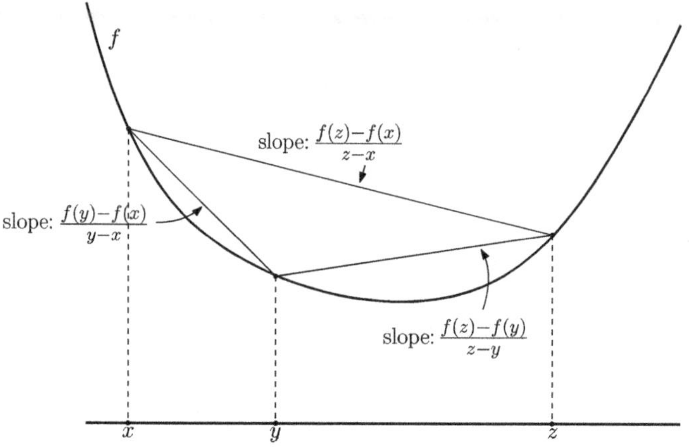

Figure 3.1 Illustration for Proposition 3.2.5.

Similarly,

$$
\begin{aligned}
\frac{f(z)-f(y)}{z-y} &= \frac{f(z)-f(\alpha x + (1-\alpha)z)}{z-\alpha x - (1-\alpha)z} \\
&\geq \frac{f(z)-\alpha f(x)-(1-\alpha)f(z)}{z-\alpha x - (1-\alpha)z} \\
&= \frac{f(z)-f(x)}{z-x}.
\end{aligned}
$$

The strict convexity implication is clear from the above.　　　　$\square$

An immediate corollary is the following relationship between the derivative of a function on the real line and convexity.

Proposition 3.2.6 *Let $f : \mathbb{R} \to \mathbb{R} \cup \{+\infty\}$ be a differentiable function over some open set $X \subseteq \mathrm{dom}(f)$. Let $C \subseteq X$ be a convex subset of X. Then f is convex over C if and only if f' is an increasing function over C, i.e., $f'(x) \leq f'(y)$ for all $x, y \in C$ with $x \leq y$. Moreover, f is strictly convex over C if and only if f' is strictly increasing over C. f is strongly convex over C with strong convexity modulus $c > 0$ if and only if $f'(x) + c(y-x) \leq f'(y)$ for all $x, y \in C$ with $x \leq y$.*

Proof ($\Rightarrow$) We may assume $x < y$; otherwise there is nothing to prove. Recall that $f'(x) = \lim_{t \to 0^+} \frac{f(x+t)-f(x)}{t}$. But for every $0 < t < y - x$, we have $\frac{f(x+t)-f(x)}{t} \leq \frac{f(y)-f(x)}{y-x}$ by Proposition 3.2.5. Thus, $f'(x) \leq \frac{f(y)-f(x)}{y-x}$. By a similar argument, we obtain $f'(y) \geq \frac{f(y)-f(x)}{y-x}$. This gives the relation.

($\Leftarrow$) If C is a singleton, there is nothing to prove. Consider any $x, z \in C$ with $x < z$ and $\alpha \in (0,1)$. Let $y = \alpha x + (1-\alpha)z$. By the Mean Value Theorem, there exists $t_1 \in [x, y]$ such that $\frac{f(y)-f(x)}{y-x} = f'(t_1)$ and $t_2 \in [y, z]$

such that $\frac{f(z)-f(y)}{z-y} = f'(t_2)$. Since $t_2 \geq t_1$ and we assume f' is increasing, then $f'(t_2) \geq f'(t_1)$. This implies that

$$\frac{f(z) - f(y)}{z - y} \geq \frac{f(y) - f(x)}{y - x}.$$

Substituting $y = \alpha x + (1 - \alpha)z$ and rearranging, we obtain that $f(\alpha x + (1 - \alpha)z) \leq \alpha f(x) + (1 - \alpha)f(z)$.

The arguments for strict and strong convexity are left as exercises for the reader. $\square$

We can now prove the main result of this subsection. A key idea behind the results below is that one can reduce testing convexity of a function on $\mathbb{R}^d$ to testing convexity of any one-dimensional "slice" of it. This enables the use of Proposition 3.2.6. More precisely:

Proposition 3.2.7 *Let $f \colon \mathbb{R}^d \to \mathbb{R} \cup \{+\infty\}$ be a function and $C \subseteq \mathbb{R}^d$ be convex. Then f is convex over C if and only if for every $\mathbf{z} \in \mathbb{R}^d$ and $\mathbf{r} \in \mathbb{R}^d \setminus \{\mathbf{0}\}$, the function $\varphi \colon \mathbb{R} \to \mathbb{R} \cup \{+\infty\}$ defined by $\varphi(t) = f(\mathbf{z} + t\mathbf{r})$ is convex over $\{t \in \mathbb{R} : \mathbf{z} + t\mathbf{r} \in C\}$.*

Proof Left as an exercise. $\square$

Theorem 3.2.8 *Let $f \colon \mathbb{R}^d \to \mathbb{R} \cup \{+\infty\}$ be differentiable over an open set $X \subseteq \mathrm{dom}(f)$, and let $C \subseteq X$ be convex. Then the following are all equivalent.*

1. f is convex over C.
2. $f(\mathbf{y}) \geq f(\mathbf{x}) + \langle \nabla f(\mathbf{x}), \mathbf{y} - \mathbf{x} \rangle$ for all $\mathbf{x}, \mathbf{y} \in C$.
3. $\langle \nabla f(\mathbf{y}) - \nabla f(\mathbf{x}), \mathbf{y} - \mathbf{x} \rangle \geq 0$ for all $\mathbf{x}, \mathbf{y} \in C$.

A characterization of strict convexity is obtained if all the above inequalities are considered strict for all $\mathbf{x} \neq \mathbf{y} \in C$. A characterization of strong convexity with modulus $c > 0$ is obtained if point 2. is replaced with $f(\mathbf{y}) \geq f(\mathbf{x}) + \langle \nabla f(\mathbf{x}), \mathbf{y} - \mathbf{x} \rangle + \frac{1}{2}c\|\mathbf{y} - \mathbf{x}\|^2$ for all $\mathbf{x}, \mathbf{y} \in C$, and point 3. is replaced with $\langle \nabla f(\mathbf{y}) - \nabla f(\mathbf{x}), \mathbf{y} - \mathbf{x} \rangle \geq c\|\mathbf{y} - \mathbf{x}\|^2$ for all $\mathbf{x}, \mathbf{y} \in C$.

Proof (1. $\implies$ 2.) Consider any $\mathbf{x}, \mathbf{y} \in C$. For every $\alpha \in (0, 1)$, convexity of f implies that $f((1 - \alpha)\mathbf{x} + \alpha\mathbf{y}) \leq (1 - \alpha)f(\mathbf{x}) + \alpha f(\mathbf{y})$. Rearranging, we obtain

$$\frac{f((1-\alpha)\mathbf{x}+\alpha\mathbf{y})-f(\mathbf{x})}{\alpha} \leq f(\mathbf{y}) - f(\mathbf{x})$$
$$\implies \frac{f(\mathbf{x}+\alpha(\mathbf{y}-\mathbf{x}))-f(\mathbf{x})}{\alpha} \leq f(\mathbf{y}) - f(\mathbf{x}).$$

Letting $\alpha \to 0$ on the left-hand side, we obtain the directional derivative $\langle \nabla f(\mathbf{x}), \mathbf{y} - \mathbf{x} \rangle$ (see Exercise 5 from Section 3.2.4) and 2. is established.

(2. $\implies$ 3.) By switching the roles of $\mathbf{x}, \mathbf{y} \in C$, we obtain the following:

$$f(\mathbf{y}) \geq f(\mathbf{x}) + \langle \nabla f(\mathbf{x}), \mathbf{y} - \mathbf{x} \rangle,$$
$$f(\mathbf{x}) \geq f(\mathbf{y}) + \langle \nabla f(\mathbf{y}), \mathbf{x} - \mathbf{y} \rangle.$$

Adding these inequalities together we obtain 3.

(3. $\implies$ 1.) Consider an arbitrary one-dimensional "slice" of f given by $\mathbf{z} \in \mathbb{R}^d$ and $\mathbf{r} \in \mathbb{R}^d \setminus \{\mathbf{0}\}$: $\varphi(t) := f(\mathbf{z} + t\mathbf{r})$. By Proposition 3.2.7, it suffices to show that φ is convex over $\{t \in \mathbb{R} : \mathbf{z} + t\mathbf{r} \in C\}$. If $\{t \in \mathbb{R} : \mathbf{z} + t\mathbf{r} \in C\}$ is a singleton, there is nothing to prove. Thus, consider any $t_1 < t_2$ in this set and define $\mathbf{x}^1 := \mathbf{z} + t_1\mathbf{r}$ and $\mathbf{x}^2 := \mathbf{z} + t_2\mathbf{r}$; thus, $\mathbf{x}^1, \mathbf{x}^2 \in C$. By the chain rule (Exercise 5 from Section 3.2.4), $\varphi'(t_1) = \langle \nabla f(\mathbf{x}^1), \mathbf{r} \rangle$ and $\varphi'(t_2) = \langle \nabla f(\mathbf{x}^2), \mathbf{r} \rangle$. Also, observe that $\mathbf{r} = \frac{1}{t_2 - t_1}(\mathbf{x}^2 - \mathbf{x}^1)$. Therefore, $\varphi'(t_2) - \varphi'(t_1) = \langle \nabla f(\mathbf{x}^2), \mathbf{r} \rangle - \langle \nabla f(\mathbf{x}^1), \mathbf{r} \rangle = \frac{1}{t_2 - t_1}\langle \nabla f(\mathbf{x}^2) - \nabla f(\mathbf{x}^1), \mathbf{x}^2 - \mathbf{x}^1 \rangle \geq 0$, using 3. with $\mathbf{x} = \mathbf{x}^1$ and $\mathbf{y} = \mathbf{x}^2$. By Proposition 3.2.6, we obtain that $\varphi(t)$ is a convex function over $\{t \in \mathbb{R} : \mathbf{z} + t\mathbf{r} \in C\}$.

The arguments for strict and strong convexity are left as exercises for the reader. $\qquad\qquad\square$

Part 2. of Theorem 3.2.8 says that $\nabla f(\mathbf{x})$ is a subgradient for f and is in the subdifferential (see Definition 3.1.15). We will see later in Theorem 3.4.4 that, in fact, when a convex function f is differentiable at a point $\mathbf{x}$, its subdifferential is a singleton consisting of the gradient $\nabla f(\mathbf{x})$.

As is evident from the proof above, Part 3. of Theorem 3.2.8 simply says that a differentiable function is convex if and only if its directional derivative in a direction $\mathbf{r} = \mathbf{y} - \mathbf{x}$ increases when going from $\mathbf{x}$ to $\mathbf{y}$, which is just the condition for convexity of the one-dimensional "slice" of the function through $\mathbf{x}$ and $\mathbf{y}$.

3.2.3 Second-Order Derivative Properties

A simple consequence of Proposition 3.2.6 for twice differentiable functions on the real line is the following.

Corollary 3.2.9 *Let $f : \mathbb{R} \to \mathbb{R} \cup \{+\infty\}$ be a twice differentiable function over an open set $X \subseteq \mathrm{dom}(f)$ and let $C \subseteq X$ be an open interval. Then f is convex over C if and only if $f''(x) \geq 0$ for all $x \in C$. If $f''(x) > 0$ for all $x \in C$, then f is strictly convex over C.*

Remark 3.2.10 From Proposition 3.2.6, we know strict convexity of f is equivalent to the condition that f' is strictly increasing. However, this is not equivalent to $f''(x) > 0$; the implication only goes in one direction. This is why we lose the other direction when discussing strict convexity in Corollary 3.2.9. As a concrete example, consider $f(x) = x^4$ which is strictly convex, but the second derivative is 0 at $x = 0$.

Corollary 3.2.9 enables one to characterize convexity of a function f in terms of its Hessian $\nabla^2 f$.

Theorem 3.2.11 *Let* $f \colon \mathbb{R}^d \to \mathbb{R} \cup \{+\infty\}$ *be a twice differentiable function over an open set* $X \subseteq \mathrm{dom}(f)$ *and let* $C \subseteq X$ *be an open convex set. Then the following are all true.*

1. *f is convex over C if and only if $\nabla^2 f(\mathbf{x})$ is positive semidefinite (PSD) for all $\mathbf{x} \in C$.*
2. *If $\nabla^2 f(\mathbf{x})$ is positive definite (PD) for all $\mathbf{x} \in C$, then f is strictly convex over C.*
3. *f is strongly convex with modulus $c > 0$ over C if and only if $\nabla^2 f(\mathbf{x}) - c I_d$ is positive semidefinite (PSD) for all $\mathbf{x} \in C$, i.e., the smallest eigenvalue of $\nabla^2 f(\mathbf{x})$ is at least c.*

Proof   1. ($\Rightarrow$) Let $\mathbf{x} \in C$, and we would like to show that $\nabla^2 f(\mathbf{x})$ is positive semidefinite. Consider any $\mathbf{r} \in \mathbb{R}^d \setminus \{\mathbf{0}\}$. Define the function $\varphi(t) = f(\mathbf{x} + t\mathbf{r})$. By Proposition 3.2.7, φ is convex over the open interval $\{t \in \mathbb{R} : \mathbf{x} + t\mathbf{r} \in C\}$ (since C is open). By Corollary 3.2.9, $0 \le \varphi''(0) = \langle \nabla^2 f(\mathbf{x})\mathbf{r}, \mathbf{r} \rangle$ where the equality follows from the chain rule (Exercise 5 from Section 3.2.4). Since the choice of $\mathbf{r}$ was arbitrary, this shows that $\nabla^2 f(\mathbf{x})$ is positive semidefinite.

    ($\Leftarrow$) Assume $\nabla^2 f(\mathbf{x})$ is positive semidefinite for all $\mathbf{x} \in C$, and consider $\bar{\mathbf{x}} \in \mathbb{R}^d$ and $\mathbf{r} \in \mathbb{R}^d \setminus \{\mathbf{0}\}$. Define the function $\varphi(t) = f(\bar{\mathbf{x}} + t\mathbf{r})$. Now $\varphi''(t) = \langle \nabla^2 f(\bar{\mathbf{x}} + t\mathbf{r})\mathbf{r}, \mathbf{r} \rangle \ge 0$ for all $t \in \mathbb{R}$ such that $\bar{\mathbf{x}} + t\mathbf{r} \in C$ (Exercise 5 from Section 3.2.4), since $\nabla^2 f(\bar{\mathbf{x}} + t\mathbf{r})$ is positive semidefinite. By Corollary 3.2.9, φ is convex over $\{t \in \mathbb{R} : \bar{\mathbf{x}} + t\mathbf{r} \in C\}$. By Proposition 3.2.7, f is convex.

2. This follows from the same construction as in 1. above, and the sufficient condition that if the second derivative of a one-dimensional function is strictly positive, then the function is strictly convex.

3. By Proposition 3.1.4, f is strongly convex with modulus $c > 0$ if and only if the function $g(\mathbf{x}) := f(\mathbf{x}) - \frac{1}{2}c\|\mathbf{x}\|^2$ is convex. Applying part 1. to g, we obtain the result.

$\qquad\qquad\qquad\qquad\qquad\qquad\qquad\qquad\qquad\qquad\qquad\qquad\qquad\qquad\quad\square$

3.2.4 Exercises

1. Show that any convex function on $\mathbb{R}^d$ that is finite valued everywhere, i.e., the domain is all of $\mathbb{R}^d$, is closed, convex.
2. Let $f : \mathbb{R}^d \to \mathbb{R}$ be a convex function. Let $\{\mathbf{x}^k\}_{k \in \mathbb{N}}$ be a sequence converging to $\mathbf{x}$ and let $\mathbf{s}^k \in \partial f(\mathbf{x}^k)$ for all $k \in \mathbb{N}$. Show that if $\mathbf{s}^k$ converges to $\mathbf{s}$ then $\mathbf{s} \in \partial f(\mathbf{x})$.
3. Complete the proof of Proposition 3.2.6 and Theorem 3.2.8 for strict and strong convexity.
4. Prove Proposition 3.2.7.
5. Let $f : \mathbb{R}^d \to \mathbb{R} \cup \{+\infty\}$ be a function and let $\mathbf{z} \in \mathbb{R}^d$ and $\mathbf{r} \in \mathbb{R}^d \setminus \{\mathbf{0}\}$. Define the function $\varphi : \mathbb{R} \to \mathbb{R} \cup \{+\infty\}$ as $\varphi(t) = f(\mathbf{z} + t\mathbf{r})$. Show that if f is differentiable at $\mathbf{z} + \bar{t}\mathbf{r}$ for some $\bar{t} \in \mathbb{R}$, then φ is differentiable at $\bar{t}$ and $\varphi'(\bar{t}) = \langle \nabla f(\mathbf{z} + \bar{t}\mathbf{r}), \mathbf{r} \rangle$. Show that if f is twice differentiable at $\mathbf{z} + \bar{t}\mathbf{r}$, then φ is twice differentiable at $\bar{t}$ and $\varphi''(\bar{t}) = \langle \nabla^2 f(\mathbf{z} + \bar{t}\mathbf{r})\mathbf{r}, \mathbf{r} \rangle$.
6. Show that the function

$$f(x) := \begin{cases} \frac{1}{x} & x > 0, \\ +\infty & x \le 0 \end{cases}$$

   is convex.
7. Show that the function $x \mapsto e^x$ is convex. Conclude that $1 + x \le e^x$ for all $x \in \mathbb{R}$.
8. **(Logarithms, Young's inequality, and the AM-GM inequality)** Show that the following function is convex.

$$f(x) := \begin{cases} -\log(x) & x > 0, \\ +\infty & x \le 0. \end{cases}$$

   Use this to derive the following inequality: For any real numbers $r_1, \ldots, r_k \ge 0$ and $\gamma_1, \ldots, \gamma_k \ge 0$ such that $\gamma_1 + \cdots + \gamma_k = 1$, we must have

$$\gamma_1 r_1 + \gamma_2 r_2 + \cdots + \gamma_k r_k \ge r_1^{\gamma_1} r_2^{\gamma_2} \cdots \cdots r_k^{\gamma_k}.$$

   The special case of $k = 2$ is known as *Young's inequality* and the special case of $\gamma_1 = \gamma_2 = \cdots = \gamma_k$ is known as the *Arithmetic Mean–Geometric Mean (AM-GM) inequality*.
9.* **(Hölder's inequality and Minkowski's inequality)** Let $p, q \ge 1$ such that $\frac{1}{p} + \frac{1}{q} = 1$ (allowing for p or q to be ∞). Show that

$$|\langle \mathbf{x}, \mathbf{y} \rangle| \le \|\mathbf{x}\|_p \|\mathbf{y}\|_q$$

for every $\mathbf{x}, \mathbf{y} \in \mathbb{R}^d$. Moreover, if $p, q > 1$ then equality holds if and only if $|\mathbf{x}_i| = |\mathbf{y}_i|^{\frac{q}{p}}$. This inequality is known as *Hölder's inequality* and it generalizes the Cauchy–Schwarz inequality (Theorem 1.1.6).

Use Hölder's inequality to derive *Minkowski's inequality*: For any $p \geq 1$ and for all $\mathbf{x}, \mathbf{y} \in \mathbb{R}^d$,

$$\|\mathbf{x} + \mathbf{y}\|_p \leq \|\mathbf{x}\|_p + \|\mathbf{y}\|_p.$$

This establishes that $\| \cdot \|_p$ is indeed a norm (Example 1.1.2).

10. Let $p, q \geq 1$ such that $\frac{1}{p} + \frac{1}{q} = 1$ (allowing for p or q to be ∞). Use Hölder's inequality (Exercise 9 above) to show that $(B_{\ell^p}(\mathbf{0}, 1))^\circ = B_{\ell^q}(\mathbf{0}, 1)$.

11. Let Q be a symmetric $d \times d$ matrix, and $\mathbf{c} \in \mathbb{R}^d$. Show that the function $f(\mathbf{x}) = \mathbf{x}^T Q \mathbf{x} + \langle \mathbf{c}, \mathbf{x} \rangle$ is convex if and only if Q is positive semidefinite.

3.3 Sublinear Functions

We will now introduce a more structured subfamily of convex functions that is easier to deal with analytically and yet has very important uses in diverse areas. Let us discuss two high-level motivations for considering this new family of functions. The first one comes from the observation that norms are convex functions (see part 3. of Example 3.1.11). The two key properties of norms are the triangle inequality and the scaling property (Definition 1.1.1). We will relax the scaling property a little bit to define this new class of functions that retains some important properties of norms but strictly generalizes to a more flexible class of functions. This will give a set of tools that have proved to be useful in all aspects of convex geometry and analysis, including optimization. The second motivation comes directly from optimization. A very important class of optimization problems is that of maximizing a linear function over closed, convex sets. If we consider the linear function as a parameter with a fixed closed, convex set, then the optimal values of these optimization problems become a function of the vector representing the linear function. It turns out that this function is convex and shares some important properties with norms. In fact, we discover the same class of convex functions that we obtain from our first construction with the relaxed scaling property. This dual perspective on this class of convex functions leads to a rich and fascinating set of results with various applications, many of which we will see in subsequent chapters. So, without further ado, let us make the central definition.

Definition 3.3.1 A function $f : \mathbb{R}^d \to \mathbb{R} \cup \{+\infty\}$ is called *sublinear* if it satisfies the following two properties:

(i) f is *positively homogeneous*, i.e., $f(\lambda \mathbf{r}) = \lambda f(\mathbf{r})$ for all $\mathbf{r} \in \mathbb{R}^d$ and $\lambda > 0$.

(ii) f is *subadditive*, i.e., $f(\mathbf{x} + \mathbf{y}) \le f(\mathbf{x}) + f(\mathbf{y})$ for all $\mathbf{x}, \mathbf{y} \in \mathbb{R}^d$.

Note how the first property is a relaxation of the scaling property of norms to consider only positive scalings. The second property is nothing but the triangle inequality. Here is the connection with convexity.

Proposition 3.3.2 *Let $f : \mathbb{R}^d \to \mathbb{R} \cup \{+\infty\}$. Then the following are equivalent:*

1. *f is sublinear.*
2. *f is convex and positively homogeneous.*
3. *$f(\lambda_1 \mathbf{x}^1 + \lambda_2 \mathbf{x}^2) \le \lambda_1 f(\mathbf{x}^1) + \lambda_2 f(\mathbf{x}^2)$ for all $\mathbf{x}^1, \mathbf{x}^2 \in \mathbb{R}^d$ and $\lambda_1, \lambda_2 > 0$.*

Proof Left as an exercise. $\square$

Positive homogeneity applied with $\mathbf{r} = \mathbf{0}$ and $\lambda = 2$ implies the following useful property of sublinear functions.

Proposition 3.3.3 *Let $f : \mathbb{R}^d \to \mathbb{R} \cup \{+\infty\}$ be a sublinear function. Then either $f(\mathbf{0}) = 0$ or $f(\mathbf{0}) = +\infty$.*

A characterization of sublinear functions via epigraphs is also possible.

Proposition 3.3.4 *Let $f : \mathbb{R}^d \to \mathbb{R} \cup \{+\infty\}$ such that $f(\mathbf{0}) = 0$. Then f is sublinear if and only if $\mathrm{epi}(f)$ is a convex cone in $\mathbb{R}^d \times \mathbb{R}$.*

Proof ($\Rightarrow$) From Proposition 3.3.2, we know that f is convex and positively homogeneous. From Proposition 3.1.7, this implies that $\mathrm{epi}(f)$ is convex. So we only need to verify that if $(\mathbf{x}, t) \in \mathrm{epi}(f)$ then $\lambda(\mathbf{x}, t) = (\lambda \mathbf{x}, \lambda t) \in \mathrm{epi}(f)$ for all $\lambda \ge 0$. If $\lambda = 0$, then the result follows from the assumption that $f(\mathbf{0}) = 0$. Now consider $\lambda > 0$. Since $(\mathbf{x}, t) \in \mathrm{epi}(f)$, we have $f(\mathbf{x}) \le t$ and by positive homogeneity of f, $f(\lambda \mathbf{x}) = \lambda f(\mathbf{x}) \le \lambda t$, and so $(\lambda \mathbf{x}, \lambda t) \in \mathrm{epi}(f)$.

($\Leftarrow$) We verify part 3. of Proposition 3.3.2. Consider any $\mathbf{x}^1, \mathbf{x}^2 \in \mathbb{R}^d$ and $\lambda_1, \lambda_2 > 0$. If $f(\mathbf{x}^1) = +\infty$ or $f(\mathbf{x}^2) = +\infty$, we are done; so we assume $f(\mathbf{x}^1), f(\mathbf{x}^2) \in \mathbb{R}$. Since $\mathrm{epi}(f)$ is a convex cone, we have that $(\lambda_1 \mathbf{x}^1 + \lambda_2 \mathbf{x}^2, \lambda_1 f(\mathbf{x}^2) + \lambda_2 f(\mathbf{x}^2)) \in \mathrm{epi}(f)$ because $(\mathbf{x}^1, f(\mathbf{x}^1)), (\mathbf{x}^2, f(\mathbf{x}^2)) \in \mathrm{epi}(f)$. By the definition of the epigraph, we obtain that $f(\lambda_1 \mathbf{x}^1 + \lambda_2 \mathbf{x}^2) \le \lambda_1 f(\mathbf{x}^1) + \lambda_2 f(\mathbf{x}^2)$ for all $\mathbf{x}^1, \mathbf{x}^2 \in \mathbb{R}^d$. $\square$

The following are some important examples of sublinear functions.

Example 3.3.5   1. Any linear function $f : \mathbb{R}^d \to \mathbb{R}$ is sublinear, i.e., the map $\mathbf{x} \mapsto \langle \mathbf{c}, \mathbf{x} \rangle$ is sublinear for any $\mathbf{c} \in \mathbb{R}^d$.

2. Any norm on $\mathbb{R}^d$ is sublinear.

3. If $f : \mathbb{R}^d \to \mathbb{R}$ is a convex function, then at any $\mathbf{x} \in \mathbb{R}^d$ the directional derivative function $f'(\mathbf{x}; \mathbf{r})$ is sublinear in $\mathbf{r}$ (see (1.3.1) in Definition 1.3.16). This will be established in Section 3.4 and will be an important application of sublinearity.

3.3.1 Nonnegative Sublinear Functions: Gauges

As mentioned earlier, any norm $N : \mathbb{R}^d \to \mathbb{R}$ is a sublinear function simply because the scaling property of norms is a strengthening of positive homogeneity, and subadditivity is nothing but triangle inequality. There are two important differences between norms and general sublinear functions. First, norms are nonnegative, but sublinear functions may take negative values (linear functions are sublinear; see part 1. of Example 3.3.5). Second, the scaling property of norms with arbitrary real scalings implies that a norm has the additional "symmetry" property that $N(\mathbf{x}) = N(-\mathbf{x})$. Sublinear functions do not have this property in general; linear functions are again a counterexample. In this subsection, we will look at nonnegative sublinear functions that are not necessarily symmetric, giving us an important family of sublinear functions that are a strict superset of norms and strict subset of all sublinear functions.

Since a sublinear function is convex (Proposition 3.3.2), and sublevel sets of convex sets are convex (Proposition 3.1.10), we immediately know that the unit norm balls $B_N(\mathbf{0}, 1) = \{\mathbf{x} \in \mathbb{R}^d : N(\mathbf{x}) \leq 1\}$ are convex sets. Because of the "symmetry property" of norms, these unit norm balls are also "symmetric" about the origin. This merits a definition.

Definition 3.3.6   A convex set $C \subseteq \mathbb{R}^d$ is said to be **0**-*symmetric* if $\mathbf{x} \in C$ implies that $-\mathbf{x} \in C$.

We therefore have the following.

Proposition 3.3.7 *Let $N : \mathbb{R}^d \to \mathbb{R}$ be a norm. Then the unit norm ball $B_N(\mathbf{0}, 1) = \{\mathbf{x} \in \mathbb{R}^d : N(\mathbf{x}) \leq 1\}$ is a **0**-symmetric, closed, convex set.*

One can actually prove a converse to the above statement, which will establish a nice one-to-one correspondence between norms and **0**-symmetric convex sets. In fact, a more general correspondence will be shown between all nonnegative sublinear functions and all closed, convex sets containing **0**. The starting point is a simple, geometric observation about norms. Given a norm N on $\mathbb{R}^d$, for any $\mathbf{x} \in \mathbb{R}^d$, $N(\mathbf{x})$ is equal to the smallest scaling factor by which we can scale the unit norm ball $B_N(\mathbf{0}, 1)$ and contain $\mathbf{x}$. For example, for the ℓ_∞ norm, $\|\mathbf{x}\|_\infty = \max_{i=1,\dots,d} |\mathbf{x}_i|$ is the smallest $\lambda > 0$ such that $\lambda[-1,1]^d$ contains $\mathbf{x}$, i.e., $\mathbf{x} \in [-\lambda, \lambda]^d$. This motivates the following definition for an arbitrary closed, convex set containing the origin.

Definition 3.3.8 Let $C \subseteq \mathbb{R}^d$ be a closed, convex set such that $\mathbf{0} \in C$. Define the following function $\gamma_C \colon \mathbb{R}^d \to \mathbb{R} \cup \{+\infty\}$ as

$$\gamma_C(\mathbf{r}) = \inf\{\lambda > 0 : \mathbf{r} \in \lambda C\}.$$

γ_C is called the *gauge* or the *Minkowski functional* of C.

The following is a useful observation for the analysis of gauge functions.

Lemma 3.3.9 *Let $C \subseteq \mathbb{R}^d$ be a closed convex set such that $\mathbf{0} \in C$, and let $\mathbf{r} \in \mathbb{R}^d$ be any vector. Then the set $\{\lambda > 0 : \mathbf{r} \in \lambda C\}$ is either empty or a convex interval of the real line of the form $(a, +\infty)$ or $[a, +\infty)$.*

Proof Define $I := \{\lambda > 0 : \mathbf{r} \in \lambda C\}$ and suppose it is nonempty. It suffices to show that if $\bar\lambda \in I$ then for all $\lambda \geq \bar\lambda$, $\lambda \in I$. This follows from the fact that $\bar\lambda \in I$ implies that $\frac{1}{\bar\lambda}\mathbf{r} \in C$. For any $\lambda \geq \bar\lambda$, we have $\frac{1}{\lambda}\mathbf{r} = \frac{\bar\lambda}{\lambda}(\frac{1}{\bar\lambda}\mathbf{r}) + (\frac{\lambda-\bar\lambda}{\lambda})\mathbf{0}$ which is in C because C is convex and $\mathbf{0} \in C$. $\qquad\square$

A useful intuition to keep in mind is that for any $\mathbf{r} \neq \mathbf{0}$, the gauge function value $\gamma_C(\mathbf{r})$ gives a factor to scale $\mathbf{r}$ so that one ends up on the boundary of C. More precisely:

Proposition 3.3.10 *Let $C \subseteq \mathbb{R}^d$ be a closed, convex set such that $\mathbf{0} \in C$. Suppose $\mathbf{r} \in \mathbb{R}^d$ such that $0 < \gamma_C(\mathbf{r}) < +\infty$. Then $\frac{1}{\gamma_C(\mathbf{r})}\mathbf{r} \in \mathrm{relbd}(C)$.*

Proof From Lemma 3.3.9, we have that for all $\lambda > \gamma_C(\mathbf{r})$, we have that $\mathbf{r} \in \lambda C$, i.e., $\frac{1}{\lambda}\mathbf{r} \in C$. Taking the limit $\lambda \downarrow \gamma_C(\mathbf{r})$ and using the fact that C is closed, we obtain that $\frac{1}{\gamma_C(\mathbf{r})}\mathbf{r} \in C$. If $\frac{1}{\gamma_C(\mathbf{r})}\mathbf{r} \in \mathrm{relint}(C)$, then we can scale $\frac{1}{\gamma_C(\mathbf{r})}\mathbf{r}$ by $\alpha > 1$ and obtain that $\frac{\alpha}{\gamma_C(\mathbf{r})}\mathbf{r} \in C$ (since both $\mathbf{0}$ and $\frac{1}{\gamma_C(\mathbf{r})}\mathbf{r}$ are in C, therefore $\mathbf{r} \in \mathrm{aff}(C)$), which would imply that $\mathbf{r} \in \frac{\gamma_C(\mathbf{r})}{\alpha}C$, contradicting the fact that $\gamma_C(\mathbf{r}) = \inf\{\lambda > 0 : \mathbf{r} \in \lambda C\}$ since $\frac{\gamma_C(\mathbf{r})}{\alpha} < \gamma_C(\mathbf{r})$. $\qquad\square$

The following theorem relates geometric properties of C with analytical properties of the gauge function. These relations prove to be quite handy.

Theorem 3.3.11 *Let $C \subseteq \mathbb{R}^d$ be a closed, convex set such that $\mathbf{0} \in C$. Then the following are all true.*

1. γ_C is a closed, nonnegative, sublinear function.
2. $C = \{\mathbf{x} \in \mathbb{R}^d : \gamma_C(\mathbf{x}) \leq 1\}$.

3. $\operatorname{rec}(C) = \{\mathbf{r} \in \mathbb{R}^d : \gamma_C(\mathbf{r}) = 0\}$.
4. *If* $\mathbf{0} \in \operatorname{relint}(C)$, *then* $\operatorname{relint}(C) = \{\mathbf{x} \in \mathbb{R}^d : \gamma_C(\mathbf{x}) < 1\}$.

Proof 1. Nonnegativity follows from the definition. Positive homogeneity of γ_C follows from the observation that for any $\mathbf{r} \in \mathbb{R}^d$ and $t > 0$,

$$\inf\{\lambda > 0 : t\mathbf{r} \in \lambda C\} = t \inf\left\{\frac{\lambda}{t} : t\mathbf{r} \in \lambda C\right\} = t \inf\left\{\frac{\lambda}{t} : \mathbf{r} \in \frac{\lambda}{t}C\right\}$$
$$= t \inf\{\lambda' > 0 : \mathbf{r} \in \lambda'C\},$$

where we used part 1. of Theorem 1.3.1. Subadditivity follows from the fact that $\mathbf{r} \in \lambda C$ and $\mathbf{r}' \in \lambda'C$ implies that $\mathbf{r} + \mathbf{r}' \in (\lambda + \lambda')C$ by Exercise 3 from Section 2.1.1. In other words,

$$\{\lambda : \mathbf{r} + \mathbf{r}' \in \lambda C\} \supseteq \{\lambda : \mathbf{r} \in \lambda C\} + \{\lambda : \mathbf{r}' \in \lambda C\},$$

where the sum on the right-hand side is a Minkowski sum of subsets of real numbers. Now, apply parts 2. and 3. of Theorem 1.3.1. We leave the property of closedness as an exercise.

2. We first observe that $C \subseteq \{\mathbf{x} \in \mathbb{R}^d : \gamma_C(\mathbf{x}) \le 1\}$ because $\mathbf{x} \in C$ implies that $1 \in \{\lambda > 0 : \mathbf{x} \in \lambda C\}$ and therefore $\inf\{\lambda > 0 : \mathbf{x} \in \lambda C\} \le 1$. Now, we verify that $\{\mathbf{x} \in \mathbb{R}^d : \gamma_C(\mathbf{x}) \le 1\} \subseteq C$. $\gamma_C(\mathbf{x}) \le 1$ implies that $\inf\{\lambda > 0 : \mathbf{x} \in \lambda C\} \le 1$, and thus $\{\lambda > 0 : \mathbf{x} \in \lambda C\}$ cannot be empty and must be an unbounded interval by Lemma 3.3.9. This means that either $1 \in \{\lambda > 0 : \mathbf{x} \in \lambda C\}$, and thus $\mathbf{x} \in C$, or $1 = \inf\{\lambda > 0 : \mathbf{x} \in \lambda C\} = \gamma_C(\mathbf{x})$. By Proposition 3.3.10, we have that $1 \cdot \mathbf{x} \in C$.

3. Since $\{\lambda > 0 : \mathbf{r} \in \lambda C\}$ is convex by Lemma 3.3.9, we observe that $\gamma_C(\mathbf{r}) = 0$ if and only if $\frac{1}{\lambda}\mathbf{r} \in C$ for all $\lambda > 0$. Since $\mathbf{0} \in C$, this is equivalent to saying that $t\mathbf{r} \in C$ for all $t \ge 0$; more explicitly, $\mathbf{0} + t\mathbf{r} \in C$ for all $t \ge 0$. This is equivalent to saying that $\mathbf{r}$ satisfies Definition 2.4.18 of being a recession direction.

4. Consider any $\mathbf{x} \in \operatorname{relint}(C)$. By definition of relative interior, there exists $\lambda > 1$ such that $\lambda\mathbf{x} \in C$ (since both $\mathbf{0}$ and $\mathbf{x}$ are in $C \subseteq \operatorname{aff}(C)$). By part 2. above, $\gamma_C(\lambda\mathbf{x}) \le 1$ and by part 1. above, γ_C is positively homogeneous, and thus, $\gamma_C(\mathbf{x}) \le \frac{1}{\lambda} < 1$.

   Now suppose $\mathbf{x} \in \mathbb{R}^d$ such that $\gamma_C(\mathbf{x}) < 1$. If $\gamma_C(\mathbf{x}) = 0$, then $\mathbf{x} \in \operatorname{rec}(C)$ by part 3. above. Since $\mathbf{0} \in \operatorname{relint}(C)$, we also have $\mathbf{x} = \mathbf{0} + \mathbf{x} \in \operatorname{relint}(C)$ (Exercise 34 from Section 2.4.4). Now suppose $0 < \gamma_C(\mathbf{x}) < 1$. By part 2. above, $\mathbf{x} \in C$. Suppose to the contrary that $\mathbf{x} \notin \operatorname{relint}(C)$. By Theorem 2.4.15, $\mathbf{x}$ is contained in a proper face F of C. Since $\mathbf{0} \in \operatorname{relint}(C)$, $\mathbf{0}$ is not contained in F. Also, $\gamma_C\left(\frac{\mathbf{x}}{\gamma_C(\mathbf{x})}\right) = 1$ by positive homogeneity of

γ_C, from part 1. above. Therefore, $\frac{\mathbf{x}}{\gamma_C(\mathbf{x})} \in C$. However, $\mathbf{x} = (1-\gamma_C(\mathbf{x}))\mathbf{0} + \gamma_C(\mathbf{x})\left(\frac{\mathbf{x}}{\gamma_C(\mathbf{x})}\right)$. Since $\gamma_C(\mathbf{x}) < 1$ and $\mathbf{0} \notin F$, this would contradict the fact that F is a face.

$\square$

We derive some immediate consequences.

Corollary 3.3.12 *Let $C \subseteq \mathbb{R}^d$ be a closed, convex set with $\mathbf{0} \in C$. Then C is compact if and only if $\gamma(\mathbf{r}) > 0$ for all $\mathbf{r} \in \mathbb{R}^d \setminus \{\mathbf{0}\}$.*

Corollary 3.3.13 *[Uniqueness of the gauge] Let C be a closed, convex set with $\mathbf{0} \in C$. Let $f : \mathbb{R}^d \to \mathbb{R} \cup \{+\infty\}$ be any nonnegative, sublinear function with $f(\mathbf{0}) = 0$. Then $C = \{\mathbf{x} \in \mathbb{R}^d : f(\mathbf{x}) \le 1\}$ if and only if $f = \gamma_C$.*

Proof The sufficiency follows from Theorem 3.3.11, part 2. For the necessity, suppose to the contrary that $f(\bar{\mathbf{x}}) \ne \gamma_C(\bar{\mathbf{x}})$ for some $\bar{\mathbf{x}} \in \mathbb{R}^d$. We first observe that $\bar{\mathbf{x}} \ne \mathbf{0}$ because $f(\mathbf{0}) = 0 = \gamma_C(\mathbf{0})$.

First suppose $f(\bar{\mathbf{x}}) > \gamma_C(\bar{\mathbf{x}}) \ge 0$. If $\gamma_C(\bar{\mathbf{x}}) = 0$, then by part 3. of Theorem 3.3.11, $\bar{\mathbf{x}} \in \mathrm{rec}(C)$ and so $\bar{\mathbf{x}} \in C$. We also must have $f(\bar{\mathbf{x}}) < +\infty$, since otherwise $\bar{\mathbf{x}} \notin C$ as $C = \{\mathbf{x} \in \mathbb{R}^d : f(\mathbf{x}) \le 1\}$. Since $\bar{\mathbf{x}} \in \mathrm{rec}(C)$, the point $\frac{2}{f(\bar{\mathbf{x}})}\bar{\mathbf{x}} \in C$. However, by positive homogeneity, $f(\frac{2\bar{\mathbf{x}}}{f(\bar{\mathbf{x}})}) = 2 > 1$ which says that $\frac{2\bar{\mathbf{x}}}{f(\bar{\mathbf{x}})} \notin C$ because of the assumption that $C = \{\mathbf{x} \in \mathbb{R}^d : f(\mathbf{x}) \le 1\}$, resulting in a contradiction. If $\gamma_C(\bar{\mathbf{x}}) > 0$, then $f(\bar{\mathbf{x}}) > \gamma_C(\bar{\mathbf{x}})$ implies that $\gamma_C(\bar{\mathbf{x}}) < +\infty$. Consider the point $\frac{1}{\gamma_C(\bar{\mathbf{x}})}\bar{\mathbf{x}}$. By Proposition 3.3.10, $\frac{1}{\gamma_C(\bar{\mathbf{x}})}\bar{\mathbf{x}} \in \mathrm{relbd}(C)$. However, since f is positively homogeneous, $f\left(\frac{1}{\gamma_C(\bar{\mathbf{x}})}\bar{\mathbf{x}}\right) = \frac{1}{\gamma_C(\bar{\mathbf{x}})}f(\bar{\mathbf{x}}) > 1$ because $f(\bar{\mathbf{x}}) > \gamma_C(\mathbf{x})$. This says $\frac{1}{\gamma_C(\bar{\mathbf{x}})}\bar{\mathbf{x}} \notin C$ because $C = \{\mathbf{x} \in \mathbb{R}^d : f(\mathbf{x}) \le 1\}$. This again leads to a contradiction.

Next suppose $f(\bar{\mathbf{x}}) < \gamma_C(\bar{\mathbf{x}})$. If $f(\bar{\mathbf{x}}) = 0$, then by positive homogeneity, $f(\lambda\bar{\mathbf{x}}) = 0$ for all $\lambda \ge 0$. Thus, $\lambda\bar{\mathbf{x}} \in C$ for all $\lambda \ge 0$ by the assumption that $C = \{\mathbf{x} \in \mathbb{R}^d : f(\mathbf{x}) \le 1\}$. This means that $\bar{\mathbf{x}} \in \mathrm{rec}(C)$. Thus, $\gamma_C(\bar{\mathbf{x}}) = 0$ by part 3. of Theorem 3.3.11, which contradicts the assumption that $f(\bar{\mathbf{x}}) < \gamma_C(\bar{\mathbf{x}})$. Finally, consider $f(\bar{\mathbf{x}}) > 0$ and the point $\frac{1}{f(\bar{\mathbf{x}})}\bar{\mathbf{x}}$. By positive homogeneity of γ_C, we obtain that $\gamma_C\left(\frac{1}{f(\bar{\mathbf{x}})}\bar{\mathbf{x}}\right) = \gamma_C\left(\frac{1}{f(\bar{\mathbf{x}})}\bar{\mathbf{x}}\right) = \frac{\gamma_C(\bar{\mathbf{x}})}{f(\bar{\mathbf{x}})} > 1$. Therefore, $\frac{1}{f(\bar{\mathbf{x}})}\bar{\mathbf{x}} \notin C$ by part 2. of Theorem 3.3.11. However, $f\left(\frac{1}{f(\bar{\mathbf{x}})}\bar{\mathbf{x}}\right) = 1$, which shows that $\frac{1}{f(\bar{\mathbf{x}})}\bar{\mathbf{x}} \in C$ because $C = \{\mathbf{x} \in \mathbb{R}^d : f(\mathbf{x}) \le 1\}$. This is a contradiction. $\square$

Corollary 3.3.13 is a fundamental result. It establishes a one-to-one correspondence between the family of closed, convex sets containing the origin and the family of closed, nonnegative, sublinear functions that take value 0 at the origin. More concretely, the gauge operation maps a closed, convex set containing the origin to a closed, nonnegative sublinear function. In the other

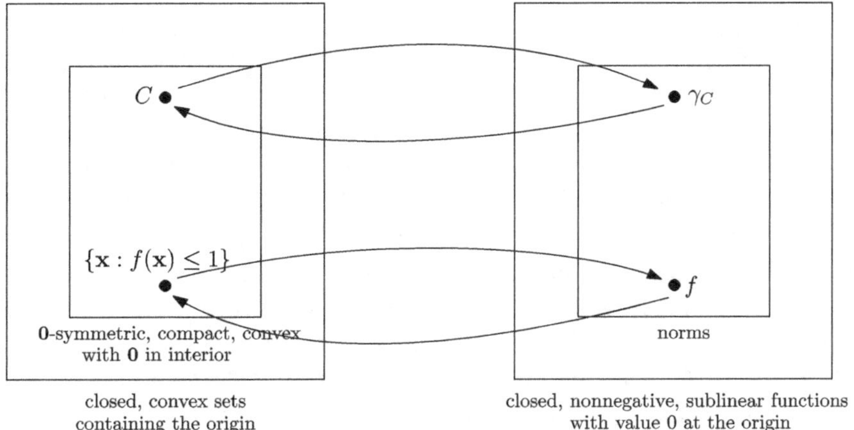

Figure 3.2 One-to-one correspondence between closed, convex sets containing the origin and closed, nonnegative, sublinear functions taking value 0 at the origin. Arrows from left to right show the gauge operation and arrows from right to left show the 1-sublevel set operation.

direction, taking 1-sublevel sets maps a closed, nonnegative sublinear function to a closed, convex set containing the origin. Moreover, these operations are inverses of each other: start with a closed, convex set C containing the origin; its gauge γ_C is a closed, nonnegative sublinear function whose 1-sublevel set is C; and start with a closed, nonnegative, sublinear function f and let C be the 1-sublevel set of f, then $\gamma_C = f$. In particular, every closed, nonnegative, sublinear function taking value 0 at the origin is the gauge of some closed, convex set containing the origin, and every such closed, convex set is the 1-sublevel set of some closed, nonnegative, sublinear function taking value 0 at the origin. See Figure 3.2. This correspondence can be used to reason about nonnegative, sublinear functions using geometric arguments about their 1-sublevel sets, and vice versa, one can study closed, convex sets (containing the origin, which can be assumed up to translation) using analytic arguments on their gauge functions.

This correspondence can be restricted to **0**-symmetric, compact convex sets on the one hand and norms on the other (see Figure 3.2).

Theorem 3.3.14 *Let $N: \mathbb{R}^d \to \mathbb{R}$ be a norm. Then $B_N(\mathbf{0}, 1) = \{\mathbf{x} \in \mathbb{R}^d : N(\mathbf{x}) \leq 1\}$ is a **0**-symmetric, compact convex set with **0** in its interior. Moreover, $\gamma_{B_N(\mathbf{0}, 1)} = N$.*

*Conversely, let B be a **0**-symmetric, compact convex set containing **0** in its interior. Then γ_B is a norm on $\mathbb{R}^d$ and $B = B_{\gamma_B}(\mathbf{0}, 1)$.*

Proof For the first part, since N is sublinear, it is convex (Proposition 3.3.2). By definition, $B_N(\mathbf{0}, 1) = \{\mathbf{x} \in \mathbb{R}^d : N(\mathbf{x}) \leq 1\}$ is a sublevel set for N, and it is thus a convex set (Proposition 3.1.10). It is closed, since N is continuous by Theorem 3.2.3. Since $N(\mathbf{x}) = N(-\mathbf{x})$, this also shows that $B_N(\mathbf{0}, 1)$ is $\mathbf{0}$-symmetric. We now show that $\mathrm{rec}(B_N(\mathbf{0}, 1)) = \{\mathbf{0}\}$; this will imply that it is compact by Theorem 2.4.22. Consider any nonzero vector $\mathbf{r}$, and let $N(\mathbf{r}) = M > 0$. Then, $\frac{2}{M}\mathbf{r} = \mathbf{0} + \frac{2}{M}\mathbf{r}$, but $N\left(\frac{2}{M}\mathbf{r}\right) = 2$. Thus, $\frac{2}{M}\mathbf{r} \notin B_N(\mathbf{0}, 1)$, and so $\mathbf{r}$ cannot be a recession direction for $B_N(\mathbf{0}, 1)$.

We next verify that $\mathbf{0} \in \mathrm{int}(B_N(\mathbf{0}, 1))$. If not, then by the supporting hyperplane theorem 2.4.5, there exists $\mathbf{a} \in \mathbb{R}^d \setminus \{\mathbf{0}\}$ and $\delta \in \mathbb{R}$ such that $B_N(\mathbf{0}, 1) \subseteq H^{\leq}(\mathbf{a}, \delta)$ and $\langle \mathbf{a}, \mathbf{0} \rangle = \delta$. Thus, $\delta = 0$. Now, since $\mathbf{a} \neq \mathbf{0}$, $N(\mathbf{a}) > 0$. Thus, $N\left(\frac{\mathbf{a}}{N(\mathbf{a})}\right) = 1$ and by definition, $\frac{\mathbf{a}}{N(\mathbf{a})} \in B_N(\mathbf{0}, 1)$. However, $\left\langle \mathbf{a}, \frac{\mathbf{a}}{N(\mathbf{a})} \right\rangle = \frac{\|\mathbf{a}\|^2}{N(\mathbf{a})} > 0$, which contradicts the fact that $B_N(\mathbf{0}, 1) \subseteq H^{\leq}(\mathbf{a}, 0)$. Finally, from Corollary 3.3.13, we obtain that $N = \gamma_{B_N(\mathbf{0}, 1)}$ since N is a nonnegative, sublinear function taking value 0 at the origin.

For the second part, we know that γ_B is nonnegative and sublinear, and since B is compact, $\gamma_B(\mathbf{r}) > 0$ for all $\mathbf{r} \neq \mathbf{0}$ by Corollary 3.3.12. Since $\mathbf{0} \in \mathrm{int}(B)$, Exercise 4 from Section 3.3.5 implies that γ_B is finite valued everywhere. To confirm that γ_B is a norm, all that remains to be checked is that $\gamma_B(\mathbf{x}) = \gamma_B(-\mathbf{x})$ for all $\mathbf{x} \neq \mathbf{0}$. Suppose to the contrary that $\gamma_B(\mathbf{x}) > \gamma_B(-\mathbf{x})$ (note that this is without loss of generality). This implies that $\gamma_B\left(\frac{1}{\gamma_B(-\mathbf{x})}\mathbf{x}\right) > 1$. Therefore, $\frac{1}{\gamma_B(-\mathbf{x})}\mathbf{x} \notin B$ by Theorem 3.3.11, part 2. However, $\gamma_B\left(-\frac{1}{\gamma_B(-\mathbf{x})}\mathbf{x}\right) = \frac{1}{\gamma_B(-\mathbf{x})}\gamma_B(-\mathbf{x}) = 1$ showing that $-\frac{1}{\gamma_B(-\mathbf{x})}\mathbf{x} \in B$ by Theorem 3.3.11, part 2. This contradicts the fact that B is $\mathbf{0}$-symmetric. Thus, γ_B is a norm on $\mathbb{R}^d$. Moreover, by Theorem 3.3.11, part 2., $B = \{\mathbf{x} \in \mathbb{R}^d : \gamma_B(\mathbf{x}) \leq 1\} = B_{\gamma_B}(\mathbf{0}, 1)$. $\qquad\square$

We now turn to some computational aspects of the gauge function. First, let us give an explicit formula for the gauge of a halfspace containing the origin.

Example 3.3.15 Let $H := H^{\leq}(\mathbf{a}, \delta)$ be a halfspace defined by some $\mathbf{a} \in \mathbb{R}^d$ and $\delta \in \mathbb{R}$ such that $\mathbf{0} \in H^{\leq}(\mathbf{a}, \delta)$. We assume that we have normalized δ to be 0 or 1. If $\delta = 0$, then

$$\gamma_H(\mathbf{r}) = \begin{cases} 0 & \text{if } \langle \mathbf{a}, \mathbf{r} \rangle \leq 0, \\ +\infty & \text{if } \langle \mathbf{a}, \mathbf{r} \rangle > 0. \end{cases}$$

If $\delta = 1$, then

$$\gamma_H(\mathbf{r}) = \max\{0, \langle \mathbf{a}, \mathbf{r} \rangle\}.$$

The above calculation, along with the next theorem, gives powerful computational tools for gauge functions.

Theorem 3.3.16 *Let C_i, $i \in I$ be a (not necessarily finite) family of closed, convex sets containing the origin, and let $C = \cap_{i\in I} C_i$. Then*

$$\gamma_C = \sup_{i\in I} \gamma_{C_i}.$$

Proof Consider any $\mathbf{r} \in \mathbb{R}^d$. Let us define $A_i = \{\lambda > 0 : \mathbf{r} \in \lambda C_i\}$ for each $i \in I$, and define $A = \{\lambda > 0 : \mathbf{r} \in \lambda C\}$. Observe that $A = \cap A_i$. If any A_i is empty, then $\gamma_{C_i}(\mathbf{r}) = +\infty$, and A is empty and therefore $\gamma_C(\mathbf{r}) = +\infty$, and the equality holds. Now suppose all A_i's are nonempty, and so by Lemma 3.3.9, each A_i is of the form (a_i, ∞) or $[a_i, \infty)$. If $A = \emptyset$, then it must mean that $a_i \to \infty$. Since $\gamma_{C_i}(\mathbf{r}) = \inf A_i = a_i$, this shows that $\sup_{i\in I} \gamma_{C_i}(\mathbf{r}) = \infty$. Moreover, $A = \emptyset$ implies that $\gamma_C(\mathbf{r}) = \inf A = +\infty$. Finally, consider the case that A is nonempty. Then since $A = \cap A_i$, A must be of the form $(a, +\infty)$ or $[a, +\infty)$ where $a := \sup_{i\in I} a_i$. Then $\gamma_C(\mathbf{r}) = a = \sup_{i\in I} a_i = \sup_{i\in I} \gamma_{C_i}(\mathbf{r})$. $\square$

This shows that gauge functions for polyhedra can be computed very easily.

Corollary 3.3.17 *Let P be a polyhedron with $\mathbf{0} \in \mathrm{int}(P)$. Thus, by scaling the normals and shifts of the defining halfspaces there exist $\mathbf{a}^1, \ldots, \mathbf{a}^m \in \mathbb{R}^d$ such that*

$$P = \{\mathbf{x} \in \mathbb{R}^d : \langle \mathbf{a}^i, \mathbf{x} \rangle \leq 1 \ \ i = 1, \ldots, m\}.$$

Then

$$\gamma_P(\mathbf{r}) = \max\{0, \langle \mathbf{a}^1, \mathbf{r} \rangle, \ldots, \langle \mathbf{a}^m, \mathbf{r} \rangle\}.$$

Proof Use the formula from Example 3.3.15 and Theorem 3.3.16. $\square$

3.3.2 Support Functions

We established in the previous subsection that nonnegative, sublinear functions are precisely the gauges of closed, convex sets containing the origin. We would like to arrive at such a geometric characterization of general sublinear functions, without imposing nonnegativity. It turns out that the one-to-one correspondence from Corollary 3.3.13 (Figure 3.2) can be expanded to a one-to-one correspondence between general sublinear functions and all closed, convex sets. For this, we will have to define a new operation on closed, convex sets on the one hand and a new operation on the family of sublinear functions on the other. Let us begin with closed, convex sets not necessarily containing the origin and extracting sublinear functions out of them. The operation is inspired by optimization of linear functions over closed, convex sets.

Definition 3.3.18 Let $S \subseteq \mathbb{R}^d$ be any set. The *support function* for S is a function on $\mathbb{R}^d$ defined as

$$\sigma_S(\mathbf{r}) = \sup_{\mathbf{x} \in S} \langle \mathbf{r}, \mathbf{x} \rangle.$$

The following is easy to verify.

Proposition 3.3.19 *Let* $S \subseteq \mathbb{R}^d$. *Then*

$$\sigma_S = \sigma_{\mathrm{cl}(S)} = \sigma_{\mathrm{conv}(S)} = \sigma_{\mathrm{cl}(\mathrm{conv}(S))}.$$

Proposition 3.3.20 *Let* $S \subseteq \mathbb{R}^d$. *Then* σ_S *is a closed, sublinear function, i.e., its epigraph is a closed, convex cone.*

Proof We first check that σ_S is sublinear. We check positive homogeneity. For any $\mathbf{r} \in \mathbb{R}^d$ and $\lambda > 0$,

$$\sigma_S(\lambda \mathbf{r}) = \sup_{\mathbf{x} \in S} \langle \lambda \mathbf{r}, \mathbf{x} \rangle = \sup_{\mathbf{x} \in S} \lambda \langle \mathbf{r}, \mathbf{x} \rangle = \lambda \sup_{\mathbf{x} \in S} \langle \mathbf{r}, \mathbf{x} \rangle = \lambda \sigma_S(\mathbf{r}).$$

We check subadditivity. Let $\mathbf{r}^1, \mathbf{r}^2 \in \mathbb{R}^d$. Then,

$$
\begin{aligned}
\sigma_S(\mathbf{r}^1 + \mathbf{r}^2) &= \sup_{\mathbf{x} \in S} \langle \mathbf{r}^1 + \mathbf{r}^2, \mathbf{x} \rangle \\
&= \sup_{\mathbf{x} \in S} (\langle \mathbf{r}^1, \mathbf{x} \rangle + \langle \mathbf{r}^2, \mathbf{x} \rangle) \\
&\leq \sup_{\mathbf{x} \in S} \langle \mathbf{r}^1, \mathbf{x} \rangle + \sup_{\mathbf{x} \in S} \langle \mathbf{r}^2, \mathbf{x} \rangle \\
&= \sigma_S(\mathbf{r}^1) + \sigma_S(\mathbf{r}^2).
\end{aligned}
$$

Since σ_S is the supremum of linear functions $\langle \mathbf{x}, \mathbf{r} \rangle$, $\mathbf{x} \in S$, $\mathrm{epi}(f)$ is the intersection of closed halfspaces, which shows that it is closed. The fact that it is a convex cone follows from Proposition 3.3.4. $\qquad\square$

Observe that if $\mathbf{0} \in C$, then the support function σ_C is nonnegative. Conversely, if $\mathbf{0} \notin C$ then σ_C takes negative values (Exercise 14 from Section 3.3.5). Given the correspondence between nonnegative sublinear functions and closed, convex sets containing the origin via the gauge operation, this suggests that there should be a connection between support functions and gauges. And indeed, there is a fundamental correspondence between gauges and support functions of closed, convex sets containing the origin via polarity.

Theorem 3.3.21 *Let* C *be a closed, convex set with* $\mathbf{0} \in C$. *Then*

$$\gamma_C = \sigma_{C^\circ}.$$

Proof Recall that $C = (C^\circ)^\circ$ by Proposition 2.4.10 part 2. Unwrapping the definitions, this says that

$$C = \{\mathbf{x} \in \mathbb{R}^d : \langle \mathbf{a}, \mathbf{x} \rangle \leq 1 \ \forall \mathbf{a} \in C^\circ\} = \cap_{\mathbf{a} \in C^\circ} H^{\leq}(\mathbf{a}, 1).$$

By Theorem 3.3.16 and Example 3.3.15, we obtain that

$$\gamma_C(\mathbf{r}) = \sup_{\mathbf{a} \in C^\circ} \gamma_{H^{\leq}(\mathbf{a},1)}(\mathbf{r}) = \sup_{\mathbf{a} \in C^\circ} \max\{0, \langle \mathbf{a}, \mathbf{r} \rangle\}.$$

Since $\mathbf{0} \in C^\circ$, the last term above can be written as $\sup_{\mathbf{a} \in C^\circ} \langle \mathbf{a}, \mathbf{r} \rangle = \sigma_{C^\circ}(\mathbf{r})$. $\square$

Generalized Cauchy–Schwarz/Hölder's inequality. Theorem 3.3.21 has wide applications in different areas of convex and functional analysis, and we will use it many times in later chapters as well. Here, we illustrate how the relationship between norms, gauges, and support functions leads to a far-reaching generalization of Hölder's inequality (and consequently, the Cauchy–Schwarz inequality); see Exercise 9 in Section 3.2.4.

Theorem 3.3.22 *Let $C \subseteq \mathbb{R}^d$ be a compact, convex set with $\mathbf{0} \in C$. Then*

$$\langle \mathbf{x}, \mathbf{y} \rangle \leq \gamma_C(\mathbf{x})\sigma_C(\mathbf{y}) \quad \forall \mathbf{x}, \mathbf{y} \in \mathbb{R}^d.$$

Proof Consider any $\mathbf{x}, \mathbf{y} \in \mathbb{R}^d$. Since C is compact, $\gamma_C(\mathbf{x}) > 0$ by Corollary 3.3.12, and $\sigma_C(\mathbf{y}) < +\infty$. If $\gamma_C(\mathbf{x}) = +\infty$, then the inequality follows immediately; so assume otherwise. By Proposition 3.3.10, $\frac{\mathbf{x}}{\gamma_C(\mathbf{x})} \in C$, and therefore,

$$\left\langle \frac{\mathbf{x}}{\gamma_C(\mathbf{x})}, \mathbf{y} \right\rangle \leq \sup_{\mathbf{z} \in C} \langle \mathbf{z}, \mathbf{y} \rangle = \sigma_C(\mathbf{y}).$$

This immediately implies $\langle \mathbf{x}, \mathbf{y} \rangle \leq \gamma_C(\mathbf{x})\sigma_C(\mathbf{y})$. $\square$

Corollary 3.3.23 *Let $C \subseteq \mathbb{R}^d$ be a compact, convex set with $\mathbf{0} \in C$. Then*

$$\langle \mathbf{x}, \mathbf{y} \rangle \leq \gamma_C(\mathbf{x})\gamma_{C^\circ}(\mathbf{y}) \quad \forall \mathbf{x}, \mathbf{y} \in \mathbb{R}^d.$$

Proof Follows from Theorems 3.3.22 and 3.3.21. $\square$

The above corollary generalizes Hölder's inequality (Exercise 9 in Section 3.2.4) because when $\frac{1}{p} + \frac{1}{q} = 1$, then the ℓ^p and ℓ^q unit balls are polars of each other (Exercise 10 in Section 3.2.4). Note that Theorem 3.3.22 and Corollary 3.3.23 have no assumption of $\mathbf{0}$-symmetric sets, so they strictly generalize the norm inequalities of Hölder and Cauchy–Schwarz. When C is a $\mathbf{0}$-symmetric, compact, convex set containing the origin in its interior, so is C° (Exercise 11 from Section 3.3.5). Then γ_C and $\gamma_{C^\circ}(= \sigma_C)$ are called *dual norms*.

3.3.3 Correspondence between Closed, Convex Sets and Closed, Sublinear Functions

Proposition 3.3.20 shows that support functions are closed, sublinear functions. Proposition 3.3.19 shows that two different sets, e.g., S and $\text{conv}(S)$, may give rise to the same sublinear function $\sigma_S = \sigma_{\text{conv}(S)}$ via the support function construction. In other words, if we consider the mapping $S \to \sigma_S$ as a mapping from the family of subsets of $\mathbb{R}^d$ to the family of closed, sublinear functions, this mapping is not injective. But if we restrict to closed, convex sets, it can be shown that this mapping is injective; see Exercise 13 from Section 3.3.5. A natural question then is whether the mapping $C \to \sigma_C$ from the family of closed, convex sets to the family of closed, sublinear functions is *onto*. The answer is yes! Thus, all closed, sublinear functions are support functions and vice versa. Moreover, there is an explicit inverse to the operation of taking the support function of a closed, convex set.

Theorem 3.3.24 *Let* $f : \mathbb{R}^d \to \mathbb{R} \cup \{+\infty\}$ *be a closed, sublinear function that is not identically* $+\infty$ *everywhere. Then the set*

$$C_f := \{\mathbf{x} \in \mathbb{R}^d : \langle \mathbf{r}, \mathbf{x} \rangle \leq f(\mathbf{r}) \ \forall \mathbf{r} \in \mathbb{R}^d\} = \cap_{\mathbf{r} \in \mathbb{R}^d} H^{\leq}(\mathbf{r}, f(\mathbf{r})) \tag{3.3.1}$$

is a nonempty, closed, convex set. Moreover, $\sigma_{C_f} = f$.

Conversely, if C *is a nonempty, closed, convex set, then* $C_{\sigma_C} = C$.

Proof First, observe that if $f(\mathbf{r}) = +\infty$ for all $\mathbf{r} \neq \mathbf{0}$, then $C_f = \mathbb{R}^d$ and the result holds since $f(\mathbf{0}) = 0$ by Proposition 3.3.3. So, in the rest of the proof we assume that f takes a finite value at some nonzero vector. Proposition 3.3.3, combined with the assumption that $\text{epi}(f)$ is a nonempty, closed set, implies that $f(\mathbf{0}) = 0$ (see Exercise 2 from Section 3.3.5). By Proposition 3.3.4, $\text{epi}(f)$ is a closed, convex cone. By parts 2. and 3. of Proposition 2.4.10, $\text{epi}(f)$ is the intersection of all halfspaces of the form $H^{\leq}((\mathbf{y}, \eta), 0)$, where $(\mathbf{y}, \eta) \in \text{epi}(f)^{\circ}$. Exercise 10 from Section 3.1.1 shows that $\eta \leq 0$ for all $(\mathbf{y}, \eta) \in \text{epi}(f)^{\circ}$. We claim that $(\mathbf{y}, \eta) \in \text{epi}(f)^{\circ}$ with $\eta < 0$ if and only if $\frac{\mathbf{y}}{-\eta} \in C_f$. This follows from the following observation.

$$
\begin{aligned}
\mathbf{y} \in C_f &\Leftrightarrow \langle \mathbf{r}, \mathbf{y} \rangle \leq f(\mathbf{r}) && \forall \mathbf{r} \in \mathbb{R}^d \\
&\Leftrightarrow \langle \mathbf{r}, \mathbf{y} \rangle \leq f(\mathbf{r}) && \forall \mathbf{r} \in \text{dom}(f) \\
&\Leftrightarrow \langle \mathbf{r}, \mathbf{y} \rangle \leq t && \forall \mathbf{r} \in \text{dom}(f), t \in \mathbb{R} \text{ such that } f(\mathbf{r}) \leq t \\
&\Leftrightarrow \langle \mathbf{r}, \mathbf{y} \rangle - t \leq 0 && \forall \mathbf{r} \in \text{dom}(f), t \in \mathbb{R} \text{ such that } f(\mathbf{r}) \leq t \\
&\Leftrightarrow \langle (\mathbf{r}, t), (\mathbf{y}, -1) \rangle \leq 0 && \forall \mathbf{r} \in \text{dom}(f), t \in \mathbb{R} \text{ such that } f(\mathbf{r}) \leq t \\
&\Leftrightarrow \langle (\mathbf{y}, -1), (\mathbf{r}, t) \rangle \leq 0 && \forall (\mathbf{r}, t) \in \text{epi}(f) \\
&\Leftrightarrow (\mathbf{y}, -1) \in \text{epi}(f)^{\circ}.
\end{aligned}
$$

$$\tag{3.3.2}$$

We next observe that C_f is nonempty. Indeed, consider any $\bar{\mathbf{r}} \neq \mathbf{0}$ with $f(\bar{\mathbf{r}}) < +\infty$. Since the point $(\bar{\mathbf{r}}, f(\bar{\mathbf{r}}) - 1)$ is not in $\text{epi}(f)$, by Theorem 2.4.2, there must exist a separating hyperplane given by $(\mathbf{y}, \eta) \in \text{epi}(f)^\circ$ such that $\langle \mathbf{y}, \bar{\mathbf{r}} \rangle + \eta f(\bar{\mathbf{r}}) \leq 0$ and $\langle \mathbf{y}, \bar{\mathbf{r}} \rangle + \eta(f(\bar{\mathbf{r}}) - 1) > 0$. Together, these inequalities imply that $\eta < 0$. By the argument made above, $\frac{\mathbf{y}}{-\eta} \in C_f$.

Finally, we show that $(\mathbf{y}, 0) \in \text{epi}(f)^\circ$ if and only if $\mathbf{y} \in \text{rec}(C_f)$. If $(\mathbf{y}, 0) \in \text{epi}(f)^\circ$, then $\mathbf{y} \in \text{rec}(C_f)$. This follows from the fact that $(\mathbf{y}, 0) \in \text{epi}(f)^\circ$ implies that $\langle \mathbf{r}, \mathbf{y} \rangle \leq 0$ for all $\mathbf{r} \in \text{dom}(f)$. Thus, for any $\mathbf{x} \in C_f$ and $\lambda \geq 0$, we have $\langle \mathbf{r}, \mathbf{x} + \lambda \mathbf{y} \rangle \leq \langle \mathbf{r}, \mathbf{x} \rangle \leq f(\mathbf{r})$ for every $\mathbf{r} \in \text{dom}(f)$, and therefore $\langle \mathbf{r}, \mathbf{x} + \lambda \mathbf{y} \rangle \leq f(\mathbf{r})$ for all $\mathbf{r} \in \mathbb{R}^d$. This implies that $\mathbf{x} + \lambda \mathbf{y} \in C_f$ for all $\mathbf{x} \in C_f$ and $\lambda \geq 0$ and so $\mathbf{y} \in \text{rec}(C_f)$. Similarly, if $\mathbf{y} \in \text{rec}(C_f)$, then one can show that $\langle \mathbf{r}, \mathbf{y} \rangle \leq 0$ for all $\mathbf{r} \in \text{dom}(f)$. Otherwise, if $\langle \bar{\mathbf{r}}, \mathbf{y} \rangle > 0$ for some $\bar{\mathbf{r}} \in \text{dom}(f)$, then for any $\mathbf{x} \in C_f$ (C_f is nonempty), $\langle \bar{\mathbf{r}}, \mathbf{x} + \lambda \mathbf{y} \rangle$ can be made as large as needed by choosing $\lambda > 0$ large enough; in particular, it can be made larger than $f(\bar{\mathbf{r}})$. This contradicts that $\mathbf{x} + \lambda \mathbf{y} \in C_f$ for all $\lambda \geq 0$. Since $\langle \mathbf{r}, \mathbf{y} \rangle \leq 0$ for all $\mathbf{r} \in \text{dom}(f)$, we have that for all $(\mathbf{r}, t) \in \text{epi}(f)$, $\langle (\mathbf{r}, t), (\mathbf{y}, 0) \rangle \leq 0$. This implies that $(\mathbf{y}, 0) \in \text{epi}(f)^\circ$. Therefore, $(\mathbf{y}, 0) \in \text{epi}(f)^\circ$ if and only if $\mathbf{y} \in \text{rec}(C_f)$.

We now establish that $\sigma_{C_f} = f$. For any $\mathbf{r} \in \mathbb{R}^d$, since $C_f \subseteq H^{\leq}(\mathbf{r}, f(\mathbf{r}))$, we must have that $\sigma_{C_f}(\mathbf{r}) = \sup_{\mathbf{x} \in C_f} \langle \mathbf{r}, \mathbf{x} \rangle \leq f(\mathbf{r})$. To show that $\sigma_{C_f}(\mathbf{r}) \geq f(\mathbf{r})$, we consider two cases.

Case 1: $f(\mathbf{r}) < +\infty$. Consider an arbitrary $\delta > 0$. Since the point $(\mathbf{r}, f(\mathbf{r}) - \delta)$ is not in $\text{epi}(f)$, by Theorem 2.4.2, there must exist a separating hyperplane given by $(\mathbf{y}, \eta) \in \text{epi}(f)^\circ$ such that $\langle \mathbf{y}, \mathbf{r} \rangle + \eta f(\mathbf{r}) \leq 0$ and $\langle \mathbf{y}, \mathbf{r} \rangle + \eta(f(\mathbf{r}) - \delta) > 0$. Together, these inequalities imply that $\eta < 0$. By the argument made above, $\frac{\mathbf{y}}{-\eta} \in C_f$. Moreover, the second inequality can be rearranged to give $\langle \frac{\mathbf{y}}{-\eta}, \mathbf{r} \rangle > f(\mathbf{r}) - \delta$. Therefore, $\sup_{\mathbf{x} \in C_f} \langle \mathbf{r}, \mathbf{x} \rangle \geq \langle \frac{\mathbf{y}}{-\eta}, \mathbf{r} \rangle > f(\mathbf{r}) - \delta$. Since $\delta > 0$ was arbitrary, this implies $\sup_{\mathbf{x} \in C_f} \langle \mathbf{r}, \mathbf{x} \rangle \geq f(\mathbf{r})$.

Case 2: $f(\mathbf{r}) = +\infty$. It suffices to show that $\sigma_{C_f}(\mathbf{r}) \geq t$ for all $t \in \mathbb{R}$, implying that $\sigma_{C_f}(\mathbf{r}) = +\infty$. Fix an arbitrary $\bar{t} \in \mathbb{R}$. Since $f(\mathbf{r}) = +\infty$, the point $(\mathbf{r}, \bar{t}) \notin \text{epi}(f)$. Thus, there exists $(\mathbf{y}, \eta) \in \text{epi}(f)^\circ$ such that $\langle \mathbf{r}, \mathbf{y} \rangle + \eta \bar{t} > 0$. If $\eta < 0$, then the arguments above show that $\frac{\mathbf{y}}{-\eta} \in C_f$. However, the inequality $\langle \mathbf{r}, \mathbf{y} \rangle + \eta \bar{t} > 0$ after rearragement and diving by $-\eta$ gives $\langle \mathbf{r}, \frac{\mathbf{y}}{-\eta} \rangle > \bar{t}$. Thus, $\sup_{\mathbf{x} \in C_f} \langle \mathbf{r}, \mathbf{x} \rangle \geq \langle \mathbf{r}, \frac{\mathbf{y}}{-\eta} \rangle > \bar{t}$ and we are done. If $\eta = 0$, then from the arguments above, $\mathbf{y} \in \text{rec}(C_f)$. Since C_f is nonempty, consider any $\bar{\mathbf{x}} \in C_f$. Thus, $\bar{\mathbf{x}} + \lambda \mathbf{y} \in C_f$ for all $\lambda \geq 0$. However, $\langle \mathbf{r}, \mathbf{y} \rangle = \langle \mathbf{r}, \mathbf{y} \rangle + \eta \bar{t} > 0$ Thus, we can make $\langle \mathbf{r}, \bar{\mathbf{x}} + \lambda \mathbf{y} \rangle$ as large as we want by choosing λ large enough. In particular, we can make it larger than $\bar{t}$.

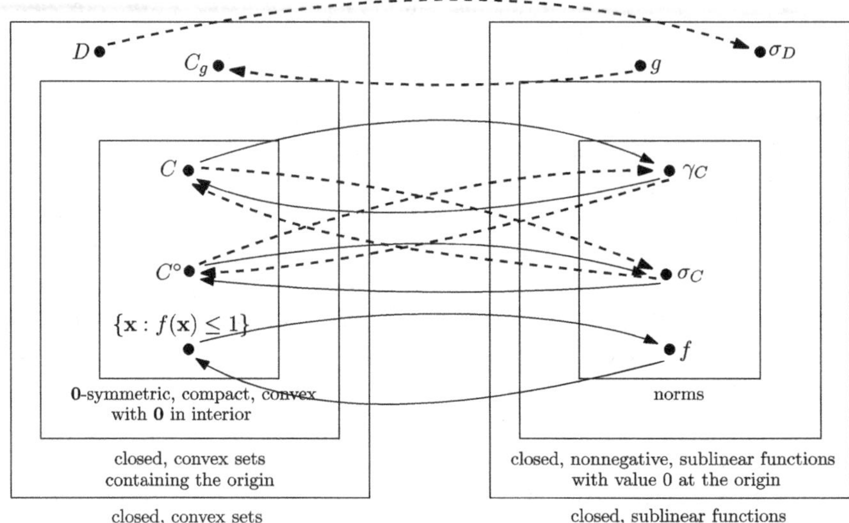

Figure 3.3 One-to-one correspondence between closed, convex sets and closed, sublinear functions, as an extension of Figure 3.2. Solid arrows have the same interpretation as before from Figure 3.2. Dashed arrows from left to right show the support function operation and dashed arrows from right to left show the operation from Theorem 3.3.24.

We now show that $C_{\sigma_C} = C$ for any closed, convex set C. Consider any $\mathbf{x} \in C$. Then $\langle \mathbf{r}, \mathbf{x} \rangle \leq \sup_{\mathbf{y} \in C} \langle \mathbf{r}, \mathbf{y} \rangle = \sigma_C(\mathbf{r})$. Therefore, $\mathbf{x} \in H^{\leq}(\mathbf{r}, \sigma(\mathbf{r}))$ for all $\mathbf{r} \in \mathbb{R}^d$. This shows that $\mathbf{x} \in C_{\sigma_C}$, and therefore, $C \subseteq C_{\sigma_C}$. To show the reverse inclusion, consider any $\mathbf{y} \notin C$. Since C is a closed, convex set, there exists a separating hyperplane $H^{=}(\mathbf{a}, \delta)$ such that $C \subseteq H^{\leq}(\mathbf{a}, \delta)$ and $\langle \mathbf{a}, \mathbf{y} \rangle > \delta$. $C \subseteq H^{\leq}(\mathbf{a}, \delta)$ implies that $\sigma_C(\mathbf{a}) = \sup_{\mathbf{x} \in C} \langle \mathbf{a}, \mathbf{x} \rangle \leq \delta$. Since C_{σ_C} has $\langle \mathbf{a}, \mathbf{x} \rangle \leq \sigma_C(\mathbf{a})$ as a defining halfspace, and $\langle \mathbf{a}, \mathbf{y} \rangle > \delta \geq \sigma_C(\mathbf{a})$, we observe that $\mathbf{y} \notin C_{\sigma_C}$. $\qquad\square$

Thanks to Theorem 3.3.24, we can now expand Figure 3.2 to include all closed, sublinear functions and all closed, convex sets to obtain Figure 3.3. The proof of Theorem 3.3.24 also gives another nice picture associated with the sublinear function f, which corresponds to the following proposition. See Figure 3.4.

Proposition 3.3.25 *Let* $f : \mathbb{R}^d \to \mathbb{R} \cup \{+\infty\}$ *be a closed, sublinear function that is not identically* $+\infty$ *everywhere, and let* C_f *be defined as in Theorem 3.3.24. Then* $\mathbf{y} \in C_f$ *if and only if* $(\mathbf{y}, -1) \in \mathrm{epi}(f)^\circ$. *In other words,* $C_f = \{\mathbf{y} \in \mathbb{R}^d : (\mathbf{y}, -1) \in \mathrm{epi}(f)^\circ\}$. *Moreover,* $\mathbf{y} \in \mathrm{rec}(C_f)$ *if and only if* $(\mathbf{y}, 0) \in \mathrm{epi}(f)^\circ$. *In other words,* $\mathrm{rec}(C_f) = \{\mathbf{y} \in \mathbb{R}^d : (\mathbf{y}, 0) \in \mathrm{epi}(f)^\circ\}$.

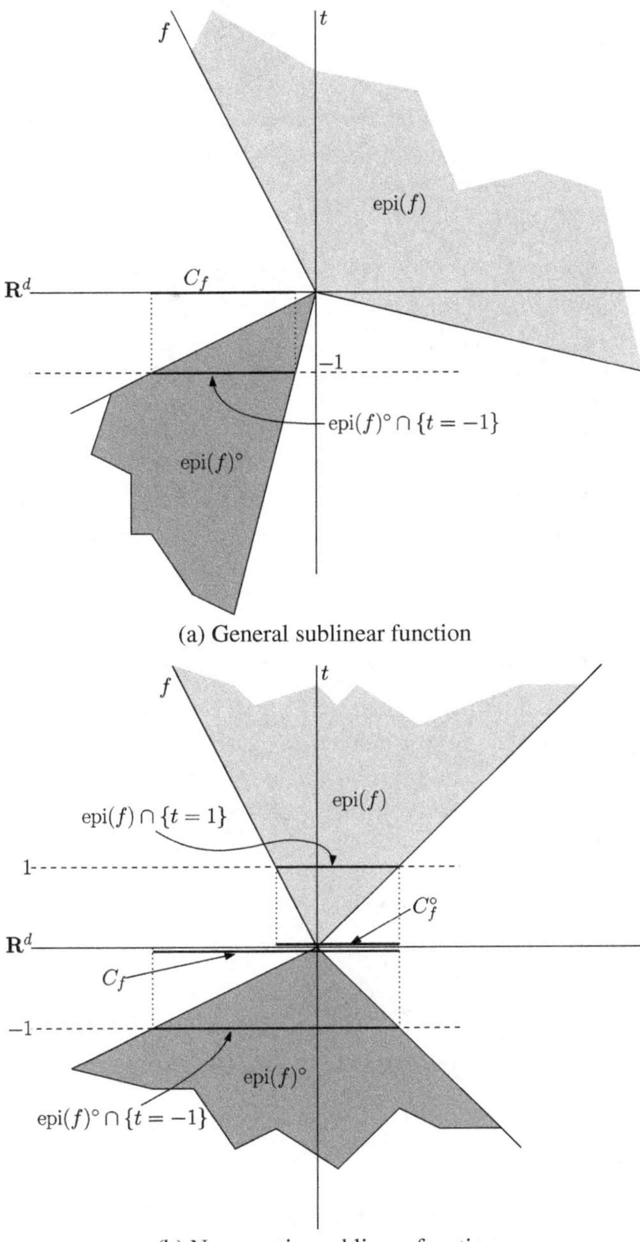

(a) General sublinear function

(b) Nonnegative sublinear function

Figure 3.4 Illustration of Propositions 3.3.25(a) and 3.3.26(b).

When f is a nonnegative sublinear function, even more can be said.

Proposition 3.3.26 *Let $f : \mathbb{R}^d \to \mathbb{R}$ be a sublinear function that is nonnegative everywhere, and let C_f be defined as in Theorem 3.3.24. Then $f = \gamma_{(C_f)^\circ}$, i.e., f is the gauge function for $(C_f)^\circ$. Consequently, $(C_f)^\circ = \{\mathbf{y} \in \mathbb{R}^d : (\mathbf{y}, 1) \in \text{epi}(f)\} = \{\mathbf{y} \in \mathbb{R}^d : f(\mathbf{y}) \le 1\}$.*

Proof   Since $f \ge 0$, $\text{epi}(f) \subseteq \{(\mathbf{r}, t) : t \ge 0\}$. Therefore, $(\mathbf{0}, -1) \in \text{epi}(f)^\circ$. By Proposition 3.3.25, $\mathbf{0} \in C_f$. Moreover, by Theorems 3.3.24 and 3.3.21, $f = \sigma_{C_f} = \gamma_{(C_f)^\circ}$. By Theorem 3.3.11 part 2., this shows that $(C_f)^\circ = \{\mathbf{y} \in \mathbb{R}^d : f(\mathbf{y}) \le 1\}$. By Problem 15 from the Section 3.1.1, we have that $(C_f)^\circ = \{\mathbf{y} \in \mathbb{R}^d : (\mathbf{y}, 1) \in \text{epi}(f)\} = \{\mathbf{y} \in \mathbb{R}^d : f(\mathbf{y}) \le 1\}$. $\square$

Figure 3.4 shows that for a nonnegative, sublinear function f, the polar of its epigraph gives the epigraph of the gauge of C_f, when the picture is inverted about the $\mathbb{R}^d$ plane. In other words, for any closed, convex set C, the polarity of C and C° can be seen as the polarity of the epigraphs of the gauge and support functions of C (or C°).

3.3.4 Representing Closed, Convex Sets as 1-Sublevel Sets

Corollary 3.3.13 showed us that given a closed, convex set C containing the origin, the gauge function is the *unique* nonnegative, sublinear function whose 1-sublevel set is C. If we allow sublinear functions that are not necessarily nonnegative, then there may exist many distinct sublinear functions whose 1-sublevel sets coincide.

Example 3.3.27   Consider the polyhedron

$$P = \{\mathbf{x} \in \mathbb{R}^2 : -\mathbf{x}_1 - \mathbf{x}_2 \le 1, \quad -\mathbf{x}_1 \le 1, \quad -\mathbf{x}_2 \le 1\}.$$

From Corollary 3.3.17, we obtain that

$$\gamma_P(\mathbf{r}) = \max\{0, -\mathbf{r}_1 - \mathbf{r}_2, -\mathbf{r}_1, -\mathbf{r}_2\},$$

and by Theorem 3.3.11 part 2., we obtain that $P = \{\mathbf{x} \in \mathbb{R}^2 : \gamma_P(\mathbf{x}) \le 1\}$. Now consider the function

$$f(\mathbf{r}) = \max\{-\mathbf{r}_1 - \mathbf{r}_2, -\mathbf{r}_1, -\mathbf{r}_2\}.$$

It turns out that $P = \{\mathbf{x} \in \mathbb{R}^2 : f(\mathbf{x}) \le 1\}$ because

$$\begin{aligned}
\mathbf{x} \in P &\Leftrightarrow -\mathbf{x}_1 - \mathbf{x}_2 \le 1, \quad -\mathbf{x}_1 \le 1, \quad -\mathbf{x}_2 \le 1 \\
&\Leftrightarrow \max\{-\mathbf{x}_1 - \mathbf{x}_2, -\mathbf{x}_1, -\mathbf{x}_2\} \le 1 \\
&\Leftrightarrow f(\mathbf{x}) \le 1.
\end{aligned}$$

Notice that $f((1,1)) = -1 \neq 0 = \gamma_P((1,1))$. Also, f is sublinear because f is the support function of the set $S = \{(-1, -1), (-1, 0), (0, -1,)\}$. See Figure 3.5.

We make a definition that will help in the following discussion.

Definition 3.3.28 Given a closed, convex set $C \subseteq \mathbb{R}^d$ with $\mathbf{0} \in C$, we say that a sublinear function $f : \mathbb{R}^d \to \mathbb{R} \cup \{+\infty\}$ is a *representation of* C if $C = \{\mathbf{x} \in \mathbb{R}^d : f(\mathbf{x}) \leq 1\}$. $\mathcal{R}_C$ will denote the set of all sublinear functions that represent C.

Example 3.3.27 shows that Corollary 3.3.13 really breaks down if the assumption of nonnegativity is removed. Even so, given a closed, convex set C, any sublinear function that represents C must match the gauge on $\mathbb{R}^d \setminus \text{int}(\text{rec}(C))$ (see Exercise 19 from Section 3.3.5). Thus, for compact, convex sets, we have a unique sublinear representation. Exercise 19 from Section 3.3.5 also shows that the function must be nonpositive at all points in $\text{rec}(C)$. Since Corollary 3.3.13 shows that for a given set C the gauge is the unique nonnegative function that represents C, this implies that the gauge is the point-wise *largest* sublinear function among all sublinear functions that represent C. It turns out there is a *smallest* such function as well and it can be described explicitly.

Theorem 3.3.29 *Let $C \subseteq \mathbb{R}^d$ be a closed, convex set $\mathbf{0} \in \text{int}(C)$. There exists a sublinear function $\rho_C \in \mathcal{R}_C$ such that $\rho_C \leq f$ everywhere for all $f \in \mathcal{R}_C$. In other words, $\rho_C \leq f \leq \gamma_C$ for all $f \in \mathcal{R}_C$. Moreover, $\rho_C = \sigma_{\widehat{C}}$ where*

$$\widehat{C} := \{\mathbf{y} \in C^\circ : \exists \mathbf{x} \in C \text{ such that } \langle \mathbf{y}, \mathbf{x} \rangle = 1\}. \tag{3.3.3}$$

Proof We first show that $\sigma_{\widehat{C}} \in \mathcal{R}_C$, i.e., $C = \{\mathbf{x} \in \mathbb{R}^d : \sigma_{\widehat{C}}(\mathbf{x}) \leq 1\}$. Since $\widehat{C} \subseteq C^\circ$, we have that $\langle \mathbf{y}, \mathbf{x} \rangle \leq 1$ for all $\mathbf{x} \in C$ and $\mathbf{y} \in \widehat{C}$. Therefore, $\sigma_{\widehat{C}}(\mathbf{x}) \leq 1$ for all $\mathbf{x} \in C$. Now, consider any $\mathbf{z} \notin C$. Let $\mathbf{a} = \mathbf{z} - \text{Proj}_C(\mathbf{z})$ and $\delta = \langle \mathbf{a}, \text{Proj}_C(\mathbf{z}) \rangle$. By Exercise 2 from Section 2.4.4, $\langle \mathbf{a}, \mathbf{x} \rangle = \delta$ is a supporting hyperplane for C. Since $\mathbf{0} \in \text{int}(C)$, we must have $0 = \langle \mathbf{a}, \mathbf{0} \rangle < \delta$. Define $\bar{\mathbf{y}} := \frac{\mathbf{a}}{\delta}$. Then, $\langle \bar{\mathbf{y}}, \mathbf{x} \rangle \leq 1$ for all $\mathbf{x} \in C$, i.e., $\bar{\mathbf{y}} \in C^\circ$ and $\langle \bar{\mathbf{y}}, \text{Proj}_C(\mathbf{z}) \rangle = 1$. Thus, $\bar{\mathbf{y}} \in \widehat{C}$. However, $\langle \bar{\mathbf{y}}, \mathbf{z} \rangle = \frac{1}{\delta} \langle \mathbf{a}, \mathbf{z} \rangle > 1$. Thus, $\sigma_{\widehat{C}}(\mathbf{z}) > 1$ and we have that $C = \{\mathbf{x} \in \mathbb{R}^d : \sigma_{\widehat{C}}(\mathbf{x}) \leq 1\}$.

Next, we show that for any $f \in \mathcal{R}_C$, we must have $\sigma_{\widehat{C}}(\mathbf{r}) \leq f(\mathbf{r})$. We first note that f is finite valued everywhere since $\mathbf{0} \in \text{int}(C)$ (Exercise 18 from Section 3.3.5). Consider any arbitrary $\bar{\mathbf{r}} \in \mathbb{R}^d$. We consider two cases based on the sign of $f(\bar{\mathbf{r}})$.

Case 1: $f(\bar{\mathbf{r}}) > 0$. Let $\bar{\mathbf{x}} = \frac{\bar{\mathbf{r}}}{f(\bar{\mathbf{r}})}$. Since $f(\bar{\mathbf{x}}) = 1$, we have $\bar{\mathbf{x}} \in C$ and therefore, $\langle \mathbf{y}, \bar{\mathbf{x}} \rangle \leq 1$ for all $\mathbf{y} \in \widehat{C} \subseteq C^{\circ}$. Therefore, $\frac{\sigma_{\widehat{C}}(\bar{\mathbf{r}})}{f(\bar{\mathbf{r}})} = \sigma_{\widehat{C}}(\bar{\mathbf{x}}) = \sup_{\mathbf{y} \in \widehat{C}} \langle \mathbf{y}, \bar{\mathbf{x}} \rangle \leq 1 = f(\bar{\mathbf{x}}) = \frac{f(\bar{\mathbf{r}})}{f(\bar{\mathbf{r}})}$, showing that $\sigma_{\widehat{C}}(\bar{\mathbf{r}}) \leq f(\bar{\mathbf{r}})$.

Case 2: $f(\bar{\mathbf{r}}) \leq 0$. By positive homogeneity of f, $t\bar{\mathbf{r}} \in C$ for all $t > 0$ since $f(t\bar{\mathbf{r}}) = tf(\bar{\mathbf{r}}) \leq 0 \leq 1$. Thus, $\bar{\mathbf{r}} \in \mathrm{rec}(C)$. Consider an arbitrary $\bar{\mathbf{y}} \in \widehat{C}$. It suffices to show that $\langle \bar{\mathbf{y}}, \bar{\mathbf{r}} \rangle \leq f(\bar{\mathbf{r}})$ since by taking a supremum over $\widehat{C}$ on the left-hand side, we would obtain $\sigma_{\widehat{C}}(\bar{\mathbf{r}}) \leq f(\bar{\mathbf{r}})$.

Since $\bar{\mathbf{y}} \in \widehat{C}$, there exists $\bar{\mathbf{x}} \in C$ such that $\langle \bar{\mathbf{y}}, \bar{\mathbf{x}} \rangle = 1$. This implies that $\bar{\mathbf{x}} \in \mathrm{bd}(C)$. Theorem 3.3.11, part 4., combined with Exercise 19 from Section 3.3.5, implies $f(\bar{\mathbf{x}}) = 1$. Since f is finite valued everywhere, Theorem 3.2.3 shows that f is continuous. Therefore, there exists $t > 0$ small enough such that $f(\bar{\mathbf{x}} + t\bar{\mathbf{r}}) > 0$. By applying Case 1., $\sigma_{\widehat{C}}(\bar{\mathbf{x}} + t\bar{\mathbf{r}}) \leq f(\bar{\mathbf{x}} + t\bar{\mathbf{r}})$. Therefore, we obtain

$$
\begin{aligned}
1 + t\langle \bar{\mathbf{y}}, \bar{\mathbf{r}} \rangle &= \langle \bar{\mathbf{y}}, \bar{\mathbf{x}} + t\bar{\mathbf{r}} \rangle \\
&\leq \sup_{\mathbf{y} \in \widehat{C}} \langle \mathbf{y}, \bar{\mathbf{x}} + t\bar{\mathbf{r}} \rangle \\
&= \sigma_{\widehat{C}}(\bar{\mathbf{x}} + t\bar{\mathbf{r}}) \\
&\leq f(\bar{\mathbf{x}} + t\bar{\mathbf{r}}) \\
&\leq f(\bar{\mathbf{x}}) + tf(\bar{\mathbf{r}}) \\
&= 1 + tf(\bar{\mathbf{r}}),
\end{aligned}
$$

where the last inequality follows from sublinearity of f. This gives us $\langle \bar{\mathbf{y}}, \bar{\mathbf{r}} \rangle \leq f(\bar{\mathbf{r}})$ as desired. $\qquad\square$

The following is an immediate consequence of Theorem 3.3.29 and strengthens it by completely characterizing which sublinear functions can represent a closed, convex set with the origin in its interior.

Corollary 3.3.30 *Let $C \subseteq \mathbb{R}^d$ be a closed, convex set $\mathbf{0} \in \mathrm{int}(C)$ and let ρ_C be the smallest representation of C as guaranteed by Theorem 3.3.29. Let $f : \mathbb{R}^d \to \mathbb{R} \cup \{+\infty\}$ be a sublinear function. Then f is a representation of C if and only if $\rho_C \leq f \leq \gamma_C$.*

Proof The necessity is the content of Theorem 3.3.29. Now consider any sublinear function f satisfying $\rho_C \leq f \leq \gamma_C$. Then, $C = \{\mathbf{x} \in \mathbb{R}^d : \gamma_C(\mathbf{x}) \leq 1\} \subseteq \{\mathbf{x} \in \mathbb{R}^d : f(\mathbf{x}) \leq 1\} \subseteq \{\mathbf{x} \in \mathbb{R}^d : \rho_C(\mathbf{x}) \leq 1\} = C$, where the first equality is from part 2. of Theorem 3.3.11 and the last equality is from Theorem 3.3.29. This shows that f is a representation of C. $\qquad\square$

Let us end our discussion with a couple of different perspectives on Theorem 3.3.29. From Theorem 3.3.24, every sublinear function $f \in \mathcal{R}_C$ is the support function of a closed, convex set. This prompts the following definition.

Definition 3.3.31 Given a closed, convex set $C \subseteq \mathbb{R}^d$, we say that a set $X \subseteq \mathbb{R}^d$ is a *prepolar of* C if C is the 1-sublevel set of σ_X, i.e., σ_X represents C.

By Proposition 3.3.19, $\rho_C = \sigma_{\widehat{C}} = \sigma_{\text{cl}(\text{conv}(\widehat{C}))}$, where $\widehat{C}$ is as defined in (3.3.3). Thus, a geometric reformulation of Theorem 3.3.29 and Corollary 3.3.30 is the following.

Theorem 3.3.32 *Let* $C \subseteq \mathbb{R}^d$ *be a closed, convex set with* $\mathbf{0} \in \text{int}(C)$. *Define* $C^{\bullet} := \text{cl}(\text{conv}(\widehat{C}))$ *where* $\widehat{C}$ *is as defined in* (3.3.3). *Then a set* $X \subseteq \mathbb{R}^d$ *is a prepolar of* C *if and only if* $C^{\bullet} \subseteq \text{cl}(\text{conv}(X)) \subseteq C^{\circ}$.

Example 3.3.33 (Example 3.3.27 continued) Consider the polyhedron P defined in Example 3.3.27. The polar P° is the square $\text{conv}\{(-1, -1), (-1, 0), (0, 0), (0, -1)\}$. Note that $\widehat{P}$ as defined in (3.3.3) simply consists of the two line segments $\text{conv}(\{(-1, -1), (-1, 0)\})$ and $\text{conv}(\{(-1, -1), (0, -1)\})$. The smallest closed, convex prepolar $P^{\bullet}$ is the triangle formed by the convex hull of these three points. See Figure 3.5.

We next observe that X is a prepolar of C if and only if $C = \{\mathbf{x} \in \mathbb{R}^d : \langle \mathbf{y}, \mathbf{x} \rangle \leq 1 \ \forall \mathbf{y} \in X\}$, i.e., $C = X^{\circ}$ (Exercise 21 from Section 3.3.5). In fact, this is the motivation behind the use of the term "prepolar." In other words, C can be described as the intersection of all halfspaces given by normals in a prepolar and shifts equal to 1. The fact $C = (C^{\circ})^{\circ}$ for every closed, convex set C can be restated as, "Every closed, convex set is the intersection of all halfspaces

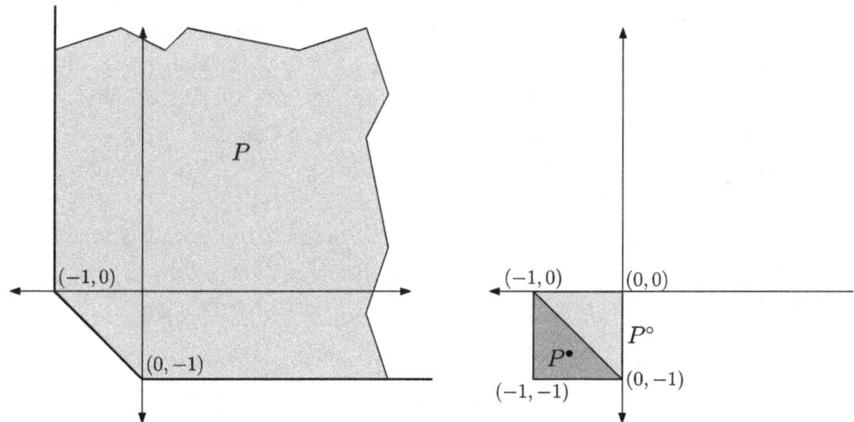

Figure 3.5 The polyhedron P from Example 3.3.27, and its polar P° and the smallest convex prepolar $P^{\bullet}$. The jagged lines for P are meant to show that P is unbounded. P° is the square $\text{conv}\{(-1, -1), (-1, 0), (0, 0), (0, -1)\}$ and $P^{\bullet}$ is the triangle shaded in dark gray.

that contain it" (Corollary 2.4.3). Thus, roughly speaking, the search for the "smallest prepolar" is equivalently a search for a "smallest" set of halfspaces whose intersection gives C (see also Section 2.4.3). One can view $\widehat{C}$ as the normals of supporting hyperplanes of C (with shift equal to 1). The fact that C is the 1-sublevel set of the support function of $\widehat{C}$ can be stated equivalently as follows.

Theorem 3.3.34 *Any closed, convex set is equal to the intersection of all its supporting halfspaces.*

3.3.5 Exercises

1. Prove Proposition 3.3.2.
2. Show that a closed, sublinear function f that is not identically $+\infty$ everywhere must have $f(\mathbf{0}) = 0$.
3. Show that the gauge of a closed, convex set containing the origin is closed, i.e., the epigraph is a closed, convex cone.
4. Let C be a closed, convex set with $\mathbf{0} \in C$. Show that γ_C is finite valued everywhere if and only if $\mathbf{0} \in \operatorname{int}(C)$.
5. Let $A, B \subseteq \mathbb{R}^d$ be closed, convex sets containing the origin. Show that $A \subseteq B$ if and only if $\gamma_A \geq \gamma_B$.
6. Let $A \subseteq \mathbb{R}^d$ and $\lambda > 0$. Show that $\gamma_{\lambda A} = \frac{1}{\lambda}\gamma_A$.
7. Prove the assertion in Example 3.3.15.
8. Let $f \colon \mathbb{R}^d \to \mathbb{R}$ be a sublinear function that satisfies the subadditivity relation at equality, i.e., $f(\mathbf{x} + \mathbf{y}) = f(\mathbf{x}) + f(\mathbf{y})$ for all $\mathbf{x}, \mathbf{y} \in \mathbb{R}^d$. Show that f is a linear function, i.e., $f(\mathbf{x}+\mathbf{y}) = f(\mathbf{x}) + f(\mathbf{y})$ for all $\mathbf{x}, \mathbf{y} \in \mathbb{R}^d$ and $f(t\mathbf{x}) = tf(\mathbf{x})$ for all $t \in \mathbb{R}$ and $\mathbf{x} \in \mathbb{R}^d$. (Note that positive homogeneity of f gives the second condition only for $t > 0$.)
9. Let $\sigma \colon \mathbb{R}^d \to \mathbb{R}$ be a sublinear function. Show that if $\sigma(\mathbf{x}) + \sigma(-\mathbf{x}) = 0$ for some $\mathbf{x} \in \mathbb{R}^d \setminus \{\mathbf{0}\}$, then $\sigma(\mathbf{x} + \mathbf{y}) = \sigma(\mathbf{x}) + \sigma(\mathbf{y})$ for all $\mathbf{y} \in \mathbb{R}^d$. Conclude that a sublinear function σ is actually *linear* if and only if $\sigma(\mathbf{x}) + \sigma(-\mathbf{x}) = 0$ for *all* $\mathbf{x} \in \mathbb{R}^d$ (see Exercise 8 above).
10. (Operations that preserve sublinearity) Prove the following.

    (i) Let $\sigma_1, \sigma_2 \colon \mathbb{R}^d \to \mathbb{R}$ be two sublinear functions. Show that $t_1\sigma_1 + t_2\sigma_2$ is a sublinear function for all $t_1, t_2 \geq 0$.

    (ii) Let $\sigma_i, i \in I$ be some family of sublinear functions. Show that $\sigma := \sup_{i \in I} \sigma_i$ is also sublinear.

11. Show that if C is a $\mathbf{0}$-symmetric, compact, convex set containing the origin in its interior, so is C°.

12. Prove Proposition 3.3.19.
13. Let C_1, C_2 be closed, convex sets. Then $\sigma_{C_1} = \sigma_{C_2}$ if and only if $C_1 = C_2$.
14. Let C be a closed, convex set. Show that σ_C is nonnegative everywhere if and only if $\mathbf{0} \in C$.
15. **Calculus of support functions.** Show the following.

   (i) Let $A, B \subseteq \mathbb{R}^d$ be closed, convex sets. Show that $A \subseteq B$ if and only if $\sigma_A \leq \sigma_B$.

   (ii) Let $A, B \subseteq \mathbb{R}^d$ be closed, convex sets, and $\lambda_1, \lambda_2 \geq 0$. Then
   $$\sigma_{\lambda_1 A + \lambda_2 B} = \lambda_1 \sigma_A + \lambda_2 \sigma_B.$$

   (iii) Let $C_i, i \in I$ be a family of closed, convex sets, and let $C = \mathrm{cl}(\mathrm{conv}(\cup_{i \in I} C_i))$. Then $\sigma_C = \sup_{i \in I} \sigma_{C_i}$.

   (iv) Let $T \colon \mathbb{R}^d \to \mathbb{R}^m$ be a linear transformation, and let $T^* \colon \mathbb{R}^m \to \mathbb{R}^d$ be its adjoint transformation; i.e., for all $\mathbf{x} \in \mathbb{R}^d$ and $\mathbf{y} \in \mathbb{R}^m$, we have $\langle \mathbf{y}, T\mathbf{x} \rangle = \langle T^*\mathbf{y}, \mathbf{x} \rangle$ (in matrix language, T^* is represented by the transpose of the matrix representing T). Show that for any set S, $\sigma_{T(S)}(\mathbf{r}) = \sigma_S(T^*\mathbf{r})$ for all $\mathbf{r} \in \mathbb{R}^m$.

16. Let P be a polyhedron and $\mathbf{v}^1, \ldots, \mathbf{v}^k$ be its extreme points. Show that

$$\sigma_P(\mathbf{r}) = \begin{cases} \max\{\langle \mathbf{v}^1, \mathbf{r} \rangle, \ldots, \langle \mathbf{v}^k, \mathbf{r} \rangle\} & \text{if } \mathbf{r} \in \mathrm{rec}(P)^\circ, \\ +\infty & \text{otherwise.} \end{cases}$$

17. Let $C \subseteq \mathbb{R}^d$ be a nonempty, closed, convex set. Show that $\{\mathbf{r} \in \mathbb{R}^d : \sigma_C(\mathbf{r}) < \infty\} \subseteq \mathrm{rec}(C)^\circ$. Thus, those linear objectives $\mathbf{r} \in \mathbb{R}^d$ for which the convex optimization problem $\sup_{\mathbf{x} \in C} \langle \mathbf{r}, \mathbf{x} \rangle$ is finite are in the polar of the recession cone of C. Construct an example where the containment is strict.
18. Show that any sublinear function that represents a closed, convex set containing the origin in its interior is finite valued everywhere.
19. Show that if $f \colon \mathbb{R}^d \to \mathbb{R}$ is a sublinear function that represents a closed convex set C containing the origin in its interior, then

$$f(\mathbf{r}) = \gamma_C(\mathbf{r}) \quad \forall \mathbf{r} \notin \mathrm{int}(\mathrm{rec}(C)).$$

Show also that $f(\mathbf{r}) \leq 0$ for all $\mathbf{r} \in \mathrm{rec}(C)$.
20. Show that in Theorem 3.3.29, one cannot relax the condition $\mathbf{0} \in \mathrm{int}(C)$ to $\mathbf{0} \in C$.
21. Let $C \subseteq \mathbb{R}^d$ be a closed, convex set containing the origin. Show that X is a prepolar of C if and only if $C = \{\mathbf{x} \in \mathbb{R}^d : \langle \mathbf{y}, \mathbf{x} \rangle \leq 1 \ \forall \mathbf{y} \in X\}$, i.e., $C = X^\circ$.

3.4 Directional Derivatives, Subgradients, and Subdifferential Calculus

It follows from the chain rule in calculus that if a function f is differentiable at a point $\mathbf{x}$, then the directional derivative (see Definition 1.3.16 and (1.3.1)) is a linear function of the direction given by the inner product with the gradient, i.e., $f'(\mathbf{x};\mathbf{r}) = \langle \nabla f(\mathbf{x}),\mathbf{r} \rangle$ for all $\mathbf{r} \in \mathbb{R}^d$. We used this to analyze gradients of differentiable convex functions in Theorem 3.2.8. It turns out that even if a convex function is not differentiable, its directional derivatives have a special structure; in particular, $f'(\mathbf{x};\mathbf{r})$ is a sublinear function, generalizing the fact that it is linear for differentiable (convex) functions. After establishing this fact in Proposition 3.4.2, we will use the tools of sublinearity developed in Section 3.3 to further analyze subgradients of convex functions (Definition 3.1.15) and their relationships with directional derivatives.

Lemma 3.4.1 *If $f\colon \mathbb{R}^d \to \mathbb{R}$ is convex, for any $\mathbf{x},\mathbf{r} \in \mathbb{R}^d$, the expression $\frac{f(\mathbf{x}+t\mathbf{r})-f(\mathbf{x})}{t}$ is a nondecreasing function of t for $t \neq 0$.*

Proof By Proposition 3.2.7, the function $\varphi(t) = f(\mathbf{x} + t\mathbf{r})$ is a convex function. By Proposition 3.2.5, we observe that $\frac{\varphi(t) - \varphi(0)}{t}$ is a nondecreasing function of t. $\qquad\square$

Proposition 3.4.2 *Let $f\colon \mathbb{R}^d \to \mathbb{R}$ be a convex function, and let $\mathbf{x} \in \mathbb{R}^d$. Then the limit in (1.3.1) exists and is finite for all $\mathbf{r} \in \mathbb{R}^d$, and the function $f'(\mathbf{x};\cdot)\colon \mathbb{R}^d \to \mathbb{R}$ is sublinear.*

Proof By Proposition 3.2.7, the function $\varphi(t) = f(\mathbf{x} + t\mathbf{r})$ is a convex function, and $f'(\mathbf{x};\mathbf{r}) = \lim_{t\downarrow 0} \frac{\varphi(t) - \varphi(0)}{t}$. By Lemma 3.4.1, we observe that $\frac{\varphi(t) - \varphi(0)}{t}$ is a nondecreasing function of t for $t \neq 0$, and restricting to $t > 0$, $\frac{\varphi(t) - \varphi(0)}{t}$ is lower bounded by the value at $t = -1$, i.e., $\frac{\varphi(-1) - \varphi(0)}{-1}$. Therefore, $\lim_{t\downarrow 0} \frac{\varphi(t) - \varphi(0)}{t}$ exists and is in fact equal to $\inf_{t>0} \frac{\varphi(t)-\varphi(0)}{t}$.

We now prove positive homogeneity of $f'(\mathbf{x};\cdot)$. For any $\mathbf{r} \in \mathbb{R}^d$ and $\lambda > 0$, we obtain that

$$
\begin{aligned}
f'(\mathbf{x};\lambda\mathbf{r}) &= \lim_{t\downarrow 0} \frac{f(\mathbf{x}+t\lambda\mathbf{r}) - f(\mathbf{x})}{t} \\
&= \lim_{t\downarrow 0} \lambda \frac{f(\mathbf{x}+t\lambda\mathbf{r}) - f(\mathbf{x})}{\lambda t} \\
&= \lambda \lim_{t\downarrow 0} \frac{f(\mathbf{x}+t\lambda\mathbf{r}) - f(\mathbf{x})}{\lambda t} \\
&= \lambda \lim_{t'\downarrow 0} \frac{f(\mathbf{x}+t'\mathbf{r}) - f(\mathbf{x})}{t'} \\
&= \lambda f'(\mathbf{x};\mathbf{r}).
\end{aligned}
$$

We next establish that $f'(\mathbf{x};\cdot)$ is subadditive. Consider any $\mathbf{r}^1,\mathbf{r}^2 \in \mathbb{R}^d$ and $\lambda \in (0,1)$.

$$
\begin{aligned}
f'(\mathbf{x}; \mathbf{r}^1 + \mathbf{r}^2) &= \lim_{t \downarrow 0} \frac{f(\mathbf{x} + t(\mathbf{r}^1 + \mathbf{r}^2)) - f(\mathbf{x})}{t} \\
&= \lim_{t \downarrow 0} \frac{2f\left(\mathbf{x} + 2t\left(\frac{\mathbf{r}^1 + \mathbf{r}^2}{2}\right)\right) - 2f(\mathbf{x})}{2t} \\
&= \lim_{\bar{t} \downarrow 0} \frac{2f\left(\mathbf{x} + \bar{t}\left(\frac{\mathbf{r}^1 + \mathbf{r}^2}{2}\right)\right) - 2f(\mathbf{x})}{\bar{t}} \\
&= \lim_{\bar{t} \downarrow 0} \frac{2f\left(\frac{\mathbf{x} + \bar{t}\mathbf{r}^1}{2} + \frac{\mathbf{x} + \bar{t}\mathbf{r}^2}{2}\right) - f(\mathbf{x}) - f(\mathbf{x})}{\bar{t}} \\
&\le \lim_{\bar{t} \downarrow 0} \frac{2\left(\frac{1}{2}f(\mathbf{x} + \bar{t}\mathbf{r}^1) + \frac{1}{2}f(\mathbf{x} + \bar{t}\mathbf{r}^2)\right) - f(\mathbf{x}) - f(\mathbf{x})}{\bar{t}} \\
&= \lim_{\bar{t} \downarrow 0} \frac{f(\mathbf{x} + \bar{t}\mathbf{r}^1) - f(\mathbf{x})}{\bar{t}} + \lim_{\bar{t} \downarrow 0} \frac{f(\mathbf{x} + \bar{t}\mathbf{r}^2) - f(\mathbf{x})}{\bar{t}} \\
&= f'\left(\mathbf{x}; \mathbf{r}^1\right) + f'\left(\mathbf{x}; \mathbf{r}^2\right),
\end{aligned}
$$

where the third equality follows from substituting $\bar{t}$ for $2t$, and the inequality follows from convexity of f. $\qquad\square$

We now establish the fundamental connection between directional derivatives of a convex function and its subgradients and subdifferentials (recall Definition 3.1.15), using the construction of the closed, convex set C_f from a sublinear function f in Theorem 3.3.24. See Figure 3.6.

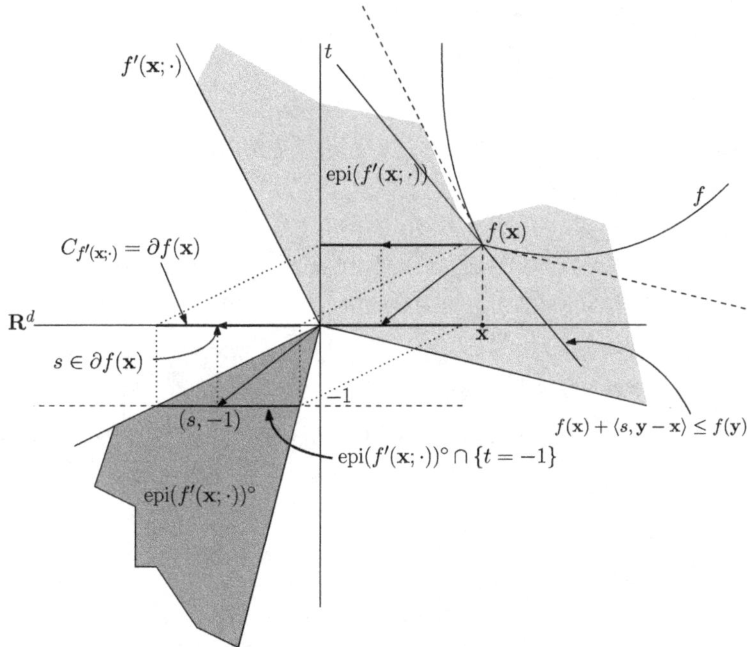

Figure 3.6 A picture illustrating the relationship between the sublinear function $f'(\mathbf{x}; \cdot)$, the set $C_{f'(\mathbf{x}; \cdot)}$, the subgradient $\partial f(\mathbf{x})$, and a supporting hyperplane to epi(f) given by an element $s \in \partial f(\mathbf{x})$. Recall the relationships from Figure 3.4.

Theorem 3.4.3 *Let $f: \mathbb{R}^d \to \mathbb{R}$ be a convex function, and let $\mathbf{x} \in \mathbb{R}^d$. Then*

$$\partial f(\mathbf{x}) = C_{f'(\mathbf{x};\cdot)}.$$

In other words, $f'(\mathbf{x};\cdot)$ is the support function for the subdifferential $\partial f(\mathbf{x})$.

Proof Recall from Definition 3.1.15 that

$$\begin{aligned}
\partial f(\mathbf{x}) &= \{\mathbf{s} \in \mathbb{R}^d : \langle \mathbf{s}, \mathbf{y} - \mathbf{x} \rangle \leq f(\mathbf{y}) - f(\mathbf{x}) \ \forall \mathbf{y} \in \mathbb{R}^d\} \\
&= \{\mathbf{s} \in \mathbb{R}^d : \langle \mathbf{s}, \mathbf{r} \rangle \leq f(\mathbf{x} + \mathbf{r}) - f(\mathbf{x}) \ \forall \mathbf{r} \in \mathbb{R}^d\}.
\end{aligned}$$

Thus, we have the following equivalences.

$$\begin{aligned}
\mathbf{s} \in \partial f(\mathbf{x}) &\Leftrightarrow \langle \mathbf{s}, \mathbf{r} \rangle \leq f(\mathbf{x} + \mathbf{r}) - f(\mathbf{x}) && \forall \mathbf{r} \in \mathbb{R}^d \\
&\Leftrightarrow \langle \mathbf{s}, t\mathbf{r} \rangle \leq f(\mathbf{x} + t\mathbf{r}) - f(\mathbf{x}) && \forall \mathbf{r} \in \mathbb{R}^d, t > 0 \\
&\Leftrightarrow \langle \mathbf{s}, \mathbf{r} \rangle \leq \frac{f(\mathbf{x} + t\mathbf{r}) - f(\mathbf{x})}{t} && \forall \mathbf{r} \in \mathbb{R}^d, t > 0 \\
&\Leftrightarrow \langle \mathbf{s}, \mathbf{r} \rangle \leq f'(\mathbf{x}; \mathbf{r}) && \forall \mathbf{r} \in \mathbb{R}^d \\
&\Leftrightarrow \mathbf{s} \in C_{f'(\mathbf{x};\mathbf{r})} && \forall \mathbf{r} \in \mathbb{R}^d,
\end{aligned}$$

where the second-to-last equivalence follows the fact that $\frac{f(\mathbf{x} + t\mathbf{r}) - f(\mathbf{x})}{t}$ is a nondecreasing function of t by Lemma 3.4.1, and the last equivalence follows from the definition of $C_{f'(\mathbf{x};\mathbf{r})}$ in (3.3.1). $\qquad\square$

A characterization of differentiability for convex functions can be obtained using these concepts.

Theorem 3.4.4 *Let $f: \mathbb{R}^d \to \mathbb{R}$ be a convex function, and let $\mathbf{x} \in \mathbb{R}^d$. Then the following are equivalent.*

(i) f is differentiable at $\mathbf{x}$.
(ii) $f'(\mathbf{x};\cdot)$ is a linear function given by $f'(\mathbf{x}, \mathbf{r}) = \langle \mathbf{a_x}, \mathbf{r} \rangle$ for some $\mathbf{a_x} \in \mathbb{R}^d$.
(iii) $\partial f(\mathbf{x})$ is a singleton, i.e., there is a unique subgradient for f at $\mathbf{x}$.

Moreover, if any of the above conditions hold then $\nabla f(\mathbf{x}) = \mathbf{a_x} = \mathbf{s}$, where $\mathbf{s}$ is the unique subgradient in $\partial f(\mathbf{x})$.

Proof *(i)* $\implies$ *(ii)*. If f is differentiable, then the chain rule from calculus says that $f'(\mathbf{x}; \mathbf{r}) = \langle \nabla f(\mathbf{x}), \mathbf{r} \rangle$; thus, setting $\mathbf{a_x} = \nabla f(\mathbf{x})$ suffices.

(ii) $\implies$ *(iii)*. By Theorem 3.4.3 and (3.3.1), we obtain that

$$\begin{aligned}
\partial f(\mathbf{x}) &= C_{f'(\mathbf{x};\cdot)} \\
&= \{\mathbf{s} \in \mathbb{R}^d : \langle \mathbf{s}, \mathbf{r} \rangle \leq f'(\mathbf{x}; \mathbf{r}) \ \forall \mathbf{r} \in \mathbb{R}^d\} \\
&= \{\mathbf{s} \in \mathbb{R}^d : \langle \mathbf{s}, \mathbf{r} \rangle \leq \langle \mathbf{a_x}, \mathbf{r} \rangle \ \forall \mathbf{r} \in \mathbb{R}^d\}.
\end{aligned}$$

We now observe that if $\langle \mathbf{s}, \mathbf{r} \rangle \leq \langle \mathbf{a_x}, \mathbf{r} \rangle$ for all $\mathbf{r} \in \mathbb{R}^d$, i.e., $\langle \mathbf{s} - \mathbf{a_x}, \mathbf{r} \rangle \leq 0$ for all $\mathbf{r} \in \mathbb{R}^d$ then we must have $\mathbf{s} = \mathbf{a_x}$. Therefore, $\partial f(\mathbf{x}) = \{\mathbf{a_x}\}$.

(iii) $\implies$ (i). Let $\mathbf{s}$ be the unique subgradient at $\mathbf{x}$. We will establish that

$$\lim_{\mathbf{h}\to\mathbf{0}} \frac{|f(\mathbf{x}+\mathbf{h}) - f(\mathbf{x}) - \langle \mathbf{s}, \mathbf{h}\rangle|}{\|\mathbf{h}\|} = 0,$$

thus showing that f is differentiable at $\mathbf{x}$ with gradient $\mathbf{s}$. In other words, given any $\delta > 0$, we must find $\epsilon > 0$ such that $\mathbf{h} \in B(\mathbf{0}, \epsilon)$ implies that $\frac{|f(\mathbf{x}+\mathbf{h}) - f(\mathbf{x}) - \langle \mathbf{s}, \mathbf{h}\rangle|}{\|\mathbf{h}\|} < \delta$.

Suppose to the contrary that for some $\delta > 0$, for every $k \geq 1$ there exists $\mathbf{h}_k$ such that $\|\mathbf{h}_k\| =: t_k \leq \frac{1}{k}$ and $\frac{|f(\mathbf{x}+\mathbf{h}_k) - f(\mathbf{x}) - \langle \mathbf{s}, \mathbf{h}_k\rangle|}{t_k} \geq \delta$. Since $\frac{\mathbf{h}_k}{t_k}$ is a sequence of unit norm vectors, by Theorem 1.3.10, there is a convergent subsequence which converges to $\mathbf{r}$ with unit norm. To keep the notation easy, we relabel indices so that $\left\{ \frac{\mathbf{h}_k}{t_k} \right\}_{k=1}^{\infty}$ is the convergent sequence. Using Theorem 3.2.3, there exists a constant $L := L(B(\mathbf{x}, 1))$ such that $|f(\mathbf{y}) - f(\mathbf{z})| \leq L\|\mathbf{y} - \mathbf{z}\|$ for all $\mathbf{y}, \mathbf{z} \in B(\mathbf{x}, 1)$. Noting that $\mathbf{x} + \mathbf{h}_k$ and $\mathbf{x} + t_k \mathbf{r}$ for all $k \geq 1$ are in the ball $B(\mathbf{x}, 1)$ (since $t_k \leq \frac{1}{k}$),

$$
\begin{aligned}
\delta &\leq \frac{|f(\mathbf{x}+\mathbf{h}_k) - f(\mathbf{x}) - \langle \mathbf{s}, \mathbf{h}_k\rangle|}{t_k} \\
&\leq \frac{|f(\mathbf{x}+\mathbf{h}_k) - f(\mathbf{x}+t_k\mathbf{r})| + |f(\mathbf{x}+t_k\mathbf{r}) - f(\mathbf{x}) - \langle \mathbf{s}, t_k\mathbf{r}\rangle| + |\langle \mathbf{s}, t_k\mathbf{r}\rangle - \langle \mathbf{s}, \mathbf{h}_k\rangle|}{t_k} \\
&\leq \frac{L\|t_k\mathbf{r}-\mathbf{h}_k\|}{t_k} + \frac{|f(\mathbf{x}+t_k\mathbf{r}) - f(\mathbf{x}) - \langle \mathbf{s}, t_k\mathbf{r}\rangle|}{t_k} + \frac{|\langle \mathbf{s}, t_k\mathbf{r}\rangle - \langle \mathbf{s}, \mathbf{h}_k\rangle|}{t_k} \\
&\leq L\|\mathbf{r} - \tfrac{\mathbf{h}_k}{t_k}\| + |\tfrac{f(\mathbf{x}+t_k\mathbf{r}) - f(\mathbf{x})}{t_k} - \langle \mathbf{s}, \mathbf{r}\rangle| + \|\mathbf{s}\|\|\mathbf{r} - \tfrac{\mathbf{h}_k}{t_k}\| \\
&= (L + \|\mathbf{s}\|)\|\mathbf{r} - \tfrac{\mathbf{h}_k}{t_k}\| + |\tfrac{f(\mathbf{x}+t_k\mathbf{r}) - f(\mathbf{x})}{t_k} - \langle \mathbf{s}, \mathbf{r}\rangle|,
\end{aligned}
$$

where the Cauchy–Schwarz inequality is used in the last inequality. We now let $k \to \infty$. The first term in the last expression above goes to 0, since $\frac{\mathbf{h}_k}{t_k}$ converges to $\mathbf{r}$. In the second term, $\frac{f(\mathbf{x}+t_k\mathbf{r}) - f(\mathbf{x})}{t_k}$ goes to its limit which is the directional derivative $f'(\mathbf{x}; \mathbf{r})$. By Theorem 3.4.3, $f'(\mathbf{x}; \mathbf{r}) = \sup_{\mathbf{y}\in\partial f(\mathbf{x})} \langle \mathbf{y}, \mathbf{r}\rangle = \langle \mathbf{s}, \mathbf{r}\rangle$, because by assumption $\partial f(\mathbf{x}) = \{\mathbf{s}\}$. Thus, the second term in the last expression above also goes to 0. This contradicts $\delta > 0$. $\qquad\square$

The following rules for manipulating subgradients and subdifferentials will be useful from an algorithmic perspective when we discuss optimization in part II of the book.

Theorem 3.4.5 Subdifferential calculus. *The following are all true.*

1. *Let $f_1, f_2 \colon \mathbb{R}^d \to \mathbb{R}$ be convex functions and let $t_1, t_2 \geq 0$. Then*

$$\partial(t_1 f_1 + t_2 f_2)(\mathbf{x}) = t_1 \partial f_1(\mathbf{x}) + t_2 \partial f_2(\mathbf{x}) \text{ for all } \mathbf{x} \in \mathbb{R}^d.$$

2. *Let $A \in \mathbb{R}^{m \times d}$ and $\mathbf{b} \in \mathbb{R}^m$ and let $T(\mathbf{x}) = A\mathbf{x} + \mathbf{b}$ be the corresponding affine map from $\mathbb{R}^d \to \mathbb{R}^m$ and let $g \colon \mathbb{R}^m \to \mathbb{R}$ be a convex function. Then*

$$\partial(g \circ T)(\mathbf{x}) = A^T \partial g(A\mathbf{x} + \mathbf{b}) \text{ for all } \mathbf{x} \in \mathbb{R}^d.$$

3. *Let $f_j \colon \mathbb{R}^d \to \mathbb{R}$, $j \in J$ be convex functions for some (possibly infinite) index set J, and let $f = \sup_{j \in J} f_j$. Then*

$$\mathrm{cl}(\mathrm{conv}(\cup_{j \in J(\mathbf{x})} \partial f_j(\mathbf{x}))) \subseteq \partial f(\mathbf{x}),$$

where $J(\mathbf{x})$ is the set of indices j such that $f_j(\mathbf{x}) = f(\mathbf{x})$.

3.4.1 Exercises

1.* Prove Theorem 3.4.5.
2. Prove Theorem 3.1.20 using Theorem 3.4.5.

3.5 Volume Relations

If one tries to visualize a convex set, the picture arises that the "central" portions of the set are "larger" than the extremities. A little more precisely, intuitively we imagine that if we intersect the convex set with parallel hyperplanes, then the intersections with the central part of the convex set should have larger $(d-1)$-dimensional volume (Section 2.2.1) than the intersections that are closer to the faces/extreme points of the convex set exposed by such hyperplanes. In other words, the volumes of these "slices" as a function of the right-hand side defining the hyperplanes should resemble a unimodal function. We will see a quantitative formalization of this intuition in this section, which goes by the name of the *Brunn–Minkowski theorem* and reveals a lot more structure than what this intuition suggests. Its applications in optimization will be explored in Chapter 6. The following notion makes the notion of "slices" of convex sets precise.

Definition 3.5.1   Given any convex set $C \subseteq \mathbb{R}^d$ and a linear subspace $L \subseteq \mathbb{R}^d$, a set of the form $C \cap (L + \mathbf{y})$, where $\mathbf{y} \in \mathbb{R}^d$ is called a *section of C parallel to L*.

The formalization begins with the following observation about sections of a convex set. Let $C \subseteq \mathbb{R}^d$ be convex and let $L \subseteq \mathbb{R}^d$ be a linear subspace. Then, for any $\mathbf{y}, \mathbf{y}^1, \mathbf{y}^2 \in \mathbb{R}^d$ such that $\mathbf{y} = \gamma \mathbf{y}^1 + (1 - \gamma)\mathbf{y}^2$, for some $\gamma \in [0, 1]$, we have

$$\gamma(C \cap (L + \mathbf{y}^1)) + (1 - \gamma)(C \cap (L + \mathbf{y}^2)) \subseteq C \cap (L + \mathbf{y})$$

(see Exercise 1 from Section 3.5.4). Thus, if we can find bounds on the volumes of Minkowski sums, we may be able to give bounds on the volume of the middle section. Such a bound is the content of the Brunn–Minkowski theorem.

3.5.1 The Brunn–Minkowski Theorem

Theorem 3.5.2 (Brunn–Minkowski inequality) *Let $X, Y \subseteq \mathbb{R}^d$ be compact sets (not necessarily convex). Then* $\operatorname{vol}(X + Y)^{1/d} \geq \operatorname{vol}(X)^{1/d} + \operatorname{vol}(Y)^{1/d}$.

We will first establish the special case when X and Y are axis-aligned "boxes."

Definition 3.5.3 An *axis-aligned box* in $\mathbb{R}^d$ is a set of the form $\{\mathbf{x} \in \mathbb{R}^d : a_i \leq \mathbf{x}_i \leq b_i \ \forall i = 1, \ldots, d\}$ for some real numbers $a_i \leq b_i, i = 1, \ldots, d$.

Lemma 3.5.4 *If $X, Y \subseteq \mathbb{R}^d$ are both axis-aligned boxes, then* $\operatorname{vol}(X+Y)^{1/d} \geq \operatorname{vol}(X)^{1/d} + \operatorname{vol}(Y)^{1/d}$.

Proof Let $X = \{\mathbf{x} \in \mathbb{R}^d : a_i \leq \mathbf{x}_i \leq b_i \ \forall i = 1, \ldots, d\}$ for some real numbers $a_i \leq b_i, i = 1, \ldots, d$ and $Y = \{\mathbf{x} \in \mathbb{R}^d : a_i' \leq \mathbf{x}_i \leq b_i' \ \forall i = 1, \ldots, d\}$. Observe that $X + Y = \{\mathbf{x} \in \mathbb{R}^d : a_i + a_i' \leq \mathbf{x}_i \leq b_i + b_i' \ \forall i = 1, \ldots, d\}$. Thus, $\operatorname{vol}(X + Y) = \prod_{i=1}^{d} \left((b_i - a_i) + (b_i' - a_i') \right)$. Also, $\operatorname{vol}(X) = \prod_{i=1}^{d}(b_i - a_i)$ and $\operatorname{vol}(Y) = \prod_{i=1}^{d}(b_i' - a_i')$. Therefore,

$$
\begin{aligned}
\frac{\operatorname{vol}(X)^{1/d} + \operatorname{vol}(Y)^{1/d}}{\operatorname{vol}(X+Y)^{1/d}} &= \left(\prod_{i=1}^{d} \frac{(b_i - a_i)}{(b_i - a_i) + (b_i' - a_i')} \right)^{1/d} + \left(\prod_{i=1}^{d} \frac{(b_i' - a_i')}{(b_i - a_i) + (b_i' - a_i')} \right)^{1/d} \\
&\leq \frac{1}{d} \left(\sum_{i=1}^{d} \frac{(b_i - a_i)}{(b_i - a_i) + (b_i' - a_i')} + \sum_{i=1}^{d} \frac{(b_i' - a_i')}{(b_i - a_i) + (b_i' - a_i')} \right) \\
&= 1,
\end{aligned}
$$

where the inequality follows from the AM-GM inequality (see Exercise 8 from Section 3.2.4). $\qquad\square$

We next use Lemma 3.5.4 to derive the inequality for finite unions of axis-aligned boxes.

Corollary 3.5.5 *If $X, Y \subseteq \mathbb{R}^d$ are both finite unions of axis aligned boxes, then* $\operatorname{vol}(X + Y)^{1/d} \geq \operatorname{vol}(X)^{1/d} + \operatorname{vol}(Y)^{1/d}$.

Proof We will prove the result by induction on the total number of axis-aligned boxes. If the total is 2, i.e., X and Y are simply axis-aligned boxes, then the result follows from Lemma 3.5.4. This takes care of the base case.

By Exercise 2 in Section 3.5.4, we may assume that both X and Y are finite unions of axis-aligned boxes whose interiors are disjoint. Since we are not in the base case, without loss of generality, we will assume X is the union of at least two axis-aligned boxes with disjoint interior. By Exercise 3 in Section 3.5.4, there exists $i \in \{1, \ldots, d\}$ and $M \in \mathbb{R}$ such that both halfspaces $\{\mathbf{x} \in \mathbb{R}^d : \mathbf{x}_i \leq M\}$ and $\{\mathbf{x} \in \mathbb{R}^d : \mathbf{x}_i \geq M\}$ contain at least one of the axis-aligned boxes from X. Let $X \cap \{\mathbf{x} \in \mathbb{R}^d : \mathbf{x}_i \leq M\}$ have volume equal to $\alpha \operatorname{vol}(X)$ for some $\alpha \in [0, 1]$. Let $M' \in \mathbb{R}$ be such that $Y \cap \{\mathbf{x} \in \mathbb{R}^d : \mathbf{x}_i \leq M'\}$

has volume equal to $\alpha \, \text{vol}(Y)$; such an M' can be found by "sweeping" $\mathbb{R}^d$ with hyperplanes parallel to $\{\mathbf{x} \in \mathbb{R}^d : \mathbf{x}_i = 0\}$. Note that both $Y_1 := Y \cap \{\mathbf{x} \in \mathbb{R}^d : \mathbf{x}_i \leq M'\}$ and $Y_2 = Y \cap \{\mathbf{x} \in \mathbb{R}^d : \mathbf{x}_i \geq M'\}$ can be expressed as a finite union of axis-aligned boxes with the number of boxes bounded above by the number of boxes used to define Y. Moreover, both $X_1 := X \cap \{\mathbf{x} \in \mathbb{R}^d : \mathbf{x}_i \leq M\}$ and $X_2 = X \cap \{\mathbf{x} \in \mathbb{R}^d : \mathbf{x}_i \geq M\}$ can be expressed as a finite union of axis-aligned boxes with the number of boxes strictly less than the number of boxes used to define X from the definition of i and M. Thus, by the induction hypothesis, $\text{vol}(X_i + Y_i)^{1/d} \geq \text{vol}(X_i)^{1/d} + \text{vol}(Y_i)^{1/d}$ for $i = 1, 2$. Moreover, $X_1 + Y_1 \subseteq \{\mathbf{x} \in \mathbb{R}^d : \mathbf{x}_i \leq M + M'\}$ and $X_2 + Y_2 \subseteq \{\mathbf{x} \in \mathbb{R}^d : \mathbf{x}_i \geq M + M'\}$. Therefore,

$$
\begin{aligned}
\text{vol}(X + Y) &\geq \text{vol}((X_1 + Y_1) \cup (X_2 + Y_2)) \\
&= \text{vol}(X_1 + Y_1) + \text{vol}(X_2 + Y_2) \\
&\geq \left(\text{vol}(X_1)^{1/d} + \text{vol}(Y_1)^{1/d}\right)^d + \left(\text{vol}(X_2)^{1/d} + \text{vol}\left(Y_2\right)^{1/d}\right)^d \\
&= \alpha\left(\text{vol}(X)^{1/d} + \text{vol}(Y)^{1/d}\right)^d + (1 - \alpha)\left(\text{vol}(X)^{1/d} + \text{vol}(Y)^{1/d}\right)^d \\
&= \left(\text{vol}(X)^{1/d} + \text{vol}(Y)^{1/d}\right)^d,
\end{aligned}
$$

where the inequality follows from Lemma 3.5.4. $\qquad\square$

We finally prove Theorem 3.5.2 by approximating the compact sets X and Y using unions of axis-aligned boxes.

Proof of Theorem 3.5.2   For each $k \in \mathbb{N}$, define the collection $\mathcal{B}_k$ of all axis-aligned boxes of the form $\{\mathbf{x} \in \mathbb{R}^d : \frac{\ell_i}{2^k} \leq \mathbf{x}_i \leq \frac{\ell_i + 1}{2^k} \ \forall i = 1, \ldots, d\}$ where $\ell_1, \ldots, \ell_d \in \mathbb{Z}$. In other words, $\mathcal{B}_k$ are all possible axis-aligned boxes with side lengths $\frac{1}{2^k}$ and vertices in the lattice $\frac{1}{2^k}\mathbb{Z}^d$. Let X_k, $k \in \mathbb{N}$ be defined as the union of all axis-aligned boxes in $\mathcal{B}_k$ that have a nonempty intersection with X. Similarly, Y_k, $k \in \mathbb{N}$ is defined as the union of all axis-aligned boxes in $\mathcal{B}_k$ that have a nonempty intersection with Y. It can be verified that $X_i \supseteq X_j \supseteq X$ for all $i \leq j$ and $\bigcap_{k=1}^{\infty} X_k = X$. Similarly, $Y_i \supseteq Y_j \supseteq Y$ for all $i \leq j$ and $\bigcap_{k=1}^{\infty} Y_k = Y$. This also implies that $X_i + Y_i \supseteq X_j + Y_j \supseteq X + Y$ for all $i \leq j$ and $\bigcap_{k=1}^{\infty}(X_k + Y_k) = X + Y$; see Exercise 4 in Section 3.5.4. The result now follows from Corollary 3.5.5, Theorem 1.3.19, and the fact that $r^{1/d}$ is a continuous function of r. $\qquad\square$

3.5.2 Sectional Volumes and Symmetrization

An equivalent formulation of the Brunn–Minkowski inequality when restricted to convex sets is the following inequality, which gives more insight into the structure of parallel sections of a convex set discussed at the beginning of this section. We will use the notion of k-dimensional volumes for k-dimensional sets discussed in Section 2.2.1.

Corollary 3.5.6 (Brunn's concavity principle) *Let $C \subseteq \mathbb{R}^d$ be a compact, convex set. Let $L \subseteq \mathbb{R}^d$ be a linear subspace of dimension $k \geq 1$. Define $f_{C,L} \colon L^\perp \to \mathbb{R}$ as $f_{C,L}(\mathbf{y}) = \mathrm{vol}_k(C \cap (L + \mathbf{y}))^{\frac{1}{k}}$. Then $f_{C,L}$ is concave over $\mathrm{Proj}_{L^\perp}(C)$ in the sense that for any $\mathbf{y}^1, \mathbf{y}^2 \in \mathrm{Proj}_{L^\perp}(C)$ and $0 \leq \lambda \leq 1$,*

$$f_{C,L}\left(\lambda \mathbf{y}^1 + (1-\lambda)\mathbf{y}^2\right) \geq \lambda f_{C,L}\left(\mathbf{y}^1\right) + (1-\lambda) f_{C,L}\left(\mathbf{y}^2\right).$$

Proof Left as an exercise – use the Brunn–Minkowski inequality (Theorem 3.5.2) on the relation from Exercise 1 from Section 3.5.4. $\qquad\square$

The above concavity result allows for a useful operation of symmetrization of a convex set with respect to an arbitrary linear subspace.

Definition 3.5.7 Let $L \subseteq \mathbb{R}^d$ be a linear subspace of dimension $k \geq 1$. For every $\mathbf{y} \in L^\perp$, define $C_{\mathbf{y}} := (L + \mathbf{y}) \cap B\left(\mathbf{y}, \frac{f_{C,L}(\mathbf{y})}{V(k)^{1/k}}\right)$, where $f_{C,L}$ is the function defined in Corollary 3.5.6 and $V(k)$ is the k-dimensional volume of the k-dimensional Euclidean unit ball. In other words, $C_{\mathbf{y}}$ is a k-dimensional Euclidean ball with center $\mathbf{y}$ and the same k-dimensional volume as the section $C \cap (L + \mathbf{y})$. The set $\bigcup_{\mathbf{y} \in \mathrm{Proj}_{L^\perp}(C)} C_{\mathbf{y}}$ is called the *spherical symmetrization of C with respect to L*. The special case when L is 1-dimensional is called *Steiner symmetrization* and that when L is $(d-1)$-dimensional is called *Schwarz symmetrization*.

Corollary 3.5.8 *Let $C \subseteq \mathbb{R}^d$ be a compact, convex set and let $L \subseteq \mathbb{R}^d$ be a linear subspace. Then the spherical symmetrization of C with respect to L is a compact, convex set with the same volume as C.*

Proof Left as an exercise. $\qquad\square$

3.5.3 The Rogers–Shephard Inequality

The Brunn–Minkowski theorem can be used to derive a relation between the volume of a convex set and the volumes of its sections and projections.

Theorem 3.5.9 (Rogers–Shephard inequality) *Let $C \subseteq \mathbb{R}^d$ be a compact, convex set and let $L \subseteq \mathbb{R}^d$ be a linear subspace of dimension k. Then, for any $\mathbf{y} \in \mathbb{R}^d$,*

$$\mathrm{vol}_d(C) \geq \tfrac{1}{2^d} \, \mathrm{vol}_k(C \cap (L + \mathbf{y})) \, \mathrm{vol}_{d-k}(\mathrm{Proj}_{L^\perp}(C)).$$

Proof We may assume $k \geq 1$ since otherwise the volumes of the 0-dimensional sections is zero. Consider the function $f_{C,L}$ defined in Corollary 3.5.6. Observe that (by Fubini's theorem on iterated integrals)

$$\mathrm{vol}_d(C) = \int_{\mathbf{y} \in L^\perp} \left(f_{C,L}(\mathbf{y})\right)^k d\mathbf{y}. \tag{3.5.1}$$

By Corollary 3.5.6, $f_{C,L}$ is concave over $\mathrm{Proj}_{L^\perp}(C)$ and by Theorem 3.2.3 it is continuous. Moreover, $\mathrm{Proj}_{L^\perp}(C)$ is compact, convex since C is compact, convex. Observe that $f_{C,L}$ takes value zero on $L^\perp \backslash \mathrm{Proj}_{L^\perp}(C)$. By Weierstrass' theorem (Theorem 1.3.13), there exists $\mathbf{y}^\star \in \mathrm{Proj}_{L^\perp}(C)$, where $f_{C,L}$ has the maximum value. Define $K := \mathbf{y}^\star + \frac{1}{2}\left(\mathrm{Proj}_{L^\perp}(C) - \mathbf{y}^\star\right) \subseteq \mathrm{Proj}_{L^\perp}(C)$. Since $f_{C,L}$ is concave and nonnegative over $\mathrm{Proj}_{L^\perp}(C)$, we must have $f_{C,L}(\mathbf{y}) \geq \frac{1}{2} f_{C,L}(\mathbf{y}^\star)$ for all $\mathbf{y} \in K$. From (3.5.1), we have

$$
\begin{aligned}
\mathrm{vol}_d(C) &\geq \int_{\mathbf{y} \in K} \left(f_{C,L}(\mathbf{y})\right)^k d\mathbf{y} \\
&\geq \int_{\mathbf{y} \in K} \tfrac{1}{2^k} \left(f_{C,L}(\mathbf{y}^\star)\right)^k d\mathbf{y} \\
&= \tfrac{1}{2^k} \left(f_{C,L}(\mathbf{y}^\star)\right)^k \mathrm{vol}_{d-k}(K) \\
&= \tfrac{1}{2^k} \left(f_{C,L}(\mathbf{y}^\star)\right)^k \cdot \tfrac{1}{2^{d-k}} \mathrm{vol}_{d-k}(\mathrm{Proj}_{L^\perp}(C)).
\end{aligned}
$$

Since $\left(f_{C,L}(\mathbf{y}^\star)\right)^k \geq \mathrm{vol}_k(C \cap (L + \mathbf{y}))$ for any $\mathbf{y} \in \mathbb{R}^d$, we have the stated result. $\qquad\square$

3.5.4 Exercises

1. Let $C \subseteq \mathbb{R}^d$ be convex and let $L \subseteq \mathbb{R}^d$ be a linear subspace. Then, for any $\mathbf{y}, \mathbf{y}^1, \mathbf{y}^2 \in \mathbb{R}^d$ such that $\mathbf{y}$ lies on the line segment $[\mathbf{y}^1, \mathbf{y}^2]$, there exists $\gamma \in [0,1]$ such that $\gamma(C \cap (L+\mathbf{y}^1)) + (1-\gamma)(C \cap (L+\mathbf{y}^2)) \subseteq C \cap (L+\mathbf{y})$.

2. If $X \subseteq \mathbb{R}^d$ is a finite union of axis-aligned boxes, show that it can be expressed as a finite union of axis-aligned boxes whose interiors are disjoint.

3. Suppose we have a finite collection of axis-aligned boxes with at least two boxes, and the interiors of all boxes are disjoint. Show that there exists $i \in \{1, \ldots, d\}$ and $M \in \mathbb{R}$ such that both halfspaces $\{\mathbf{x} \in \mathbb{R}^d : \mathbf{x}_i \leq M\}$ and $\{\mathbf{x} \in \mathbb{R}^d : \mathbf{x}_i \geq M\}$ contain at least one of these axis-aligned boxes.

4. For each $k \in \mathbb{N}$, define $\mathcal{B}_k$ to be the set of all axis-aligned boxes of the form $\{\mathbf{x} \in \mathbb{R}^d : \frac{\ell_i}{2^k} \leq \mathbf{x}_i \leq \frac{\ell_i+1}{2^k} \ \forall i = 1, \ldots, d\}$ where $\ell_1, \ldots, \ell_d \in \mathbb{Z}$. Prove the following.

    (a) Let $X \subseteq \mathbb{R}^d$ be a closed set. Define $X_k, k \in \mathbb{N}$ to be the union of all axis-aligned boxes in $\mathcal{B}_k$ that have a nonempty intersection with X. Show that $X_i \supseteq X_j \supseteq X$ for all $i \leq j$ and $\bigcap_{k=1}^\infty X_k = X$.

    (b) Let $X, Y \subseteq \mathbb{R}^d$ be compact sets. Define X_k and $Y_k, k \in \mathbb{N}$ be defined as the union of all axis aligned boxes in $\mathcal{B}_k$ that have a nonempty intersection with X and Y, respectively. Show that $X_i + Y_i \supseteq X_j + Y_j \supseteq X + Y$ for all $i \leq j$ and $\bigcap_{k=1}^\infty (X_k + Y_k) = X + Y$.

5. Prove Corollaries 3.5.6 and 3.5.8.
6.* Strengthen the Rogers–Shephard inequality in Theorem 3.5.9 to

$$\mathrm{vol}_d(C) \geq \binom{d}{k}^{-1} \mathrm{vol}_k(C \cap (L + \mathbf{y}))\, \mathrm{vol}_{d-k}(\mathrm{Proj}_{L^\perp}(C)).$$

3.6 Notes and Related Literature

The material presented in Sections 3.1–3.4 is classical convex analysis, with the exception of Section 3.3.4. As in Chapter 2, we have emphasized those aspects that we consider most relevant for Part II of the book. Interested readers can find much more in [124, 140, 199]. We have used proofs from these texts in a few places in this chapter. Section 3.2.1 is based on lecture notes prepared by Dr. Joseph Paat when he was a graduate student. The author is grateful to Dr. Paat for allowing the use of his notes.

A few words about Section 3.4 are in order. We restricted our attention to finite, real-valued convex functions. This simplifies things but many of the insights can be extended to functions taking value $+\infty$; see [140, 199]. In the subdifferential calculus of Theorem 3.4.5, part 1. needs to be modified to say that the subdifferential of the nonnegative combination is a *subset* of the nonnegative combination (in the Minkowski sum sense) of the individual subdifferentials. The containment can be strict for functions that take $+\infty$ as value. In part 3. of Theorem 3.4.5, one can establish equality when the index set J has a compact metric space structure and the map $j \mapsto f_j(\mathbf{x})$ is continuous for every $\mathbf{x} \in \mathbb{R}^d$.

The topic of minimal representations of convex sets as 1-sublevel sets of sublinear functions from Section 3.3.4 is a relatively recent strain of investigation. This question traces its motivations to optimization with integer-constrained variables; in particular, to *cutting-plane* theory within the theory of optimization with integer variables. We do not deal with this literature in this book but instead refer the reader to chapter 6 of the textbook [74] and the survey [29] for an introduction, a sample of the main results in cutting-plane theory that use the technology of sublinear functions and minimal representations, and further references to the literature. Our motivation for including the topic of minimal representations in this book comes from our view that it is a useful and fundamental insight in convex analysis that is not covered in standard textbooks. Minimal representations were first considered in [33] inspired by a fundamental cutting-plane result by Dey and Wolsey [103] and studied in further depth in [73]. The proof of Theorem 3.3.29 is an adaptation of the proof from [73]. The strict containment example in

Exercise 17 from Section 3.3.5 was suggested to me by An Wang, a Ph.D. student in my department.

Brunn–Minkowski theory and its ramifications form the backbone of many developments in modern convex geometry and its connections to functional analysis and Banach space theory. A classical reference that goes much deeper than our exposition in Section 3.5.1 is [210]; see also [124]. Our proof of Theorem 3.5.2 is based on the exposition in [173]. We also greatly benefited from the lectures notes by Rothvoss [200], especially from his exposition of the Rogers–Shepard inequality, and highly recommend the reader to look into them for more on the topic. An interesting theorem about the "average" sectional volume of ellipsoids goes back to the work of Busemann [62] and Furstenberg–Tzkoni [119], which has now come to be known as the *Furstenberg–Tzkoni formula*; see also [130, 171]. The *Busemann–Petty problem* and *Shepard's problem* are two related questions in convex geometry.

4

Geometry of Numbers

In the early part of the twentieth century, Hermann Minkowski pioneered a novel geometric approach to several questions in number theory. This approach developed into a field called the *geometry of numbers* and it had an influence on fields outside number theory as well, particularly functional analysis and the study of Banach spaces, and more recently on cryptography and discrete optimization. Minkowski's main insight was to convert certain number-theoretic problems into questions involving convex sets and points in $\mathbb{R}^d$ with integer coordinates. In fact, several of the concepts we studied earlier, such as the gauge of a convex set (Definition 3.3.8), were first introduced by Minkowski while developing this field. Our main interest in the subject comes from the fact that the algorithmic versions of some of these questions have proved to be of fundamental importance for mathematical optimization.

4.1 Lattices in Euclidean Space

Definition 4.1.1   A subset $\Lambda \subseteq \mathbb{R}^d$ is called a *lattice* if the following properties hold:

  (i) $\mathbf{0} \in \Lambda$.

  (ii) For any $\mathbf{x}, \mathbf{y} \in \Lambda$, $\mathbf{x} + \mathbf{y} \in \Lambda$.

  (iii) For any $\mathbf{x} \in \Lambda$, $-\mathbf{x} \in \Lambda$.

  (iv) There exists $\epsilon > 0$ such that $B(\mathbf{x}, \epsilon) \cap \Lambda = \{\mathbf{x}\}$ for all $\mathbf{x} \in \Lambda$. In other words, all the points in Λ are at least ϵ distance away from each other.

The *rank* of a lattice Λ is the dimension of span(Λ). With a mild abuse of notation, we use rk(Λ) to denote the rank of Λ. A lattice with rank d is said to be *full rank*.

Example 4.1.2   1. The set $\mathbb{Z}^d$, i.e., all points in $\mathbb{R}^d$ with integral coordinates, is a lattice.

2. For any linear subspace $L \subseteq \mathbb{R}^d$, $L \cap \mathbb{Z}^d$ is a lattice.

3. Consider any set of linearly independent vectors $\{\mathbf{b}^1, \ldots, \mathbf{b}^k\}$ in $\mathbb{R}^d$. The set $Z(\{\mathbf{b}^1, \ldots, \mathbf{b}^k\}) := \{\mu_1 \mathbf{b}^1 + \cdots + \mu_k \mathbf{b}^k : \mu_1, \ldots, \mu_k \in \mathbb{Z}\}$ is a lattice. We will use the shorthand $Z(B)$ to also denote $Z(\{\mathbf{b}^1, \ldots, \mathbf{b}^k\})$, where $B \in \mathbb{R}^{d \times k}$ is the matrix with $\mathbf{b}^1, \ldots, \mathbf{b}^k$ as columns.

It turns out that part 3. of Example 4.1.2 is generic in the sense that every lattice is of that form. This is a fundamental property that we establish below that allows an algorithmic perspective on lattices, because one can represent a lattice by specifying a finite set of linearly independent vectors and, consequently, perform computations on it.

Theorem 4.1.3 *Let $\Lambda \subseteq \mathbb{R}^d$ be a lattice. There exist linearly independent vectors $\mathbf{b}^1, \ldots, \mathbf{b}^k$ in $\mathbb{R}^d$ such that $Z(\{\mathbf{b}_1, \ldots, \mathbf{b}_k\}) = \Lambda$.*

Theorem 4.1.3 motivates the following definition.

Definition 4.1.4   Given a lattice Λ, any set of linearly independent vectors $\mathbf{b}^1, \ldots, \mathbf{b}^k$ in $\mathbb{R}^d$ such that $Z(\{\mathbf{b}_1, \ldots, \mathbf{b}_k\}) = \Lambda$ is called *a basis of the lattice* Λ, and Λ is said to be *generated* by $\mathbf{b}^1, \ldots, \mathbf{b}^k$.

We introduce some useful concepts to facilitate the proof of Theorem 4.1.3.

Definition 4.1.5   Let $\Lambda \subseteq \mathbb{R}^d$ be a lattice. A linear subspace $L \subseteq \mathbb{R}^d$ is said to be a *lattice subspace of* Λ if $\mathrm{span}(L \cap \Lambda) = L$. Similarly, an affine subspace $H \subseteq \mathbb{R}^d$ is said to be an *affine lattice subspace of* Λ if $\mathrm{aff}(H \cap \Lambda) = H$. A hyperplane that is an affine lattice subspace is called a *lattice hyperplane*.

Definition 4.1.6   Let $\Lambda \subseteq \mathbb{R}^d$ be a lattice. A *sublattice* of Λ is a subset $\Lambda' \subseteq \Lambda$ that is also a lattice (condition (iv) in Definition 4.1.1 is automatically satisfied since $\Lambda' \subseteq \Lambda$, so conditions (i), (ii), and (iii) are the key properties to check here). A sublattice Λ' defines an equivalence relation on Λ: $\mathbf{x}, \mathbf{y} \in \Lambda$ are related if $\mathbf{x} - \mathbf{y} \in \Lambda'$ (see Exercise 6 from Section 4.1.1). An equivalence class of this relation is called a *coset of Λ with respect to Λ'*.

The following lemma shows that the affine lattice subspaces parallel to a lattice subspace are in one-to-one correspondence with cosets with respect to the corresponding sublattice.

Lemma 4.1.7 *Let $\Lambda \subseteq \mathbb{R}^d$ be a lattice and let $L \subseteq \mathbb{R}^d$ be a lattice subspace of Λ. Then $\Lambda' := L \cap \Lambda$ is a sublattice of Λ (see Exercise 4 from Section 4.1.1). Let $H \subseteq \mathbb{R}^d$ be an affine subspace parallel to L (see Definition 2.2.9). Then the following are equivalent.*

1. *H is an affine lattice subspace.*
2. *$H \cap \Lambda$ is a coset of Λ with respect to Λ'.*
3. *$H = \mathrm{aff}(S)$ for some coset S of Λ with respect to Λ'.*

Proof Left as an exercise. $\qquad\square$

We next show a useful result that will be the core idea behind the proof of Theorem 4.1.3 and will have other important applications later.

Proposition 4.1.8 *Let $\Lambda \subseteq \mathbb{R}^d$ be a lattice and let $L \subseteq \mathbb{R}^d$ be a lattice subspace of Λ. Then $\mathrm{Proj}_{L^\perp}(\Lambda)$ is a lattice (recall that $L^\perp$ denotes the orthogonal complement of L (see Definition 2.4.7)).*

In other words, the projection of a lattice onto the orthogonal complement of a lattice subspace is again a lattice.

Proof Since the projection onto a linear subspace is a linear transformation, it is easy to verify that properties (i), (ii), and (iii) in Definition 4.1.1 are satisfied for $\mathrm{Proj}_{L^\perp}(\Lambda)$. So it remains to check property (iv).

We first observe that $\mathrm{Proj}_{L^\perp}(\Lambda)$ is in one-to-one correspondence with the cosets of Λ with respect to the sublattice $\Lambda' := L \cap \Lambda$, since each coset projects to a single point in $L^\perp$ and different cosets project to distinct points. Therefore, using Lemma 4.1.7, $\mathrm{Proj}_{L^\perp}(\Lambda)$ is in one-to-one correspondence with the affine lattice subspaces of Λ that are parallel to L. Since $L = \mathrm{span}(\Lambda')$, there exist linearly independent vectors $\mathbf{v}^1, \ldots, \mathbf{v}^k \in \Lambda'$ such that $L = \mathrm{span}\{\mathbf{v}^1, \ldots, \mathbf{v}^k\}$. Define the bounded set

$$G := \{\lambda_1 \mathbf{v}^1 + \cdots + \lambda_k \mathbf{v}^k + \mathbf{v} : 0 \le \lambda_i < 1$$
$$\forall i = 1, \ldots, k, \ \mathbf{v} \in L^\perp \text{ such that } \|\mathbf{v}\| \le 1\}.$$

See Figure 4.1 for an illustration. Observe that if an affine subspace H parallel to L does not intersect G, then $\|\mathrm{Proj}_{L^\perp}(H)\| > 1$. Thus, it suffices to show that only finitely many affine lattice subspaces parallel to L have nonempty intersection with G. Consider any such affine lattice subspace H and let $\mathbf{z} \in H \cap \Lambda$ (which exists since $H \cap \Lambda \ne \emptyset$). Then $\mathbf{z}$ has a unique decomposition of the form $\mathbf{z} = \gamma_1 \mathbf{v}^1 + \cdots + \gamma_k \mathbf{v}^k + \mathbf{w}$, where $\mathbf{w} \in L^\perp$. Note that $\|\mathbf{w}\| \le 1$ since $H \cap G \ne \emptyset$. Define $\bar{\mathbf{z}} = \{\gamma_1\}\mathbf{v}^1 + \cdots + \{\gamma_k\}\mathbf{v}^k + \mathbf{w}$, which is in the same coset as $\mathbf{z}$ and belongs to G. Thus, for any affine lattice subspace parallel to L that has nonempty intersection with G, there exists a lattice point from the corresponding coset in G. Since G is bounded, it contains only finitely many lattice points (see Exercise 1 from Section 4.1.1). In other words, there are only finitely many cosets intersecting G and consequently, only finitely many points in $\mathrm{Proj}_{L^\perp}(\Lambda)$ have norm at most 1. Setting $\epsilon > 0$ strictly less than

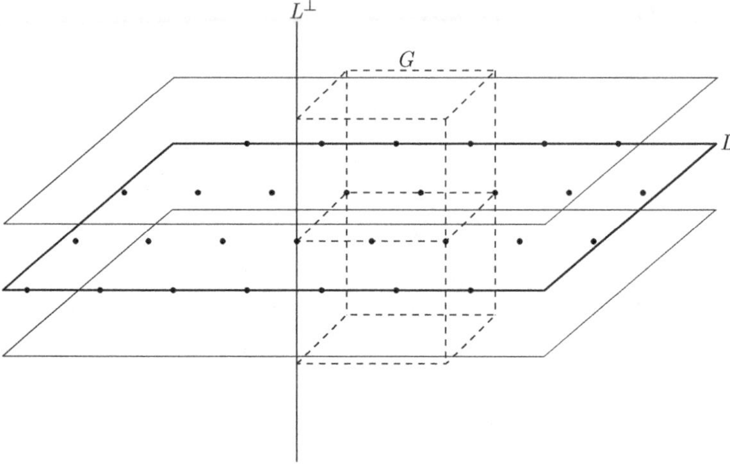

Figure 4.1 Illustration for the construction of G in the proof of Theorem 4.1.8. The subspace L and some points in $L \cap \Lambda$ are shown. G is the box shown in dashed lines. Two other lattice hyperplanes parallel to L are also shown.

the minimum norm of all the nonzero points in $\mathrm{Proj}_{L^\perp}(\Lambda)$ gives us property (iv) in Definition 4.1.1. $\qquad\square$

Proof of Theorem 4.1.3 We prove the result by induction on the rank of the lattice. If the lattice has rank 1, then a shortest nonzero vector in the lattice is a basis for the lattice (see Exercises 11 and 12 from Section 4.1.1). Consider a lattice Λ of rank $k > 1$. Let $\mathbf{u} \in \Lambda$ be a shortest nonzero lattice vector (which exists because bounded sets contain finitely many lattice points by Exercise 1 from Section 4.1.1). Then $L := \mathrm{span}(\{\mathbf{u}\})$ is a lattice subspace. By Proposition 4.1.8, $\mathrm{Proj}_{L^\perp}(\Lambda)$ is a lattice. Moreover, rank of $\mathrm{Proj}_{L^\perp}(\Lambda)$ is strictly less than k since $\mathrm{Proj}_{L^\perp}(\Lambda) \subseteq L^\perp \cap \mathrm{span}(\Lambda) \subsetneq \mathrm{span}(\Lambda)$, since $\mathbf{u} \notin L^\perp$. By the induction hypothesis, $\mathrm{Proj}_{L^\perp}(\Lambda)$ has a basis $\{\mathbf{v}^1, \dots, \mathbf{v}^m\}$, and by the definition of projection, there exist corresponding vectors $\bar{\mathbf{v}}^1, \dots, \bar{\mathbf{v}}^m \in \Lambda$ such that $\mathrm{Proj}_{L^\perp}(\bar{\mathbf{v}}^i) = \mathbf{v}^i$ for $i = 1, \dots, m$. One can now verify that $\{\bar{\mathbf{v}}^1, \dots, \bar{\mathbf{v}}^m, \mathbf{u}\}$ is a basis for Λ (Exercise 13 from Section 4.1.1). $\qquad\square$

Theorem 4.1.3 shows the existence of a basis for a lattice Λ, but it does not say that there is a unique one. In fact, a lattice has infinitely many bases if Λ has rank at least 2 (a lattice with rank 1 has exactly two bases; see Exercise 11 from Section 4.1.1). See Figure 4.2 for two bases for $\mathbb{Z}^2$. It can be verified that any two bases of a lattice must have the same number of elements (Exercise 14 from Section 4.1.1). Moreover, there is another fundamental relation between

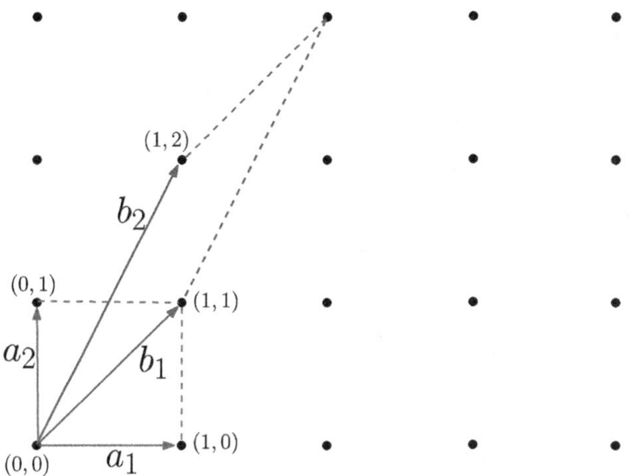

Figure 4.2 Two bases for $\mathbb{Z}^2$: $\{a_1 = (1,0), a_2 = (0,1)\}$ and $\{b_1 = (1,1), b_2 = (1,2)\}$.

any two bases of a lattice which leads to a key parameter associated with a lattice.

Definition 4.1.9 A $d \times d$ matrix U is *unimodular* if all its entries are integers and $\det(U) = \pm 1$, where $\det(U)$.

Proposition 4.1.10 *Let $A, B \in \mathbb{R}^{d \times k}$. The columns of A and the columns of B generate the same lattice, i.e., $Z(A) = Z(B)$ if and only if there exists a $k \times k$ unimodular matrix U such that $B = AU$ (and consequently $A = BU^{-1}$).*

Proof ($\Leftarrow$) If there exists a unimodular matrix U such that $A = BU$, then every column $\mathbf{a}^i$, $i = 1, \ldots, k$ of A is an integral combination of the columns of B, and $B = AU^{-1}$ implies that every column $\mathbf{b}^i$, $i = 1, \ldots, k$ is an integral combination of the columns of A. This immediately shows that the two sets generate the same lattice.

($\Rightarrow$) Suppose the columns of A and the columns of B generate the same lattice Λ. Since each $\mathbf{b}^i \in \Lambda$, $\mathbf{b}^i$ can be written as an integral combination of the columns of A. Therefore, there exist $\mathbf{x}^i \in \mathbb{Z}^k$ such that $\mathbf{b}^i = A\mathbf{x}_i$ for all $i = 1, \ldots, k$. Let $X \in \mathbb{Z}^{k \times k}$ be the matrix with columns $\mathbf{x}^1, \ldots, \mathbf{x}^k$. So $B = AX$. By the same argument, there exists a matrix Y with integral entries such that $A = BY$. This implies that $B = AX = BYX$, and since B has linear independent columns, this implies that $YX = I$. Hence, $\det(Y)\det(X) = 1$. Since X and Y are both integral matrices, their determinants are integers and this implies that $|\det(X)| = |\det(Y)| = 1$ and therefore X and Y are unimodular. We take $U = X$ to be our unimodular matrix. $\qquad\square$

Combined with the multiplicative property of determinants, Proposition 4.1.10 implies

Theorem 4.1.11 *Let $\Lambda \subseteq \mathbb{R}^d$ be a lattice and let the columns of $A, B \in \mathbb{R}^{d \times k}$ form two different bases of Λ. Then $\det(A^T A) = \det(B^T B)$.*

We are thus led to a fundamental measure of a lattice.

Definition 4.1.12   Let $\Lambda \subseteq \mathbb{R}^d$ be a lattice. The *determinant of* Λ is defined as $\det(\Lambda) := \sqrt{\det(A^T A)}$ for any matrix A whose columns form a basis of Λ.

These results can be interpreted in geometric terms.

Definition 4.1.13   Let $\mathbf{a}^1, \ldots, \mathbf{a}^k \in \mathbb{R}^d$ be linearly independent vectors. The set

$$\{\mu_1 \mathbf{a}^1 + \cdots + \mu_k \mathbf{a}^k : 0 \le \mu_i < 1 \ \forall i = 1, \ldots, k\}$$

is called the *fundamental parallelepiped* corresponding to this set of vectors. Given any lattice Λ, the fundamental parallelepiped corresponding to any basis of Λ is called *a* fundamental parallelepiped of Λ.

Remark 4.1.14   The k-dimensional volume of the parallelepiped formed by the columns of a matrix $A \in \mathbb{R}^{d \times k}$ of rank k is precisely $\sqrt{\det(A^T A)}$ (see Section 2.2.1). Thus, for any lattice Λ of rank k, the k-dimensional volume of any fundamental parallelepiped of Λ is equal to $\det(\Lambda)$.

4.1.1 Exercises

1. Let $\Lambda \subseteq \mathbb{R}^d$ be a lattice. Then $|A \cap \Lambda|$ is finite for any bounded set $A \subseteq \mathbb{R}^d$.
2. Show that the image of a lattice under any invertible linear transformation is a lattice.
3. Check that all the examples in Example 4.1.2 satisfy the properties listed in Definition 4.1.1.
4. Let $\Lambda \subseteq \mathbb{R}^d$ be a lattice and let $L \subseteq \mathbb{R}^d$ be any linear subspace. Show that $L \cap \Lambda$ is a sublattice of Λ.
5. Let $\Lambda \subseteq \mathbb{R}^d$ be a lattice. Let $H \subseteq \mathbb{R}^d$ be an affine subspace such that $H \cap \Lambda \ne \emptyset$ and L be the linear subspace parallel to it (see Definition 2.2.9). Show that H is an affine lattice subspace if and only if L is a lattice subspace.
6. Show that the relation defined in Definition 4.1.6 is an equivalence relation.

7. Prove Lemma 4.1.7.

8. Is it true that if $\Lambda \subseteq \mathbb{R}^d$ is a lattice and L is a lattice subspace of Λ, then $\mathrm{Proj}_L(\Lambda)$ is a lattice? Compare with Theorem 4.1.8.

9. Let $\Lambda \subseteq \mathbb{R}^d$ be a lattice, L be a lattice subspace of Λ, and $X \subseteq \mathbb{R}^d$ be an arbitrary subset. Show that the number of affine lattice subspaces parallel to L that intersect X equals $\#(\mathrm{Proj}_{L^\perp}(X) \cap \mathrm{Proj}_{L^\perp}(\Lambda))$.

10. Let $\Lambda \subseteq \mathbb{R}^d$ be a lattice and let L be a lattice subspace. Show that $\mathrm{rk}(\mathrm{Proj}_{L^\perp}(\Lambda)) + \mathrm{rk}(L \cap \Lambda) = \mathrm{rk}(\Lambda)$.

11. Let $\Lambda \subseteq \mathbb{R}^d$ be a lattice. Show that the following are equivalent:

    (a) Λ has rank 1.
    (b) Λ has exactly two bases.
    (c) Λ has finitely many bases.

12. Let $\Lambda \subseteq \mathbb{R}^d$ be a lattice and $\mathbf{w} \in \Lambda$. Let $L = \mathrm{span}(\{\mathbf{w}\})$. Show that the following are equivalent.

    (a) $\mathbf{w}$ is a shortest lattice vector in $L \cap \Lambda$ (with respect to any norm).
    (b) There is no point from Λ on the line segment $[\mathbf{0}, \mathbf{w}]$ different from $\mathbf{0}$ and $\mathbf{w}$.
    (c) $\{\mathbf{w}\}$ is a basis of $L \cap \Lambda$.

    $\mathbf{w}$ is called a *primitive lattice vector* if it satisfies the above three (equivalent) conditions.

13. Complete the proof of Theorem 4.1.3 by verifying that $\{\bar{\mathbf{v}}^1, \ldots, \bar{\mathbf{v}}^m, \mathbf{u}\}$ is a basis for Λ.

14. Show that the size of any basis for a lattice $\Lambda \subseteq \mathbb{R}^d$ equals $\mathrm{rk}(\Lambda)$.

15. Let $\Lambda \subseteq \mathbb{R}^d$ be a lattice and let $L \subseteq \mathbb{R}^d$ be a lattice subspace and let $\mathbf{b}^1, \ldots, \mathbf{b}^k$ be a basis for the lattice $L \cap \Lambda$. Show that Λ has a basis that includes $\mathbf{b}^1, \ldots, \mathbf{b}^k$.

16. Let $U \in \mathbb{Z}^{d \times d}$. Show that U is unimodular if and only if U^{-1} is unimodular.

17. (Hermite Normal Form)

    Consider the following *elementary column operations* for a matrix $A \in \mathbb{R}^{d \times n}$

    i. exchanging two columns.
    ii. multiplying a column by -1.
    iii. adding an integral multiple of one column to another column.

    (a) Let A be any $d \times n$ matrix and let A' be obtained from A by performing a finite number of sequential elementary operations as above. Show that A can be obtained from A' by performing elementary column operations and the set of integer linear combinations of the columns of A is precisely the set of integer linear combinations of the columns of A'.

(b) Show that any $d \times n$ matrix A with integer entries and rank equal to d (so $n \geq d$) can be converted into the form $[B, 0]$ using only the above elementary column operations such that all of the following hold:

- B is a nonsingular, lower triangular matrix with nonzero entries on the main diagonal.
- B has only nonnegative entries.
- In each row of B there is a unique maximum element, which sits on the main diagonal of B.

  $[B, 0]$ is called the *Hermite Normal Form (HNF)* of A.

(c) Let U be a $d \times d$ unimodular matrix. Show that the HNF of U is the identity matrix. Conclude that the lattice generated by the columns of a unimodular matrix is $\mathbb{Z}^d$.

18.* (Smith Normal Form) Let $\Lambda, \Lambda' \subseteq \mathbb{R}^d$ be two lattices such that $\mathrm{rk}(\Lambda') = \mathrm{rk}(\Lambda)$. Show that Λ' is a sublattice of Λ if and only if Λ has a basis $\mathbf{b}^1, \ldots, \mathbf{b}^k$ and Λ' has a basis $\mathbf{b}'^1, \ldots, \mathbf{b}'^k$ such that for every $i = 1, \ldots, k$, $\mathbf{b}'^i = u_i \mathbf{b}^i$ for some integer $u_i \geq 1$. Thus, conclude that the number of cosets of Λ with respect to Λ' is equal to $u_1 \cdots \cdots u_k = \frac{\det(\Lambda')}{\det(\Lambda)}$ (in particular, $\det(\Lambda)$ always divides $\det(\Lambda')$).

19.* Let $\mathbf{a}^1, \ldots, \mathbf{a}^n \subseteq \mathbb{Z}^d$ such that the matrix with these vectors as columns has rank equal to d (therefore, $n \geq d$). Show that the set of all integer linear combinations of $\mathbf{a}^1, \ldots, \mathbf{a}^n$ is a sublattice of $\mathbb{Z}^d$ with determinant equal to the greatest common divisor of the determinants of all $d \times d$ submatrices of the matrix with $\mathbf{a}^1, \ldots, \mathbf{a}^n$ as columns.

20. Let $\mathbf{a}^1, \ldots, \mathbf{a}^n$ be a set of vectors in $\mathbb{R}^d$. We *do not* assume that the vectors are linearly independent or that $n \leq d$. Is the set of all integer linear combinations of $\mathbf{a}^1, \ldots, \mathbf{a}^n$ always a lattice of $\mathbb{R}^d$? What if we restrict $\mathbf{a}^1, \ldots, \mathbf{a}^n$ to be rational vectors, as opposed to having arbitrary real entries?

21. Let Π be a fundamental parallelepiped of a lattice Λ. Show that $(\Pi + \mathbf{z}) \cap (\Pi + \mathbf{z}') = \emptyset$ if $\mathbf{z}, \mathbf{z}'$ are distinct elements of Λ, and $\mathrm{span}(\Lambda) = \bigcup_{\mathbf{z} \in \Lambda} (\Pi + \mathbf{z})$. In other words, Π "tiles" $\mathrm{span}(\Lambda)$.

22. Let $\Lambda \subseteq \mathbb{R}^d$ be a lattice and $S \subseteq \Lambda$. Show that either $\mathrm{vol}(\mathrm{conv}(S)) = 0$ or at least $\frac{1}{d!} \det(\Lambda)$.

4.2 Minkowski's Convex Body Theorem

Perhaps the most well-known result from the field of geometry of numbers is Minkowski's theorem about lattice points contained in convex bodies that are symmetric about the origin. This theorem has had myriad applications in mathematics and we will see two striking applications later in this chapter (Theorems 4.3.14 and 4.4.4). We proceed to develop the circle of ideas around this remarkable result.

We begin with an intuitive lemma about volumes in $\mathbb{R}^d$. It says that the standard method of computing the volume of a set in $\mathbb{R}^d$ by integrating its characteristic function can also be done "modulo a lattice."

Definition 4.2.1 Let $\Lambda \subseteq \mathbb{R}^d$ be a full rank lattice. For any set $X \subseteq \mathbb{R}^d$, define $\varphi_{X,\Lambda}(\mathbf{x}) = \#(\{\mathbf{z} \in \Lambda : \mathbf{x} + \mathbf{z} \in X\}) = \#(X \cap (\Lambda + \mathbf{x}))$ for any $\mathbf{x} \in \mathbb{R}^d$.

$\varphi_{X,\Lambda}$ is easily verified to be periodic with respect to the lattice, i.e., $\varphi_{X,\Lambda}(\mathbf{x} + \mathbf{z}) = \varphi_{X,\Lambda}(\mathbf{x})$ for all $\mathbf{x} \in \mathbb{R}^d$ and $\mathbf{z} \in \Lambda$. One can think of it as the characteristic function of X, modulo the lattice.

Lemma 4.2.2 *Let $\Lambda \subseteq \mathbb{R}^d$ be a full rank lattice and let Π be a fundamental parallelepiped of Λ. Let $X \subseteq \mathbb{R}^d$ be a bounded, Lebesgue measurable set.[1] Then*

$$\mathrm{vol}(X) = \int_{\Pi} \varphi_{X,\Lambda}(\mathbf{x})d\mathbf{x},$$

where $\mathrm{vol}(X)$ denotes the volume (Lebesgue measure) of X.

Proof By Exercise 21 from Section 4.1.1 and countable additivity of volume/measure, $\mathrm{vol}(X) = \sum_{\mathbf{z} \in \Lambda} \mathrm{vol}(X \cap (\Pi + \mathbf{z}))$. Since volume is preserved under translations, $\mathrm{vol}(X \cap (\Pi + \mathbf{z})) = \mathrm{vol}((X - \mathbf{z}) \cap \Pi)$. Therefore, $\mathrm{vol}(X) = \sum_{\mathbf{z} \in \Lambda} \mathrm{vol}((X + \mathbf{z}) \cap \Pi)$. Thus,

$$
\begin{aligned}
\mathrm{vol}(X) &= \sum_{\mathbf{z} \in \Lambda} \mathrm{vol}((X + \mathbf{z}) \cap \Pi) \\
&= \sum_{\mathbf{z} \in \Lambda} \int_{\Pi} \mathbb{1}_{X+\mathbf{z}}(\mathbf{x})d\mathbf{x} \\
&= \int_{\Pi} \sum_{\mathbf{z} \in \Lambda} \mathbb{1}_{X+\mathbf{z}}(\mathbf{x})d\mathbf{x} \\
&= \int_{\Pi} \varphi_{X,\Lambda}(\mathbf{x})d\mathbf{x},
\end{aligned}
$$

where we can switch the sum and the integral in the third equality since X is bounded and so $\mathrm{vol}(X \cap (\Pi + \mathbf{z}))$ is nonzero for finitely many $\mathbf{z} \in \Lambda$, i.e., we actually have a finite sum in the first equality. $\square$

---

[1] "Lebesgue measurable" simply means that there is a well-defined way of assigning volume to the set. We will not need any machinery from measure theory. We use the language for mathematical formalism and correctness. The interested reader may refer to [204] for an introduction to basic measure theory.

Sets of the form $\Lambda + \mathbf{x}$ are called *cosets of $\mathbb{R}^d$ modulo the lattice*; equivalently, these are the equivalence classes under the equivalence relation where $\mathbf{x}$ and $\mathbf{y}$ are related if and only if $\mathbf{x} - \mathbf{y} \in \Lambda$. A useful interpretation of Lemma 4.2.2 is that $\frac{\text{vol}(X)}{\det(\Lambda)} = \frac{\int_\Pi \varphi_{X,\Lambda}(\mathbf{x})d\mathbf{x}}{\text{vol}(\Pi)} = \frac{\int_\Pi \#(X \cap (\Lambda + \mathbf{x}))}{\text{vol}(\Pi)}$ is the average number of elements X catches from a coset of $\mathbb{R}^d$. This leads to the consequence that any set with large volume can be translated to contain many lattice points.

Theorem 4.2.3 (Blichfeldt's theorem) *Let $\Lambda \subseteq \mathbb{R}^d$ be a full rank lattice. Let $X \subseteq \mathbb{R}^d$ be a bounded, Lebesgue measurable set. Then there exists $\mathbf{x} \in \mathbb{R}^d$ such that $\#(X \cap (\Lambda + \mathbf{x})) \geq \left\lceil \frac{\text{vol}(X)}{\det(\Lambda)} \right\rceil$.*

Proof Let Π be a fundamental parallelepiped of Λ. We show that $\varphi_{X,\Lambda}(\mathbf{x}) \geq \left\lceil \frac{\text{vol}(X)}{\det(\Lambda)} \right\rceil$ for some $\mathbf{x} \in \Pi$. Suppose to the contrary, which implies that $\varphi_{X,\Lambda}(\mathbf{x}) < \frac{\text{vol}(X)}{\det(\Lambda)}$ for all $\mathbf{x} \in \Pi$ since $\varphi_{X,\Lambda}(\mathbf{x})$ is an integer valued function. Then, by Lemma 4.2.2 and Remark 4.1.14, $\text{vol}(X) = \int_\Pi \varphi_{X,\Lambda}(\mathbf{x})d\mathbf{x} < \int_\Pi \frac{\text{vol}(X)}{\det(\Lambda)}d\mathbf{x} < \text{vol}(X)$, which is a contradiction. $\square$

We can finally state Minkowski's famous convex body theorem.

Theorem 4.2.4 (Minkowski's convex body theorem) *Let $\Lambda \subseteq \mathbb{R}^d$ be a full rank lattice. Let $C \subseteq \mathbb{R}^d$ be a $\mathbf{0}$-symmetric, convex set (Definition 3.3.6). Then C contains at least $\left\lceil \frac{\text{vol}(C)}{2^d \det(\Lambda)} \right\rceil - 1$ pairs of points $\pm\mathbf{z} \in \Lambda \setminus \{\mathbf{0}\}$.*

Proof We prove the result for bounded C and leave the unbounded case as an exercise. Define $X := \frac{1}{2}C$. Then $\text{vol}(X) = \text{vol}(C)/2^d$ and by Theorem 4.2.3, there exists $\mathbf{x} \in \mathbb{R}^d$ such that $\#(X \cap (\Lambda + \mathbf{x})) \geq \left\lceil \frac{\text{vol}(C)}{2^d \det(\Lambda)} \right\rceil$. Let $X' \subseteq X \cap (\Lambda + \mathbf{x})$ be any finite set with size at least $\left\lceil \frac{\text{vol}(C)}{2^d \det(\Lambda)} \right\rceil$ and let $\mathbf{v} \in X'$ be any vertex of $\text{conv}(X')$, which is a polytope by the Minkowski–Weyl theorem (Theorem 2.5.16). Define $Z := \{\mathbf{x}' - \mathbf{v} : \mathbf{x}' \in X' \setminus \{\mathbf{v}\}\}$ and we observe that $Z \subseteq \Lambda \setminus \{\mathbf{0}\}$ with $\#(Z) \geq \left\lceil \frac{\text{vol}(C)}{2^d \det(\Lambda)} \right\rceil - 1$. Moreover, for all $\mathbf{z} \in Z$, $-\mathbf{z} \notin Z$ because otherwise, $\mathbf{v}$ can be expressed as the convex combination of $\mathbf{v} + \mathbf{z}, \mathbf{v} - \mathbf{z} \in X'$. Since $\mathbf{v} \in X' \subseteq X$, we must have $-\mathbf{v} \in X$ as well, since C is $\mathbf{0}$-symmetric. Consider any $\mathbf{z} \in Z$ with $\mathbf{z} = \mathbf{x}' - \mathbf{v}$ for $\mathbf{x}' \in X' \subseteq X$. By convexity of X, $\frac{1}{2}\mathbf{x}' + \frac{1}{2}(-\mathbf{v}) = \frac{1}{2}\mathbf{z} \in X$ and therefore $\mathbf{z} \in 2X = C$. Since C is $\mathbf{0}$-symmetric, for each $\mathbf{z} \in Z$ we also have $-\mathbf{z} \in C$. $\square$

The following slightly weaker version of Minkowski's theorem is perhaps more well known.

Corollary 4.2.5 *Let* $\Lambda \subseteq \mathbb{R}^d$ *be a full rank lattice. Let* $C \subseteq \mathbb{R}^d$ *be a* **0***-symmetric, convex set with* $\mathrm{vol}(C) > 2^d \det(\Lambda)$. *Then* C *contains a nonzero lattice point from* Λ.

A consequence of Minkowski's theorem is the following converse of Proposition 4.1.8.

Proposition 4.2.6 *Let* $\Lambda \subseteq \mathbb{R}^d$ *be a lattice and let* $L \subseteq \mathrm{span}(\Lambda)$ *be a subspace. If* $\mathrm{Proj}_{L^\perp}(\Lambda)$ *is a lattice, then* L *is a lattice subspace.*

We make an important observation that will be the key in proving Proposition 4.2.6.

Lemma 4.2.7 *Let* $\Lambda \subseteq \mathbb{R}^d$ *be a full rank lattice and let* $\mathbf{v} \in \mathbb{R}^d$ *be a unit norm vector such that* $L := \mathrm{span}(\mathbf{v})$ *is not a lattice subspace. Then* $\mathrm{Proj}_{L^\perp}(\Lambda)$ *is not a lattice. In particular, for any* $\epsilon > 0$, *there exists a nonzero point in* $B(\mathbf{0}, \epsilon) \cap \mathrm{Proj}_{L^\perp}(\Lambda)$.

Proof Consider an arbitrary $\epsilon > 0$ and let $V > 0$ be the $(d-1)$-dimensional volume of $B(\mathbf{0}, \epsilon) \cap L^\perp$. Let $M > \frac{2^d \det(\Lambda)}{V}$. Then the cylindrical set $C := \{\mathbf{x} + \lambda \mathbf{v} : \mathbf{x} \in B(\mathbf{0}, \epsilon) \cap L^\perp,\ \lambda \in [-M, M]\}$ is a **0**-symmetric, convex set with volume $2MV$ which is strictly greater than $2^d \det(\Lambda)$. By Corollary 4.2.5, there exists nonzero $\mathbf{z} \in C \cap \Lambda$. Since the line L is not a lattice subspace, it cannot contain any nonzero point from Λ. Therefore, $\mathrm{Proj}_{L^\perp}(\mathbf{z}) \neq \mathbf{0}$. Moreover, $\mathbf{z} \in C$ implies that $\mathbf{z}$ is of the form $\mathbf{x} + \lambda \mathbf{v}$ for some $\mathbf{x} \in B(\mathbf{0}, \epsilon) \cap L^\perp$ and $\lambda \in [-M, M]$. Thus, $\mathbf{z}$ has distance at most ϵ from L and therefore $\mathrm{Proj}_{L^\perp}(\mathbf{z})$ is at distance at most ϵ from $\mathbf{0}$. $\qquad\square$

Proof of Proposition 4.2.6 We can work in $\mathrm{span}(\Lambda)$ and so we assume Λ is a full rank lattice. We prove the contrapositive. Since $L \cap \Lambda$ is a lattice (Exercise 4 from Section 4.1.1), there exists $\bar{\epsilon} > 0$ such that $L \cap \Lambda \cap B(\mathbf{0}, \bar{\epsilon}) = \{\mathbf{0}\}$. We will now show that for any $0 < \epsilon < \bar{\epsilon}$, there exists a point in $\mathrm{Proj}_{L^\perp}(\Lambda)$ with norm at most ϵ. This will show that $\mathrm{Proj}_{L^\perp}(\Lambda)$ is not a lattice, violating condition (iv) in Definition 4.1.1.

Since we assume L is not a lattice subspace, there exists $\mathbf{v} \in L \backslash \{\mathbf{0}\}$ such that $\mathbf{v}$ is orthogonal to $\mathrm{span}(L \cap \Lambda)$. Consequently, $\bar{L} := \mathrm{span}(\mathbf{v}) \subseteq L$ is not a lattice subspace and, by Lemma 4.2.7, for any $0 < \epsilon < \bar{\epsilon}$, there exists nonzero $\mathbf{z}_\epsilon \in B(\mathbf{0}, \epsilon) \cap \mathrm{Proj}_{\bar{L}^\perp}(\Lambda)$. Let $\mathbf{z}'_\epsilon \in \Lambda$ be such that $\mathbf{z}_\epsilon = \mathrm{Proj}_{\bar{L}^\perp}(\mathbf{z}'_\epsilon)$. Since $\epsilon < \bar{\epsilon}$, $\mathbf{z}'_\epsilon \notin L$. Since $\bar{L} \subseteq L$, we have $L^\perp \subseteq \bar{L}^\perp$ (see Proposition 2.4.10, part 4. and Exercise 9b in Section 2.4.4). By Exercise 3 in Section 2.3.1, $\mathrm{Proj}_{L^\perp}(\mathbf{z}'_\epsilon) = \mathrm{Proj}_{L^\perp}(\mathrm{Proj}_{\bar{L}^\perp}(\mathbf{z}'_\epsilon)) = \mathrm{Proj}_{L^\perp}(\mathbf{z}_\epsilon)$. Since $\mathbf{z}'_\epsilon \notin L$, $\mathrm{Proj}_{L^\perp}(\mathbf{z}'_\epsilon) \neq \mathbf{0}$.

Since $\text{Proj}_{L^\perp}(\cdot)$ is a nonexpansive map (Proposition 2.3.3), $\|\text{Proj}_{L^\perp}(\mathbf{z}'_\epsilon)\| = \|\text{Proj}_{L^\perp}(\mathbf{z}_\epsilon) - \mathbf{0}\| \leq \|\mathbf{z}_\epsilon - \mathbf{0}\| \leq \epsilon.$ $\qquad\square$

4.2.1 Exercises

1. Let $\Lambda \subseteq \mathbb{R}^d$ be a lattice. Show that if $X \subseteq \mathbb{R}^d$ is any set (not necessarily convex) of volume strictly less than $\det(\Lambda)$, then there is a translate of X that does not contain any point from Λ.

2. Finish the proof of Theorem 4.2.4 in the case of unbounded C.

3. Show that, in Corollary 4.2.5, we can change the hypothesis by adding a compactness assumption on C and relaxing the volume condition to $\text{vol}(C) \geq 2^d \det(\Lambda)$.

4. Let $\Lambda \subseteq \mathbb{R}^d$ be a full rank lattice. Show that for any norm N on $\mathbb{R}^d$, there is a nonzero lattice vector with norm at most $2 \left(\frac{\det(\Lambda)}{\text{vol}(B_N(\mathbf{0},1))} \right)^{1/d}$. In particular, there exists a lattice vector $\mathbf{z} \in \Lambda \setminus \{\mathbf{0}\}$ such that $\|\mathbf{z}\|_\infty \leq \det(\Lambda)^{1/d}$ and a lattice vector $\|\mathbf{z}\|_2 \leq \sqrt{\frac{8d}{\pi e}} \det(\Lambda)^{1/d} < \sqrt{d} \det(\Lambda)^{1/d}$.

5. Theorem 4.2.4 gives a lower bound on the number of nonzero lattice points in a convex set.

   (i) Is it tight or can one prove a better universal lower bound that applies to all pairs of lattices and convex sets?

   (ii) Is the lower bound also an upper bound on the number of nonzero lattice points for full-dimensional convex sets, up to constant factors?

6. Strengthen Lemma 4.2.7 as follows. Let $\Lambda \subseteq \mathbb{R}^d$ be a full rank lattice and let $\mathbf{v} \in \mathbb{R}^d$ such that $L := \text{span}(\mathbf{v})$ is not a lattice subspace. Then, for any $\bar{\lambda} > 0$ and $\epsilon > 0$, there exists $\mathbf{z} \in \Lambda$ such that $\text{Proj}_{L^\perp}(\mathbf{z})$ has norm at most ϵ and $\mathbf{z} = \text{Proj}_{L^\perp}(\mathbf{z}) + \lambda\mathbf{v}$ with $\lambda \geq \bar{\lambda}$. In other words, there are points arbitrarily close to L that are arbitrarily far away.

7.* Prove Dirichlet's approximation theorem: given any $\alpha_1, \ldots, \alpha_d \in \mathbb{R}$ and $\epsilon > 0$, there exist integers $p_1, \ldots, p_d \in \mathbb{Z}$ and a natural number $1 \leq q \leq \left(\frac{1}{\epsilon} \right)^d$ such that

$$\left| \alpha_i - \frac{p_i}{q} \right| \leq \frac{\epsilon}{q} \qquad \forall i = 1, \ldots, d.$$

In other words, one can approximate a given set of real numbers by rationals whose denominators can be controlled (as a function of the error desired).

8.* Show that if $\Lambda \subseteq \mathbb{R}^d$ is a lattice and $L \subseteq \mathbb{R}^d$ is an arbitrary linear subspace, then $\mathrm{Proj}_L(\Lambda)$ is of the form $\Lambda' + D$ where $\Lambda' \subseteq L$ is a lattice and $D \subseteq L$ is a dense subset of $\mathrm{span}(D)$.

4.3 Packing, Covering, SVP, and CVP

We turn our attention to a central set of questions in geometry of numbers that has been particularly influential in cryptography and optimization. We will study the algorithmic aspects in Section 6.3.1. Here we study some geometric aspects. We start with a notion naturally associated with Minkowski's convex body theorem.

Definition 4.3.1 Let $\Lambda \subseteq \mathbb{R}^d$ be a lattice and N be a norm on $\mathbb{R}^d$. The *shortest lattice vector(s) (in the norm N)* is defined simply as $SV(N,\Lambda) :=$ $\mathrm{argmin}\{N(\mathbf{z}) : \mathbf{z} \in \Lambda \setminus \{\mathbf{0}\}\}$. Note that this is well defined by Exercise 1 from Section 4.1.1. The *shortest lattice vector problem (SVP)* is the algorithmic problem of computing a shortest lattice vector.

Exercise 4 from Section 4.2.1 provides upper bounds on the norm of a shortest lattice vector. Using the correspondence between norms and $\mathbf{0}$-symmetric, compact convex sets with $\mathbf{0}$ in the interior established in Theorem 3.3.14, a related geometric notion arises.

Definition 4.3.2 Let $\Lambda \subseteq \mathbb{R}^d$ be a lattice. Let $C \subseteq \mathbb{R}^d$ be a compact, convex set with $\mathbf{0}$ in its interior. The *packing radius of C with respect to the lattice Λ* is defined as

$$\rho(C,\Lambda) := \sup\{r > 0 : \mathrm{int}(rC) \cap (\mathrm{int}(rC) + \mathbf{z}) = \emptyset \ \forall \mathbf{z} \in \Lambda\}.$$

Equivalently, $\rho(C,\Lambda)$ is the largest $r > 0$ such that the interiors of two lattice translates of rC do not intersect.

The following relation follows from the definitions.

Lemma 4.3.3 *Let $\Lambda \subseteq \mathbb{R}^d$ be a lattice. For any norm N on $\mathbb{R}^d$, $N(SV(N,\Lambda)) = 2\rho(B_N(\mathbf{0},1),\Lambda)$. Equivalently, if $C \subseteq \mathbb{R}^d$ is a $\mathbf{0}$-symmetric, compact, convex set with $\mathbf{0}$ in its interior then $\rho(C,\Lambda) = \frac{1}{2}N_C(SV(N_C,\Lambda))$, where N_C is the norm associated with C (see Theorem 3.3.14).*[2]

---

[2] The notation $\lambda_1(C,\Lambda)$ is often used in the literature to denote the length of the shortest lattice vector with respect to the norm N_C, which is also called the *homogenous minimum of C with respect to Λ*. This is related to the so-called *successive minima* defined by Minkowski that provide powerful generalizations of Theorem 4.2.4. See the notes at the end of this chapter.

A closely related notion to the shortest lattice vector is the following.

Definition 4.3.4　Let $\Lambda \subseteq \mathbb{R}^d$ be a lattice and N be a norm on $\mathbb{R}^d$. For $\mathbf{t} \in \mathbb{R}^d$, the *closest lattice vector(s) to* $\mathbf{t}$ *(in the norm N)* is defined as $CV(\mathbf{t}, \Lambda) := \operatorname{argmin}\{d_N(\mathbf{t}, \mathbf{z}) : \mathbf{z} \in \Lambda\}$. This is well defined by Exercise 1 in Section 4.1.1. The *closest lattice vector problem (CVP)* is the algorithmic problem of computing a closest lattice vector.

Observe that $CV(\mathbf{t}, \Lambda) = CV(\operatorname{Proj}_{\operatorname{span}(\Lambda)}(\mathbf{t}), \Lambda)$. Thus, in general, one may restrict attention to $\operatorname{span}(\Lambda)$. It turns out to be useful to also consider the "farthest" point from the lattice.

Lemma 4.3.5 *Let $\Lambda \subseteq \mathbb{R}^d$ be a lattice and N be a norm on $\mathbb{R}^d$. Then*

$$\sup_{\mathbf{t} \in \operatorname{span}(\Lambda)} \min\{d_N(\mathbf{t}, \mathbf{z}) : \mathbf{z} \in \Lambda\} = \inf\{r > 0 : B_N(\mathbf{0}, r) + \Lambda \supseteq \operatorname{span}(\Lambda)\}.$$

Equivalently, for any $\mathbf{t} \in \operatorname{span}(\Lambda)$, its distance to a closest lattice vector is at most the smallest radius of a ball (in the norm N) whose lattice translates cover $\operatorname{span}(\Lambda)$. Moreover, there exists $\mathbf{t}^\star \in \operatorname{span}(\Lambda)$ that achieves this upper bound.

Proof　Left as an exercise.　　　　　　　　　　　　　　　　　　$\square$

Lemma 4.3.5 motivates the following definition.

Definition 4.3.6　Let $\Lambda \subseteq \mathbb{R}^d$ be a lattice. For any compact, convex set $C \subseteq \operatorname{span}(\Lambda)$, we define the *covering radius* or *inhomogeneous minimum of C with respect to Λ* as

$$\mu(C, \Lambda) := \inf\{r > 0 : rC + \Lambda = \operatorname{span}(\Lambda)\}.$$

Using the correspondence between $\mathbf{0}$-symmetric, compact, convex sets with $\mathbf{0}$ in the interior and norms on $\mathbb{R}^d$ established in Theorem 3.3.14, we can restate Lemma 4.3.5 as saying the following for full rank lattices: the covering radius of any such set is the maximum distance of any point in $\mathbb{R}^d$ to its closest lattice vector in the corresponding norm. The following interpretation of the covering radius proves useful in applications.

Lemma 4.3.7 *Let $\Lambda \subseteq \mathbb{R}^d$ be a lattice. For any compact, convex set $C \subseteq \operatorname{span}(\Lambda)$, $C + \Lambda \neq \operatorname{span}(\Lambda)$ if and only if there exists $\mathbf{t} \in \operatorname{span}(\Lambda)$ such that $(C + \mathbf{t}) \cap \Lambda = \emptyset$. In other words, there is a translate of C within $\operatorname{span}(\Lambda)$ that contains no lattice point if and only if the lattice translates of C do not cover $\operatorname{span}(\Lambda)$. Consequently,*

$$\mu(C, \Lambda) = \sup\{r > 0 : \exists \mathbf{t} \in \operatorname{span}(\Lambda) \text{ such that } (rC + \mathbf{t}) \cap \Lambda = \emptyset\}.$$

Proof Left as an exercise. $\qquad\square$

Shortest and closest lattice vector problems will feature as important optimization problems to be studied in Part II of this book. Our main interest in studying the corresponding notions of packing and covering radii is a theorem of Khinchine (Theorem 4.4.6) that will also be a key ingredient when we design optimization algorithms in Section 6.3.2. To formulate this deep result, we need to study how many parallel lattice hyperplanes intersect a convex set (see Definition 4.1.5).

Definition 4.3.8 Let $\Lambda \subseteq \mathbb{R}^d$ be a lattice and let L be a lattice subspace. For any convex set $C \subseteq \mathbb{R}^d$, define the *L-intersection number $iw_\Lambda(L,C)$* as the number of affine lattice subspaces parallel to L that have nonempty intersection with C (possibly infinite).

To get a computational handle on the L-intersection numbers, we relate them to solutions to certain optimization problems. This also leads us naturally to an important notion in lattice theory.

Definition 4.3.9 Let $\Lambda \subseteq \mathbb{R}^d$ be a lattice. The *dual* or *polar lattice* is defined as

$$\Lambda^* := \{\mathbf{w} \in \operatorname{span}(\Lambda) : \langle \mathbf{w}, \mathbf{z} \rangle \in \mathbb{Z} \quad \forall \mathbf{z} \in \Lambda\}.$$

A useful picture to keep in mind is that affine lattice subspaces parallel to a lattice subspace L are in one-to-one correspondence with $\operatorname{Proj}_{L^\perp}(\Lambda)$ (cf. proof of Theorem 4.1.8). Exercises 6 and 10 from Section 4.3.3 show that $\operatorname{Proj}_{L^\perp}(\Lambda) = (L^\perp \cap \Lambda^*)^*$ and so dual lattices provide another way to describe projected lattices. Thus, dual lattices can be used to investigate properties of affine lattice subspaces, in particular, counting affine lattice hyperplanes intersecting a convex set (cf. Exercise 9 from Section 4.1.1). This is made precise in Lemma 4.3.10 and Proposition 4.3.11.

Lemma 4.3.10 *Let $\Lambda \subseteq \mathbb{R}^d$ be a full rank lattice and let L be a lattice subspace of dimension $d - 1$. For any full-dimensional, compact, convex set $C \subseteq \mathbb{R}^d$, the L-intersection number $iw_\Lambda(L,C)$ equals either $\lfloor N_D(\mathbf{w}^\star) \rfloor$ or $\lfloor N_D(\mathbf{w}^\star) \rfloor + 1$, where $D = (C - C)^\circ$, N_D is the norm associated with D, and $\mathbf{w}^\star$ is a primitive vector of $L^\perp \cap \Lambda^*$ (cf. Exercise 5 from Section 2.2.3, Exercise 12 in Section 4.1.1, and Exercise 10 from Section 4.3.3).*

Proof Let $\mathcal{H}$ be the set of affine lattice subspaces parallel to L. Following the proof of Theorem 4.1.8, $\mathcal{H}$ is in one-to-one correspondence with $\operatorname{Proj}_{L^\perp}(\Lambda)$, which is a lattice of rank 1 by Exercise 10 from Section 4.1.1. By Exercise 9 from Section 4.1.1, $iw_\Lambda(L,C)$ is the number of elements in

$\text{Proj}_{L^\perp}(C) \cap \text{Proj}_{L^\perp}(\Lambda)$. Let $\mathbf{w}$ be a primitive vector of $\text{Proj}_{L^\perp}(\Lambda)$ and $\mathbf{w}^\star$ be a primitive vector of $L^\perp \cap \Lambda^\star$. By Exercise 10 from Section 4.3.3 and Exercise 12 from Section 4.1.1, $\mathbf{w}^\star$ is a basis for $(\text{Proj}_{L^\perp}(\Lambda))^\star$ and $\mathbf{w}$ is a basis of $\text{Proj}_{L^\perp}(\Lambda)$. Exercise 7 from Section 4.3.3 shows that $\langle \mathbf{w}^\star, \mathbf{w} \rangle = 1$, i.e., $\mathbf{w} = \frac{\mathbf{w}^\star}{\|\mathbf{w}^\star\|^2}$. Since C is compact and projections are linear maps, $\text{Proj}_{L^\perp}(C)$ is compact. By Exercise 6 from Section 2.3.1, $\text{Proj}_{L^\perp}(C) = \{\gamma \mathbf{w} : \ell \leq \gamma \leq u\}$, where $\ell = \inf_{\mathbf{x} \in C} \langle \mathbf{w}^\star, \mathbf{x} \rangle$ and $u = \sup_{\mathbf{x} \in C} \langle \mathbf{w}^\star, \mathbf{x} \rangle$. Since $\mathbf{w}$ is a basis for $\text{Proj}_{L^\perp}(\Lambda)$, we have $\text{Proj}_{L^\perp}(\Lambda) = \{\gamma \mathbf{w} : \gamma \in \mathbb{Z}\}$. By counting the number of integers in the interval bounded by ℓ and u, we obtain that $iw_\Lambda(L,C)$ is either $\lfloor u - \ell \rfloor$ or $\lfloor u - \ell \rfloor + 1$. We now observe that (using the notation σ_X from Section 3.3.2 to denote the support function of a set X):

$$
\begin{aligned}
u - \ell &= \sup_{\mathbf{x} \in C} \langle \mathbf{w}^\star, \mathbf{x} \rangle - \inf_{\mathbf{x} \in C} \langle \mathbf{w}^\star, \mathbf{x} \rangle \\
&= \sup_{\mathbf{x} \in C} \langle \mathbf{w}^\star, \mathbf{x} \rangle + \sup_{\mathbf{x} \in C} \langle \mathbf{w}^\star, -\mathbf{x} \rangle \\
&= \sup_{\mathbf{x} \in C} \langle \mathbf{w}^\star, \mathbf{x} \rangle + \sup_{\mathbf{y} \in -C} \langle \mathbf{w}^\star, \mathbf{y} \rangle \\
&= \sigma_C(\mathbf{w}^\star) + \sigma_{-C}(\mathbf{w}^\star) \\
&= \sigma_{C-C}(\mathbf{w}^\star) \\
&= N_D(\mathbf{w}^\star),
\end{aligned}
$$

where the second last equality follows from Exercise 15 in Section 3.3.5, and the last equality follows from Theorem 3.3.21. $\qquad\square$

Proposition 4.3.11 *Let $\Lambda \subseteq \mathbb{R}^d$ be a full rank lattice. For any full-dimensional, compact, convex set $C \subseteq \mathbb{R}^d$,*

$$\min\{iw_\Lambda(L,C) : L \subseteq \mathbb{R}^d \text{ lattice subspace of dimension } d-1\}$$

$$\leq \min_{\mathbf{w} \in \Lambda^\star \setminus \{\mathbf{0}\}} N_D(\mathbf{w}) + 1,$$

where $D = (C - C)^\circ$ and N_D is the norm associated with D (see Theorem 3.3.14).

Proposition 4.3.11 tells us that the minimum number of lattice hyperplanes that intersect a convex set is given by, up to an additive error of at most 1, the length of the shortest lattice vector in the dual lattice with respect to the norm associated with the polar of the difference body of C (see Exercise 5 from Section 2.2.3). Recall the simple relation between shortest lattice vectors and packing radii from Lemma 4.3.3. Lemma 4.3.7 connects the covering radius and lattice-free translates of a convex set. If we can find relationships connecting packing radii and covering radii, we will have a set of relationships connecting the minimum number of lattice hyperplanes intersecting a convex set and lattice-free translates of the set. This is precisely the content of Khinchine's theorem (Theorem 4.4.6). We now proceed to establish this connection between packing and covering radii.

4.3.1 Bounds for Packing and Covering Radii

It is intuitively apparent that if the shortest lattice vector in a lattice Λ is not very small in length, then the distance between parallel affine lattice hyperplanes cannot be too small either. Using the connection with dual lattices established earlier, this means the shortest lattice vectors in Λ and Λ^* cannot both be "large." This is made precise below for norms induced by positive definite matrices. Recall that the unit balls with respect to such a norm is an ellipsoid (cf. Section 2.7).

Theorem 4.3.12 *Let $\Lambda \subseteq \mathbb{R}^d$ be a full rank lattice and let $A \in \mathbb{R}^{d \times d}$ be a positive definite matrix. The product of the lengths of the shortest vectors in Λ and Λ^* with respect to N_A and $N_{A^{-1}}$ respectively is at most d. Equivalently,*

$$\rho(E(A, \mathbf{0}), \Lambda)\rho(E(A^{-1}, \mathbf{0}), \Lambda^*) \leq \frac{d}{4},$$

where we recall that $E(A^{-1}, \mathbf{0})$ is the ellipsoid that forms the unit ball with respect to the norm N_A induced by A (cf. Section 2.7).

Proof Let λ be the length of a shortest vector in Λ with respect to N_A and let λ^* be the length of a shortest vector in Λ^* with respect to $N_{A^{-1}}$. By Exercise 4 from Section 4.2.1,

$$\lambda \leq 2 \left(\frac{\det(\Lambda)}{\mathrm{vol}(E(A, \mathbf{0}))} \right)^{1/d}, \quad \lambda^* \leq 2 \left(\frac{\det(\Lambda^*)}{\mathrm{vol}(E(A^{-1}, \mathbf{0}))} \right)^{1/d}.$$

Using Exercise 7 from Section 4.3.3, $\det(\Lambda) \det(\Lambda^*) = 1$ and from Theorem 2.7.3, we obtain that

$$\lambda \lambda^* \leq \frac{4}{\mathrm{vol}(B(\mathbf{0}, 1))^{2/d}}.$$

Using Theorem 2.7.4, $\mathrm{vol}(B(\mathbf{0}, 1)) \geq \left(\frac{2}{\sqrt{d}} \right)^d$ and so $\lambda \lambda^* \leq d$.

The second statement follows from Lemma 4.3.3. $\qquad \square$

The above result can be generalized to an arbitrary norm and its dual norm (see the discussion at the end of Section 3.3.2), using a deep result of Bourgain and Milman; see the notes at the end of this chapter. The following implication of Theorem 4.3.12 for lattices that are not full rank will be useful.

Corollary 4.3.13 *Let $\Lambda \subseteq \mathbb{R}^d$ be a lattice. The product of the lengths of the shortest vectors in Λ and Λ^* with respect to the standard Euclidean norm is at most $\mathrm{rk}(\Lambda)$. Equivalently,*

$$\rho(B(\mathbf{0}, 1), \Lambda)\rho(B(\mathbf{0}, 1), \Lambda^*) \leq \frac{\mathrm{rk}(\Lambda)}{4}.$$

An induction argument that builds on Corollary 4.3.13 helps to make the connection between packing and covering radii.

Theorem 4.3.14 *Let $\Lambda \subseteq \mathbb{R}^d$ be a lattice and $E \subseteq \mathbb{R}^{d \times d}$ be an ellipsoid centered at the origin (cf. Section 2.7). Then,*

$$\mu(E, \Lambda)\rho(E^\circ, \Lambda^*) \leq \frac{\mathrm{rk}(\Lambda)^{3/2}}{4}.$$

Proof Let $A \in \mathbb{R}^{d \times d}$ be an invertible matrix such that $E = A(B(\mathbf{0}, 1))$ (see Theorem 2.7.1 and Exercise 3 from Section 2.7.2). By Exercises 9b and 10 from Section 2.4.4, $E^\circ = A^{-T}(B(\mathbf{0}, 1))$. Also, the covering and packing radii do not change under linear transformations, i.e., $\mu(E, \Lambda) = \mu(B(\mathbf{0}, 1), A^{-1}(\Lambda))$ and $\rho(E^\circ, \Lambda^*) = \rho(B(\mathbf{0}, 1), A^T(\Lambda^*))$. By Exercise 9 from Section 4.3.3, $(A^{-1}(\Lambda))^* = A^T(\Lambda^*)$. Thus, we see that it suffices to prove the theorem for the special case of $E = B(\mathbf{0}, 1)$.

We prove it by induction on $\mathrm{rk}(\Lambda)$. If $\mathrm{rk}(\Lambda) = 0$, i.e., $\Lambda = \{\mathbf{0}\} = \Lambda^*$, then $\mu(B(\mathbf{0}, 1), \Lambda) = \rho(B(\mathbf{0}, 1), \Lambda^*) = 0$ and we are done. Assume the statement holds for all lattices of rank up to k and consider a lattice Λ with $k + 1$. By Lemma 4.3.5, $\mu(B(\mathbf{0}, 1), \Lambda)$ is the distance of the farthest point in $\mathrm{span}(\Lambda)$ from Λ. Let $\mathbf{t}$ be this point. We decompose the distance between $\mathbf{t}$ and Λ into two parts. Let $\mathbf{u}$ be a shortest vector in Λ (with respect to the standard Euclidean norm) and let $L = \mathrm{span}(\{\mathbf{u}\})$. Define $\Lambda' = \mathrm{Proj}_{L^\perp}(\Lambda)$ which is a lattice of rank k by Proposition 4.1.8 and Exercise 10 in Section 4.1.1. Express $\mathbf{t} = \mathbf{t}' + \lambda\mathbf{u}$, where $\mathbf{t}' \in L^\perp$ and $\lambda \in \mathbb{R}$. Let $\mathbf{z}' \in \Lambda'$ be the closest lattice vector to $\mathbf{t}'$. Consider the line $\{\mathbf{z}' + \gamma\mathbf{u} : \gamma \in \mathbb{R}\}$. Since $\mathbf{z}'$ is in the projected lattice, this line contains lattice points of the form $\mathbf{p} + \gamma\mathbf{u}$, $\gamma \in \mathbb{Z}$, for some $\mathbf{p} \in \Lambda$. Consider the lattice point of this form closest to $\mathbf{z}' + \lambda\mathbf{u}$. It must be of the form $\mathbf{z} := \mathbf{z}' + \bar{\lambda}\mathbf{u} \in \Lambda$, where $|\bar{\lambda} - \lambda| \leq \frac{1}{2}$. Therefore,

$$\begin{aligned}
\mu(B(\mathbf{0}, 1), \Lambda)^2 &\leq \|\mathbf{t} - \mathbf{z}\|^2 \\
&= \|\mathbf{t}' + \lambda\mathbf{u} - \mathbf{z}' - \bar{\lambda}\mathbf{u}\|^2 \\
&= \|\mathbf{t}' - \mathbf{z}'\|^2 + |\lambda - \bar{\lambda}|^2\|\mathbf{u}\|^2 \\
&\leq \mu(B(\mathbf{0}, 1), \Lambda')^2 + \rho(B(\mathbf{0}, 1), \Lambda)^2,
\end{aligned}$$

where the last inequality follows from Lemmas 4.3.3 and 4.3.5 and the fact that $|\lambda - \bar{\lambda}| \leq \frac{1}{2}$. By Exercise 10 from Section 4.3.3, $(\Lambda')^* = \mathrm{Proj}_{L^\perp}(\Lambda)^* = L^\perp \cap \Lambda^* \subseteq \Lambda^*$. Combined with Lemma 4.3.3, this implies that $\rho(B(\mathbf{0}, 1), \Lambda^*) \leq \rho(B(\mathbf{0}, 1), (\Lambda')^*)$. Therefore,

$$
\begin{aligned}
\mu(B(\mathbf{0},1),\Lambda)^2 \rho(B(\mathbf{0},1),\Lambda^*)^2
&\leq (\mu(B(\mathbf{0},1),\Lambda')^2 + \rho(B(\mathbf{0},1),\Lambda)^2)\rho(B(\mathbf{0},1),\Lambda^*)^2 \\
&\leq \mu(B(\mathbf{0},1),\Lambda')^2 \rho(B(\mathbf{0},1),(\Lambda')^*)^2 + \rho(B(\mathbf{0},1),\Lambda)^2 \rho(B(\mathbf{0},1),\Lambda^*)^2 \\
&\leq \frac{k^3}{16} + \frac{(k+1)^2}{16} \\
&\leq \frac{k(k+1)^2}{16} + \frac{(k+1)^2}{16} \\
&= \frac{(k+1)^3}{16},
\end{aligned}
$$

where the third inequality follows from the induction hypothesis and Corollary 4.3.13. This completes the induction step. $\square$

4.3.2 Dirichlet–Voronoi Cells

An important geometric object associated with shortest and closest lattice vectors is the following.

Definition 4.3.15 Let $\Lambda \subseteq \mathbb{R}^d$ be a lattice. The set

$$
\mathcal{V}(\Lambda) := \{\mathbf{x} \in \mathbb{R}^d : \|\mathbf{x}\| \leq \|\mathbf{z} - \mathbf{x}\| \ \forall \mathbf{z} \in \Lambda\}
$$

of points that are closer to $\mathbf{0}$ compared to any other lattice point is called the *Dirichlet–Voronoi cell of Λ*.

The set of closest lattice vectors to a point in $\mathbb{R}^d$ can be characterized in terms of the Dirichlet–Voronoi cell; see Figure 4.3.

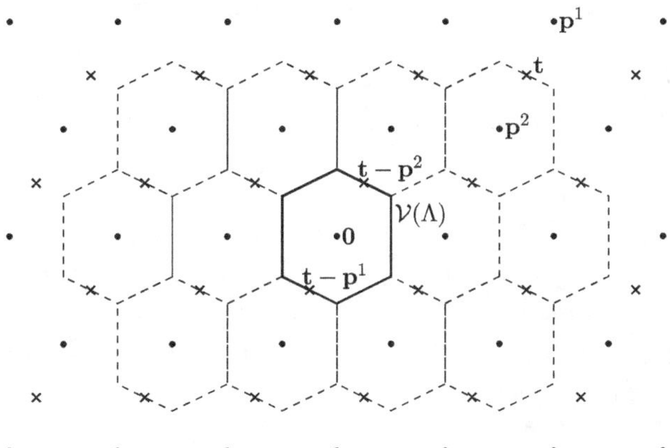

Figure 4.3 A lattice and its Voronoi cell illustrating Proposition 4.3.16. The lattice Λ is shown in solid disks. The hexagon with solid lines is the Dirichlet–Voronoi cell $\mathcal{V}(\Lambda)$. The hexagons with dashed lines are lattice translates of $\mathcal{V}(\Lambda)$. A point $\mathbf{t}$ and its coset $\Lambda + \mathbf{t}$ are shown using crosses. $\mathbf{p}^1$ and $\mathbf{p}^2$ are the closest lattice vectors to $\mathbf{t}$.

Proposition 4.3.16 *Let $\Lambda \subseteq \mathbb{R}^d$ be a lattice and let $\mathbf{t} \in \mathbb{R}^d$. Then, $(\Lambda + \mathbf{t}) \cap \mathcal{V}(\Lambda) = \{\mathbf{t} - \mathbf{p} : \mathbf{p} \in CV(\mathbf{t}, \Lambda)\}$ (where the closest lattice vector is computed in the standard Euclidean norm). Further, $(\Lambda + \mathbf{t}) \cap \mathcal{V}(\Lambda)$ is precisely the set of points from the coset $\Lambda + \mathbf{t}$ with smallest Euclidean norm.*

Proof Consider an arbitrary $\mathbf{p} \in \Lambda$. Then

$$
\begin{aligned}
\mathbf{p} \in CV(\mathbf{t}, \Lambda) &\Leftrightarrow \|\mathbf{t} - \mathbf{p}\| \le \|\mathbf{t} - \mathbf{z}\| \quad \forall \mathbf{z} \in \Lambda \\
&\Leftrightarrow \|\mathbf{t} - \mathbf{p}\| \le \|\mathbf{t} - \mathbf{p} + \mathbf{p} - \mathbf{z}\| \quad \forall \mathbf{z} \in \Lambda \\
&\Leftrightarrow \|\mathbf{t} - \mathbf{p}\| \le \|\mathbf{t} - \mathbf{p} - \mathbf{z}'\| \quad \forall \mathbf{z}' \in \Lambda \\
&\Leftrightarrow \mathbf{t} - \mathbf{p} \in \mathcal{V}(\Lambda).
\end{aligned}
$$

This establishes that $(\Lambda + \mathbf{t}) \cap \mathcal{V}(\Lambda) = \{\mathbf{t} - \mathbf{p} : \mathbf{p} \in CV(\mathbf{t}, \Lambda)\}$. We observe that $\Lambda + \mathbf{t} = \{\mathbf{t} - \mathbf{p} : \mathbf{p} \in \Lambda\}$ and therefore the set of points in $\Lambda + \mathbf{t}$ with smallest Euclidean norm is precisely $\{\mathbf{t} - \mathbf{p} : \mathbf{p} \in CV(\mathbf{t}, \Lambda)\} = (\Lambda + \mathbf{t}) \cap \mathcal{V}(\Lambda)$. $\square$

Note that for any $\mathbf{z} \in \Lambda$, the constraint $\|\mathbf{x}\| \le \|\mathbf{z} - \mathbf{x}\|$ gives a halfspace, i.e.,

$$
H_{\mathbf{z}} := \{\mathbf{x} \in \mathbb{R}^d : \|\mathbf{x}\| \le \|\mathbf{z} - \mathbf{x}\|\} = \left\{\mathbf{x} \in \mathbb{R}^d : \langle \mathbf{z}, \mathbf{x} \rangle \le \frac{\|\mathbf{z}\|^2}{2}\right\}. \tag{4.3.1}
$$

In other words, $\mathcal{V}(\Lambda) = \bigcap_{\mathbf{z} \in \Lambda} H_{\mathbf{z}}$. While this intersection is an intersection over an infinite family of halfspaces (one for each lattice point), we will now see that all but finitely many of them are redundant. In particular, $\mathcal{V}(\Lambda)$ is a polyhedron. We begin with some important observations.

Proposition 4.3.17 *Let $\Lambda \subseteq \mathbb{R}^d$ be a lattice. The following are all true.*

1. *$\mathcal{V}(\Lambda)$ is full dimensional and $\mathrm{int}(\mathcal{V}(\Lambda)) = \{\mathbf{x} \in \mathbb{R}^d : \|\mathbf{x}\| < \|\mathbf{z} - \mathbf{x}\| \; \forall \mathbf{z} \in \Lambda\}$.*
2. *$\mathcal{V}(\Lambda)$ is $\mathbf{0}$-symmetric.*
3. *For all $\mathbf{p} \in \Lambda$, $\{\mathbf{x} \in \mathbb{R}^d : \|\mathbf{p} - \mathbf{x}\| \le \|\mathbf{z} - \mathbf{x}\| \; \forall \mathbf{z} \in \Lambda\} = \mathcal{V}(\Lambda) + \mathbf{p}$. In other words, the set of points closest to $\mathbf{p}$ compared to all other lattice points is given by the translate of the Dirichlet–Voronoi cell by $\mathbf{p}$.*
4. *For any $\mathbf{p}^1, \mathbf{p}^2 \in \Lambda$, $(\mathcal{V}(\Lambda) + \mathbf{p}^1) \cap (\mathcal{V}(\Lambda) + \mathbf{p}^2)$ is a common face of $\mathcal{V}(\Lambda) + \mathbf{p}^1$ and $\mathcal{V}(\Lambda) + \mathbf{p}^2$.*
5. *$\bigcup_{\mathbf{p} \in \Lambda} \mathcal{V}(\Lambda) + \mathbf{p} = \mathbb{R}^d$.*

Proof Left as an exercise. $\square$

Theorem 4.3.18 *Let $\Lambda \subseteq \mathbb{R}^d$ be a lattice. There exists a finite set $R(\Lambda) \subseteq \Lambda$ such that $\mathcal{V}(\Lambda) = \bigcap_{\mathbf{z} \in R(\Lambda)} H_{\mathbf{z}}$ and for any set $R \subseteq \Lambda$ such that $\mathcal{V}(\Lambda) = \bigcap_{\mathbf{z} \in R} H_{\mathbf{z}}$, we have $R(\Lambda) \subseteq R$. In other words, there exists a unique minimal set $R(\Lambda)$ such that the Dirichlet–Voronoi cell of Λ is described by the halfspaces corresponding to the points in $R(\Lambda)$.*

Proof We first establish the result when Λ is full rank. Let $\mathbf{b}^1, \ldots, \mathbf{b}^d$ be a basis for Λ. Consider the bounded polytope $P := \bigcap_{i=1}^{d} H_{\mathbf{b}^i} \cap H_{-\mathbf{b}^i}$ (see Exercise 6 from Section 2.5.6) and observe that $\mathcal{V}(\Lambda) \subseteq P$. Let M be the maximum norm of a vertex of P. Consider any $\mathbf{z}$ with $\|\mathbf{z}\| > 2M$. By the Cauchy–Schwarz inequality, every vertex $\mathbf{v}$ of P satisfies $\langle \mathbf{z}, \mathbf{v} \rangle \leq M\|\mathbf{z}\| < \frac{\|\mathbf{z}\|^2}{2}$. Thus, $P \subseteq H_{\mathbf{z}}$ and every point in P satisfies the corresponding inequality $\langle \mathbf{z}, \mathbf{x} \rangle \leq \frac{\|\mathbf{z}\|^2}{2}$ strictly. This implies that $H_{\mathbf{z}}$ is redundant for $\mathcal{V}(\Lambda)$ and, therefore, $\mathcal{V}(\Lambda) = P \cap \bigcap_{\mathbf{z} \in \Lambda : \|\mathbf{z}\| \leq 2M} H_{\mathbf{z}}$. This shows that $\mathcal{V}(\Lambda)$ is a polyhedron. By Proposition 4.3.17, part 1., $\mathcal{V}(\Lambda)$ is full-dimensional, and so by Theorem 2.5.31, $\mathcal{V}(\Lambda)$ is described by a unique system of inequalities up to scaling and permutation. However, $H_{\mathbf{z}} \neq H_{\mathbf{z}'}$ for $\mathbf{z} \neq \mathbf{z}'$ (see Exercise 15 from Section 4.3.3) and thus there exists a unique minimal subset $R(\Lambda) \subseteq \Lambda$ such that $\mathcal{V}(\Lambda) = \bigcap_{\mathbf{z} \in R(\Lambda)} H_{\mathbf{z}}$.

The case where Λ is not full rank is left as an exercise. $\qquad\square$

Definition 4.3.19 Let $\Lambda \subseteq \mathbb{R}^d$ be a lattice. A vector from the set $R(\Lambda)$ from Theorem 4.3.18 is called a *Voronoi relevant vector for* Λ.

The following characterization of Voronoi relevant vectors is very useful.

Theorem 4.3.20 *Let* $\Lambda \subseteq \mathbb{R}^d$ *be a lattice and let* $\mathbf{r} \in \Lambda \setminus \{\mathbf{0}\}$. *The following are equivalent.*

1. $\mathbf{r}$ *is a Voronoi relevant vector.*
2. $\langle \mathbf{r}, \mathbf{y} \rangle \leq \frac{\|\mathbf{r}\|^2}{2}$ *defines a facet* $F \subseteq \mathcal{V}(\Lambda)$ *that is symmetric about* $\frac{\mathbf{r}}{2}$, *i.e.,* $\mathbf{y} \in F$ *if and only if* $\mathbf{r} - \mathbf{y} \in F$.
3. *There exists a facet* F *of* $\mathcal{V}(\Lambda)$ *that is symmetric about* $\frac{\mathbf{r}}{2}$, *i.e.,* $\mathbf{y} \in F$ *if and only if* $\mathbf{r} - \mathbf{y} \in F$.
4. $CV(\mathbf{r}, 2\Lambda) = \{\mathbf{0}, 2\mathbf{r}\}$.
5. $(2\Lambda + \mathbf{r}) \cap \mathcal{V}(2\Lambda) = \{+\mathbf{r}, -\mathbf{r}\}$.
6. $\{+\mathbf{r}, -\mathbf{r}\}$ *is precisely the set of shortest vectors (in Euclidean norm) in the coset of* Λ *with respect to* 2Λ *containing* $\mathbf{r}$.

Proof (1. $\Rightarrow$ 2.) By definition of a Voronoi relevant vector, $\langle \mathbf{r}, \mathbf{y} \rangle \leq \frac{\|\mathbf{r}\|^2}{2}$ defines a facet $F \subseteq \mathcal{V}(\Lambda)$. We now show that F is symmetric about $\frac{\mathbf{r}}{2}$.

If $\langle \mathbf{r}, \mathbf{y} \rangle = \frac{\|\mathbf{r}\|^2}{2}$, then $\langle \mathbf{r}, \mathbf{r} - \mathbf{y} \rangle = \|\mathbf{r}\|^2 - \frac{\|\mathbf{r}\|^2}{2} = \frac{\|\mathbf{r}\|^2}{2}$. Thus, it suffices to show that $\mathbf{r} - \mathbf{y} \in \mathcal{V}(\Lambda)$. Observe also that $\|\mathbf{r} - \mathbf{y}\|^2 = \|\mathbf{r}\|^2 - 2\langle \mathbf{r}, \mathbf{y} \rangle + \|\mathbf{y}\|^2 = \|\mathbf{y}\|^2$.

Consider any $\mathbf{z} \in \Lambda$. Then $\mathbf{r} - \mathbf{z} \in \Lambda$. Since $\mathbf{y} \in F \subseteq \mathcal{V}(\Lambda)$, we have that $\|\mathbf{r} - \mathbf{y}\| = \|\mathbf{y}\| \leq \|\mathbf{y} - (\mathbf{r} - \mathbf{z})\| = \|(\mathbf{r} - \mathbf{y}) - \mathbf{z}\|$. Thus, $\mathbf{r} - \mathbf{y} \in \mathcal{V}(\Lambda)$.

(2. $\Rightarrow$ 3.) Nothing to show.

(3. $\Rightarrow$ 4.) Since F is symmetric about $\frac{\mathbf{r}}{2}$, this point must be in the relative interior of F. This implies that for any other $\mathbf{z} \in \Lambda \setminus \{\mathbf{r}, \mathbf{0}\}$, $\langle \mathbf{z}, \frac{\mathbf{r}}{2} \rangle < \frac{\|\mathbf{z}\|^2}{2}$. In other words, $\frac{\mathbf{r}}{2}$ is strictly closer to $\mathbf{0}$ compared to all other lattice points, except $\mathbf{r}$ and $\mathbf{0}$, i.e., $CV(\frac{\mathbf{r}}{2}, \Lambda) = \{\mathbf{0}, \mathbf{r}\}$. Scaling Λ by a factor of 2 gives $CV(\mathbf{r}, 2\Lambda) = \{\mathbf{0}, 2\mathbf{r}\}$.

(4. $\Rightarrow$ 1.) Scaling by a factor of 2, $CV(\mathbf{r}, 2\Lambda) = \{\mathbf{0}, 2\mathbf{r}\}$ implies that the closest lattice points in Λ to $\frac{\mathbf{r}}{2}$ are precisely $\mathbf{0}, \mathbf{r}$. Thus, for any $\mathbf{z} \in \Lambda \setminus \{\mathbf{0}, \mathbf{r}\}$, $\|\frac{\mathbf{r}}{2}\| < \|\mathbf{z} - \frac{\mathbf{r}}{2}\|$, which implies that $\langle \mathbf{z}, \frac{\mathbf{r}}{2} \rangle < \frac{\|\mathbf{z}\|^2}{2}$. On the other hand, since $\langle \mathbf{r}, \frac{\mathbf{r}}{2} \rangle = \frac{\|\mathbf{r}\|^2}{2}$, we find that $\langle \mathbf{r}, \mathbf{y} \rangle \leq \frac{\|\mathbf{r}\|^2}{2}$ defines a facet of $\mathcal{V}(\Lambda)$ with $\frac{\mathbf{r}}{2}$ in the relative interior.

The equivalence of 4., 5., and 6. follows from Proposition 4.3.16. $\quad\square$

Corollary 4.3.21 *Let $\Lambda \subseteq \mathbb{R}^d$ be a lattice of rank r. Then $|R(\Lambda)| \leq 2(2^r - 1)$, i.e., $\mathcal{V}(\Lambda)$ has at most $2(2^r - 1)$ facets.*

Proof From part 6. of Theorem 4.3.20, if $\mathbf{r} \in \Lambda \setminus \{\mathbf{0}\}$ is a Voronoi relevant vector then it is one of the two shortest vectors in the coset of Λ with respect to 2Λ containing $\mathbf{r}$. By Exercise 18 from Section 4.1.1, the number of cosets of Λ with respect to 2Λ is 2^r. Leaving out the coset corresponding to 2Λ (since $\mathbf{r} \neq \mathbf{0}$), we have $2^r - 1$ cosets, each contributing at most two Voronoi relevant vectors. $\quad\square$

4.3.3 Exercises

1.* Let $B \in \mathbb{R}^{d \times r}$ be a basis for a lattice $\Lambda \subseteq \mathbb{R}^d$. Show that the length of the shortest lattice vector in the standard Euclidean norm is lower bounded by the smallest nonzero singular value of B.

2. Show that the supremum is attained in the definition of packing radius and show Lemma 4.3.3.

3. Prove Lemma 4.3.5 and show that the supremum in the LHS of the equality and the infimum in the RHS of the equality are both attained.

4. Prove Lemma 4.3.7.

5.* Let $\Lambda \subseteq \mathbb{R}^d$ be a lattice and $C \subseteq \mathbb{R}^d$ be a full-dimensional, compact, convex set such that $\mathrm{int}(C) \cap (\mathrm{int}(C) + \mathbf{z}) = \emptyset$ for all $\mathbf{z} \in \Lambda$. Show that λC contains at most $(\lambda + 1)^d$ points from $\Lambda + \mathbf{t}$ for any $\mathbf{t} \in \mathbb{R}^d$.

6.* Show that Λ^* is a lattice for any lattice Λ with $\mathrm{rk}(\Lambda^*) = \mathrm{rk}(\Lambda)$, and $(\Lambda^*)^* = \Lambda$.

7. Let $\Lambda \subseteq \mathbb{R}^d$ be a lattice. Show that the following are equivalent.

(a) The columns of $B \in \mathbb{R}^{d \times k}$ form a basis for Λ^*.

(b) $B^T A$ is a unimodular matrix for every matrix $A \in \mathbb{R}^{d \times k}$ whose columns form a basis for Λ.

(c) There exists a matrix $A \in \mathbb{R}^{d \times k}$ whose columns form a basis for Λ such that $B^T A$ is a unimodular matrix.

8. Show that $(\mathbb{Z}^d)^* = \mathbb{Z}^d$.

9. Show that for any lattice $\Lambda \subseteq \mathbb{R}^d$ and any invertible matrix $A \in \mathbb{R}^{d \times d}$, $(A(\Lambda))^* = A^{-T}(\Lambda^*)$ (see Exercise 2 in Section 4.1.1).

10. Let $\Lambda \subseteq \mathbb{R}^d$ be a lattice and let L be a lattice subspace. Show that

(a) $(\mathrm{Proj}_{L^{\perp}}(\Lambda))^* = L^{\perp} \cap \Lambda^*$.

(b) $L^{\perp} \cap \mathrm{span}(\Lambda)$ is a lattice subspace of Λ^*.

11. Let $\Lambda \subseteq \mathbb{R}^d$ be a full rank lattice and let $L \subseteq \mathbb{R}^d$ be a subspace. Show that L is a lattice subspace of Λ if and only if $L^{\perp}$ is a lattice subspace of Λ^*.

12. Prove Proposition 4.3.11.

13. In Theorem 4.3.12, can one prove a lower bound on the product of the shortest vectors in Λ and Λ^* with respect to N_A and $N_{A^{-1}}$, respectively, that works for all lattices?

14. Verify Corollary 4.3.13.

15. Consider the halfspace defined in (4.3.1) corresponding to $\mathbf{z} \in \Lambda$. Show that if $\mathbf{z} \neq \mathbf{z}'$ then $H_{\mathbf{z}} \neq H_{\mathbf{z}'}$.

16. Let $\Lambda \subseteq \mathbb{R}^d$ be a lattice and let $\mathbf{t} \in \mathbb{R}^d$. Proposition 4.3.16 implies that $(\Lambda + \mathbf{t}) \cap \mathcal{V}(\Lambda)$ is nonempty. Show that $(\Lambda + \mathbf{t}) \cap \mathcal{V}(\Lambda)$ is a singleton if and only if $(\Lambda + \mathbf{t}) \cap \mathrm{int}(\mathcal{V}(\Lambda)) \neq \emptyset$.

17. Let $\Lambda \subseteq \mathbb{R}^d$ be a lattice. Show that for any $\mu \geq 0$, $\mathcal{V}(\mu\Lambda) = \mu\mathcal{V}(\Lambda)$.

18. Prove Proposition 4.3.17.

19. Show that the volume of the Dirichlet–Voronoi cell of any full rank lattice is equal to the determinant of the lattice.

20. Show that for any lattice $\Lambda \subseteq \mathbb{R}^d$, $\mathrm{int}(2\mathcal{V}(\Lambda)) \cap \Lambda = \{\mathbf{0}\}$.

21. Complete the proof of Theorem 4.3.18 when Λ is not full rank.

22. Let $\Lambda \subseteq \mathbb{R}^d$ be a lattice, and let $\mathbf{p}^1, \mathbf{p}^2 \in \Lambda$. Show that the translates $\mathcal{V}(\Lambda) + \mathbf{p}^1$ and $\mathcal{V}(\Lambda) + \mathbf{p}^2$ share a common facet if and only if $\mathbf{p}^1 - \mathbf{p}^2$ is a Voronoi relevant vector.

23. Let $\Lambda \subseteq \mathbb{R}^d$ be a lattice. Let $\mathbf{p}, \mathbf{q} \in \Lambda$ such that $\mathcal{V}(\Lambda) + \mathbf{p}$ and $\mathcal{V}(\Lambda) + \mathbf{q}$ have a nonempty intersection. Show that there is a sequence of lattice vectors $\mathbf{p}^0 := \mathbf{p}, \mathbf{p}^1, \ldots, \mathbf{p}^k := q$ such that $\mathcal{V}(\Lambda) + \mathbf{p}^i$ and $\mathcal{V}(\Lambda) + \mathbf{p}^{i+1}$ share a common facet for $i = 0, \ldots, k - 1$.

4.4 Lattice-Free Convex Sets

A key algorithmic question in the geometry of numbers that is of central importance for the optimization problems to be considered in Part II is the following. Given a convex set $C \subseteq \mathbb{R}^d$ and a lattice $\Lambda \subseteq \mathbb{R}^d$, decide if $C \cap \Lambda = \emptyset$. In fact, the study of convex sets that have no lattice point in them has been a prominent theme within the traditional geometry of numbers literature as well. The notion of covering radius is intimately linked to this question (see Lemma 4.3.7) and has been around long before applications to optimization were discovered. In this section, we investigate properties of convex sets that contain no lattice point. To handle sets that are closed and those that are not in a unified way, we will employ the following definition of "lattice-free-ness."

Definition 4.4.1 Let $\Lambda \subseteq \mathbb{R}^d$ be a lattice. A convex set $C \subseteq \mathbb{R}^d$ is said to be Λ-*free* (or simply *lattice-free* if the lattice is clear from context) if $\mathrm{int}(C) \cap \Lambda = \emptyset$. A convex set is *maximal lattice-free* if it is lattice-free and maximal with respect to set inclusion, i.e., it is not strictly contained in another lattice-free set.

Example 4.4.2 Here are some examples of lattice-free and maximal lattice-free sets. See Figure 4.4 for illustrations.

1. Let $\Lambda \subseteq \mathbb{R}^d$ be a lattice. For any $\mathbf{w} \in \Lambda^*$, the set $\{\mathbf{x} \in \mathbb{R}^d : K \leq \langle \mathbf{w}, \mathbf{x} \rangle \leq K + 1\}$ where $K \in \mathbb{Z}$ is called a *split set* and is lattice-free. It is maximal lattice-free if and only if $\mathbf{w}$ is a primitive vector of Λ^* (cf. Exercise 12 from Section 4.1.1).
2. Let $\Lambda \subseteq \mathbb{R}^d$ be a full rank lattice. Let $\mathbf{w}^1, \ldots, \mathbf{w}^d$ be a basis for Λ^*. The simplex

$$\left\{ \mathbf{x} \in \mathbb{R}^d : \begin{array}{l} \langle \mathbf{w}^1, \mathbf{x} \rangle \geq 0 \ \ i = 1, \ldots, d \\ \langle \mathbf{w}^1 + \cdots + \mathbf{w}^d, \mathbf{x} \rangle \leq d \end{array} \right\}$$

   is a maximal lattice-free set.

Theorem 4.4.3 *For any lattice-free convex set C, there exists a maximal lattice-free convex set that contains it.*

Proof Left as an exercise. $\qquad\qquad\qquad\qquad\qquad\qquad\qquad\qquad\qquad\Box$

4.4.1 Maximal Lattice-Free Convex Sets

In this subsection, we prove a result that is useful as a tool to reduce the study of full-dimensional lattice-free sets to the bounded case.

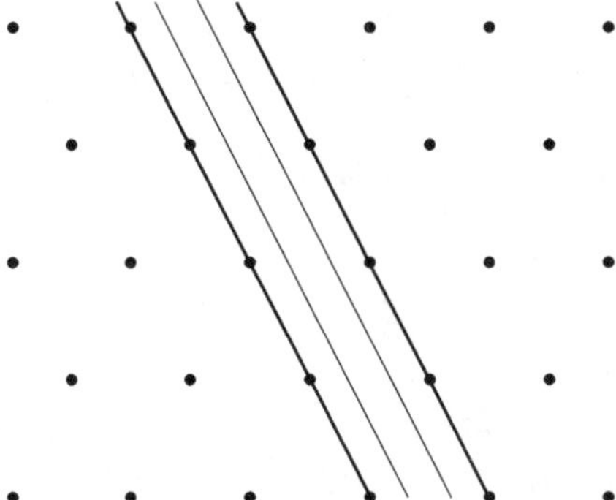

(a) A split set contained in a maximal split set (see Example 4.4.2, part 1.)

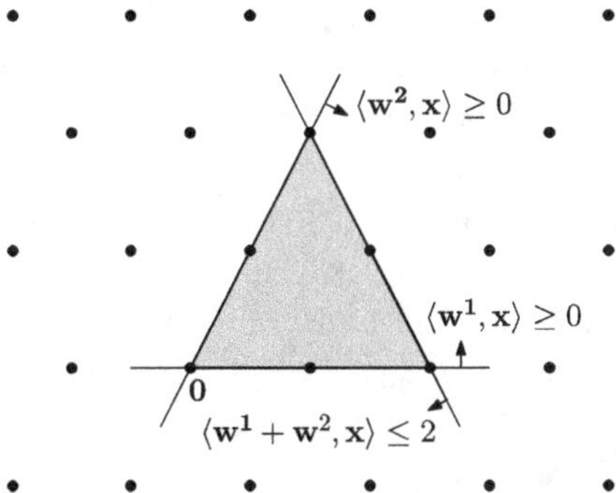

(b) The simplex from part 2. of Example 4.4.2

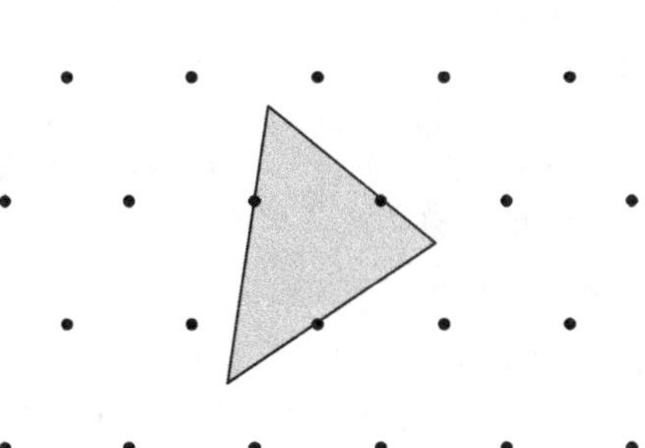

(c) Another example of a maximal lattice-free triangle in two dimensions

Figure 4.4 Illustration of maximal lattice-free sets.

Theorem 4.4.4 *Let $\Lambda \subseteq \mathbb{R}^d$ be a full rank lattice. A set $C \subseteq \mathbb{R}^d$ is a maximal Λ-free set if and only if it satisfies one of the following conditions.*

1. *C is a hyperplane that is not an affine lattice subspace.*
2. *C is a full-dimensional lattice-free polyhedron of the form $C = P + L$ where L is a lattice subspace with $\dim(L) < d$ and $P \subseteq L^{\perp}$ is a polytope of dimension $\dim(L^{\perp})$, and every facet of C contains a point from Λ in its relative interior.*

Proof ($\Leftarrow$) Suppose C is a hyperplane that is not an affine lattice subspace. By Exercise 13 in Section 2.2.3, $\text{int}(C) = \emptyset$ and therefore C is lattice-free. Consider any C' that contains C strictly. We will show that C' contains a lattice point in its interior. First, we may assume C' is closed because otherwise we may work with $\text{cl}(C')$ (observe that $\text{int}(\text{cl}(C')) = \text{int}(C')$). If $C' = \mathbb{R}^d$, then we are done. Otherwise, since $C \subseteq C'$, we have $\text{lin}(C) \subseteq \text{lin}(C')$ (see Corollary 2.4.21 from Section 2.4.4). $\text{lin}(C)$ is of dimension $d - 1$ since C is a hyperplane. Let $\mathbf{x} \in C' \backslash C$. Then $\mathbf{x} + \text{lin}(C) \subseteq C'$ and $\mathbf{x} + \text{lin}(C)$ is an affine subspace parallel to C that is disjoint from C. Let $\epsilon > 0$ be the distance between these two affine subspaces. Thus, the projection $\text{Proj}_{\text{lin}(C)^{\perp}}(C')$ contains a line segment of length ϵ. By Exercise 8 from Section 4.2.1 and Proposition 4.2.6, we see the $\text{Proj}_{\text{lin}(C)^{\perp}}(\Lambda)$ is a dense subset of $\text{lin}(C)^{\perp}$, since $\text{lin}(C)$ is not a lattice subspace by assumption. This implies that there exists $\mathbf{z}' \in \text{Proj}_{\text{lin}(C)^{\perp}}(\Lambda) \cap \text{int}(\text{Proj}_{\text{lin}(C)^{\perp}}(C'))$. Since $\mathbf{z}' + \text{lin}(C) \subseteq \text{int}(C')$, this implies the existence of a point from Λ in $\text{int}(C')$.

Next, consider C that satisfies the second part of the statement. Since C is already assumed to be lattice-free, we just need to show the maximality. Consider any C' that contains C strictly. We will show that C' contains a lattice point in its interior. Let $\mathbf{x} \in C' \backslash C$ and let F be a facet of C such that the corresponding facet defining inequality for C separates $\mathbf{x}$ from C. By assumption, there exists $\mathbf{z} \in \text{relint}(F) \cap \Lambda$. One can verify that $\mathbf{z} \in \text{int}(C')$ – see Exercise 3 from Section 4.4.3.

($\Rightarrow$) We break the proof into two cases.

Case 1: C is not full-dimensional. By Exercise 13 in Section 2.2.3, there exists a hyperplane H containing C. If H is an affine lattice hyperplane, then one can consider a split set (see part 1. from Example 4.4.2) with H as one of the bounding hyperplanes which would contradict the maximality of C. Therefore, H is not an affine lattice hyperplane. Since $\text{int}(H) = \emptyset$ (by Exercise 13 in Section 2.2.3 again), we must have that H is lattice-free. By the assumption of maximality, $C = H$ and C satisfies part 1. of the statement.

Case 2: C is full-dimensional. We first show that $\text{rec}(C) = \text{lin}(C)$. Since $\text{lin}(C) = \text{rec}(C) \cap -\text{rec}(C)$, this is equivalent to showing that

$C + (-\operatorname{rec}(C)) = C$. Since $C + (-\operatorname{rec}(C)) = \cup_{\mathbf{r} \in \operatorname{rec}(C)}(C + \operatorname{span}(\mathbf{r}))$, it suffices to show that $C + \operatorname{span}(\mathbf{r}) = C$ for all $\mathbf{r} \in \operatorname{rec}(C)$. Consider any $\mathbf{r} \in \operatorname{rec}(C)$. We will demonstrate that $C + \operatorname{span}(\mathbf{r})$ is lattice-free, and since C is a maximal lattice-free set contained in $C + \operatorname{span}(\mathbf{r})$, we shall have that $C + \operatorname{span}(\mathbf{r}) = C$. Suppose to the contrary that there exists $\mathbf{z} \in \Lambda \cap \operatorname{int}(C + \operatorname{span}(\mathbf{r}))$. By Exercises 36 and 37 in Section 2.4.4, $\mathbf{z} \in \operatorname{int}(C) + \operatorname{span}(\mathbf{r})$. This implies that there exists $\bar{\lambda} > 0$ such that $\mathbf{z} + \bar{\lambda}\mathbf{r} \in \operatorname{int}(C)$. Since $\mathbf{r} \in \operatorname{rec}(C)$, there exists $\epsilon > 0$ such that $B(\mathbf{z}, \epsilon) + \lambda \mathbf{r} \subseteq C$ for all $\lambda \geq \bar{\lambda}$. If $\operatorname{span}(\mathbf{r})$ is a lattice subspace, then there exist lattice points in the set $\{\mathbf{z} + \lambda \mathbf{r} : \lambda \geq \bar{\lambda}\}$. Such lattice points would lie in the interior of C contradicting that it is lattice-free. If $\operatorname{span}(\mathbf{r})$ is not a lattice subspace, by Exercise 6 in Section 4.2.1, there exist lattice points at distance less than ϵ from the set $\{\mathbf{z} + \lambda \mathbf{r} : \lambda \geq \bar{\lambda}\}$. Such lattice points would again lie in the interior of C violating the hypothesis that C is lattice-free.

Next, we observe that $L := \operatorname{rec}(C) = \operatorname{lin}(C)$ has dimension strictly less than d, otherwise $C = \mathbb{R}^d$ which is not Λ-free. Now, we show that $L = \operatorname{rec}(C) = \operatorname{lin}(C)$ is a lattice subspace. By Exercise 8 from Section 4.2.1, $\operatorname{Proj}_{L^{\perp}}(\Lambda) = \Lambda' + D$, where $\Lambda' \subseteq L^{\perp}$ is a lattice and $D \subseteq L^{\perp}$ is dense in $\operatorname{span}(D)$. By Exercise 38 in Section 2.4.4, $\widehat{C} := \operatorname{Proj}_{L^{\perp}}(C) = C \cap L^{\perp}$ is a compact set with dimension $\dim(L^{\perp})$. We observe that $\widehat{C} + \operatorname{span}(D)$ cannot contain any point from $\operatorname{Proj}_{L^{\perp}}(\Lambda) = \Lambda' + D$ in its relative interior. Indeed, $\operatorname{relint}(\widehat{C} + \operatorname{span}(D)) = \operatorname{relint}(\widehat{C}) + \operatorname{span}(D)$ by Exercise 37 in Section 2.4.4. If there exists $\mathbf{z} \in \operatorname{Proj}_{L^{\perp}}(\Lambda)$ in $\operatorname{relint}(\widehat{C} + \operatorname{span}(D)) = \operatorname{relint}(\widehat{C}) + \operatorname{span}(D)$, then $(\mathbf{z} + \operatorname{span}(D)) \cap \operatorname{relint}(\widehat{C})$ is nonempty. Since D is dense in $\operatorname{span}(D)$, we will obtain a point from $\operatorname{Proj}_{L^{\perp}}(\Lambda)$ in the relative interior of $\widehat{C}$. Since $C = \widehat{C} + L$, this implies that there is a point from Λ in the interior of C which contradicts the hypothesis that C is lattice-free. Thus, $\operatorname{relint}(\widehat{C} + \operatorname{span}(D))$ cannot contain any point from $\operatorname{Proj}_{L^{\perp}}(\Lambda) = \Lambda' + D$. This implies that $\widehat{C} + \operatorname{span}(D) + L = C + \operatorname{span}(D)$ is lattice-free. By the maximality assumption, $C + \operatorname{span}(D) = C$. Since $\operatorname{span}(D) \subseteq L^{\perp}$, this implies that $\operatorname{span}(D)$ has dimension 0, i.e., $D = \{\mathbf{0}\}$. Therefore, $\operatorname{Proj}_{L^{\perp}}(\Lambda) = \Lambda' + D = \Lambda'$ is a lattice. By Proposition 4.2.6, L is a lattice subspace.

We next show that $\widehat{C}$ is a polytope. The arguments above show that $\operatorname{relint}(\widehat{C}) \cap \operatorname{Proj}_{L^{\perp}}(\Lambda) = \emptyset$. Since $\operatorname{Proj}_{L^{\perp}}(\Lambda)$ is a lattice with $k := \operatorname{rk}(\operatorname{Proj}_{L^{\perp}}(\Lambda)) = \dim(L^{\perp})$ (see Exercise 10 from Section 4.1.1), there exist $\mathbf{b}^1, \ldots, \mathbf{b}^k \in L^{\perp}$ such that these vectors form a basis for $\operatorname{Proj}_{L^{\perp}}(\Lambda)$ and $L^{\perp}$. Consider the invertible linear transformation $T : L^{\perp} \to \mathbb{R}^k$ defined by $T(\gamma_1 \mathbf{b}^1 + \cdots + \gamma_k \mathbf{b}^k) = (\gamma_1, \ldots, \gamma_k)$. $\widehat{C}$ is mapped to a full-dimensional compact set $Q \subseteq \mathbb{R}^k$ and $\operatorname{Proj}_{L^{\perp}}(\Lambda)$ is mapped to $\mathbb{Z}^k$. $\operatorname{relint}(\widehat{C}) \cap \operatorname{Proj}_{L^{\perp}}(\Lambda) = \emptyset$ implies that $\operatorname{int}(Q) \cap \mathbb{Z}^k = \emptyset$. Since Q is compact, there exists $R > 0$ such that $Q \subseteq [-R, R]^k$. There are finitely many points $\mathbf{z}^1, \ldots, \mathbf{z}^m \in [-R, R]^k \cap \mathbb{Z}^k$, none of which are in $\mathrm{int}(Q)$. By the separating and supporting hyperplane theorems (Theorems 2.4.2 and 2.4.5), there exist halfspaces $H_i \subseteq \mathbb{R}^k$, $i = 1, \ldots, m$ that contain Q, but do not contain $\mathbf{z}^i$ in their interior. Then $X = [-R, R]^k \cap H_1 \cap \cdots \cap H_m$ is a $\mathbb{Z}^k$-free polytope containing Q. $T^{-1}(X)$ is a polytope $P \subseteq L^\perp$ such that $\mathrm{relint}(P) \cap \mathrm{Proj}_{L^\perp}(\Lambda) = \emptyset$ and $\widehat{C} \subseteq P$. $\mathrm{relint}(P) \cap \mathrm{Proj}_{L^\perp}(\Lambda) = \emptyset$ implies that $P + L$ is Λ-free and contains $\widehat{C} + L = C$. By maximality of C, we must have $C = P + L$. In particular, C is a full-dimensional polyhedron.

We finally show that every facet of C has a point from Λ is its relative interior. Suppose to the contrary. Then $P = C \cap L^\perp = \mathrm{Proj}_{L^\perp}(C)$ has a facet F that has no points from $\mathrm{Proj}_{L^\perp}(\Lambda)$ in its relative interior. Let this facet be given by the inequality $\langle \mathbf{a}, \mathbf{x} \rangle \leq \delta$ for some $\mathbf{a} \in \mathbb{R}^d, \delta \in \mathbb{R}$. Define $\tilde{P}$ as the set obtained by intersecting all other facets defining halfspaces for P with $\{\mathbf{x} \in \mathbb{R}^d : \delta \leq \langle \mathbf{a}, \mathbf{x} \rangle \leq \delta + 1\}$. Then, $\tilde{P}$ is bounded, since the recession cone of $\tilde{P}$ is a subset of the recession cone of P (see Exercise 4 from Section 2.5.6). This implies that $X := \{\mathbf{z} \in \mathrm{Proj}_{L^\perp}(\Lambda) \cap \tilde{P} : \delta < \langle \mathbf{a}, \mathbf{x} \rangle\}$ is a finite set; let $\delta^\star = \min\{\langle \mathbf{a}, \mathbf{z} \rangle : \mathbf{z} \in X\} > \delta$. Note that since the facet F does not contain any points from $\mathrm{Proj}_{L^\perp}(\Lambda)$ in its relative interior, the set P' obtained from P by replacing $\langle \mathbf{a}, \mathbf{x} \rangle \leq \delta$ with $\langle \mathbf{a}, \mathbf{x} \rangle \leq \delta^\star$ is $\mathrm{Proj}_{L^\perp}(\Lambda)$-free and is a strict superset of P. This implies that $P' + L$ is Λ-free and strictly contains C. This contradicts the maximality of C. $\qquad\square$

4.4.2 Khinchine's Flatness Theorem

We now complete the chain of ideas introduced in the previous section on packing and covering radii, and their connection to lattice-free sets and lattice width (see the discussion just above Section 4.3.1).

It is clear to intuition that a lattice-free set cannot be very "large." However, one has to be careful about which measure to use to determine the "largeness" of a convex set. For instance, it is not true that there is an upper bound on the volume of lattice-free sets: split sets (see Example 4.4.2, part 1.) are lattice-free but have infinite volume. It turns out that the right measure is based on the L-intersection numbers $iw_\Lambda(L, C)$ introduced in Definition 4.3.8 for $d - 1$-dimensional lattice subspaces L. In this regard, we first make a definition.

Definition 4.4.5 Let $\Lambda \subseteq \mathbb{R}^d$ be a full rank lattice. For any convex set C, define the *lattice-width of C with respect to Λ*, or simply Λ-*width of C* as $\min\{iw_\Lambda(L, C) : L \subseteq \mathbb{R}^d \text{ lattice subspace of dimension } d - 1\}$.

Theorem 4.4.6 *Let $\Lambda \subseteq \mathbb{R}^d$ be a full rank lattice and let $C \subseteq \mathbb{R}^d$ be a full-dimensional Λ-free convex set. Then the Λ-width of C is at most $d^{5/2} + 1$. For the special case when C is an ellipsoid, the Λ-width is bounded by $d^{3/2} + 1$.*

Proof We first prove the result when C is bounded. Since the Λ-width of C is less than or equal to the Λ-width of its closure and C is Λ-free is and only if its closure is Λ-free (see Exercise 1 from Section 4.4.3), we may assume C is closed, i.e., C is compact. By Theorem 2.7.7, there exists an ellipsoid $E \subseteq \mathbb{R}^d$ centered at some point $\mathbf{c} \in \mathbb{R}^k$ such that $\mathbf{c} + \frac{1}{d}(E - \mathbf{c}) \subseteq C \subseteq E$. Since C is Λ-free, so is the ellipsoid $E' = \mathbf{c} + \frac{1}{d}(E - \mathbf{c})$. By Lemma 4.3.7, $\mu(E' - \mathbf{c}, \Lambda) \geq 1$. By Theorem 4.3.14, $\rho((E' - \mathbf{c})^\circ, \Lambda^*) \leq \frac{d^{3/2}}{4}$. By Exercise 9a in Section 2.4.4, we obtain $\rho(d(E - \mathbf{c})^\circ, \Lambda^*) \leq \frac{d^{3/2}}{4}$. From the definition of the packing radius, $\rho(d(E - \mathbf{c})^\circ, \Lambda^*) = \frac{1}{d}\rho((E - \mathbf{c})^\circ, \Lambda^*)$ which implies that $\rho((E - \mathbf{c})^\circ, \Lambda^*) \leq \frac{d^{5/2}}{4}$. Note that since $E - \mathbf{c}$ is $\mathbf{0}$-symmetric, the difference body $E - E = 2(E - \mathbf{c})$. In other words, $(E - E)^\circ = \frac{1}{2}(E - \mathbf{c})^\circ$. By Proposition 4.3.11, the Λ-width of E is at most 1 plus the length of the shortest lattice vector in Λ^* with respect to the norm induced by $(E - E)^\circ$, which is $2\rho(\frac{1}{2}(E - \mathbf{c})^\circ, \Lambda^*) = 4\rho((E - \mathbf{c})^\circ, \Lambda^*) \leq d^{5/2}$. Since $C \subseteq E$, the Λ-width of C is at most $d^{5/2} + 1$ as well.

When C is an ellipsoid, we have $E' = E$ and the above argument gives the bound $d^{3/2} + 1$.

To handle the case when C is unbounded, we appeal to the characterization of maximal lattice-free sets in Theorem 4.4.4. By Theorem 4.4.3, there exists a maximal lattice-free set $K \supseteq C$. Since the Λ-width of C is at most the Λ-width of K, it suffices to prove the result for K. By Theorem 4.4.4, $K = P + L$ where L is a lattice subspace and $P \subseteq L^\perp$ is a polytope of dimension $\dim(L^\perp) \leq d$. One can now make the arguments from the bounded case on P and $\mathrm{Proj}_{L^\perp}(\Lambda)$. We leave the details for the reader. $\qquad\square$

4.4.3 Exercises

1. Let $\Lambda \subseteq \mathbb{R}^d$ be a lattice. Show that a convex set C is Λ-free if and only if $\mathrm{cl}(C)$ is Λ-free.
2.* Prove Theorem 4.4.3.
3. Let $Q \subsetneq Q' \subseteq \mathbb{R}^d$ be two full-dimensional polyhedra. Let F be a facet of Q defined by an inequality that is not valid for Q'. Show that $\mathrm{relint}(F) \subseteq \mathrm{int}(Q')$.
4. Show that one cannot drop the assumption of Λ being full rank in Theorem 4.4.4. State and prove an extension for lattices that are not full rank.
5.* Show that any full-dimensional maximal lattice-free set in $\mathbb{R}^d$ has at most 2^d facets.

6. Show that one cannot drop the assumption in Theorem 4.4.6 that C is full-dimensional.
7. Fill in the details for the unbounded case in the proof of Theorem 4.4.6.

4.5 Notes and Related Literature

The literature on the geometry of numbers is vast. The books [67, 217] are very accessible introductions and the monograph [125] is encyclopedic. The chapter on geometry of numbers from [124] and chapters VII and VIII from [24] are also good references that cover some modern topics not found in [67, 125, 217].

As mentioned in the footnote to Lemma 4.3.3, there is a powerful generalization of Minkowski's convex body theorem (Theorem 4.2.4 and Corollary 4.2.5) obtained by extending the concept of the shortest lattice vector in a given norm to the so-called *successive minima* of a norm with respect to a given lattice Λ. Given a norm N_C in $\mathbb{R}^d$ induced by a convex body C and a lattice $\Lambda \subseteq \mathbb{R}^d$, for every $k \geq 1$ define $\lambda_k(C, \Lambda)$ to be the smallest $\lambda \geq 0$ such that there are k linearly independent lattice vectors with norm less than or equal to λ (this can be generalized to general gauge functions that are not norms). Thus, $\lambda_1(C, \Lambda)$ is precisely the length of the shortest nonzero lattice vector with respect to the norm N_C. The so-called *Minkowski's second theorem in geometry of numbers* states that

$$\lambda_1(C, \Lambda)\lambda_2(C, \Lambda) \cdots \lambda_d(C, \Lambda) \leq \frac{2^d \det(\Lambda)}{\mathrm{vol}(C)}.$$

Since $\lambda_1(C, \Lambda) \leq \lambda_2(C, \Lambda) \leq \cdots \leq \lambda_d(C, \Lambda)$, this immediately gives $\lambda_1(C, \Lambda) \leq 2 \left(\frac{\det(\Lambda)}{\mathrm{vol}(C)} \right)^{1/d}$, which is the content of Corollary 4.2.5 (see Exercise 4 from Section 4.1.1). See the references above for a proof and more on successive minima.

As discussed in this chapter, the concept of the L-intersection number goes back to Khinchine's flatness theorem (Theorem 4.4.6), which is a statement about the L-intersection number of lattice-free convex bodies when L is a hyperplane. The idea of using L-intersection numbers for general subspaces goes back to Kannan [150]. An algorithmic version of the flatness theorem was used by Lenstra to design an algorithm that decides if a polytope is lattice-free [165]. This algorithm, in its modern variant, will be one of the highlights in Part II of the book. Kannan used the idea of L-intersection numbers for general L to improve the running time of Lenstra's algorithm significantly. Subsequently, some deep results related to L-intersection numbers were obtained in a landmark paper by Kannan and Lovasz [151] and these ideas have been pursued over the past decades in various forms; an excellent survey can be found in [80]. A recent breakthrough in these methods was obtained by Reis and Rothvoss in [197]. The central result of this paper sharpens one of the main results in the paper by Kannan and Lovasz [151] mentioned above. One consequence of this is a much improved bound for the flatness theorem compared to previous work – the lattice width of any lattice-free convex body is upper bounded by $O(d \log^3(2d))$. The best known bound previously was $O(d^{4/3} \operatorname{poly}(\log(d)))$ due to results by Rudelson [205] and Banaszczyk et al. [21, 22]. Another consequence of Reis and Rothvoss' work is an algorithm with the best known running time for the problem of deciding if a given convex body is lattice-free. More on this topic will be discussed in Chapter 6.

The proof of the flatness theorem we present crucially depends on Theorem 4.3.14. This result was proved by Lagarias, Lenstra, and Schnorr [159]; our proof follows the exposition in [24, section VII.7]. Theorem 4.3.14 in turn depends on Theorem 4.3.12 which is a fairly straightforward consequence of Minkowski's convex body theorem (Theorem 4.2.4 and Corollary 4.2.5). Based on a deep result about the relation between the volumes of a convex set and its polar due to Bourgain and Milman, Theorem 4.3.12 can be generalized to arbitrary norms. Bourgain and Milman [58] showed that for any $\mathbf{0}$-symmetric, full-dimensional, compact convex set $C \subseteq \mathbb{R}^d$,

$$\operatorname{vol}(C) \operatorname{vol}(C^\circ) \geq \frac{a^d}{d^d}$$

for some universal constant $a > 0$ (an upper bound of $\frac{(a')^d}{d^d}$ for some constant $a' > 0$ was known for a while due to results by Blaschke [50] and Santaló [207]). Following the proof of Theorem 4.3.12, one then obtains

$$\rho(C, \Lambda)\rho(C^\circ, \Lambda^*) \leq \frac{2d}{a}$$

for any $\mathbf{0}$-symmetric, full-dimensional, compact convex set $C \subseteq \mathbb{R}^d$. Equivalently, for any norm, the product of the length of the shortest nonzero vector in a lattice in that norm and the length of the shortest nonzero vector in the dual lattice in the dual norm is at most $\frac{8d}{a}$.

The characterization of maximal lattice-free sets presented in Theorem 4.4.4 is due to Lovász [169]. Several proofs and extensions of this result have been subsequently found and used in optimization [14, 15, 27, 28, 103, 176]; see also the survey [29]. The proof of Theorem 4.4.4 presented in this book is based on the proof by Averkov in [15].

Dirichlet–Voronoi cells have been widely used for various kinds of lattice problems. Our main application in this book will be in algorithms for computing shortest and closest lattice vectors in Part II (Section 6.3.1). We recommend the chapter on geometry of numbers in the book [124] for a good introduction to this area. That chapter also discusses the problem of covering and tiling Euclidean space by lattice translates of convex bodies, an example of which is given by the Dirichlet–Voronoi cells (see Proposition 4.3.17, part 5.). Dirichlet–Voronoi cells have also been employed outside the context of lattices in a variety of applications, especially related to computational problems in discrete geometry. See [90, 96] for more.

PART II

Optimization

5

Ingredients of Mathematical Optimization

A mathematical optimization problem is typically defined as follows. We are given an *objective function* $f(x_1, \ldots, x_d)$ of d *decision variables* and one needs to select values $x_1, \ldots, x_d \in \mathbb{R}$ to minimize (or maximize) this function such that the corresponding point $\mathbf{x} \in \mathbb{R}^d$ with these coordinates lies in some *constraint set* $C \subseteq \mathbb{R}^d$. More formally, one wants to solve $\inf\{f(\mathbf{x}) : \mathbf{x} \in C\}$, where $f : \mathbb{R}^d \to \mathbb{R}$ and $C \subseteq \mathbb{R}^d$. While we will also spend most of the subsequent chapters investigating this concrete mathematical problem, we believe it is beneficial to define the general optimization problem in a "coordinate-free" manner. This is what we will proceed to do in this chapter. In our opinion, this has several advantages, which we discuss below.

First, real-world applications where one uses the tools of mathematical optimization seldom come with a set of predefined coordinates. Rather, one *models* the situation by assigning real-valued decision variables to represent certain quantitative aspects of the problem at hand and then selecting objective functions and constraint sets involving these decision variables such that the solution to the corresponding mathematical optimization problem as defined above can be *interpreted* as giving a solution to the original problem.[1] Further, there may be multiple ways of modeling the problem that use different numbers of decision variables and different types of objective functions and constraint sets. This means that the same underlying problem is mapped to very different

---

[1] Consider, for example, the task of assigning job requests to a cloud computing cluster. Given a set of jobs, one may wish to find an assignment of jobs to servers in the cluster so that the total processing time or the maximum of the processing times is minimized. One can model this by introducing a decision variable for every pair of job and server and constrain this variable to take value 1 or 0: the value 1 models the situation when this job is assigned to this server, and 0 otherwise. The objective function can be defined in terms of these decision variables if one knows the processing times of each pair of job and machine. Apart from the 0/1 constraint discussed above, one may have constraints of the form that a particular job cannot be assigned to more than 1 machine, or there may be a capacity constraint on how many jobs get assigned to a particular machine.

173

mathematical optimization problems. We believe it is worthwhile to start our discussions with a more abstract, coordinate-free definition of a mathematical optimization problem that does not commit upfront to any such modeling decisions.

Second, defining optimization problems in a coordinate-free manner allows for the possibility of designing algorithms for the problem that do not use any coordinates at all. This is actually quite common when tackling optimization problems of a combinatorial nature. For readers who are familiar, one may cite many classical optimization problems on graphs (e.g., shortest path problems) for which there are algorithms that do not work with coordinates. On the other hand, numerical optimization or continuous optimization problems almost invariably start with coordinates, objective functions, and constraint sets. The coordinate-free approach allows for a seamless discussion of both kinds of optimization problems – combinatorial or numerical – under a unifying umbrella.

Third, in our experience, the abstraction away from coordinates seems to allow for an especially transparent discussion of the *complexity* of solving optimization problems. The notion of complexity is often handled somewhat differently in the combinatorial/discrete optimization and numerical/continuous optimization research communities. We believe that starting from a coordinate-free definition allows for a smoother and unified discussion of the complexity of optimization algorithms.

5.1 The General Optimization Problem

Definition 5.1.1 (General optimization problem) An *optimization problem class* is given by a set $\mathcal{I}$ of *instances*, a set X of *possible solutions*, and a *solution operator*

$$\mathcal{S} \colon \mathcal{I} \times \mathbb{R}_+ \to 2^X,$$

which satisfies the following property: for any two nonnegative real numbers $\epsilon_1 \leq \epsilon_2$ and any $I \in \mathcal{I}$, we have $\mathcal{S}(I, \epsilon_1) \subseteq \mathcal{S}(I, \epsilon_2)$.

The interpretation of the above definition is the following: $\mathcal{I}$ is the set of optimization problem instances we wish to study, X is the space of solutions to the problems, and for any instance $I \in \mathcal{I}$ and $\epsilon \geq 0$, $\mathcal{S}(I, \epsilon)$ is the set of ϵ-*approximate solutions* for the instance I. The advantage of this definition is that there is no need to assume any structure in the set X; for example, it could be some Euclidean space, or it could just as well be some combinatorial set like the set of edges in a graph. Linear programming in $\mathbb{R}^d$ would set $X = \mathbb{R}^d$ while the Traveling Salesperson Problem on n cities corresponds to setting X to be all tours in the complete graph K_n. It is also not hard to encode optimization problems in varying "dimensions" in this framework, e.g., X is allowed to be $\bigcup_{d \in \mathbb{N}} \mathbb{R}^d$. Also, the notion of "$\epsilon$-approximate" does not require any kind of norm or distance structure on X. The condition imposed on the solution operator simply requires that as we allow more error, we obtain more solutions. Thus, Definition 5.1.1 captures a wide variety of optimization problems in a unified framework while allowing for a flexible notion of an "ϵ-approximate" solution for $\epsilon \geq 0$. $S(I, 0)$ is often said to be the *set of (exact) optimal solutions for instance I* (but not always – see part 3. of Example 5.1.2). We note that $\mathcal{S}(I, \epsilon)$ is allowed to be the empty set, which is interpreted as "there is no solution to the instance I with error ϵ." This possibility can arise due to several reasons; see the examples below.

Example 5.1.2 We now formulate several classical optimization problems in this framework.

1. *Traveling Salesperson Problem (TSP).* For any natural number $n \in \mathbb{N}$, the (symmetric) traveling salesperson problem for n cities seeks to find a tour of minimum length that visits all cities exactly once, given pairwise (nonnegative) intercity distances. One way to represent this problem in the above formalism is as follows. The set of instances $\mathcal{I}$ is clear: we have an instance for every $n \in \mathbb{N}$ and every collection of $\binom{n}{2}$ nonnegative numbers d_{ij} for the distance between cities $i, j \in \{1, \ldots, n\}$. $X = \cup_{n \in \mathbb{N}} \Sigma_n$, where Σ_n is set of all permutations of $\{1, \ldots, n\}$. $\mathcal{S}(I, \epsilon)$ can be taken to be the set of all permutations $\sigma \in \Sigma_n$, where n is the number of cities in the instance I, such that $\sum_{i=1}^{n-1} d_{\sigma(i)\sigma(i+1)} + d_{\sigma(n)\sigma(1)}$ is within an additive ϵ error or within a multiplicative $(1 + \epsilon)$ factor of the optimal tour length in I.

2. *Mixed-integer linear optimization (MILO).*

   (a) (Fixed dimension) Let $n, d \in \mathbb{N}$ be fixed. The *mixed-integer linear optimization (MILO)* problem is the following:

   $$\sup\{\langle \mathbf{c}_1, \mathbf{x} \rangle + \langle \mathbf{c}_2, \mathbf{y} \rangle \ : \ A\mathbf{x} + B\mathbf{y} \leq \mathbf{b}, \ \mathbf{x} \in \mathbb{Z}^n, \mathbf{y} \in \mathbb{R}^d\},$$

   where $A \in \mathbb{R}^{m \times n}, B \in \mathbb{R}^{m \times d}$, and $\mathbf{b} \in \mathbb{R}^m, \mathbf{c}_1 \in \mathbb{R}^n, \mathbf{c}_2 \in \mathbb{R}^d$, and m can be chosen to be any natural number. These parameters are often called the "input data" of the problem. We say the problem is *infeasible* if there is no $(\mathbf{x}, \mathbf{y}) \in \mathbb{Z}^n \times \mathbb{R}^d$ such that $A\mathbf{x} + B\mathbf{y} \leq \mathbf{b}$. We say the problem is *unbounded* if the supremum is $+\infty$. $(\mathbf{x}^\star, \mathbf{y}^\star) \in \mathbb{R}^n \times \mathbb{R}^d$ is said to be an *optimal solution* to the problem if $A\mathbf{x}^\star + B\mathbf{y}^\star \leq \mathbf{b}, \mathbf{x}^\star \in \mathbb{Z}^n$

and $\langle \mathbf{c}_1, \mathbf{x}^\star \rangle + \langle \mathbf{c}_2, \mathbf{y}^\star \rangle \geq \langle \mathbf{c}_1, \mathbf{x}' \rangle + \langle \mathbf{c}_2, \mathbf{y}' \rangle$ for all $(\mathbf{x}', \mathbf{y}') \in \mathbb{Z}^n \times \mathbb{R}^d$ such that $A\mathbf{x}' + B\mathbf{y}' \leq \mathbf{b}$.

Let us put this in the formalism of Definition 5.1.1. Here an instance $I \in \mathcal{I}$ is given by a choice of $m \in \mathbb{N}$, $A, B, \mathbf{b}, \mathbf{c}_1$ and $\mathbf{c}_2$, and $X = \mathbb{R}^n \times \mathbb{R}^d$. Several options exist for $\mathcal{S}(I, \epsilon)$:

- $\mathcal{S}(I, \epsilon)$ may be defined to be all $(\mathbf{x}, \mathbf{y}) \in X$ such that $A\mathbf{x} + B\mathbf{y} \leq \mathbf{b}$, $\mathbf{x} \in \mathbb{Z}^n$ and $\langle \mathbf{c}_1, \mathbf{x} \rangle + \langle \mathbf{c}_2, \mathbf{y} \rangle \geq \langle \mathbf{c}_1, \mathbf{x}' \rangle + \langle \mathbf{c}_2, \mathbf{y}' \rangle - \epsilon$ for all $(\mathbf{x}', \mathbf{y}') \in X$ such that $A\mathbf{x}' + B\mathbf{y}' \leq \mathbf{b}$, $\mathbf{x}' \in \mathbb{Z}^n$.
- $\mathcal{S}(I, \epsilon)$ may be defined to be the set of $(\mathbf{x}, \mathbf{y}) \in X$ such that $A\mathbf{x} + B\mathbf{y} \leq \mathbf{b}$, $\mathbf{x} \in \mathbb{Z}^n$ and there is an optimal solution to I within ϵ distance to $(\mathbf{x}, \mathbf{y})$.
- Another possibility is to define $\mathcal{S}(I, \epsilon)$ as the $(\mathbf{x}, \mathbf{y}) \in X$ such that $A\mathbf{x} + B\mathbf{y} \leq \mathbf{b} + \epsilon \mathbf{1}$, $\mathbf{x} \in \mathbb{Z}^n$ and $\langle \mathbf{c}_1, \mathbf{x} \rangle + \langle \mathbf{c}_2, \mathbf{y} \rangle \geq \langle \mathbf{c}_1, \mathbf{x}' \rangle + \langle \mathbf{c}_2, \mathbf{y}' \rangle - \epsilon$ for all $(\mathbf{x}', \mathbf{y}') \in X$ such that $A\mathbf{x}' + B\mathbf{y}' \leq \mathbf{b}$, $\mathbf{x}' \in \mathbb{Z}^n$.

Note that if an instance I is infeasible or unbounded, $\mathcal{S}(I, 0) = \emptyset$ in all three definitions above of the solution operator. $\mathcal{S}(I, \epsilon)$ may be empty for strictly positive ϵ as well in all three cases; see the exercises below in Section 5.1.2.

(b) (Variable dimension) We can consider the family $\mathcal{I}$ to be all MILO problems with a fixed number of integer variables, but allowing for any number of continuous variables. Here $n \in \mathbb{N}$ is fixed and $X = \bigcup_{d \in \mathbb{N}} (\mathbb{R}^n \times \mathbb{R}^d)$. Everything else is defined as above. Similarly, we may also allow the number of integer variables to vary by letting $X = \bigcup_{n \in \mathbb{N}, d \in \mathbb{N}} (\mathbb{R}^n \times \mathbb{R}^d)$.

Fixing $n = 0$ in the above settings gives us *(pure) linear optimization* or *linear programming*.

3. *Nonlinear optimization.* Nonlinear optimization problems are of the form

$$\inf\{ f(\mathbf{x}) : g_i(\mathbf{x}) \leq 0 \ i = 1, \ldots, m, \ \mathbf{x} \in \mathbb{R}^d \}, \tag{5.1.1}$$

where $f, g_1, \ldots, g_m$ are real-valued functions defined on $\mathbb{R}^d$. The function f is called the *objective function*; $g_1, \ldots, g_m$ are called the *constraints*; and $\{\mathbf{x} \in \mathbb{R}^d : g_i(\mathbf{x}) \leq 0 \ i = 1, \ldots, m\}$ is called the *feasible region*. We say the problem is *infeasible* if there is no $\mathbf{x} \in \mathbb{R}^d$ such that $g_i(\mathbf{x}) \leq 0$ for all $i = 1, \ldots, m$. We say the problem is *unbounded* if the infimum is $-\infty$. An *optimal solution* is a point $\mathbf{x}^\star \in \mathbb{R}^d$ such that $g_i(\mathbf{x}^\star) \leq 0$ for all $i = 1, \ldots, m$ and $f(\mathbf{x}^\star) \leq f(\mathbf{x}')$ for all $\mathbf{x}'$ satisfying the constraints.

The class $\mathcal{I}$ may restrict the structure of the objective and constraint functions (e.g., convex, twice continuously differentiable, nondifferentiable but Lipschitz continuous, etc.). $X = \mathbb{R}^d$ (if d is fixed) or $X = \bigcup_{d \in \mathbb{N}} \mathbb{R}^d$. Several options for $\mathcal{S}(I, \epsilon)$ are studied in nonlinear optimization:

- $\mathcal{S}(I, \epsilon)$ may be defined as the set of all $\mathbf{x} \in \mathbb{R}^d$ such that $g_i(\mathbf{x}) \leq 0$ for all $i = 1, \ldots, m$ and $f(\mathbf{x}) \leq f(\mathbf{x}') + \epsilon$ for all $\mathbf{x}'$ satisfying the constraints.
- $\mathcal{S}(I, \epsilon)$ may be defined as the set of all $\mathbf{x} \in \mathbb{R}^d$ satisfying the constraints within ϵ distance of the set of optimal solutions (with respect to some chosen norm on $\mathbb{R}^d$).
- One could allow points $\mathbf{x} \in \mathbb{R}^d$ such that $g_i(\mathbf{x}) \leq \epsilon$ in the above two bullet points, instead of requiring the constraints to be satisfied exactly.
- When there are no constraints g_i, $\mathcal{S}(I, \epsilon)$ may be defined as the set of points where the norm of the gradient ∇f is at most ϵ. Such points are often called ϵ-*stationary points*.

5.1.1 Solution Operators and Loss Functions

There is an alternate view of the quality of a solution to an optimization problem, which is more prevalent in statistics and economics. Given the set of instances $\mathcal{I}$ and the set of solutions X, instead of a solution operator $S(I, \epsilon)$, one defines a *loss function*

$$\mathcal{L}: \mathcal{I} \times X \to \mathbb{R}_+ \cup \{+\infty\}.$$

From a loss function $\mathcal{L}$, one can extract a "dual" solution operator

$$\mathcal{S}_{\mathcal{L}}(I, \epsilon) := \{x \in X : \mathcal{L}(I, x) \leq \epsilon\}, \tag{5.1.2}$$

which is seen to satisfy the condition in Definition 5.1.1. Moreover, given a solution operator $\mathcal{S}$, there exists a "dual" loss function defined as

$$\mathcal{L}_{\mathcal{S}}(I, x) := \inf\{\epsilon \geq 0 : x \in \mathcal{S}(I, \epsilon)\}. \tag{5.1.3}$$

A loss function should not be confused with *objective function*, although the two concepts are often related (see Exercise 6 in Section 5.1.2). Note that our general optimization problem (Definition 5.1.1) makes no mention of an explicit "objective function."

5.1.2 Exercises

1. For each of the settings in Example 5.1.2, verify that the solution operator as defined satisfies the condition of Definition 5.1.1.
2. Construct an instance of the MILO problem (part 2. of Example 5.1.2) where the supremum is finite but not attained. Can it happen for $n = 0$?

Show that if A, B, and $\mathbf{b}$ have rational entries, then the supremum is always attained if it is finite. What does this mean for $\mathcal{S}(I,\epsilon)$ for the different operators defined in part 2. of Example 5.1.2?

3. Suppose that in part 2. of Example 5.1.2, $\mathcal{S}(I,\epsilon)$ is defined as in the third bullet point. Let I be an instance that is not unbounded. Show that there exists $\epsilon_0 \geq 0$ such that $S(I,\epsilon) = \emptyset$ for all $0 \leq \epsilon < \epsilon_0$, and $S(I,\epsilon)$ is nonempty for all $\epsilon \geq \epsilon_0$. Construct an instance where this happens for a strictly positive ϵ_0.

4. In the MILO problem (part 2. of Example 5.1.2), decide if the following definition of the solution operator satisfies the condition in Definition 5.1.1. $\mathcal{S}(I,\epsilon)$ is the set of $(\mathbf{x},\mathbf{y}) \in X$ such that $A\mathbf{x} + B\mathbf{y} \leq \mathbf{b} + \epsilon\mathbf{1}$, $\mathbf{x} \in \mathbb{Z}^n$ and $\langle \mathbf{c}_1,\mathbf{x}\rangle + \langle \mathbf{c}_2,\mathbf{y}\rangle \geq \langle \mathbf{c}_1,\mathbf{x}'\rangle + \langle \mathbf{c}_2,\mathbf{y}'\rangle - \epsilon$ for all $(\mathbf{x}',\mathbf{y}') \in X$ such that $A\mathbf{x}' + B\mathbf{y}' \leq \mathbf{b} + \epsilon\mathbf{1}$, $\mathbf{x}' \in \mathbb{Z}^n$.

5. In part 3. of Example 5.1.2, decide if the following definition of the solution operator satisfies the condition in Definition 5.1.1. $\mathcal{S}(I,\epsilon)$ is the set of $\mathbf{x} \in \mathbb{R}^d$ such that $g_i(\mathbf{x}) \leq \epsilon$ for all $i = 1,\ldots,m$ and $f(\mathbf{x}) \leq f(\mathbf{x}') + \epsilon$ for all $\mathbf{x}' \in \mathbb{R}^d$ such that $g_i(\mathbf{x}') \leq \epsilon$ for all $i = 1,\ldots,m$.

6. For each of the settings in Example 5.1.2, what is the dual loss function associated with the solution operator $\mathcal{S}$? Is it the same as the "objective function"?

7. Let $\mathcal{I}$ be a set of instances, X be a set of solutions, and $\mathcal{L}$ be a loss function. Is the dual loss function to the dual solution operator to $\mathcal{L}$ always the same as $\mathcal{L}$?

8. Let $(\mathcal{I}, X, \mathcal{S})$ be an optimization problem class as in Definition 5.1.1. Is the dual solution operator to the dual loss function to $\mathcal{S}$ always the same as $\mathcal{S}$?

9. Let $\mathcal{I}$ be a set of instances and X be a set of solutions. Is every loss function the dual loss function of some solution operator? Is every solution operator the dual solution operator to a loss function? Which concept is more general – the solution operator or loss function?

5.2 Algorithmic Optimization

A study of general optimization methods in the 1970s and 1980s led to the insight that all such procedures (whether combinatorial or analytical in nature) can be understood in a unified framework. The overall idea, roughly speaking, is that one gathers *information* about an optimization problem which tells the optimizer which instance needs to be solved within a problem class, and then computations are performed on the information gathered to arrive at a solution (possibly approximate, with guaranteed bounds on error). The act of gathering information is formalized by the notion of an oracle introduced in Definition 1.4.3.

Definition 5.2.1 An *oracle for an optimization problem class* $\mathcal{I}$ is an oracle for the set $\mathcal{I}$ in the sense of Definition 1.4.3.

Example 5.2.2 We now consider some standard oracles for the settings considered in Example 5.1.2.

1. For the TSP, the typical oracle uses two types of queries. One is the dimension query $q^{\dim}$, which returns the number $q^{\dim}(I)$ of cities in the instance I, and the queries $q_{ij}(I)$ which returns the intercity distance between cities i, j (with appropriate error exceptions if i or j are not in the range $\{1, \ldots, q^{\dim}(I)\}$).
2. For MILO, the typical oracle uses the following queries: the dimension queries for n and d (unless one or both of them are fixed and known), a query $q_{ij}^A(I)$ that reports the entry of matrix A in row i and column j for the instance I, similar queries $q_{ik}^B, q_i^{\mathbf{b}}, q_j^{\mathbf{c}_1}, q_k^{\mathbf{c}_1}$ for the matrix B, and vectors $\mathbf{b}, \mathbf{c}_1, \mathbf{c}_2$ (with appropriate error exceptions if the queried index is out of bounds).
3. For nonlinear optimization, the most commonly used oracles return function values, gradient/subgradient values, and Hessian or higher-order derivative values at a queried point. Thus, we have queries such as $q_0^{obj,\mathbf{x}}(I)$ which returns $f(\mathbf{x})$ for the objective function f in an instance of (5.1.1) where $\mathbf{x}$ is a point in the appropriate domain of f, or the query $q_1^{obj,\mathbf{x}}(I)$, which returns the gradient $\nabla f(\mathbf{x})$. Similarly, one has queries for the constraints.

   Within nonlinear optimization, if the problem class $\mathcal{I}$ has an algebraic form, e.g., the functions $f, g_1, \ldots, g_m$ are all polynomials, then the oracle queries may be set up to return the values of the coefficients appearing in any of these polynomials.

We now have everything in place to define what we mean by an optimization algorithm/procedure. An optimization algorithm will query oracles accessible to it to gather information about the instance and then perform its computations. To be able to pose the queries from $\mathcal{Q}$, process the query responses in H and compute solutions in X, it is necessary that these sets be representable in the model of computation (Definition 1.4.1). Thus, we come to the following definition.

Definition 5.2.3 Let $(\mathcal{I}, X, \mathcal{S})$ be an optimization problem class and let $(\mathcal{Q}, H)$ be an oracle for $\mathcal{I}$. Consider any model of computation such that X has a representation in this model (Definition 1.4.1) and the oracle (Q, H) is compatible with this model (Definition 1.4.3). For any $\epsilon \geq 0$, an *ϵ-approximation algorithm for* $(\mathcal{I}, X, \mathcal{S})$ *using* (Q, H) is an oracle-based algorithm (Definition 1.4.3) in this model of computation that, for any $I \in \mathcal{I}$, ends its computation with an element of $\mathcal{S}(I, \epsilon)$, when it receives the answer $q(I)$ for any query $q \in Q$ it poses to the oracle.

If $\mathcal{A}$ is such an ϵ-approximation algorithm, we define the *total complexity* $\mathrm{comp}_{\mathcal{A}}(I)$ to be the number of elementary operations performed by $\mathcal{A}$ during its run on an instance I (meaning that it receives $q(I)$ as the answers to any query q it poses to the oracle). If $\mathcal{A}$ makes k queries during its run on an instance I, we say the *information complexity* $\mathrm{icomp}_{\mathcal{A}}(I)$ is k.

The *(worst-case) total complexity* of the algorithm $\mathcal{A}$ for the problem class $(\mathcal{I}, X, \mathcal{S})$ is defined as

$$\mathrm{comp}_{\mathcal{A}} := \sup_{I \in \mathcal{I}} \; \mathrm{comp}_{\mathcal{A}}(I),$$

and the *(worst-case) information complexity* is defined as

$$\mathrm{icomp}_{\mathcal{A}} := \sup_{I \in \mathcal{I}} \; \mathrm{icomp}_{\mathcal{A}}(I).$$

Remark 5.2.4   One can assume that any algorithm that poses a query q reads and uses the answer $q(I)$ in at least one elementary step of its computations. Indeed, if it ignores the answer to any query completely, then one can simply skip the step in the algorithm that poses that query. This implies that the information complexity $\mathrm{icomp}_{\mathcal{A}}(I)$ is less than or equal to the total complexity $\mathrm{comp}_{\mathcal{A}}(I)$ of any algorithm $\mathcal{A}$ running on any instance I.

One should bear in mind the following aspect of an oracle-based algorithm in a particular model of computation. The compatibility condition for the oracle means that Q has a representation in the model of computation. This makes sense because the algorithm poses queries based on its computations, and therefore it should only have access to those queries that it can represent in its model of computation. Consider, for instance, the setting of general nonlinear optimization from part 3. of Example 5.1.2 and the corresponding oracle from Example 5.2.2. In the arithmetic model of computation, it makes sense to query the gradient oracle, for example, at a point with irrational coordinates, since such a point has a representation in this model of computation. In the Turing machine model, the algorithm cannot produce irrational numbers in its computation.[2] Consequently, it cannot query function/gradient/Hessian

---

[2] It is possible to encode a *countable* list of irrational numbers, a priori, with (finite) binary strings. This type of computation is done, for example, in the open source software `SageMath` (www.sagemath.org/). Nonetheless, there are still uncountably many irrational numbers which

oracles at nonrational points. What about the responses to such queries? Even at rational query points, the gradient or function value may not be rational. In the arithmetic model of computation, this is again not a problem because we have the representations. In the Turing machine model such responses are not allowed; in particular, H does not have a representation in this model of computation. Thus, one must restrict oneself to oracles of the form $q_0^{obj,\mathbf{x},\epsilon}$, indexed by all pairs of *rational* points $\mathbf{x}$ in the domain of the objective function f and positive rationals ϵ, such that $q_0^{obj,\mathbf{x},\epsilon}(I)$ is a *rational* number satisfying $|q_0^{obj,\mathbf{x},\epsilon}(I) - f(\mathbf{x})| \le \epsilon$, and similarly for gradient or Hessian queries.

It is worth emphasizing the fact that the information complexity $\mathrm{icomp}_A(I)$ (consequently, the total complexity as well) of an algorithm on instance I depends very much on the type of oracle access one has. Consider the MILO example in part 2. of Example 5.1.2. One kind of oracle was described in Example 5.2.2. However, another natural oracle is one whose queries return only a certain bit of the binary encoding of a queried entry of $A, B, \mathbf{b}, \mathbf{c}_1$, or $\mathbf{c}_2$. Even if we restrict ourselves to instances with rational data, i.e., all these matrices and vectors have rational entries, the information complexity will be different when computed with respect to these different oracles because, in the first case, one query returns the entire entry and, in the second case, one has to make several queries to get the entire entry.[3] Moreover, depending on the model of computation the difference may or may not be washed out in the total complexity of the algorithm. In the arithmetic model, the first kind of oracle may, and usually will, result in a smaller total complexity, compared to the second oracle. However, in the Turing machine model of computation, since one has to read and process the entire response of a query from the first oracle, the algorithm effectively will read all the bits of a response, and so there will be no difference compared to the second kind of oracle as far as total complexity is concerned.

5.2.1 Direct Encodings and Oracle-Based Access

For several important classes of optimization problems the problem class $\mathcal{I}$ is directly representable in the model of computation used to design an

---

cannot be obtained/computed during the execution of an algorithm in the Turing machine model.

[3] Recall that we count the number of queries in defining information complexity. It can be argued that one should instead add up the encoding lengths (in the appropriate model of computation – see Definition 1.4.1) of all the responses during the execution of the algorithm. We prefer to take the approach of counting simply the number of queries. If one wants to keep track of the encoding lengths of the responses, one can simply look at the "finer" oracle that returns the appropriate element (bit or real number) of the sequence that encodes the response. In our opinion, this approach of outsourcing the complexity to the type of oracle is the cleanest way to proceed.

algorithm in the sense of Definition 1.4.1. Problems of an algebraic nature, such as mixed-integer linear optimization (part 2. of Example 5.1.2) or polynomial optimization (part 3. of Example 5.1.2 where $f, g_1, \ldots, g_m$ are all polynomials) are of this type, and the vast majority of problems with a combinatorial or discrete nature also fall under this category (e.g., part 1. of Example 5.1.2). As mentioned in the discussion under Definition 1.4.3, the oracle becomes especially simple in this case: there is a single query in $\mathcal{Q}$ that returns the encoding of the problem instance. The algorithm needs to make a single query at the beginning and then proceeds with its computations with no other oracle queries needed. However, this view of computation is already seen to be insufficient for nonlinear optimization with more general functions (part 3. of Example 5.1.2). The function value and gradient-based oracles need to be queried multiple times and usually in an adaptive manner to figure out what instance one is dealing with. Both situations are handled seamlessly in the oracle-based view of computation. Defining optimization problems in a "coordinate-free" manner as in Definition 5.1.1 makes it easy to talk about oracle-based algorithms for optimization at a sufficiently general level.

Remark 5.2.5 The oracles we discussed so far for mathematical optimization are deterministic: given an instance $I \in \mathcal{I}$ and a query $q \in \mathcal{Q}$, there is a unique response $q(I)$ that one obtains (this is baked into Definition 1.4.3 of an oracle because the queries in $\mathcal{Q}$ are well-defined functions on $\mathcal{I}$). However, in many situations, the oracles are not deterministic; rather, the responses are stochastic or random. This is imperative to capture many applications, e.g., in statistical inference, decision-making under uncertainty, and artificial intelligence, to name a few. We will not deal with stochastic oracles in this book. The reader can find a formal definition of stochastic oracles in [182, chapter 1] (the monograph has a slightly narrower definition of a general optimization problem and the nature of queries of the oracle, but their definition of a stochastic oracle can be easily adapted to the more general setting above).

Remark 5.2.6 An important notion of complexity that we will not discuss in depth in this book is that of *space complexity*. This is defined as the maximum amount of information (from the oracle queries) and auxiliary computational memory that is maintained by the algorithm during its entire run. As in Definition 5.2.3, one can define the *total space complexity* and the *information space complexity*. Both notions can be quite different from $\text{comp}_{\mathcal{A}}$ and $\text{icomp}_{\mathcal{A}}$, respectively, since one keeps track of only the amount of information and auxiliary data held in memory at any given stage of the computation, as opposed to the overall amount of information or auxiliary memory used. In many optimization algorithms, it is not necessary to maintain all of the answers to previous queries or computations in memory. A classic example of this is the (sub)gradient descent algorithm (see Section 6.5.2), where only the current function values and (sub)gradients are stored in memory for computations, and they are not needed in subsequent iterations.

5.3 Bounding the Complexity

The criteria of having low information and total complexity have guided the study and design of algorithms for mathematical optimization. However, by no means are these the only criteria that is used to judge the performance of algorithms. A careful discussion of what makes an optimization algorithm "good" is too vast of a topic and is out of the scope of this book. Appealing to the fact that theoretical complexity analyses are a central aspect in the study of algorithms, and with good reason, we will focus on obtaining the tightest bounds on information and total complexity. Formally, we will focus on the following question.

> Let $(\mathcal{I}, X, \mathcal{S})$ be an optimization problem class (Definition 5.1.1) and let $(\mathcal{Q}, H)$ be an oracle for $\mathcal{I}$. Consider any model of computation such that X has a representation in this model (Definition 1.4.1) and the oracle $(\mathcal{Q}, H)$ is compatible with this model (Definition 1.4.3). Give bounds on
>
> $\inf\{\text{icomp}_{\mathcal{A}} : \mathcal{A} \text{ oracle-based algorithm in this model of computation using } (\mathcal{Q}, H)\}$
>
> and
>
> $\inf\{\text{comp}_{\mathcal{A}} : \mathcal{A} \text{ oracle-based algorithm in this model of computation using } (\mathcal{Q}, H)\}$.

Upper bounds on the above quantities are obtained by designing an algorithm and analyzing its complexity. The question of lower bounds is a little more subtle.

5.3.1 Oracle Ambiguity and Lower Bounds on Complexity

For many settings, especially problems in numerical optimization, a finite number of oracle queries may not pin down the exact problem one is facing. For example, consider $(\mathcal{I}, X, \mathcal{S})$ to be the problem class of the form (5.1.1) where $f, g_1, \ldots, g_m$ can be any convex, continuously differentiable functions, and suppose the oracle $(\mathcal{Q}, H)$ allows function evaluation and gradient queries. Given any finite number of queries, there are infinitely many instances that give the same answers to those queries.

Definition 5.3.1 Let $(\mathcal{I}, X, \mathcal{S})$ be an optimization problem class and let $(\mathcal{Q}, H)$ be an oracle for $\mathcal{I}$. For any subset $Q \subseteq \mathcal{Q}$ of queries, define an equivalence relation on $\mathcal{I}$ as follows: $I \sim_Q I'$ if $q(I) = q(I')$ for all $q \in Q$. For any instance I, let $V(I, Q)$ denote the equivalence class that I falls in, i.e.,

$$V(I, Q) = \{I' : q(I) = q(I') \ \forall q \in Q\}.$$

The above definition formalizes the fact that if one only knows the answers to queries in Q, then one has a course-grained view of $\mathcal{I}$. This is why the notion of an ϵ-approximate solution becomes especially pertinent. The following theorem gives a necessary condition on the nature of queries used by any ϵ-approximation algorithm.

Theorem 5.3.2 *Let $(\mathcal{I}, X, \mathcal{S})$ be an optimization problem class and let $(\mathcal{Q}, H)$ be an oracle for $\mathcal{I}$. If $\mathcal{A}$ is an ϵ-approximation algorithm for this optimization problem for some $\epsilon \geq 0$, then*

$$\bigcap_{I' \in V(I, Q(I))} \mathcal{S}(I', \epsilon) \neq \emptyset \qquad \forall I \in \mathcal{I},$$

where $Q(I)$ is the set of queries used by $\mathcal{A}$ when processing instance I.

Proof Left as an exercise. $\square$

Thus, one has to find a strategy that adaptively poses queries to the oracle such that the above condition holds for any instance $I \in \mathcal{I}$. This leads us to the following definition.

Definition 5.3.3 Let $(\mathcal{I}, X, \mathcal{S})$ be an optimization problem class and let $(\mathcal{Q}, H)$ be an oracle for $\mathcal{I}$. Let $(\mathcal{Q} \times H)^*$ denote the collection of all *finite* sequences of pairs $(q, h) \in \mathcal{Q} \times H$ (including the empty sequence).

A *query strategy* is a function $D: (\mathcal{Q} \times H)^* \to \mathcal{Q}$. The *transcript* $\Pi(D, I)$ *of a strategy D on an instance I* is the sequence of query and response pairs $(q_i, q_i(I))$, $i = 1, 2, \ldots$ obtained when one applies D on I, i.e., $q_1 = D(\emptyset)$ and $q_i = D(((q_1, q_1(I)), \ldots, (q_{i-1}, q_{i-1}(I))))$ for $i \geq 2$. $\Pi_k(D, I)$ will denote the truncation of $\Pi(D, I)$ to the first k terms, $k \in \mathbb{N}$. We will use $Q(D, I)$ (and $Q_k(D, I)$) to denote the set of queries in the transcript $\Pi(D, I)$ (and $\Pi_k(D, I)$). The ϵ-*information complexity of an instance I for a query strategy D* is defined as

$$\text{icomp}_\epsilon(D, I) := \inf \left\{ k \in \mathbb{N} : \bigcap_{I' \in V(I, Q_k(D, I))} \mathcal{S}(I', \epsilon) \neq \emptyset \right\}$$
$$= \inf \left\{ k \in \mathbb{N} : \bigcap_{I' : \Pi_k(D, I') = \Pi_k(D, I)} \mathcal{S}(I', \epsilon) \neq \emptyset \right\}.$$

The ϵ-*information complexity of a query strategy* D *for the problem class* $(\mathcal{I}, X, \mathcal{S})$ is defined as

$$\mathrm{icomp}_\epsilon(D) := \sup_{I \in \mathcal{I}} \ \mathrm{icomp}_\epsilon(D, I).$$

The ϵ-*information complexity of the problem class* $(\mathcal{I}, X, \mathcal{S})$ *with respect to oracle* $(\mathcal{Q}, H)$ is defined as

$$\mathrm{icomp}_\epsilon := \inf_D \ \mathrm{icomp}_\epsilon(D),$$

where the infimum is over all possible query strategies.

We can now formally state the results for lower bounding algorithmic complexity. Remark 5.2.4 and Theorem 5.3.2 imply the following.

Corollary 5.3.4 *Let* $(\mathcal{I}, X, \mathcal{S})$ *be an optimization problem class and let* $(\mathcal{Q}, H)$ *be an oracle for* $\mathcal{I}$. *If* $\mathcal{A}$ *is an* ϵ-*approximation algorithm for* $(\mathcal{I}, X, \mathcal{S})$ *using* $(\mathcal{Q}, H)$ *for some* $\epsilon \geq 0$, *then*

$$\mathrm{icomp}_\epsilon \leq \mathrm{icomp}_\mathcal{A} \leq \mathrm{comp}_\mathcal{A}.$$

Both inequalities in Corollary 5.3.4 can be strict, even if one minimizes $\mathrm{icomp}_\mathcal{A}$ and $\mathrm{comp}_\mathcal{A}$ over all ϵ-approximation algorithms. In Definition 5.3.3, there is no restriction on the kinds of functions that are allowed as query strategies. In particular, one allows for query strategies D that are not computable, i.e., there does not exist any algorithm in a model of computation compatible with $(\mathcal{Q}, H)$ that takes as input a sequence s of query-response pairs and outputs $D(s)$. One advantage of this is that one does not have to rely on any complexity theory assumptions such as $P \neq NP$ and the lower bounds are unconditional.

In the literature, icomp_ϵ is often referred to as the *analytical complexity* of the problem class (see, e.g., [184]). We prefer the phrase *information complexity* since we wish to have a unified framework for continuous and discrete optimization and "analytical" suggests problems that are numerical in nature or involve the continuum. Another term that is used in the literature is *oracle complexity*. This might suggest the complexity of implementing the oracle, rather than the number of queries needed. Since icomp_ϵ is inspired by information theory ideas, we follow the trend [60, 66, 181, 182, 222] of using the term *information complexity (with respect to an oracle)*. $\mathrm{comp}_\mathcal{A}$ is sometimes referred to as *arithmetic complexity* [184], *computational complexity* [13], or *combinatorial complexity* [222] of $\mathcal{A}$. We prefer using *(worst-case) total complexity* of the algorithm $\mathcal{A}$.

5.3.2 Exercises

1. Prove Theorem 5.3.2. Is the converse true? In other words, is it true that if for some $\epsilon \geq 0$, an oracle-based algorithm $\mathcal{A}$ has a query strategy such that for every instance $I \in \mathcal{I}$, $\bigcap_{I' \in V(I, Q(I))} \mathcal{S}(I', \epsilon) \neq \emptyset$, $\mathcal{A}$ is an ϵ-approximation algorithm?

2. Let $(\mathcal{I}, X, \mathcal{S})$ be an optimization problem class and let $(\mathcal{Q}, H)$ be an oracle for $\mathcal{I}$. Show that the ϵ-information complexity of $(\mathcal{I}, X, \mathcal{S})$ with respect to $(\mathcal{Q}, H)$ is a nonincreasing function of ϵ.

3. Show that if $\mathcal{S}, \mathcal{S}'$ are two different solution operators for $\mathcal{I}, X$ such that $\mathcal{S}(I, \epsilon) \subseteq \mathcal{S}'(I, \epsilon)$ for all $I \in \mathcal{I}$ and $\epsilon \geq 0$, i.e., the operator $\mathcal{S}$ is stricter than $\mathcal{S}'$, then the complexity measures with respect to $\mathcal{S}$ are at least as large as the corresponding measures with respect to $\mathcal{S}'$.

5.4 Parameterized Complexity

The notions of complexity defined so far are either too fine or too coarse. At one extreme are the instance-dependent notions $\text{comp}_{\mathcal{A}}(I)$, $\text{icomp}_{\mathcal{A}}(I)$, and $\text{icomp}_{\epsilon}(D, I)$, and at the other extreme are the worst-case notions $\text{comp}_{\mathcal{A}}$, $\text{icomp}_{\mathcal{A}}$, and $\text{icomp}_{\epsilon}(D)$. It is almost always impossible to give a fine-tuned analysis of the instance-based complexity notions; on the other hand, the worst-case notions give too little information, at best as a function of ϵ, and in the worst case, these values are actually ∞ for most problem classes of interest. Typically, a middle path is taken where a hierarchy within the problem class is defined and the complexity is analyzed as a function of the levels in the hierarchy.

Definition 5.4.1   Let $(\mathcal{I}, X, \mathcal{S})$ be an optimization problem class. A *(real valued) parameterization of* $\mathcal{I}$ is simply a function $P : \mathcal{I} \to \mathbb{R}_+ \cup \{+\infty\}$.

Let $P_1, \ldots, P_k$ be k different parameterizations of $\mathcal{I}$. The (worst-case) complexity of any algorithm $\mathcal{A}$ for the problem, with respect to these parameterizations, is defined as

$$\text{comp}_{\mathcal{A}}(\lambda_1, \ldots, \lambda_k) := \sup\{\text{comp}_{\mathcal{A}}(I) : P_i(I) \leq \lambda_i, \ i = 1, \ldots, k\},$$

$$\forall \lambda_1, \ldots, \lambda_k \in \mathbb{R}_+ \cup \{+\infty\}.$$

Similarly, the (worst-case) information complexity of any algorithm $\mathcal{A}$ for the problem, with respect to these parameterizations, is defined as

$$\text{icomp}_{\mathcal{A}}(\lambda_1, \ldots, \lambda_k) := \sup\{\text{icomp}_{\mathcal{A}}(I) : P_i(I) \leq \lambda_i, \ i = 1, \ldots, k\},$$

and the (worst-case) ϵ-information complexity of the problem class, with respect to these parameterizations, is defined as

$$\mathrm{icomp}_\epsilon(\lambda_1, \ldots, \lambda_k) := \inf_D \ \sup\{\mathrm{icomp}_{\mathcal{A}}(I) : P_i(I) \le \lambda_i, \ i = 1, \ldots, k\},$$

where the infimum is taken over all adaptive query strategies D.

Example 5.4.2 Let us look at some standard parameterizations in optimization literature.

1. If $\mathcal{I}$ has a direct encoding in the model of computation (see 5.2.1), then a widely used parameterization is the following: $P(I)$ is the encoding size of the instance $I \in \mathcal{I}$ (Definition 1.4.1). As mentioned before, this takes care of large classes of optimization problems with a combinatorial or algebraic structure. Complexity is then defined as a "one-dimensional" function of this single parameter, i.e., encoding size.

   Note that when different choices of encodings are used, this leads to different parameterizations. For example, in the TSP problem class defined in part 1. of Example 5.1.2, one typically encodes the intercity distances. If one focuses on the so-called *Euclidean TSP* instances, then one can define a different parameterization based on encodings of the coordinates of the cities instead of the distances. Another classical example where different encodings are considered is the optimization problem class of *knapsack problems*. Here, one is given n items with weights $w_1, \ldots, w_n \in \mathbb{Z}_+$ and values $v_1, \ldots, v_n \in \mathbb{Z}_+$, and the goal is to find the subset of items with maximum value with total weight bounded by a given budget $W \in \mathbb{Z}$. Consider the Turing machine model of computation. The standard "binary encoding" maps these numbers to their binary representations. However, one can also encode the problems by representing $w_1, \ldots, w_n$ and $v_1, \ldots, v_n$ in binary, but representing W as a string of W ones – the so-called *unary* representation. The well-known dynamic programming-based algorithm $\mathcal{A}$ has complexity $\mathrm{comp}_{\mathcal{A}}(\lambda)$ which is exponential in λ with respect to the first parameterization, while it is polynomial in λ with respect to the second parameterization.

   Moreover, other parameterizations may also be considered even in cases where direct encodings of $\mathcal{I}$ are available. For example, in part 1. of Example 5.1.2, one may simply define $P(I)$ to be the number of cities in the instance I. Similarly, for MILO problems from part 2. of Example 5.1.2, instead of considering the size of encoding of all the entries of A, B, b, c, one may simply define $P(I) := m(n + d)$, or even simply as $n + d$.

2. For nonlinear optimization problems of the form (5.1.1), direct encodings are not available. Consider, for example, the setting where the functions $f, g_1, \ldots, g_m$ are convex. Unless the functions have some algebraic form (e.g., polynomials), it is not clear how to encode them in the arithmetic or the Turing machine model of computations. The following four parameters are widely used instead.

   (a) A parameter P_1 capturing the dimension: $P_1(I) = d$, where d is the ambient dimension of I.
   (b) A parameter P_2 capturing a bound on the feasible region: $P_2(I)$ is the infimum over all $R \in \mathbb{R}_+$ such that the feasible region
   $\{\mathbf{x} \in \mathbb{R}^d : g_i(\mathbf{x}) \leq 0 \ i = 1, \ldots, m\}$ is contained in the box
   $\{\mathbf{x} \in \mathbb{R}^d : \|\mathbf{x}\|_\infty \leq R\}$.
   (c) A parameter P_3 capturing the "thickness" of the feasible region: $P_3(I)$ is the infimum over all $\lambda > 0$ such that the feasible region contains a ball of radius $1/\lambda$ inside it.
   (d) A parameter P_4 that controls the variability of the objective function f: $P_4(I)$ is the infimum over all $M \in \mathbb{R}_+$ such that f is Lipschitz continuous with Lipschitz constant M over the feasible region.

   One can then define the complexity measures $\mathrm{comp}_{\mathcal{A}}(\lambda_1, \lambda_2, \lambda_3, \lambda_4)$ and $\mathrm{icomp}_{\mathcal{A}}(\lambda_1, \lambda_2, \lambda_3, \lambda_4)$ for any ϵ-approximation algorithm $\mathcal{A}$, and the algorithm independent complexity measure $\mathrm{icomp}_{\epsilon}(\lambda_1, \lambda_2, \lambda_3, \lambda_4)$, as functions of these four parameters as well as ϵ. In the next chapter we will proceed to do precisely this.

5.5 Notes and Related Literature

The perspective on optimization problems presented here via oracles is heavily influenced by the monograph [182] and, more generally, by the paradigm of *information-based complexity* [222, 223]. The monograph [182] is concerned primarily with numerical optimization problems with convex functions and constraints (with a somewhat narrower concept of an oracle compared to our definition). As we hope to have conveyed in this chapter, the oracle-based view of optimization seamlessly encompasses combinatorial and algebraic problems and thus seems to be a more flexible language for studying optimization. The area of information-based complexity is much broader in its scope beyond optimization and addresses virtually every computational problem in its generality. We have found its application in optimization, as presented in this chapter, especially helpful in the study of complexity of optimization algorithms.

Classical computational complexity deals primarily with problems that have direct encodings (part 1. of Example 5.4.2) and studies complexity as a function of encoding size. In contrast, complexity analysis in numerical and continuous optimization has traditionally looked at complexity as a function of multiple parameters of interest, e.g., part 2. of Example 5.4.2. The relatively recent developments of *parameterized complexity* and *fixed parameter tractability* study complexity of combinatorial and algebraic problems using multiple parameters beyond just encoding size even when direct encodings are available. These trends comprise a very active area of current research in computational complexity [108, 109].

6

Complexity of Convex Optimization
with Integer Variables

6.1 Problem Setup

In this chapter, we will consider an optimization problem class that is very rich in its modeling capabilities. It is a common generalization of both parts 2. and 3. (with convex constraints) of Example 5.1.2. Moreover, the TSP problem (part 1. of Example 5.1.2) and most combinatorial optimization problems can also be modeled using this framework. The problem class is called *mixed-integer convex optimization*:

$$\inf\left\{ f(\mathbf{x},\mathbf{y}) : (\mathbf{x},\mathbf{y}) \in C, (\mathbf{x},\mathbf{y}) \in \mathbb{Z}^n \times \mathbb{R}^d \right\}, \tag{6.1.1}$$

where $f : \mathbb{R}^n \times \mathbb{R}^d \to \mathbb{R} \cup \{+\infty\}$ is a closed, convex (possibly nondifferentiable) function such that $\mathrm{dom}(f)$ is a closed set, and $C \subseteq \mathbb{R}^n \times \mathbb{R}^d$ is a closed, convex set. Let us formulate using the notation and terminology from Chapter 5. An instance I is given by f, C and $\mathcal{I}$ is the set of all such instances for varying $n, d \in \mathbb{Z}_+$ with $n + d \geq 1$. We have $X = \bigcup_{n,d \in \mathbb{Z}_+} \mathbb{R}^n \times \mathbb{R}^d$ and we consider the solution operator $\mathcal{S}(I, \epsilon)$ to be all feasible solutions in $C \cap \mathrm{dom}(f)$ that have value at most ϵ more than the optimal value. More precisely, we define

$$\mathcal{S}(I,\epsilon) := \left\{ (\mathbf{x},\mathbf{y}) \in C \cap \mathrm{dom}(f) \cap \left(\mathbb{Z}^n \times \mathbb{R}^d\right) : f(\mathbf{x},\mathbf{y}) \leq f(\mathbf{x}',\mathbf{y}') + \epsilon, \right.$$
$$\left. \forall (\mathbf{x}',\mathbf{y}') \in C \cap \mathrm{dom}(f) \cap \left(\mathbb{Z}^n \times \mathbb{R}^d\right) \right\}.$$

The set $C \cap \mathrm{dom}(f) \cap (\mathbb{Z}^n \times \mathbb{R}^d)$ is called the *feasible region* and the points in this set are called *feasible solutions*. We say that $\mathbf{x}_1, \ldots, \mathbf{x}_n$ are the *integer-valued decision variables* or simply the *integer variables* of the problem, and $\mathbf{y}_1, \ldots, \mathbf{y}_d$ are called the *real-valued decision variables* or simply the *continuous variables*. When $n = 0$, i.e., there are no integer variables, the problem is called a *pure continuous convex optimization* problem.

190

When $d = 0$, i.e., there are no continuous variables, the problem is called a *pure integer convex optimization* problem. Recall from Section 5.1 that $S(I, \epsilon)$ is called the set of ϵ-approximate solutions and $S(I, 0)$ is the set of all (exact) optimal solutions. As noted in Section 5.1, $S(I, \epsilon)$ may be empty for certain $\epsilon \geq 0$. For example, this can happen if $C \cap \mathrm{dom}(f) \cap (\mathbb{Z}^n \times \mathbb{R}^d) = \emptyset$, i.e., there are no feasible solutions. Another possibility is that the infimum in (6.1.1) is not attained and therefore $S(I, 0) = \emptyset$.

6.1.1 First-Order Oracles

We now define the oracle we will consider for this problem (see Section 5.2).

Definition 6.1.1 A *first-order oracle* for (6.1.1) has three different types of queries.

1. *Dimension query.* We have a query $q^{\dim}$ such that $q^{\dim}(I) = (n, d) \in \mathbb{Z}_+ \times \mathbb{Z}_+$ where the instance I is defined in $\mathbb{R}^n \times \mathbb{R}^d$.
2. *Separation queries.* For every $\mathbf{z} \in \mathbb{R}^n \times \mathbb{R}^d$, there is a query $q_{\mathbf{z}}^{\mathrm{sep}}$ such that for any instance $I = (f, C)$, $q_{\mathbf{z}}^{\mathrm{sep}}(I)$ returns YES if $\mathbf{z} \in C$, and if $\mathbf{z} \notin C$ then it returns $\mathbf{a} \in \mathbb{R}^d \setminus \{\mathbf{0}\}, \delta \in \mathbb{R}$ such that $H^=(\mathbf{a}, \delta)$ is a separating hyperplane for C and $\mathbf{z}$ (see Theorem 2.4.2). The set of all separation queries is also called a *separation oracle*.
3. *Functional queries.* For every $\mathbf{z} \in \mathbb{R}^n \times \mathbb{R}^d$, there is a query $q_{\mathbf{z}}^{\mathrm{fun}}$ such that for any instance $I = (f, C)$, $q_{\mathbf{z}}^{\mathrm{fun}}(I) = (f(\mathbf{z}), \mathbf{s})$, where $\mathbf{s} \in \partial f(\mathbf{z})$, if $f(\mathbf{z}) < +\infty$. If $f(\mathbf{z}) = +\infty$, then $q_{\mathbf{z}}^{\mathrm{fun}}(I)$ returns a separating hyperplane for $\mathrm{dom}(f)$ and $\mathbf{z}$. The set of all function queries is also called a *first-order functional oracle*.

Remark 6.1.2 We make some clarifying remarks.

1. It is to be understood that if the point $\mathbf{z}$ and the instance I are not in the same ambient space $\mathbb{R}^n \times \mathbb{R}^d$, the query returns an exception. Note that the responses to the queries are of different types, e.g., "YES," $(\mathbf{a}, \delta)$, $(f(\mathbf{z}), \mathbf{s})$, "Exception." Thus, the response set H is the union of all these range spaces.
2. Observe that there exist multiple first-order oracles since separating hyperplanes and subgradients are not necessarily unique. Thus, for every choice of a separating hyperplane or subgradient at a point (depending on the instance), we would obtain a different separation or functional query corresponding to that point.
3. When a functional query is made corresponding to a point outside the domain of the objective function, we will consider this as a separation query. Thus, we will assume without further comment that functional queries always return a finite value for the function and a subgradient.

Since $X = \bigcup_{n,d \in \mathbb{Z}_+} \mathbb{R}^n \times \mathbb{R}^d$ and first-order oracles as defined in Definition 6.1.1 are representable only in the arithmetic model of computation, all the results will be presented in the arithmetic model. In Section 6.6, we provide references on what is known in the Turing machine model of computation.

6.1.2 Parameterization of the Instances

As mentioned in Section 5.4, we need parameterizations to be able to give concrete bounds on the information and total complexity.

Definition 6.1.3   A *fiber box* in $\mathbb{Z}^n \times \mathbb{R}^d$ is a set of the form $\{\mathbf{x}\} \times [\ell_1, u_1] \times \cdots \times [\ell_d, u_d]$ where $\mathbf{x} \in \mathbb{Z}^n$ and $\ell_i, u_i \in \mathbb{R} \cup \{-\infty, +\infty\}$ for $i = 1, \ldots, d$. A fiber box is the empty set if $u_i < \ell_i$ for some $i = 1, \ldots, d$. The *length* of a nonempty fiber box in coordinate j is $u_j - \ell_j$. The *width* of such a fiber box is the minimum of $u_j - \ell_j$, $j = 1, \ldots, d$. If $n = 0$, a fiber box is called a *hypercuboid* in $\mathbb{R}^d$. If all $\ell_i = -\infty$ and all $u_i = \infty$, then the set is simply called a *fiber* over $\mathbf{x}$.

Definition 6.1.4   We introduce five parameterizations for the set of all instances in the problem class (6.1.1).

1. $P_1(I)$ is the number of integer variables in the instance.
2. $P_2(I)$ is the number of continuous variables in the instance.
3. $P_3(I)$ is the infimum over all $R \in \mathbb{R}_+$ such that $C \cap \text{dom}(f)$ is contained in the box $\{\mathbf{z} \in \mathbb{R}^n \times \mathbb{R}^d : \|\mathbf{z}\|_\infty \leq R\}$. Equivalently, $P_3(I) = \sup\{\|\mathbf{z}\|_\infty : \mathbf{z} \in C \cap \text{dom}(f)\}$.
4. $P_4(I)$ is defined as follows. If $d \geq 1$ and $S(I, 0) \neq \emptyset$, i.e., there are (exact) optimal solutions, $P_4(I)$ is the supremum over all $\rho \in \mathbb{R}_+$ such that there exists an optimal solution $(\mathbf{x}^\star, \mathbf{y}^\star)$ and some $\hat{\mathbf{y}} \in \mathbb{R}^d$ satisfying $\{(\mathbf{x}^\star, \mathbf{y}) : \|\mathbf{y} - \hat{\mathbf{y}}\|_\infty \leq \rho\} \subseteq C \cap \text{dom}(f)$. In other words, there is a point $(\mathbf{x}^\star, \hat{\mathbf{y}})$ in the same fiber as the optimum $(\mathbf{x}^\star, \mathbf{y}^\star)$ that is "strictly feasible" in the continuous variables with a fiber box of width ρ around $(\mathbf{x}^\star, \hat{\mathbf{y}})$ contained in $C \cap \text{dom}(f)$. If $S(I, 0) = \emptyset$ or $d = 0$, then $P_4(I)$ is defined to be $+\infty$.
5. $P_5(I)$ is the infimum over all $M \in \mathbb{R}_+$ such that f is Lipschitz continuous with respect to the $\| \cdot \|_\infty$-norm with Lipschitz constant M on $\{\mathbf{x}\} \times [-P_3(I), P_3(I)]^d$ for all $\mathbf{x} \in [-P_3(I), P_3(I)]^n \cap \mathbb{Z}^n$ where we use the convention that if f is identically $+\infty$ on $\{\mathbf{x}\} \times [-P_3(I), P_3(I)]^d$, then any M works for this fiber box. In other words, for any $(\mathbf{x}, \mathbf{y}), (\mathbf{x}, \mathbf{y}') \in (\mathbb{Z}^n \times \mathbb{R}^d) \cap ([-P_3(I), P_3(I)]^n \times [-P_3(I), P_3(I)]^d)$ with $\|\mathbf{y} - \mathbf{y}'\|_\infty \leq P_3(I)$, $|f(\mathbf{x}, \mathbf{y}) - f(\mathbf{x}, \mathbf{y}')| \leq M \|\mathbf{y} - \mathbf{y}'\|_\infty$ with the convention that $\infty - \infty = 0$.

Table 6.1 *Parameters of the problem instance used to state the complexity bounds*

Parameter	Meaning
n	Number of integer variables
d	Number of continuous variables
R	Boundedness parameter for the feasible region
ρ	Strict feasibility parameter for the feasible region
M	Lipschitz constant for the objective function

Table 6.1 gives a synopsis. Note that if $C \cap \mathrm{dom}(f) \cap (\mathbb{Z}^n \times \mathbb{R}^d) = \emptyset$, i.e., the instance I has no feasible points, then $P_4(I) = +\infty$. If $C \cap \mathrm{dom}(f)$ does contain a point from $\mathbb{Z}^n \times \mathbb{R}^d$, then for all instances with finite $P_3(I)$, there is an optimal solution since f is closed, convex and $C \cap \mathrm{dom}(f) \cap (\mathbb{Z}^n \times \mathbb{R}^d)$ is nonempty and compact (see Exercise 1 in Section 6.2.4). In this case $P_4(I)$ is a (finite) nonnegative real number (for $d \geq 1$).

We introduce the notation

$$\mathcal{I}_{n,d,R,\rho,M} := \{I \in \mathcal{I} : P_1(I) \leq n, P_2(I) \leq d, P_3(I) \leq R, P_4(I) \geq \rho, P_5(I) \leq M\},$$

and define

$$\mathrm{comp}_{\mathcal{A}}(n,d,R,\rho,M) := \sup\{\mathrm{comp}_{\mathcal{A}}(I) : I \in \mathcal{I}_{n,d,R,\rho,M}\},$$
$$\mathrm{icomp}_{\mathcal{A}}(n,d,R,\rho,M) := \sup\{\mathrm{icomp}_{\mathcal{A}}(I) : I \in \mathcal{I}_{n,d,R,\rho,M}\}$$

for any oracle-based algorithm $\mathcal{A}$, and

$$\mathrm{icomp}_{\epsilon}(n,d,R,\rho,M) := \inf_D \ \sup\{\mathrm{icomp}_{\epsilon}(D,I) : I \in \mathcal{I}_{n,d,R,\rho,M}\}$$

where the infimum is taken over all adaptive strategies D.[1]

6.1.3 Approximate Solutions

The following lemma is a very important tool that will be used extensively in our complexity analyses. It establishes the existence of a full-dimensional ball of ϵ-approximate solutions under certain conditions on the feasible region. This forms the basis of many subsequent assertions about ϵ-approximate solutions.

---

[1] The parameterized complexity functions are defined here slightly differently compared to the general definition in Definition 5.4.1 since we use the condition $P_4(I) \geq \rho$ as opposed to $P_4(I) \leq \rho$. This is a superficial difference since we could define the alternate parameter $\lambda = 1/\rho$ and we would have no discrepancy. We choose to have things different in this chapter to be consistent with the literature on the subject, where ρ is taken as the more natural parameter.

Lemma 6.1.5 *Let $1 \leq p \leq \infty$. Let $C \subseteq \mathbb{R}^k$ be a closed, convex set such that $\{\mathbf{z} \in \mathbb{R}^k : \|\mathbf{z} - \mathbf{a}\|_p \leq \rho\} \subseteq C \subseteq \{\mathbf{z} \in \mathbb{R}^k : \|\mathbf{z}\|_p \leq R\}$ for some $R, \rho \in \mathbb{R}_+$ and $\mathbf{a} \in \mathbb{R}^k$. Let $f : \mathbb{R}^k \to \mathbb{R} \cup \{+\infty\}$ be a convex function that is Lipschitz continuous over $\{\mathbf{z} \in \mathbb{R}^k : \|\mathbf{z}\|_p \leq R\}$ with respect to the $\|\cdot\|_p$-norm with Lipschitz constant M, with the convention that $\infty - \infty = 0$. For any $0 \leq \epsilon \leq 2MR$ and for any $\mathbf{z}^\star \in C$, the set $\{\mathbf{z} \in C : f(\mathbf{z}) \leq f(\mathbf{z}^\star) + \epsilon\}$ contains an $\|\cdot\|_p$ ball of radius $\frac{\epsilon\rho}{2MR}$, with center lying on the line segment between $\mathbf{z}^\star$ and $\mathbf{a}$.*

Proof Since $C \subseteq \{\mathbf{z} : \|\mathbf{z}\|_p \leq R\}$, we must have $C \subseteq \{\mathbf{z} : \|\mathbf{z} - \mathbf{z}^\star\|_p \leq 2R\}$. By convexity of C and the fact that $0 \leq \frac{\epsilon}{2MR} \leq 1$, $\mathbf{z}^\star + \frac{\epsilon}{2MR}(C - \mathbf{z}^\star) \subseteq C$. Hence,

$$\mathbf{z}^\star + \frac{\epsilon}{2MR}(C - \mathbf{z}^\star) \subseteq \left\{\mathbf{z} \in C : \|\mathbf{z} - \mathbf{z}^\star\|_p \leq \frac{\epsilon}{M}\right\} \subseteq \{\mathbf{z} \in C : f(\mathbf{z}) \leq f(\mathbf{z}^\star) + \epsilon\},$$

where the second containment follows from the Lipschitz property of f. Since C contains an $\|\cdot\|_p$ ball of radius ρ centered at $\mathbf{a}$, the set $\mathbf{z}^\star + \frac{\epsilon}{2MR}(C - \mathbf{z}^\star)$ (i.e., the $\frac{\epsilon}{2MR}$ scaling of C about $\mathbf{z}^\star$) must contain a ball of radius $\frac{\epsilon\rho}{2MR}$ centered at a point on the line segment between $\mathbf{z}^\star$ and $\mathbf{a}$. $\qquad\square$

6.2 Information Complexity

In this section, we establish lower and upper bounds on the ϵ-information complexity of (6.1.1). The complexity measures are defined with respect to a choice of first-order oracle (see item 2. in Remark 6.1.2), but this is not made explicit for ease of notation. This should be kept in mind when reading the theorem statements below.

Theorem 6.2.1 *Using the definitions from Definition 6.1.4, the following relations can be shown, assuming $R \geq 1$.*

Lower bounds. *There exists a first-order oracle such that the following hold.*

- *If $d \geq 1$,*

$$\mathrm{icomp}_\epsilon(n, d, R, \rho, M) \in \Omega\left(d2^n \log\left(\frac{MR}{\min\{\rho, 1\}\epsilon}\right)\right).$$

- *If $d = 0$,*

$$\mathrm{icomp}_\epsilon(n, d, R, \rho, M) \in \Omega\left(2^n \log(R)\right).$$

Upper bounds.

- *If $n, d \geq 1$, there exists an ϵ-approximate algorithm $\mathcal{A}$ such that for any first-order oracle,*

$$\text{icomp}_{\mathcal{A}}(n,d,R,\rho,M) \in O\left((n+d)d2^n \log\left(\tfrac{MR}{\min\{\rho,1\}\epsilon}\right)\right).$$

- *If $d = 0$, there exists an ϵ-approximate algorithm $\mathcal{A}$ such that for any first-order oracle,*

$$\text{icomp}_{\mathcal{A}}(n,d,R,\rho,M) \in O\left(n2^n \log(R)\right).$$

- *If $n = 0$, there exists an ϵ-approximate algorithm $\mathcal{A}$ such that for any first-order oracle,*

$$\text{icomp}_{\mathcal{A}}(n,d,R,\rho,M) \in O\left(d \log\left(\tfrac{MR}{\min\{\rho,1\}\epsilon}\right)\right).$$

Note that when $n = 0$, i.e., there are no integer variables, we have $\text{icomp}_{\epsilon}(n,d,R,\rho,M) = \Theta\left(d \log\left(\tfrac{MR}{\min\{\rho,1\}\epsilon}\right)\right)$, giving a tight characterization of the complexity. Moreover, when $R < 1$, there is at most one integer fiber corresponding to $\mathbf{x} = \mathbf{0}_n \in \mathbb{Z}^n$ that can contribute feasible solutions. Thus, it reduces to the continuous problem with $n = 0$.

For $d = 0$, i.e., there are no continuous variables, our upper and lower bounds are off by a linear factor in the dimension, which is of much lower order compared to the dominating term of $2^n \log(R)$. Put another way, both bounds are $2^{O(n)} \log(R)$. Additionally, since the strict feasibility assumption is vacuous and for small enough $\epsilon > 0$, $\mathcal{S}(I,\epsilon)$ is the set of exact optimum solutions, M, ϵ, and ρ do not play a role in the upper and lower bounds; in particular, they are the bounds for obtaining exact solutions ($\epsilon = 0$) as well.

There seems to be scope for nontrivial improvement in the upper bound for the mixed-integer case, i.e., $n,d \geq 1$. When one plugs in $n = 0$ in the mixed-integer upper bound ($n,d \geq 1$), one does not recover a tight upper bound for $n = 0$; instead, the bound is off by a factor of d. We believe this can likely be improved, for example, if Conjecture 6.2.6 in Section 6.2.3 is proved to be true in the future. Then one would have an upper bound of $O\left((n+d)2^n \log\left(\tfrac{MR}{\rho\epsilon}\right)\right)$ in the mixed-integer case that more accurately generalizes both the pure continuous ($n = 0$) and pure integer ($d = 0$) upper bounds.

6.2.1 Proof of the Lower Bounds in Theorem 6.2.1

The general strategy is the following: given any adaptive query sequence D, we will construct two instances $(f_1, C_1), (f_2, C_2) \in \mathcal{I}_{n,d,R,\rho,M}$ such that the transcripts $\Pi_k(D, (f_1, C_1))$ and $\Pi_k(D, (f_2, C_2))$ are equal for any k less than the lower bound, but $\mathcal{S}((f_1, C_1), \epsilon) \cap \mathcal{S}((f_2, C_2), \epsilon) = \emptyset$.

The case with $d \geq 1$. We will show that

$$\mathrm{icomp}_\epsilon(n, d, R, \rho, M) \geq \max\left\{ d2^n \log_3\left(\tfrac{2R}{3\rho}\right), d2^n \left\lfloor \log_8\left(\tfrac{MR}{2\epsilon}\right) \right\rfloor \right\}$$

$$\in \Omega\left(d2^n \log\left(\tfrac{MR}{\min\{\rho, 1\}\epsilon}\right) \right).$$

Case 1: The feasibility lower bound. We show that $\mathrm{icomp}_\epsilon(n, d, R, \rho, M) \geq d2^n \log_3\left(\tfrac{2R}{3\rho}\right)$. We construct $C_1, C_2 \subseteq [-R, R]^{n+d}$ such that $C_1 \cap C_2 \cap (\mathbb{Z}^n \times \mathbb{R}^d) = \emptyset$, both sets satisfy the strict feasibility condition dictated by ρ, and any separation oracle query from D on C_1 and C_2 has the same answer. Our instances will consist of these two sets as feasible regions and $f_1 = f_2$ as constant functions, and any functional oracle query in D will simply have as response this constant value and $\mathbf{0}$ as the subgradient. Since there is no common feasible point, $\mathcal{S}((f_1, C_1), \epsilon) \cap \mathcal{S}((f_2, C_2), \epsilon) = \emptyset$ as required.

The construction of C_1 and C_2 goes as follows. Since the functional oracle calls are superfluous, we focus on the separation queries. Let there be $k < 2^n \cdot d \log_3\left(\tfrac{2R}{3\rho}\right)$ such (adaptive) separation queries. Define $X_0 := [0, 1]^n \times [-R, R]^d$. We create a nested sequence $X_0 \supseteq X_1 \supseteq X_2 \supseteq \cdots \supseteq X_k$ of compact, convex sets such that $X_i \cap (\{\mathbf{x}\} \times \mathbb{R}^d)$ is a fiber box for any $\mathbf{x} \in \{0, 1\}^n$ and any $i = 1, \ldots, k$. This sequence will depend on the queries that are made adaptively and will determine our responses to the queries. More precisely, X_i and the response to the ith query $q_{\mathbf{z}^i}$ is defined inductively from X_{i-1} for $i = 1, \ldots, k$.

If the queried point $\mathbf{z}^i \notin X_{i-1}$, then we simply report any hyperplane separating $\mathbf{z}^i$ from X_{i-1} as the answer to $q_{\mathbf{z}^i}$ and define $X_i = X_{i-1}$. If $\mathbf{z}^i \in X_{i-1} \backslash (\mathbb{Z}^n \times \mathbb{R}^d)$, we define $X_i = X_{i-1}$ and the answer to the query $q_{\mathbf{z}^i}$ is that $\mathbf{z}^i$ is in the set.

Suppose now $\mathbf{z}^i = (\mathbf{x}^i, \mathbf{y}^i) \in X_{i-1} \cap (\mathbb{Z}^n \times \mathbb{R}^d)$. Let $B := X_{i-1} \cap (\{\mathbf{x}^i\} \times \mathbb{R}^d)$. Since B is a fiber box, there exist $\ell_j, u_j, j = 1, \ldots, d$ such that $B = \{\mathbf{x}^i\} \times [\ell_1, u_1] \times \cdots \times [\ell_d, u_d]$.

If B has width strictly less than 3ρ, then consider a halfspace H that (strictly) separates $X_{i-1} \cap (\{\mathbf{x}^i\} \times \mathbb{R}^d)$ from the rest of the fiber boxes $X_{i-1} \cap (\{\mathbf{x}\} \times \mathbb{R}^d)$ for $\mathbf{x} \neq \mathbf{x}_i$, as well as the queried points $\mathbf{z}^\ell$ that were reported to be inside X_ℓ for $\ell \leq i - 1$. Report that $\mathbf{z}^i$ is not in the set and return H as the separating halfspace. Define $X_i = X_{i-1} \cap H$.

If B has width at least 3ρ, choose the coordinate $j^\star \in \{1, \ldots, d\}$ such that B has the maximum length in that coordinate. If $\mathbf{y}^i_{j^\star} \leq \tfrac{\ell_{j^\star} + u_{j^\star}}{2}$, set $\hat{H} := \left\{ (\mathbf{x}, \mathbf{y}) : \mathbf{y}_{j^\star} \geq \tfrac{\ell_{j^\star} + 2u_{j^\star}}{3} \right\}$, and if $\mathbf{y}^i_{j^\star} > \tfrac{\ell_{j^\star} + u_{j^\star}}{2}$, set $\hat{H} := \left\{ (\mathbf{x}, \mathbf{y}) : \mathbf{y}_{j^\star} \leq \tfrac{2\ell_{j^\star} + u_{j^\star}}{3} \right\}$. Note that $\hat{H}$ is a halfspace that reduces the length of the fiber box B by one-third in the coordinate $j^\star$ (see Figure 6.1).

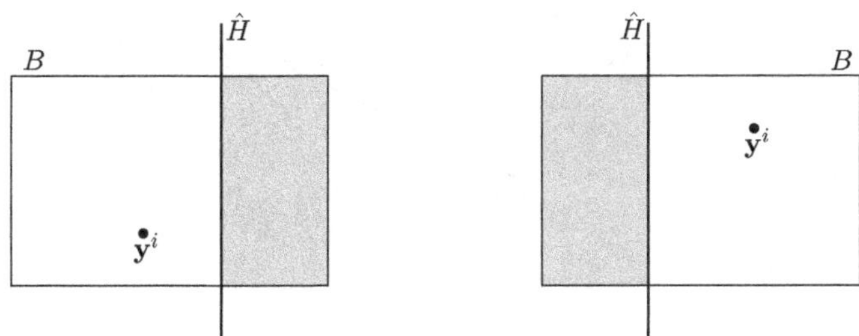

Figure 6.1 A picture illustrating B and $\hat{H}$. The light gray region is the updated fiber box.

We now "rotate" this halfspace $\hat{H}$ to obtain a halfspace H such that three conditions are satisfied (see Exercise 3 from Section 6.2.4):

1. For all $\mathbf{x} \in \{0,1\}^n \setminus \{\mathbf{x}^i\}$, $X_{i-1} \cap (\{\mathbf{x}\} \times \mathbb{R}^d) \subseteq H$.
2. $X_{i-1} \cap (\{\mathbf{x}^i\} \times \mathbb{R}^d) \cap H = X_{i-1} \cap (\{\mathbf{x}^i\} \times \mathbb{R}^d) \cap \hat{H}$.
3. For all $\ell \leq i - 1$ such that $\mathbf{z}^\ell$ was reported to be inside $X_{\ell-1}$, we have $\mathbf{z}^\ell \in H$.

The response to the query $q_{\mathbf{z}^i}$ is that $\mathbf{z}^i$ is not in the set and H is the separating halfspace. Define $X_i = X_{i-1} \cap H$. Observe that X_i has the same fiber boxes as X_{i-1} for all $\mathbf{x} \in \{0,1\}^n \setminus \{\mathbf{x}^i\}$, and in the fiber corresponding to $\mathbf{x}^i$, the fiber box has its length reduced by one-third in the coordinate $j^\star$. Also, the above construction ensures inductively that for any $i \in \{1, \ldots, k\}$, the set $X_i \cap (\{\mathbf{x}\} \times \mathbb{R}^d)$ is a fiber box for $\mathbf{x} \in \{0,1\}^n$.

Since $k < 2^n \cdot d \log_3 \left(\frac{2R}{3\rho}\right)$, we observe that X_k contains a fiber box B of width at least 3ρ (see Exercise 4 from Section 6.2.4). Thus, we can select two fiber boxes $B_1, B_2 \subseteq B$ such that $B_1 \cap B_2 = \emptyset$, and B_1 and B_2 have width ρ. For $i = 1, 2$, define C_i to be the convex hull of B_i and all the points queried by D that were reported to be in the set. We observe that $C_i \cap (\mathbb{Z}^n \times \mathbb{R}^d) = B_i$ for $i = 1, 2$ and thus we have no common feasible points in C_1, C_2. The construction also ensures that both instances f_1, C_1 and f_2, C_2 will give the same responses to all oracle queries.

Case 2: The optimality lower bound. We next show that $\mathrm{icomp}_\epsilon(n, d, R, \rho, M) \geq d2^n \left\lfloor \log_8 \left(\frac{MR}{2\epsilon}\right) \right\rfloor$. We begin with a construction of a family of convex functions over the real line.

Definition 6.2.2 Let $M, R > 0$. We define a family of convex functions defined on $\mathbb{R}$ indexed by all finite binary strings (including the empty string) that are Lipschitz continuous with Lipschitz constant M on their domains.

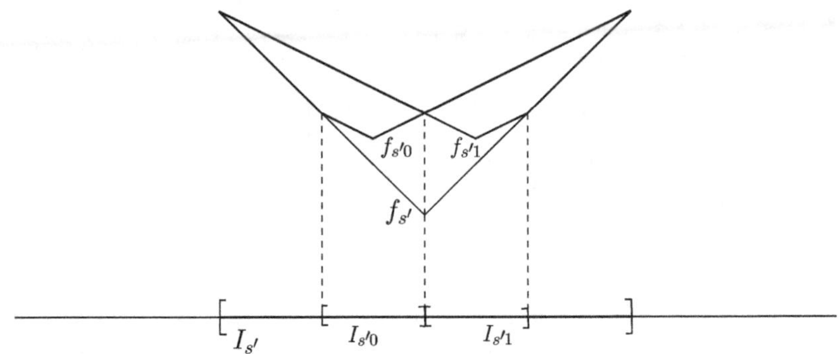

Figure 6.2 A picture illustrating the construction of f_s, I_s from $f_{s'}, I_{s'}$.

The definition proceeds inductively on the length of the binary strings, starting with the empty string. We define the interval $I_{\mathrm{empty}} = [-R, R]$ and define $f_{\mathrm{empty}}(x) = M|x|$ for $x \in [-R, R]$ and $f_{\mathrm{empty}}(x) = +\infty$ if $x \notin [-R, R]$. Next, consider s to be a binary string of length at least 1. We define I_s and f_s in the following way (see Figure 6.2). We use $|s|$ to denote the length of the string.

If $s = s'0$, i.e., the last bit in s is 0 with prefix s', define $I_s = \left[\frac{3\ell_{s'}+u_{s'}}{4}, \frac{\ell_{s'}+u_{s'}}{2} \right]$, where $\ell_{s'} \le u_{s'}$ are the endpoints of $I_{s'}$. Define

$$f_s(x) := \max \left\{ f_{s'}(x), f_{s'}\left(\frac{3\ell_{s'} + u_{s'}}{4} \right) - \frac{M}{2^{|s|}} \left(x - \frac{3\ell_{s'} + u_{s'}}{4} \right), \right.$$
$$\left. f_{s'}(u_{s'}) + \frac{M}{2^{|s|}}(x - u_{s'}) \right\}.$$

If $s = s'1$, i.e., the last bit in s is 1 with prefix s', then set $I_s = \left[\frac{\ell_{s'}+u_{s'}}{2}, \frac{\ell_{s'}+3u_{s'}}{4} \right]$, where $\ell_{s'} \le u_{s'}$ are the endpoints of $I_{s'}$. Define

$$f_s(x) := \max \left\{ f_{s'}(x), f_{s'}(\ell_{s'}) - \frac{M}{2^{|s|}}(x - \ell_{s'}), \right.$$
$$\left. f_{s'}\left(\frac{\ell_{s'} + 3u_{s'}}{4} \right) + \frac{M}{2^{|s|}} \left(x - \frac{\ell_{s'} + 3u_{s'}}{4} \right) \right\}.$$

We next extend this definition to d dimensions. Given d binary strings $s_1, \ldots, s_d$, define $f_{s_1, s_2, \ldots, s_d} : \mathbb{R}^d \to \mathbb{R}$ and $B_{s_1, \ldots, s_d} \subseteq \mathbb{R}^d$ as follows.

$$f_{s_1, s_2, \ldots, s_d}(\mathbf{y}) := \max\{f_{s_1}(\mathbf{y}_1), f_{s_2}(\mathbf{y}_2), \ldots, f_{s_d}(\mathbf{y}_d)\}, \quad B_{s_1, \ldots, s_d} := I_{s_1} \times \cdots \times I_{s_d}.$$
$$(6.2.1)$$

Now let us come back to the construction of the two instances with disjoint solutions that will give the same responses to an adaptive query. We will construct two distinct functions f_1, f_2 and take $C_1 = C_2 = [0,1]^n \times [-R,R]^d$. Any separation query will therefore return the same response. Let there be $k < d2^n \lfloor \log_8 \left(\frac{MR}{2\epsilon} \right) \rfloor$ (adaptive) functional queries at $\mathbf{z}^i = (\mathbf{x}^i, \mathbf{y}^i)$, $i = 1,\ldots,k$. Similar to Case 1 (feasibility), we will create a nested sequence of polyhedra $Y_0 \supseteq Y_1 \supseteq \cdots \supseteq Y_k$ contained in $[0,1]^n \times [-R,R]^d \times \mathbb{R}$ which will depend on the queries made adaptively and will determine our responses $(v^i, \mathbf{s}^i) \in \mathbb{R} \times (\mathbb{R}^n \times \mathbb{R}^d)$ of functional value and subgradient pairs. In addition, we will keep some more data along with Y_i at every iteration i: for every $\mathbf{x} \in \{0,1\}^n$, we will keep track of d binary strings $s_1^{i,\mathbf{x}}, \ldots, s_d^{i,\mathbf{x}}$ of length at most $\lfloor \log_8 \left(\frac{MR}{2\epsilon} \right) \rfloor$. We will maintain the following iterative invariants for all $i = 1,\ldots,k$:

1. The recession cone of Y_i is $\{(\mathbf{0},\mathbf{0},\lambda) \in \mathbb{R}^n \times \mathbb{R}^d \times \mathbb{R} : \lambda \geq 0\}$.
2. For any $(\mathbf{x},\mathbf{y}) \in [0,1]^n \times [-R,R]^d$, there exists $\bar{t} \in \mathbb{R}$ such that $(\mathbf{x},\mathbf{y},\bar{t}) \in Y_i$.
3. For every $\mathbf{x} \in \{0,1\}^n$, $\{\mathbf{x}\} \times \mathrm{epi}(f_{t_1,\ldots,t_d})$ is contained in Y_i for all binary strings $t_1,\ldots,t_d$ of length at least $\lfloor \log_8 \left(\frac{MR}{2\epsilon} \right) \rfloor$ such that $s_j^{i,\mathbf{x}}$ is a prefix of t_j for all $j = 1,\ldots,d$.
4. For all $\ell \leq i$, the halfspace $\{(\mathbf{z},t) \in (\mathbb{R}^n \times \mathbb{R}^d) \times \mathbb{R} : \langle \mathbf{s}^\ell, \mathbf{z} \rangle - t \leq \langle \mathbf{s}^\ell, \mathbf{z}^\ell \rangle - v^\ell\}$ is a supporting halfspace for Y_i at $(\mathbf{z}^\ell, v^\ell)$. If the query point $\mathbf{z}^\ell = (\mathbf{x}^\ell, \mathbf{y}^\ell) \in \{0,1\}^n \times [-R,R]^d$, then it is a supporting halfspace for $\{\mathbf{x}^\ell\} \times \mathrm{epi}(f_{t_1,\ldots,t_d})$ as well (at $(\mathbf{z}^\ell, v^\ell)$) for all binary strings $t_1,\ldots,t_d$ such that $s_j^{\ell,\mathbf{x}^\ell}$ is a prefix of t_j for all $j = 1,\ldots,d$.

The set Y_k will be used to construct the epigraphs of f_1 and f_2.

Let $m = \lfloor \log_8 \left(\frac{MR}{2\epsilon} \right) \rfloor$. Let $\mathcal{F} := \{f_{s_1,s_2,\ldots,s_d} : s_i$ binary string of length m $\forall i = 1,\ldots,d\}$. By Exercise 6d in Section 6.2.4, all functions in $\mathcal{F}$ have a common minimum value $f^\star$. Define $Y_0 := [0,1]^n \times [-R,R]^d \times \{t \in \mathbb{R} : t \geq f^\star\}$. Define $s_j^{0,\mathbf{x}}$ to be the empty string for all $\mathbf{x} \in \{0,1\}^n$ and $j = 1,\ldots,d$. We now consider different cases depending on the functional query point $\mathbf{z}^i = (\mathbf{x}^i, \mathbf{y}^i)$.

If $\mathbf{z}^i \notin [0,1]^n \times [-R,R]^d$, we report a function value of $+\infty$ and return a separating hyperplane between $\mathbf{z}^i$ and $[0,1]^n \times [-R,R]^d$. We define $Y_i := Y_{i-1}$ and for every $\mathbf{x} \in \{0,1\}^n$, $s_j^{i,\mathbf{x}} = s_j^{i-1,\mathbf{x}}$ for all $j = 1,\ldots,d$.

If $\mathbf{x}^i \in [0,1]^n \setminus \{0,1\}^n$, we return as function value $v^i := \min\{t \in \mathbb{R} : (\mathbf{x}^i,\mathbf{y}^i,t) \in Y_{i-1}\}$, which exists by part 2. of the iterative invariant that is maintained for Y_{i-1}. From part 1. of the iterative invariant, it follows that there exists a supporting hyperplane at $(\mathbf{x}^i,\mathbf{y}^i,v^i)$ of the form $(\mathbf{a}^1,\mathbf{a}^2, -1) \in \mathbb{R}^n \times \mathbb{R}^d \times \mathbb{R}$ (see Exercises 9, 12, and 13 from Section 3.1.1). We return

$\mathbf{s}^i := (\mathbf{a}^1, \mathbf{a}^2)$ as the subgradient response to this functional query. We define $Y_i := Y_{i-1}$. Define, for every $\mathbf{x} \in \{0,1\}^n$, $s_j^{i,\mathbf{x}} = s_j^{i-1,\mathbf{x}}$ for all $j = 1, \ldots, d$.

Now suppose $\mathbf{x}^i \in \{0,1\}^n$. We first define the strings $s_j^{i,\mathbf{x}}$, depending on a case analysis. Let $j^\star \in \{1, \ldots, d\}$ such that $f_{s_1^{i-1,\mathbf{x}^i}, \ldots, s_d^{i-1,\mathbf{x}^i}}(\mathbf{y}^i) = f_{s_{j^\star}^{i-1,\mathbf{x}}}(\mathbf{y}_{j^\star}^i)$.

If $s_{j^\star}^{i-1,\mathbf{x}^i}$ has length m, then $Y_i = Y_{i-1}$ and $s_j^{i,\mathbf{x}} = s_j^{i-1,\mathbf{x}}$ for all $\mathbf{x} \in \{0,1\}^n$ and $j \in \{1, \ldots, d\}$.

If $\mathbf{y}_{j^\star}^i \notin I_{s_{j^\star}^{i-1,\mathbf{x}^i}}$, define for every $\mathbf{x} \in \{0,1\}^n$, $s_j^{i,\mathbf{x}} = s_j^{i-1,\mathbf{x}}$ for all $j = 1, \ldots, d$.

If $\mathbf{y}_{j^\star}^i \in I_{s_{j^\star}^{i-1,\mathbf{x}^i}}$, we set $b := 0$ if $\mathbf{y}_{j^\star}^i$ is greater than or equal to the midpoint of $I_{s_{j^\star}^{i-1,\mathbf{x}}}$; otherwise we set $b = 1$. For every $\mathbf{x} \in \{0,1\}^n \setminus \{\mathbf{x}^i\}$, $s_j^{i,\mathbf{x}} = s_j^{i-1,\mathbf{x}}$ for all $j = 1, \ldots, d$. Further, define $s_j^{i,\mathbf{x}^i} = s_j^{i-1,\mathbf{x}^i}$ for all $j \neq j^\star$. Finally, define $s_{j^\star}^{i,\mathbf{x}^i} = s_{j^\star}^{i-1,\mathbf{x}^i} b$, i.e., $s_{j^\star}^{i,\mathbf{x}^i}$ is obtained by appending the bit b to $s_{j^\star}^{i-1,\mathbf{x}^i}$.

Now we define Y_i and the response to the query. Let $v := f_{s_1^{i,\mathbf{x}^i}, \ldots, s_d^{i,\mathbf{x}^i}}(\mathbf{y}^i)$ and $\mathbf{q} := \gamma \mathbf{e}^{j^\star}$ where γ is a subgradient value for $f_{s_{j^\star}^{i,\mathbf{x}^i}}$ at $\mathbf{y}_{j^\star}^i$. Define the halfspace $\hat{H} := \{(\mathbf{x}, \mathbf{y}, t) \in \mathbb{R}^n \times \mathbb{R}^d \times \mathbb{R} : \langle \mathbf{q}, \mathbf{y} \rangle - t \leq \langle \mathbf{q}, \mathbf{y}^i \rangle - v\}$ (restricted to the the fiber over $\mathbf{x}$, $\hat{H}$ is a supporting halfspace to the epigraph of $f_{s_1^{i,\mathbf{x}^i}, \ldots, s_d^{i,\mathbf{x}^i}}$ at $(\mathbf{y}^i, v)$). Similar to Case 1 (feasibility), we now "rotate" this halfspace to obtain a halfspace H that satisfies the following conditions (see Exercise 3 from Section 6.2.4). More precisely, let $\mathbf{p} \in \mathbb{R}^n$ be such that $H := \{(\mathbf{x}, \mathbf{y}, t) \in \mathbb{R}^n \times \mathbb{R}^d \times \mathbb{R} : \langle \mathbf{p}, \mathbf{x} \rangle + \langle \mathbf{q}, \mathbf{y} \rangle - t \leq \langle \mathbf{p}, \mathbf{x}^i \rangle + \langle \mathbf{q}, \mathbf{y}^i \rangle - v\}$ satisfies the following:

1. For all $\mathbf{x} \in \{0,1\}^n \setminus \{\mathbf{x}^i\}$, $Y_{i-1} \cap (\{\mathbf{x}\} \times \mathbb{R}^d \times \mathbb{R}) \subseteq H$.
2. $Y_{i-1} \cap (\{\mathbf{x}^i\} \times \mathbb{R}^d \times \mathbb{R}) \cap H = Y_{i-1} \cap (\{\mathbf{x}^i\} \times \mathbb{R}^d \times \mathbb{R}) \cap \hat{H}$.
3. For all $\ell \leq i - 1$ such that $\mathbf{x}^\ell \in [0,1]^n \setminus \{0,1\}^n$, we have $(\mathbf{x}^\ell, \mathbf{y}^\ell, v^\ell) \in H$.

Define $Y_i = Y_{i-1} \cap H$. Exercise 7 from Section 6.2.4 verifies that the iterative invariant is maintained if the above definitions are followed for Y_i and the strings $s_j^{i,\mathbf{x}}$, $\mathbf{x} \in \{0,1\}^n$ and $j \in \{1, \ldots, d\}$. Observe that H is a supporting halfspace for Y_i. As response to the query at $(\mathbf{x}^i, \mathbf{y}^i)$, we return $v^i := v$ as the function value and $\mathbf{s}^i := (\mathbf{p}, \mathbf{q})$ as the subgradient.

Since $k < 2^n \cdot d \lfloor \log_8 \left(\frac{MR}{2\epsilon} \right) \rfloor$, there must exist $\mathbf{x}^\star \in \{0,1\}^n$ and $j^\star \in \{1, \ldots, d\}$ such that there are strictly less than $\lfloor \log_8 \left(\frac{MR}{2\epsilon} \right) \rfloor = m$ functional queries at $(\mathbf{x}^i, \mathbf{y}^i)$ such that $\mathbf{x}^i = \mathbf{x}^\star$ and $f_{s_1^{i-1,\mathbf{x}^i}, \ldots, s_d^{i-1,\mathbf{x}^i}}(\mathbf{y}^i) = f_{s_{j^\star}^{i-1,\mathbf{x}}}(\mathbf{y}_{j^\star}^i)$. Since the length of the strings increases by at most 1 in every step, $s_{j^\star}^{i-1,\mathbf{x}^\star}$

has length strictly less than m. Thus, there exist distinct binary strings s, s' of length m such that $s_{j^\star}^{i-1,\mathbf{x}^\star}$ is a prefix of both s and s'. For $j \neq j^\star$, define $s_j^{\text{fin},\mathbf{x}^\star}$ to be any binary string of length m such that $s_j^{i-1,\mathbf{x}^\star}$ is a prefix of $s_j^{\text{fin},\mathbf{x}^\star}$. Define $\tilde{f}_1 = f_{s_1^{\text{fin}},\ldots,s,\ldots,s_d^{\text{fin}}}$ and $\tilde{f}_2 = f_{s_1^{\text{fin}},\ldots,s',\ldots,s_d^{\text{fin}}}$, where s and s' appear in the $j^\star$ position. The key thing to observe here is that both $\tilde{f}_1$ and $\tilde{f}_2$ are consistent with all the responses to queries on the fiber corresponding to $\mathbf{x}^\star$, by Exercise 7d from Section 6.2.4. Finally, for $\mathbf{x} \neq \mathbf{x}^\star$, define $s_j^{\text{fin},\mathbf{x}}$ to be any string of length at least $m + 1$ such that $s_j^{k,\mathbf{x}}$ is a prefix of $s_j^{\text{fin},\mathbf{x}}$ for every $j = 1, \ldots, d$.

Let $I := \{ i \in \{1, \ldots, k\} : \mathbf{x}^i \in [0,1]^n \setminus \{0,1\}^n \}$. Define

$$E_1 = \text{cl}\left(\text{conv}\left(\left(\{\mathbf{x}^\star\} \times \text{epi}(\tilde{f}_1) \right) \cup \bigcup_{\mathbf{x} \neq \mathbf{x}^\star} \left(\{\mathbf{x}\} \times \text{epi}\left(f_{s_1^{\text{fin},\mathbf{x}},\ldots,s_d^{\text{fin},\mathbf{x}}} \right) \right) \cup \bigcup_{i \in I} \left\{ (\mathbf{x}^i, \mathbf{y}^i, t) : t \geq v^i \right\} \right) \right)$$

and

$$E_2 = \text{cl}\left(\text{conv}\left(\left(\{\mathbf{x}^\star\} \times \text{epi}(\tilde{f}_2) \right) \cup \bigcup_{\mathbf{x} \neq \mathbf{x}^\star} \left(\{\mathbf{x}\} \times \text{epi}\left(f_{s_1^{\text{fin},\mathbf{x}},\ldots,s_d^{\text{fin},\mathbf{x}}} \right) \right) \cup \bigcup_{i \in I} \left\{ (\mathbf{x}^i, \mathbf{y}^i, t) : t \geq v^i \right\} \right) \right).$$

By parts 1., 3., and 4. of the iterative invariant, E_1 and E_2 are both closed convex sets contained in Y_k, and for any $i \in \{1, \ldots, k\}$ we have that $\min\{t \in \mathbb{R} : (\mathbf{x}^i, \mathbf{y}^i, t) \in E_1\} = \min\{t \in \mathbb{R} : (\mathbf{x}^i, \mathbf{y}^i, t) \in E_1\} = v^i$ and the halfspace $\{(\mathbf{z}, t) \in (\mathbb{R}^n \times \mathbb{R}^d) \times \mathbb{R} : \langle \mathbf{s}^i, \mathbf{z} \rangle - t \leq \langle \mathbf{s}^i, \mathbf{z}^i \rangle - v^i \}$ is a supporting halfspace for both E_1 and E_2 at $(\mathbf{x}^i, \mathbf{y}^i, v^i)$. By Exercise 9 from Section 3.1.1, there exist convex functions f_1, f_2 with domain $[0,1]^n \times [-R,R]^d$ and with epigraphs E_1 and E_2, respectively. Therefore, for any $i \in \{1, \ldots, k\}$, $(v^i, \mathbf{s}^i)$ are valid function value and subgradient pairs at $\mathbf{z}^i$ for both f_1 and f_2. Exercise 6d from Section 6.2.4 shows that the minimum of both f_1 and f_2 are on the fiber over $\mathbf{x}^\star$ (since on all other fibers we used strings of length at least $m + 1$). Exercises 6e and 6f from Section 6.2.4 then imply that $\mathcal{S}((f_1, C_1), \epsilon) \cap \mathcal{S}((f_2, C_2), \epsilon) = \emptyset$ (recall $C_1 = C_2 = [0,1]^n \times [-R,R]^d$).

The case with $d = 0$. The proof proceeds in a similar manner to Case 1 (feasibility lower bound) of the $d \geq 1$ setting, with $X_0 = [0,1]^{n-1} \times [-\lfloor R \rfloor, \lfloor R \rfloor] \subseteq \mathbb{R}^n$. The "fibers" are now $\{\mathbf{x}\} \times \{-\lfloor R \rfloor, -\lfloor R \rfloor + 1, \ldots, \lfloor R \rfloor\}$ for $\mathbf{x} \in \{0,1\}^{n-1}$. If $k < 2^{n-1} \log_2(2\lfloor R \rfloor + 1)$, one can again construct $C_1, C_2 \subseteq X_0$ such that $C_1 \cap C_2 \cap \mathbb{Z}^n = \emptyset$ by an iterative argument based on the queries from D, and take f_1, f_2 as constant functions. We leave the details as an exercise (Exercise 8 from Section 6.2.4).

6.2.2 A Geometric Detour: Centerpoints

The idea of the upper bound hinges on a geometric concept that has appeared in several different areas, including convex geometry, statistics, and computer science. This concept can be viewed as a high-dimensional version of the median and is also a useful way to formalize a "center" for an arbitrary convex set.

Definition 6.2.3 For any $S \subseteq \mathbb{Z}^n \times \mathbb{R}^d$ with $d \geq 1$, $\nu(S)$ will denote the *mixed-integer volume* of S, i.e.,

$$\nu(S) := \sum_{\mathbf{x} \in \mathbb{Z}^n} \mathrm{vol}_d \left(S \cap (\{\mathbf{x}\} \times \mathbb{R}^d) \right),$$

where vol_d is the d-dimensional volume in $\mathbb{R}^d$ (see also the discussion on volumes in Section 2.2.1). If $d = 0$, we overload notation and use $\nu(S)$ to denote the number of integer points in S, i.e., the counting measure on $\mathbb{Z}^n$.

Note that if $S = C \cap (\mathbb{Z}^n \times \mathbb{R}^d)$ for a compact convex set $C \subseteq \mathbb{R}^n \times \mathbb{R}^d$, then $\nu(S)$ is finite.

Definition 6.2.4 For any $S \subseteq \mathbb{Z}^n \times \mathbb{R}^d$ and $\mathbf{x} \in \mathbb{R}^n \times \mathbb{R}^d$, define

$$h_S(\mathbf{x}) := \inf\{\nu(S \cap H) : \text{halfspace } H \subseteq \mathbb{R}^n \times \mathbb{R}^d \text{ such that } \mathbf{x} \in H\}.$$

The set of *centerpoints of S* is defined as $C(S) := \mathrm{argmax}_{\mathbf{x} \in S}\, h_S(\mathbf{x})$.

The motivation behind the concept of a centerpoint is to define a mixed-integer point that is "deep inside" of S. The intuition is that if we make a separation or functional query at a centerpoint, then no matter what the response is, we can reduce the search space by a significant amount. For this purpose, we would like to have *lower bounds* on the score $h_S(\mathbf{x})$ of a centerpoint $\mathbf{x} \in S$. The next theorem quantifies this.

Theorem 6.2.5 *Let $n, d \in \mathbb{Z}_+$ with $n + d \geq 1$. Let $C \subseteq \mathbb{R}^n \times \mathbb{R}^d$ be any compact, convex set and let $S = C \cap (\mathbb{Z}^n \times \mathbb{R}^d)$. Then $C(S)$ is nonempty and $h_S(\hat{\mathbf{x}}) \geq \frac{1}{2^n (d+1)} \nu(S)$ for any centerpoint $\hat{\mathbf{x}}$. If $n = 0$, then*

$$h_S(\hat{\mathbf{x}}) \geq \left(\frac{d}{d+1} \right)^d \nu(S) \geq \frac{1}{e} \nu(S), \text{ where } e \text{ is Euler's constant.}$$

The first bound in Theorem 6.2.5 was first established in [186] and is a special case of a result involving Helly numbers [38, Theorem 3.3]. The second bound ($n = 0$) is due to Grünbaum [126]. There is clearly a gap in the two cases and the following sharper lower bound is conjectured to be true [38, 186]; a matching upper bound is given by $C = [0, 1]^n \times \Delta_d$, where Δ_d is the standard d-dimensional simplex (see Exercise 9 from Section 6.2.4).

Conjecture 6.2.6 Under the hypothesis of Theorem 6.2.5, $h_S(\hat{\mathbf{x}}) \geq \frac{1}{2^n}\left(\frac{d}{d+1}\right)^d v(S) \geq \frac{1}{2^n}\frac{1}{e}v(S)$ for any centerpoint $\hat{\mathbf{x}}$, where e is Euler's constant.

Proof of Theorem 6.2.5 We split the proof into two parts, as suggested by the statement of the theorem.

The case with $n \geq 1$. Given any $0 \leq \alpha \leq 1$, let $\mathcal{H}_\alpha$ be the set of all halfspaces H such that $v(H \cap S) \geq \alpha v(S)$. Since C is compact, $D_\alpha := \cap_{H \in \mathcal{H}_\alpha} H$ is a compact, convex subset of C (every halfspace that contains C belongs to $\mathcal{H}_\alpha$ for any $\alpha \in [0, 1]$).

We now claim that for any $\mathbf{x} \in D_\alpha$, we have $h_S(\mathbf{x}) \geq (1-\alpha)v(S)$. To see this, consider any halfspace $H = \{\mathbf{y} \in \mathbb{R}^d : \langle \mathbf{a}, \mathbf{y} \rangle \leq \delta\}$ that contains $\mathbf{x} \in D_\alpha$. Since the halfspace $\bar{H} := \{\mathbf{y} \in \mathbb{R}^d : \langle \mathbf{a}, \mathbf{y} \rangle \geq \delta + \epsilon\}$ does not contain $\mathbf{x}$ for any $\epsilon > 0$, $v(S \cap \bar{H}) < \alpha v(S)$. Therefore, $v(S \cap \{\mathbf{y} \in \mathbb{R}^d : \langle \mathbf{a}, \mathbf{y} \rangle < \delta + \epsilon\}) \geq (1 - \alpha)v(S)$ for all $\epsilon > 0$. Taking the limit $\epsilon \to 0$, we obtain that $v(S \cap H) \geq (1 - \alpha)v(S)$.

Let $\alpha^\star = 1 - \frac{1}{2^n(d+1)}$. It suffices to show that $S \cap D_{\alpha^\star + \epsilon}$ is nonempty for every $\epsilon > 0$, because using compactness and the fact that $D_\alpha \subseteq D_\beta$ when $0 \leq \alpha \leq \beta \leq 1$, we would have $\cap_{\epsilon > 0}(S \cap D_{\alpha^\star + \epsilon})$ is nonempty (see Theorem 1.3.11), and any point $\mathbf{x}$ in this set will satisfy $h_S(\mathbf{x}) \geq \left(\frac{1}{2^n(d+1)} - \epsilon\right)v(S)$ for every $\epsilon > 0$, i.e., $h_S(\mathbf{x}) \geq \frac{1}{2^n(d+1)}v(S)$.

Fix $\epsilon > 0$. We want to show that $S \cap D_{\alpha^\star + \epsilon}$ is nonempty. Consider the family of $(\mathbb{Z}^n \times \mathbb{R}^d)$-convex sets (recall Definition 2.6.6) of the form $S \cap H$, $H \in \mathcal{H}_{\alpha^\star + \epsilon}$. We have $S \cap D_{\alpha^\star + \epsilon} = \bigcap_{H \in \mathcal{H}_{\alpha^\star + \epsilon}}(S \cap H)$. Suppose to the contrary that $\bigcap_{H \in \mathcal{H}_{\alpha^\star + \epsilon}}(S \cap H) = \emptyset$. By Theorem 2.6.14 and Proposition 2.6.8, there exist halfspaces $H_1, \ldots, H_{2^n(d+1)} \in \mathcal{H}_{\alpha^\star + \epsilon}$ such that $(S \cap H_1) \cap \cdots \cap (S \cap H_{2^n(d+1)}) = \emptyset$. This implies that $S \subseteq H_1^c \cup \cdots H_{2^n(d+1)}^c$, where H_i^c denotes the complement $\mathbb{R}^d \setminus H_i$ for $i = 1, \ldots, 2^n(d+1)$. Since $v(S \cap H_i) \geq \left(1 - \frac{1}{2^n(d+1)} + \epsilon\right)v(S)$, we must have $v(S \cap H_i^c) \leq \left(\frac{1}{2^n(d+1)} - \epsilon\right)v(S)$. Then,

$$\begin{aligned}
v(S) &= v\left(\bigcup_{i=1}^{2^n(d+1)}(S \cap H_i^c)\right) \\
&\leq \sum_{i=1}^{2^n(d+1)} v(S \cap H_i^c) \leq 2^n(d+1)\left(\frac{1}{2^n(d+1)} - \epsilon\right)v(S) \\
&= (1 - 2^n(d+1)\epsilon)v(S) \\
&< v(S),
\end{aligned}$$

which yields a contradiction.

The case with $n = 0$. We now have $S = C$. If $\mathrm{vol}(C) = 0$, then any point in C is a centerpoint. Thus, we consider the case when C is full dimensional with nonzero volume. Consider the *centroid* of C, defined as

$$\mathbf{x}^{\mathrm{cen}} := \frac{\int_C \mathbf{x}\, d\mathbf{x}}{\mathrm{vol}(C)}.$$

We claim that $h_S(\mathbf{x}^{\mathrm{cen}}) \geq \left(\frac{d}{d+1}\right)^d \mathrm{vol}(C)$, which will finish the proof.

Consider any halfspace $H^{\leq}(\mathbf{a}, \delta)$ that contains $\mathbf{x}^{\mathrm{cen}}$, where $\mathbf{a} \in \mathbb{R}^d \setminus \{\mathbf{0}\}$ is of unit Euclidean norm. Let $\widetilde{C}$ be the Schwarz symmetrization of C with respect to $L := \mathrm{span}(\mathbf{a})^{\perp}$ (see Definition 3.5.7). Corollary 3.5.8 shows that $\widetilde{C}$ is a compact, convex set with $\mathrm{vol}(\widetilde{C}) = \mathrm{vol}(C)$ and $\mathrm{vol}(\widetilde{C} \cap H^{\leq}(\mathbf{a}, \delta)) = \mathrm{vol}(C \cap H^{\leq}(\mathbf{a}, \delta))$. Thus, it suffices to show that $\mathrm{vol}(\widetilde{C} \cap H^{\leq}(\mathbf{a}, \delta)) \geq \left(\frac{d}{d+1}\right)^d \mathrm{vol}(\widetilde{C})$. Observe also that the centroid of $\widetilde{C}$ is $\langle \mathbf{a}, \mathbf{x}^{\mathrm{cen}} \rangle \mathbf{a}$ and it is contained in $H^{\leq}(\mathbf{a}, \delta)$. Thus, it suffices to show that $\mathrm{vol}\left(\widetilde{C} \cap H^{\leq}(\mathbf{a}, \langle \mathbf{a}, \mathbf{x}^{\mathrm{cen}} \rangle)\right) \geq \left(\frac{d}{d+1}\right)^d \mathrm{vol}(\widetilde{C})$.

Let $\ell \leq \langle \mathbf{a}, \mathbf{x}^{\mathrm{cen}} \rangle$ be such that the convex hull of $\ell \mathbf{a}$ and the $(d-1)$-dimensional ball $\widetilde{C} \cap H^{=}(\mathbf{a}, \langle \mathbf{a}, \mathbf{x}^{\mathrm{cen}} \rangle)$ has volume equal to $\mathrm{vol}\left(\widetilde{C} \cap H^{\leq}(\mathbf{a}, \langle \mathbf{a}, \mathbf{x}^{\mathrm{cen}} \rangle)\right)$. Let $u = \ell + \left(\frac{\mathrm{vol}(\widetilde{C})}{\mathrm{vol}(\widetilde{C} \cap H^{\leq}(\mathbf{a}, \langle \mathbf{a}, \mathbf{x}^{\mathrm{cen}} \rangle))}\right)^{1/d} (\langle \mathbf{a}, \mathbf{x}^{\mathrm{cen}} \rangle - \ell)$. Consider the affine function $p(t)$ over the interval $[\ell, u]$ whose value at ℓ is 0 and the value at $\langle \mathbf{a}, \mathbf{x}^{\mathrm{cen}} \rangle$ is the radius of $\widetilde{C} \cap H^{=}(\mathbf{a}, \langle \mathbf{a}, \mathbf{x}^{\mathrm{cen}} \rangle)$. What we have formally done here is the following. Consider the compact, convex set P formed by the convex hull of $\ell \mathbf{a}$ and $\{\mathbf{x} \in \mathbb{R}^d : \langle \mathbf{a}, \mathbf{x} \rangle = u, \; \|\mathbf{x} - u\mathbf{a}\|_2 \leq p(u)\}$ (such a set is called a *pyramid*). Then the following are all true (see Figure 6.3):

1. The radius of the section $P \cap H^{=}(\mathbf{a}, t)$ equals $p(t)$ for all $t \in [\ell, u]$.
2. $\mathrm{vol}(P) = \mathrm{vol}(\widetilde{C})$.
3. $\mathrm{vol}(P \cap H^{\leq}(\mathbf{a}, \langle \mathbf{a}, \mathbf{x}^{\mathrm{cen}} \rangle)) = \mathrm{vol}\left(\widetilde{C} \cap H^{\leq}(\mathbf{a}, \langle \mathbf{a}, \mathbf{x}^{\mathrm{cen}} \rangle)\right)$.

Exercise 10 shows that the centroid of the pyramid P is at $\bar{t}\mathbf{a}$, with $\bar{t} \leq \langle \mathbf{a}, \mathbf{x}^{\mathrm{cen}} \rangle$, i.e., the centroid of P is to the left of the centroid of $\widetilde{C}$ in Figure 6.3. Exercise 11 shows that $\mathrm{vol}(P \cap H^{\leq}(\mathbf{a}, \langle \mathbf{a}, \mathbf{x}^{\mathrm{cen}} \rangle)) \geq \left(\frac{d}{d+1}\right)^d \mathrm{vol}(P)$. This completes the proof because of properties 2. and 3. above. $\square$

6.2.3 Proof of the Upper Bounds in Theorem 6.2.1

We now proceed with the proof of the upper bounds in Theorem 6.2.1.

The mixed-integer case with $n, d \geq 1$. We consider the following adaptive search strategy D. For any finite (possibly empty) sequence $T \in (\mathcal{Q} \times H)^*$ of query and response pairs, where $\mathcal{Q}$ is the set of all possible functional or separation oracle queries in $[-R, R]^{n+d}$ and H is the set of possible responses to such queries, define the next query $D(T)$ as follows. Let $\mathbf{z}^1, \dots, \mathbf{z}^q$ be the points queried in T where either a functional oracle call to the objective

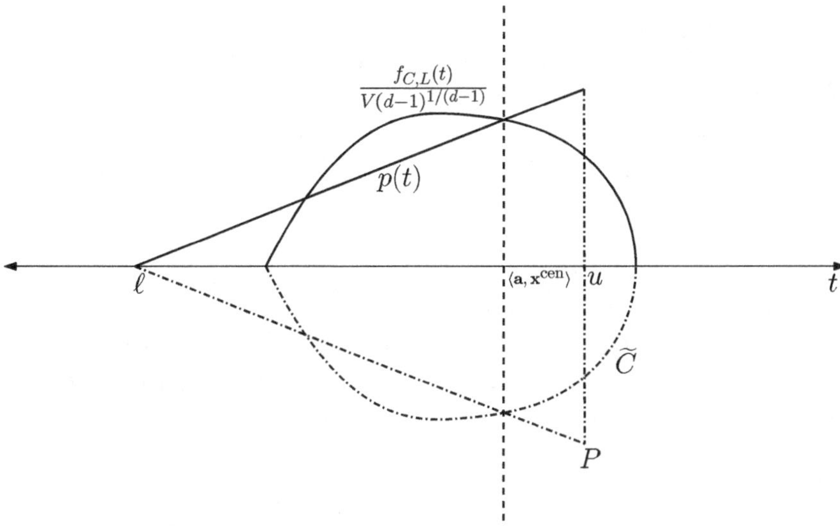

Figure 6.3 A picture illustrating the construction of $\ell, u, p(t)$ and the pyramid P. $f_{C,L}$ is the function defined in Corollary 3.5.6 with $L = \mathrm{span}(\mathbf{a})^{\perp}$ and $V(d-1)$ is the volume of the Euclidean unit ball; $\dfrac{f_{C,L}(t)}{V(d-1)^{1/(d-1)}}$ is thus the radius of the $(d-1)$-dimensional ball $\widetilde{C} \cap H^{=}(\mathbf{a}, t)$. If we "rotate" the graph of this function about the t-axis, we obtain $\widetilde{C}$. If we "rotate" the graph of $p(t)$ about the t-axis, we obtain P.

function was made, or a separation oracle call was made that returned a separating hyperplane. Let $\mathbf{s}^{j}$ be the subgradient or normal vector to the separating hyperplane returned at $\mathbf{z}^{j}$, $j = 1, \ldots, q$. Define $v^{\min}$ to be the minimum function value seen so far ($+\infty$ if T has no functional query).

The next query for D will be at the centerpoint $\hat{\mathbf{z}}$ (see Definition 6.2.4) of the set

$$\left\{ \mathbf{z} \in \mathbb{Z}^{n} \times \mathbb{R}^{d} : \begin{array}{l} \langle \mathbf{s}^{i}, \mathbf{z} - \mathbf{z}^{i}\rangle \leq 0 \ i = 1, \ldots, q, \\ \|\mathbf{z}\|_{\infty} \leq R \end{array} \right\}.$$

If T is the transcript on an instance f, C, then the *search space* polyhedron P defined by the above inequalities contains $C \cap \mathrm{dom}(f) \cap \{\mathbf{z} : f(\mathbf{z}) \leq v^{\min}\}$. D now queries the centerpoint of $P \cap (\mathbb{Z}^{n} \times \mathbb{R}^{d})$ so that any separating hyperplane or subgradient inequality can remove a guaranteed fraction of the mixed-integer volume of the current search space. More formally, D first queries the separation oracle for $C \cap \mathrm{dom}(f)$ at $\hat{\mathbf{z}}$. If the separation oracle says $\hat{\mathbf{z}}$ is in $C \cap \mathrm{dom}(f)$, then D queries the first-order oracle for f at $\hat{\mathbf{z}}$. Now we analyze the information complexity of this adaptive query strategy (Definition 5.3.3).

Consider any instance $I = (f, C) \in \mathcal{I}_{n,d,R,M,\rho}$. Define $\bar{\rho} = \min\{\rho, 1\}$. Recall that we assume $C \cap \mathrm{dom}(f)$ contains a fiber box of width at least $\rho \geq \bar{\rho}$ in the same fiber as the optimum solution if $C \cap \mathrm{dom}(f) \cap (\mathbb{Z}^n \times \mathbb{R}^d)$ is nonempty. Consider any natural number

$$k \geq 2 \cdot \left(\log_b \left(\left(\frac{2R+1}{\bar{\rho}} \right)^{n+d} \right) + \log_b \left(\left(\frac{M(2R+1)}{\epsilon} \right)^{n+d} \right) \right),$$

where $b = \frac{2^n(d+1)}{2^n(d+1)-1}$. We claim that either the instance has no feasible solutions or at least one functional oracle query appears in the transcript $\Pi_k(D, I)$ and $\mathbf{z}^{\min} \in S(I, \epsilon)$, where $\mathbf{z}^{\min}$ is a point queried in the transcript $\Pi_k(D, I)$ with the minimum function value among all points queried with a functional query on f in $\Pi_k(D, I)$. In other words, we have $\mathbf{z}^{\min} \in C \cap \mathrm{dom}(f)$ and $f(\mathbf{z}^{\min}) \leq OPT + \epsilon$ where OPT is the minimum value of f on C. This will prove the result since

$$
\begin{aligned}
2 \cdot \left(\log_b \left(\left(\tfrac{2R+1}{\bar{\rho}} \right)^{n+d} \right) + \log_b \left(\left(\tfrac{M(2R+1)}{\epsilon} \right)^{n+d} \right) \right) &= 2(n+d) \log_b \left(\frac{M(2R+1)^2}{\bar{\rho}\epsilon} \right) \\
&= 2(n+d) \ln \left(\frac{M(2R+1)^2}{\bar{\rho}\epsilon} \right) \Big/ \ln(b) \\
&\leq 2(n+d) 2^n (d+1) \ln \left(\frac{M(2R+1)^2}{\bar{\rho}\epsilon} \right) \\
&\in O\left((n+d)d2^n \log \left(\frac{MR}{\min\{\rho, 1\}\epsilon} \right) \right).
\end{aligned}
$$

First, let k' be the number of queries in $\Pi_k(D, I)$ that were either functional queries on f or separation queries on $C \cap \mathrm{dom}(f)$ that returned a separating hyperplane, i.e., we ignore the separation queries on points inside $C \cap \mathrm{dom}(f)$. Observe that $k' \geq k/2 \geq \log_b \left(\left(\frac{2R+1}{\bar{\rho}} \right)^{n+d} \right) + \log_b \left(\left(\frac{M(2R+1)}{\epsilon} \right)^{n+d} \right)$ since a separation query on any point in $C \cap \mathrm{dom}(f)$ is immediately followed by a functional query on the same point. Theorem 6.2.5 implies that each of these k' queries reduces the mixed-integer volume of the current search space P by at least $1/b$. Recall that we start with a mixed-integer volume of at most $(2R+1)^{n+d}$ and $C \cap \mathrm{dom}(f)$ contains a fiber box of mixed-integer volume at least $\bar{\rho}^d \geq \bar{\rho}^{n+d}$ (since $\bar{\rho} \leq 1$) if there exist feasible solutions. Thus, if more than $\log_b \left(\left(\frac{2R+1}{\bar{\rho}} \right)^{n+d} \right)$ separation oracle queries are made we would (correctly) conclude that the instance has no feasible solutions. Let us then consider the case where we have at least $\log_b \left(\left(\frac{M(2R+1)}{\epsilon} \right)^{n+d} \right)$ functional queries to f at points inside $C \cap \mathrm{dom}(f) \cap (\mathbb{Z}^n \times \mathbb{R}^d)$. Let $k'' \geq 1$ denote the number of such queries, queried at $\mathbf{z}^1, \ldots, \mathbf{z}^{k''}$ with responses $\mathbf{s}^1, \ldots, \mathbf{s}^{k''}$ as the subgradients and $v^1, \ldots, v^{k''}$ as the function values. Let $v^{\min}$ be the minimum of these function values, corresponding to the query point

$\mathbf{z}^{\min}$. Observe that all the points $\mathbf{z}^1, \ldots, \mathbf{z}^{k''}$, including $\mathbf{z}^{\min}$, are in $C \cap \mathrm{dom}(f)$. We now verify that $f(\mathbf{z}^{\min}) \leq OPT + \epsilon$ where OPT is the minimum value of f on $C \cap \mathrm{dom}(f) \cap (\mathbb{Z}^n \times \mathbb{R}^d)$ attained at, say, $\mathbf{z}^\star = (\mathbf{x}^\star, \mathbf{y}^\star)$.

Let $C' = C \cap (\{\mathbf{x}^\star\} \times \mathbb{R}^d)$ be the intersection of C with the fiber containing $\mathbf{z}^\star$. Consider the polyhedron

$$\tilde{P} := \left\{ \mathbf{z} : \langle \mathbf{s}^\ell, \mathbf{z} - \mathbf{z}^\ell \rangle \leq 0 \ \ell = 1, \ldots, k'' \right\}.$$

Since we have been reducing the mixed-integer volume at a rate of $1/b$, $C \cap \mathrm{dom}(f) \cap \tilde{P}$ has mixed-integer volume at most $(2R+1)^{n+d}/b^{k'}$ and therefore $C' \cap \mathrm{dom}(f) \cap \tilde{P}$ has d-dimensional volume at most $(2R+1)^{n+d}/b^{k'}$. Since $k' \geq \log_b\left(\left(\frac{2R+1}{\bar{\rho}}\right)^{n+d}\right) + \log_b\left(\left(\frac{M(2R+1)}{\epsilon}\right)^{n+d}\right)$, we must have $b^{k'} \geq \left(\frac{2R+1}{\bar{\rho}}\right)^{n+d} \cdot \left(\frac{M(2R+1)}{\epsilon}\right)^{n+d}$. Thus, $C' \cap \mathrm{dom}(f) \cap \tilde{P}$ has d-dimensional volume at most $\left(\frac{\bar{\rho}\epsilon}{M(2R+1)}\right)^{n+d}$. We may assume $\frac{\epsilon}{2MR} \leq 1$; otherwise, any feasible solution is an ϵ approximate solution, and so is $\mathbf{z}^{\min}$. Since $\bar{\rho} \leq 1$ as well, this means $\frac{\bar{\rho}\epsilon}{M(2R+1)} \leq 1$. Therefore, $\left(\frac{\bar{\rho}\epsilon}{M(2R+1)}\right)^{n+d} \leq \left(\frac{\bar{\rho}\epsilon}{M(2R+1)}\right)^{d} < \left(\frac{\bar{\rho}\epsilon}{2MR}\right)^{d}$. From Lemma 6.1.5, $\{\mathbf{z} \in C' \cap \mathrm{dom}(f) : f(\mathbf{z}) \leq f(\mathbf{z}^\star) + \epsilon\}$ has volume at least $\left(\frac{\bar{\rho}\epsilon}{2MR}\right)^{d}$. Thus, at least one point $\hat{\mathbf{z}}$ in $\{\mathbf{z} \in C' \cap \mathrm{dom}(f) : f(\mathbf{z}) \leq OPT + \epsilon\}$ must be outside $C' \cap \mathrm{dom}(f) \cap \tilde{P}$. Such a point must violate one of the subgradient inequalities defining $\tilde{P}$, say, corresponding to index $\tilde{\ell} \in \{1, \ldots, k''\}$. In other words, $\langle \mathbf{s}^{\tilde{\ell}}, \hat{\mathbf{z}} - \mathbf{z}^{\tilde{\ell}} \rangle > 0$. This means $f(\hat{\mathbf{z}}) \geq f(\mathbf{z}^{\tilde{\ell}}) + \langle \mathbf{s}^{\tilde{\ell}}, \hat{\mathbf{z}} - \mathbf{z}^{\tilde{\ell}} \rangle > f(\mathbf{z}^{\tilde{\ell}})$. Thus, $f(\mathbf{z}^{\min}) \leq f(\mathbf{z}^{\tilde{\ell}}) < f(\hat{\mathbf{z}}) \leq OPT + \epsilon$.

The pure integer case with $d = 0$. The proof proceeds in a very similar manner except that one can stop when we have at most one integer point left in the polyhedral search space. Thus, we start from the box $[-R, R]^n$ containing $(2R+1)^n$ integer points and end with at most a single integer point, removing at least $\frac{1}{2n}$ fraction of integer points every time by Theorem 6.2.5.

The pure continuous case with $n = 0$. The proof is very similar and the only difference is that we can use the stronger bound on the centerpoints due to Grünbaum from Theorem 6.2.5. In other words, b can be taken to be $\frac{e}{e-1}$, where e is Euler's constant while mimicking the proof of the $n, d \geq 1$ case above. $\qquad\square$

Remark 6.2.7 The upper and lower bounds achieved above are roughly a consequence of the concept of Helly numbers discussed in Section 2.6. By Theorem 2.6.14, there exists a family of halfspaces $H_1, \ldots, H_{2^n(d+1)}$ such that the family $H_i \cap (\mathbb{Z}^n \times \mathbb{R}^d)$, $i = 1, \ldots, 2^n(d+1)$ of $\mathbb{Z}^n \times \mathbb{R}^d$-convex sets are a critical family. Now consider the family of $2^n(d+1)$ polyhedra $\cap_{j \neq i} H_j$ for $i = 1, \ldots, 2^n(d+1)$, along with the polyhedron $\cap_{j=1}^{2^n(d+1)} H_j$. If one makes less than $2^n(d+1)$ separation oracle queries, then every time we can simply report the halfspace H_j that does not contain a mixed-integer query point (such a halfspace exists since $\cap_{j=1}^{2^n(d+1)} H_j \cap (\mathbb{Z}^n \times \mathbb{R}^d) = \emptyset$), and if the query point is not in $\mathbb{Z}^n \times \mathbb{R}^d$, we truthfully report if it is in $\cap_{j=1}^{2^n(d+1)} H_j$ or not. The intersection of these reported halfspaces still contains a point from $\mathbb{Z}^n \times \mathbb{R}^d$ since it is a critical family and we have less than $2^n(d+1)$ queries. Therefore, we are unable to distinguish between the case $\cap_{j=1}^{2^n(d+1)} H_j$ which has no point from $\mathbb{Z}^n \times \mathbb{R}^d$ and the nonempty case. This gives a lower bound of $2^n(d+1)$. As we saw in the proof of the upper bound above, the key result is Theorem 6.2.5 which is based on Helly numbers again.

6.2.4 Exercises

1. Show that if f is closed, convex, and $C \cap \mathrm{dom}(f)$ is compact, then the infimum is attained in the mixed-integer convex optimization problem (6.1.1) if the problem has feasible solutions.

2. Verify that $\max\left\{ d2^n \log_3\left(\frac{2R}{3\rho}\right), d2^n \left\lfloor \log_8\left(\frac{MR}{\epsilon}\right) \right\rfloor \right\} \in \Omega\left(d2^n \log\left(\frac{MR}{\min\{\rho, 1\}\epsilon}\right)\right)$.

3. Let $\bar{\mathbf{x}} \in \{0, 1\}^n$, $\mathbf{a} \in \mathbb{R}^d \setminus \{\mathbf{0}\}$, $R > 0$ and $\delta \in \mathbb{R}$. Let $\hat{H} := \{\mathbf{y} \in \mathbb{R}^d : \langle \mathbf{a}, \mathbf{y} \rangle \leq \delta\}$. Let $S_1, \ldots, S_m \subseteq [0, 1]^n \times [-R, R]^d$ be compact sets such that $S_i \cap (\{\bar{\mathbf{x}}\} \times \mathbb{R}^d) = \emptyset$, i.e., no S_i intersects the fiber over $\bar{\mathbf{x}}$. Show that there exists a halfspace $H \subseteq \mathbb{R}^n \times \mathbb{R}^d$ such that

   (a) $S_i \subseteq H$ for all $i = 1, \ldots, m$.
   (b) $(\{\bar{\mathbf{x}}\} \times \mathbb{R}^d) \cap H = (\{\bar{\mathbf{x}}\} \times \mathbb{R}^d) \cap \hat{H}$.

4. Consider the construction of $X_1, \ldots, X_k$ in Case 1 of the proof (feasibility arguments) of the lower bound for information complexity for the $d \geq 1$ setting. Show that if $k < 2^n \cdot d \log_3\left(\frac{2R}{3\rho}\right)$, X_k contains a fiber box B of width at least 3ρ.

5. Given $M, R > 0$, consider the family of functions and intervals indexed by finite binary strings from Definition 6.2.2. Show the following properties hold for this family:

   (a) Every function has Lipschitz constant bounded by M.
   (b) The length of $I_s = \frac{2R}{4^{|s|}}$ for any binary string s.
   (c) The midpoint of I_s is the global minimizer of f_s over $\mathbb{R}$.

(d) All functions indexed by strings of the same length have the same minimum value. Moreover, strings of strictly longer length imply a strictly higher value of the minimum.

(e) The difference between the maximum value and minimum value of f_s over I_s is exactly $\frac{MR}{8^{|s|}}$.

(f) $\text{int}(I_{s_1}) \cap \text{int}(I_{s_2}) = \emptyset$ for two distinct binary strings s_1, s_2 of the same length.

6. Given $M, R > 0$, consider the family of functions $f_{s_1,\ldots,s_d}$ and sets $B_{s_1,\ldots,s_d}$ defined in (6.2.1), where $s_1, \ldots, s_d$ are finite binary strings. Show the following properties:

(a) Every function $f_{s_1,\ldots,s_d}$ has Lipschitz constant with respect to the ℓ_∞ norm bounded by M.

(b) For binary strings $s_1, \ldots, s_d$, $B_{s_1,\ldots,s_d}$ is a hypercuboid with length $\frac{2R}{4^{|s_j|}}$ in coordinate j.

(c) The center of $B_{s_1,\ldots,s_d}$ is a global minimizer of $f_{s_1,\ldots,s_d}$ over $\mathbb{R}^d$, where the center is defined as the cartesian product of the midpoints of $I_{s_1}, \ldots, I_{s_d}$.

(d) Let $k \in \mathbb{Z}_+$. All functions $f_{s_1,\ldots,s_d}$ such that $s_1, \ldots, s_d$ all have length k have the same minimum value. Moreover, for $k_1 < k_2$, the minimum corresponding to k_1 is strictly smaller than the minimum corresponding to k_2.

(e) The difference between the maximum value and minimum value of $f_{s_1,\ldots,s_d}$ over $B_{s_1,\ldots,s_d}$ is at least $\min_i \frac{MR}{8^{|s_i|}}$.

(f) Let $s_1, \ldots, s_d$ and $t_1, \ldots, t_d$ be binary strings all of the same length such that there exists i such that $s_i \neq t_i$. Then $\text{int}(B_{s_1,\ldots,s_d}) \cap \text{int}(B_{t_1,\ldots,t_d}) = \emptyset$.

7. Given $M, R > 0$, family of functions $f_{s_1,\ldots,s_d}$ and sets $B_{s_1,\ldots,s_d}$ defined in (6.2.1). For any binary string s and $y \in [-R, R]$, define $\text{depth}(s; y) := \max\{m \in \mathbb{N} : y \in I_{s|_m}\}$, where $s|_m$ denotes the prefix of s of length m. For any binary strings $s_1, \ldots, s_d$ and $\mathbf{y} \in [-R, R]^d$, define $\text{depth}(s_1, \ldots, s_d; \mathbf{y}) := \text{argmin}\{\text{depth}(s_j; y_j) : j \in \{1, \ldots, d\}\}$. Show the following.

(a) For any binary strings s and t such that s is a prefix of t, and any $y \in [-R, R]$, $\text{depth}(s; \mathbf{y}) \leq \text{depth}(t; \mathbf{y})$.

(b) Let $s_1, \ldots, s_d$ and $t_1, \ldots, t_d$ be binary strings such that s_i is a prefix of t_i for all $i = 1, \ldots, d$. Let $\mathbf{y} \in [-R, R]$. Suppose $j \in \text{depth}(s_1, \ldots, s_d; \mathbf{y})$ and $\text{depth}(s_j; y_j) = \text{depth}(t_j; y_j)$. Show that $j \in \text{depth}(t_1, \ldots, t_d; \mathbf{y})$.

(c) The following are equivalent:

    i. $f_{s_1,\ldots,s_d}(\mathbf{y}) = f_{s_j}(\mathbf{y}_j)$.

    ii. $\gamma \mathbf{e}_j \in \partial f_{s_1,\ldots,s_d}(\mathbf{y})$, where γ is the derivative of f_{s_j} at $\mathbf{y}_j$.

    iii. $j \in \mathrm{depth}(s_1,\ldots,s_d;\mathbf{y})$.

(d) Let $s_1,\ldots,s_d$ be binary strings. Let $\mathbf{y} \in [-R,R]^d$. Let $j^\star \in \{1,\ldots,d\}$ such that $f_{s_1,\ldots,s_d}(\mathbf{y}) = f_{s_{j^\star}(\mathbf{y}_{j^\star})}$. If $\mathbf{y}_{j^\star} \in I_{s_{j^\star}}$, we set $b := 0$ if $\mathbf{y}_{j^\star}$ is greater than or equal to the midpoint of $I_{s_{j^\star}}$; otherwise we set $b = 1$. Define $s'_j = s_j$ for all $j \neq j^\star$ and define $s'_{j^\star} = s_{j^\star}b$, i.e., $s'_{j^\star}$ is obtained by appending the bit b to $s_{j^\star}$. Show that for any binary strings $t_1,\ldots,t_d$ such that s'_i is a prefix of t_i, $f_{t_1,\ldots,t_d}(\mathbf{y}) = f_{s'_1,\ldots,s'_d}(\mathbf{y}) = f_{s'_{j^\star}(\mathbf{y}_{j^\star})}$ and $\mathbf{e}^{j^\star} \in \partial f_{t_1,\ldots,t_d}(\mathbf{y})$.

Conclude that the iterative invariants are maintained during the iterative constructions in Case 2. (optimality lower bound) of the proof of the lower bound for information complexity for the $d \geq 1$ setting.

8. Complete the proof of the lower bound on information complexity in the pure integer case ($d = 0$).

9. Let $n,d \in \mathbb{Z}_+$ such that $n + d \geq 1$. Let $C = [0,1]^n \times \Delta_d$, where Δ_d is the standard d-dimensional simplex and $S = C \cap (\mathbb{Z}^n \times \mathbb{R}^d)$. Show that $h_S(\mathbf{z}) \leq \frac{1}{2^n}\left(\frac{d}{d+1}\right)^d v(S)$ for all $\mathbf{z} \in S$.

10.* Show that the centroid of the pyramid P constructed in the proof of Theorem 6.2.5 is at $\bar{t}\mathbf{a}$ with $\bar{t} \leq \langle \mathbf{a}, \mathbf{x}^{\mathrm{cen}} \rangle$.

11. Let $B \subseteq \mathbb{R}^d$ be any compact, convex set of dimension $d-1$ and let $\mathbf{v} \in \mathbb{R}^d$ be a point that is not on the affine hull of B. Define $P := \mathrm{conv}(\{\mathbf{v}\} \cup B)$. Show that if H is a halfspace whose bounding hyperplane is parallel to the affine hull of B and H contains both $\mathbf{v}$ and the centroid of P, then $\mathrm{vol}(P \cap H) \geq \left(\frac{d}{d+1}\right)^d \mathrm{vol}(P)$.

6.3 Algorithmic Complexity

In this section, we present upper bounds on the algorithmic complexity of mixed-integer convex optimization. For the general problem (6.1.1), the best known upper bounds on the total worst-case complexity are worse than the information complexity bounds from Section 6.2. Moreover, the information complexity of state-of-the-art algorithms that achieve these upper bounds are also worse than the information complexity bounds established for the general problem. This leads to some of the most challenging open questions in mathematical optimization:

- Does there exist an algorithm for the general mixed-integer convex optimization problem (6.1.1) whose worst-case *computational* complexity is the same as the information complexity of mixed-integer convex optimization established in Section 6.2?
- Are there algorithms for the general mixed-integer convex optimization problem (6.1.1) whose worst-case *information* complexity is the same as the information complexity of mixed-integer convex optimization established in Section 6.2 and whose computational complexity is no worse than the best known state-of-the-art algorithms?

  Recall that the upper bound on the information complexity of the problem presented in Section 6.2.3 was based on an algorithm that needs to compute mixed-integer centerpoints. This is computationally a very hard problem and even using the best known algorithms for computing mixed-integer centerpoints will result in the overall computational complexity of the centerpoint algorithm being worse than the state-of-the-art algorithms.

However, we will see below that for the case with no integer variables (pure continuous convex optimization), the state-of-the-art algorithms have worst-case computational complexity comparable to the information complexity of pure continuous convex optimization.

We will start by designing algorithms for some special cases of the problem. We will use these as building blocks to put together an algorithm that tackles the general mixed-integer convex optimization problem (6.1.1).

6.3.1 The Unconstrained Quadratic Case

Consider the special case of (6.1.1), where $f(\mathbf{z}) = \frac{1}{2}\mathbf{z}^T Q \mathbf{z} - \mathbf{c}^T \mathbf{z}$ for some positive definite matrix $Q \in \mathbb{R}^{(n+d)\times(n+d)}$ and $\mathbf{c} \in \mathbb{R}^n \times \mathbb{R}^d$, and $C = \mathbb{R}^n \times \mathbb{R}^d$. This is equivalent to computing the closest mixed-integer point to $\mathbf{t} := Q^{-1}\mathbf{c}$ in the norm induced by Q (Exercise 2 from Section 6.3.3). This special case is called *unconstrained quadratic minimization* because the objective function is a degree two polynomial of the coordinates of $\mathbf{z}$. Since this is a special case with an algebraic structure, we will assume oracle access to the entries of Q and coordinates of $\mathbf{c}$, which is stronger than a first-order functional oracle discussed at the beginning of the chapter. Further, because we have this algebraic structure, we do not need to assume any *a priori* bound R on the feasible region and can allow $C = \mathbb{R}^n \times \mathbb{R}^d$. This is because the norm of the solution can be bounded in terms of some properties of Q and $\mathbf{c}$, and we will prove upper bounds on the algorithmic complexity as a function of these properties.

Two separate ingredients are needed to solve this problem: one coming from the case $n = 0$ (no integer variables) and the other coming from the case $d = 0$ (no continuous variables).

6.3.1.1 Inner Products, Positive Definite Matrices, and Quadratic Functions

We will first collect some useful notions and facts involving positive definite matrices and quadratic functions.

Definition 6.3.1 An *inner product on* $\mathbb{R}^k$ is any function $\langle \cdot, \cdot \rangle : \mathbb{R}^k \times \mathbb{R}^k \to \mathbb{R}$ that satisfies the following conditions.

1. (Nonnegativity) $\langle \mathbf{x}, \mathbf{x} \rangle \geq 0$ for all $\mathbf{x} \in \mathbb{R}^k$. Moreover, $\langle \mathbf{x}, \mathbf{x} \rangle = 0$ if and only if $\mathbf{x} = 0$.
2. (Symmetry) $\langle \mathbf{x}, \mathbf{y} \rangle = \langle \mathbf{y}, \mathbf{x} \rangle$ for all $\mathbf{x}, \mathbf{y} \in \mathbb{R}^k$.
3. (Bilinearity – vector addition) $\langle \mathbf{x} + \mathbf{y}, \mathbf{z} \rangle = \langle \mathbf{x}, \mathbf{z} \rangle + \langle \mathbf{y}, \mathbf{z} \rangle$ for all $\mathbf{x}, \mathbf{y}, \mathbf{z} \in \mathbb{R}^k$.
4. (Bilinearity – scalar multiplication) $\langle \alpha\mathbf{x}, \mathbf{y} \rangle = \alpha\langle \mathbf{x}, \mathbf{y} \rangle$ for all $\alpha \in \mathbb{R}$ and $\mathbf{x}, \mathbf{y} \in \mathbb{R}^k$.

Example 6.3.2 Let $Q \in \mathbb{R}^{k \times k}$ be a positive definite matrix. Then $\langle \mathbf{x}, \mathbf{y} \rangle_Q := \mathbf{x}^T Q \mathbf{y}$ is an inner product. If Q is the identity matrix, then we recover the standard inner product on $\mathbb{R}^k$. Also, $\langle \mathbf{x}, \mathbf{x} \rangle_Q = N_Q(\mathbf{x})$ (see Exercise 1 from Section 1.2.3).

Theorem 6.3.3 *Any inner product on* $\mathbb{R}^k$ *is of the form* $\langle \mathbf{x}, \mathbf{y} \rangle_Q$ *for some positive definite matrix* Q.

Proof Consider an arbitrary inner product $\langle \cdot, \cdot \rangle$ on $\mathbb{R}^k$. Define the matrix $Q_{ij} := \langle \mathbf{e}^i, \mathbf{e}^j \rangle$, where $\mathbf{e}^i$, $i = 1, \ldots, k$ denotes the standard unit vector. Bilinearity of the inner product shows that $\langle \mathbf{x}, \mathbf{y} \rangle = \mathbf{x}^T Q \mathbf{y}$ for all $\mathbf{x}, \mathbf{y} \in \mathbb{R}^k$, and nonnegativity and symmetry of the inner product implies that Q is positive definite. $\qquad\square$

In the following, we will always index an inner product with a positive definite matrix Q, unless we mean the standard inner product on $\mathbb{R}^k$, i.e., Q is the identity matrix. In this case, we will not have any subscript.

Definition 6.3.4 (Orthogonality) Any inner product on $\mathbb{R}^k$ gives a notion of "orthogonal vectors." We define the following:

1. We say that $\mathbf{x}, \mathbf{y} \in \mathbb{R}^k$ are *orthogonal with respect to the inner product* $\langle \cdot, \cdot \rangle_Q$ if $\langle \mathbf{x}, \mathbf{y} \rangle_Q = 0$. We also say that $\mathbf{x}$ and $\mathbf{y}$ are *Q-conjugate*.
2. Let $L \subseteq \mathbb{R}^k$ be a linear subspace. Then the *orthogonal complement of L with respect to* $\langle \cdot, \cdot \rangle_Q$ is defined as $L_Q^{\perp} := \{ \mathbf{y} \in \mathbb{R}^k : \langle \mathbf{x}, \mathbf{y} \rangle_Q = 0 \;\; \forall \mathbf{x} \in L \}$. It can be verified that $L_Q^{\perp}$ is a linear subspace (see Exercise 1 from Section 6.3.3). We also say that $L_Q^{\perp}$ is the *Q-conjugate subspace* of L.

The following decomposition result is a useful consequence of the notion of Q-conjugacy.

Theorem 6.3.5 *Let $Q \in \mathbb{R}^{k \times k}$ be a positive definite matrix and $\mathbf{c} \in \mathbb{R}^k$. Let L be an arbitrary linear subspace of $\mathbb{R}^k$. Any $\mathbf{z} \in \mathbb{R}^k$ can be uniquely expressed as $\mathbf{z} = \mathbf{p} + \mathbf{q}$, where $\mathbf{p} \in L$ and $\mathbf{q} \in L_Q^{\perp}$.*

Moreover, let $f(\mathbf{z}) = \frac{1}{2}\mathbf{z}^T Q \mathbf{z} - \mathbf{c}^T \mathbf{z} = \frac{1}{2}\langle \mathbf{z}, \mathbf{z} \rangle_Q - \mathbf{c}^T \mathbf{z}$. Then, $f(\mathbf{z}) = f(\mathbf{p}) + f(\mathbf{q})$ and $N_Q(\mathbf{z})^2 = N_Q(\mathbf{p})^2 + N_Q(\mathbf{q})^2$.

Proof   Left as an exercise. $\square$

Definition 6.3.6   Given $\mathbf{z} \in \mathbb{R}^k$, a linear subspace $L \subseteq \mathbb{R}^k$ and an inner product $\langle \cdot, \cdot \rangle_Q$, the vector $\mathbf{p} \in L$ from Theorem 6.3.5 is called the *projection of $\mathbf{z}$ on L with respect to* $\langle \cdot, \cdot \rangle_Q$.

Lemma 6.3.7 *Let $\mathbf{s}^1, \ldots, \mathbf{s}^p$ be pairwise Q-conjugate for some positive definite matrix $Q \in \mathbb{R}^{k \times k}$. The following are both true.*

1. $\mathbf{s}^1, \ldots, \mathbf{s}^p$ are all linearly independent.
2. For any $\mathbf{z} \in \mathbb{R}^k$, the projection of $\mathbf{z}$ on the subspace $\mathrm{span}(\{\mathbf{s}^1, \ldots, \mathbf{s}^p\})$ with respect to $\langle \cdot, \cdot \rangle_Q$ is given by $\frac{\langle \mathbf{z}, \mathbf{s}^1 \rangle_Q}{\langle \mathbf{s}^1, \mathbf{s}^1 \rangle_Q}\mathbf{s}^1 + \frac{\langle \mathbf{z}, \mathbf{s}^2 \rangle_Q}{\langle \mathbf{s}^2, \mathbf{s}^2 \rangle_Q}\mathbf{s}^2 + \cdots + \frac{\langle \mathbf{z}, \mathbf{s}^p \rangle_Q}{\langle \mathbf{s}^p, \mathbf{s}^p \rangle_Q}\mathbf{s}^p$. In particular, if $\mathbf{z} \in \mathrm{span}(\{\mathbf{s}^1, \ldots, \mathbf{s}^p\})$, then $\mathbf{z}$ is the sum of the projections of $\mathbf{x}$ onto the linear subspaces spanned by $\mathbf{s}^1, \ldots, \mathbf{s}^p$.

Theorem 6.3.8 (Gram–Schmidt Orthogonalization) . *Let $Q \in \mathbb{R}^{k \times k}$ be a positive definite matrix and let $\mathbf{z}^1, \ldots, \mathbf{z}^p$ ($1 \leq p \leq k$) be a set of linearly independent vectors in $\mathbb{R}^k$. Then one can compute a set of vectors $\mathbf{s}^1, \ldots, \mathbf{s}^p$ that are pairwise Q-conjugate such that for all $i = 1, \ldots, p$, $\mathrm{span}(\{\mathbf{s}^1, \ldots, \mathbf{s}^i\}) = \mathrm{span}(\{\mathbf{z}^1, \ldots, \mathbf{z}^i\})$. The algorithm has worst-case complexity $O(p^2 k + pk^2)$.*

Proof   Set $\mathbf{s}^1 = \mathbf{z}^1$. For $i = 1, \ldots, p - 1$, set

$$\mathbf{s}^{i+1} := \mathbf{z}^{i+1} - \sum_{j=1}^{i} \frac{\langle \mathbf{z}^{i+1}, \mathbf{s}^j \rangle_Q}{\langle \mathbf{s}^j, \mathbf{s}^j \rangle_Q} \mathbf{s}^j.$$

In other words, $\mathbf{s}^{i+1}$ is the difference between $\mathbf{z}^{i+1}$ and the projection of $\mathbf{z}^{i+1}$ on the linear subspace spanned by $\mathbf{s}^1, \ldots, \mathbf{s}^i$ with respect to the inner product

$\langle \cdot, \cdot \rangle_Q$. It can be verified that $\mathbf{s}^1, \ldots, \mathbf{s}^p$ so defined satisfy the condition stated in the theorem (Exercise 5 from Section 6.3.3).

The complexity comes from making the following observation. After computing $\mathbf{s}^{i+1}$, one can compute the vector $\tilde{\mathbf{s}}^{i+1} := Q\mathbf{s}^{i+1}$. Then all subsequent inner products can be obtained by taking the standard inner product with the $\tilde{\mathbf{s}}^i$ vectors. Computing the matrix-vector product $Q\mathbf{s}^{i+1}$ is $O(k^2)$ and this is repeated p times. The iterative steps take $O(p^2 k)$ time. $\qquad\square$

Remark 6.3.9 Theorem 6.3.8 and Lemma 6.3.7 can be used to compute the projection of any point $\mathbf{z} \in \mathbb{R}^k$ on any linear subspace $L \subseteq \mathbb{R}^k$ with respect to an inner product $\langle \cdot, \cdot \rangle_Q$ if we have access to a basis for L.

6.3.1.2 The Pure Continuous Case: Conjugate Gradients

We now consider the problem of minimizing $f(\mathbf{z}) = \frac{1}{2}\mathbf{z}^T Q \mathbf{z} - \mathbf{c}^T \mathbf{z}$ with $n = 0$, i.e., there are no integer variables ($Q \in \mathbb{R}^{d \times d}$ and $\mathbf{c} \in \mathbb{R}^d$). From the perspective that we are trying to find the mixed-integer point closest to $Q^{-1}\mathbf{c}$ (Exercise 2 from Section 6.3.3), the solution in this case is easy: report $\mathbf{z} = Q^{-1}\mathbf{c}$ since there is no integrality to worry about. However, in the Turing machine model of computation, inverting a matrix is prone to numerical instability when working with roundings of real or rational numbers. We will present another popular method called the *Conjugate Gradient method* which circumvents this issue and is also faster in practice for large-scale problems. Consequently, it is also often the method of choice for solving systems of linear equations involving positive definite matrices. In the arithmetic model of computation, this is not a major concern and both the inversion operation and the Conjugate Gradient method have the same complexity of $O(d^3)$.

The main observation is that if $\{\mathbf{s}^1, \ldots, \mathbf{s}^d\}$ are any pairwise Q-conjugate vectors, then the problem becomes easier to solve in this basis. In particular, write $\mathbf{z} = \lambda_1 \mathbf{s}^1 + \cdots + \lambda_d \mathbf{s}^d$. The objective then becomes

$$\frac{1}{2}\langle \mathbf{z}, \mathbf{z} \rangle_Q - \mathbf{c}^T \mathbf{z} = \lambda_1^2 \frac{\langle \mathbf{s}^1, \mathbf{s}^1 \rangle_Q}{2} + \lambda_2^2 \frac{\langle \mathbf{s}^2, \mathbf{s}^2 \rangle_Q}{2} + \cdots + \lambda_d^2 \frac{\langle \mathbf{s}^d, \mathbf{s}^d \rangle_Q}{2}$$
$$- \lambda_1(\mathbf{c}^T \mathbf{s}^1) - \lambda_2(\mathbf{c}^T \mathbf{s}^2) - \cdots - \lambda_d(\mathbf{c}^T \mathbf{s}^d). \qquad (6.3.1)$$

Thus, we are now minimizing

$$\min_{\lambda_1, \ldots, \lambda_d} \frac{a_1}{2}\lambda_1^2 + \cdots + \frac{a_d}{2}\lambda_d^2 - b_1 \lambda_1 - \cdots - b_d \lambda_d,$$

where $a_1, \ldots, a_d > 0$ and $b_1, \ldots, b_d \in \mathbb{R}$. This is an easy problem to solve because the λ_i's do not interact with each other. In other words, we can set $\bar{\lambda}_i = \frac{b_i}{a_i}$ and the solution is $\bar{\mathbf{z}} = \bar{\lambda}_1 \mathbf{s}^1 + \cdots + \bar{\lambda}_d \mathbf{s}^d$.

We have thus reduced the problem to finding Q-conjugate vectors. By the Gram–Schmidt Orthogonalization process (Theorem 6.3.8), we can start with an arbitrary basis of $\mathbb{R}^d$ – for example, the standard unit vectors $\mathbf{e}^1, \ldots, \mathbf{e}^d$ – and produce a set of Q-conjugate vectors using the Gram–Schmidt process. But this involves taking many matrix-vector products. We will now show that if we pick a suitably chosen set of linear independent vectors, then significant savings in computation can be achieved. This is based on the following two lemmas.

Lemma 6.3.10 *Let $\mathbf{z}^0$ be an arbitrary point in $\mathbb{R}^d$ and let L be any linear subspace of $\mathbb{R}^d$. Let $\bar{\mathbf{z}}$ be the minimizer of the function $f(\mathbf{z}) = \frac{1}{2}\mathbf{z}^T Q\mathbf{z} - \mathbf{c}^T\mathbf{z}$ restricted to $\mathbf{z}^0 + L$. Then $\nabla f(\bar{\mathbf{z}}) = Q\bar{\mathbf{z}} - \mathbf{c}$ is orthogonal to L with respect to the standard inner product on $\mathbb{R}^d$.*

Lemma 6.3.11 *Let $\mathbf{z}^0$ be an arbitrary point in $\mathbb{R}^d$ and let L be any linear subspace of $\mathbb{R}^d$. Let $\bar{\mathbf{z}}$ be the minimizer of the function $f(\mathbf{z}) = \frac{1}{2}\mathbf{z}^T Q\mathbf{z} - \mathbf{c}^T\mathbf{z}$ restricted to $\mathbf{z}^0 + L$. Let $\mathbf{s} \in \mathbb{R}^d$ be any vector that is Q-conjugate to L. Let $\hat{\mathbf{z}}$ be the minimizer of the one-dimensional function obtained by restricting f to the line $\{\bar{\mathbf{z}} + \alpha\mathbf{s} : \alpha \in \mathbb{R}\}$. Then $\hat{\mathbf{z}}$ is the minimizer of f restricted to $\mathbf{z}^0 + \mathrm{span}(L \cup \{\mathbf{s}\})$.*

We now describe a specialized, iterative way of choosing the Q-conjugate vectors.

Conjugate Gradient method: Basic.

1. Initialize an arbitrary $\mathbf{z}^0 \in \mathbb{R}^d$. Let $\mathbf{s}^0 = \mathbf{r}^0 = \mathbf{c} - Q\mathbf{z}^0 = -\nabla f(\mathbf{z}^0)$.
2. For $i = 0, 1, \ldots, d-1$,

   (a) Minimize the one-dimensional function $\varphi(\alpha) = f(\mathbf{z}^i + \alpha\mathbf{s}^i)$; let α_i be the minimizer.
   (b) Set $\mathbf{z}^{i+1} := \mathbf{z}^i + \alpha_i\mathbf{s}^i$.
   (c) Set $\mathbf{r}^{i+1} := -\nabla f(\mathbf{z}^{i+1}) = \mathbf{c} - Q\mathbf{z}^{i+1}$.
   (d) Set $\mathbf{s}^{i+1} := \mathbf{r}^{i+1} - \sum_{j=0}^{i} \frac{\langle \mathbf{r}^{i+1}, \mathbf{s}^j \rangle_Q}{\langle \mathbf{s}^j, \mathbf{s}^j \rangle_Q} \mathbf{s}^j$.

Lemma 6.3.12 *Let $\{\mathbf{z}^0, \ldots, \mathbf{z}^{k+1}\}$, $\{\mathbf{r}^0, \ldots, \mathbf{r}^{k+1}\}$, and $\{\mathbf{s}^0, \ldots, \mathbf{s}^{k+1}\}$ be the sequence of vectors generated in the algorithm above for some $0 \le k \le d-1$. Then the following hold:*

(i) $\mathrm{span}(\{\mathbf{r}^0, \ldots, \mathbf{r}^k\}) = \mathrm{span}(\{\mathbf{s}^0, \ldots, \mathbf{s}^k\})$.
(ii) $\mathbf{z}^{k+1}$ is minimizer of f restricted to $\mathbf{z}^0 + \mathrm{span}(\{\mathbf{r}^0, \ldots, \mathbf{r}^k\})$.

Proof The proof is by induction on k. For $k = 0$, part (i) holds by the initialization step. By Steps 2(a) and 2(b), $\mathbf{z}^1$ is by definition the minimizer claimed in part (ii). For $k \ge 1$, by the induction hypothesis, $\mathbf{z}^k$ is the minimizer restricted to $\mathbf{z}^0 + \mathrm{span}(\{\mathbf{r}^0, \ldots, \mathbf{r}^{k-1}\})$. By Step 2(c) and Lemma 6.3.10, $\mathbf{r}^k = -\nabla f(\mathbf{z}^k)$ is orthogonal to the span of $\mathbf{r}^0, \ldots, \mathbf{r}^{k-1}$, which is the same as the span of $\mathbf{s}^0, \ldots, \mathbf{s}^{k-1}$ by the induction hypothesis. Part (i) then follows from the Gram–Schmidt procedure in Step 2(d). Steps 2(a) and 2(b), combined with Lemma 6.3.11, imply that $\mathbf{z}^{k+1}$ is the minimizer over $\mathbf{z}^0 + \mathrm{span}(\{\mathbf{r}^0, \ldots, \mathbf{r}^k\})$.

$\qquad\qquad\square$

The next lemma helps to greatly reduce the amount of calculation in Step 2(d).

Lemma 6.3.13 *Let* $\{\mathbf{z}^0, \ldots, \mathbf{z}^{k+1}\}$, $\{\mathbf{r}^0, \ldots, \mathbf{r}^{k+1}\}$, *and* $\{\mathbf{s}^0, \ldots, \mathbf{s}^{k+1}\}$ *be the sequence of vectors generated in the algorithm above for some* $1 \le k \le d-1$. *Then* $\langle \mathbf{r}^{k+1}, \mathbf{s}^j \rangle_Q = 0$ *for all* $0 \le j \le k-1$.

Proof By Lemma 6.3.12 (i), $\mathrm{span}(\{\mathbf{r}^0, \ldots, \mathbf{r}^j\}) = \mathrm{span}(\{\mathbf{s}^0, \ldots, \mathbf{s}^j\})$ for all $0 \le j \le k$. Let this linear space be denoted by L_j. By Lemma 6.3.12 (i), $\mathbf{z}^{k+1}$ is the minimizer for f restricted to $\mathbf{z}^0 + L_k$. By Lemma 6.3.10, $\mathbf{r}^{k+1}$ is orthogonal to L_k with respect to the standard inner product. In other words, $(\mathbf{r}^{k+1})^T \mathbf{r}^j = 0$ for all $0 \le j \le k$. Now consider

$$\begin{aligned}
\langle \mathbf{r}^{k+1}, \mathbf{s}^j \rangle_Q &= \left(\mathbf{r}^{k+1}\right)^T Q \mathbf{s}^j \\
&= \tfrac{1}{\alpha_j}\left(\mathbf{r}^{k+1}\right)^T Q\left(\mathbf{z}^{j+1} - \mathbf{z}^j\right) \\
&= \tfrac{1}{\alpha_j}\left(\mathbf{r}^{k+1}\right)^T \left((\mathbf{c} - Q\mathbf{z}^j) - (\mathbf{c} - Q\mathbf{z}^{j+1})\right) \\
&= \tfrac{1}{\alpha_j}\left(\mathbf{r}^{k+1}\right)^T \left(\mathbf{r}^j - \mathbf{r}^{j+1}\right).
\end{aligned} \qquad (6.3.2)$$

The last term is 0 for all $j \le k-1$.

$\qquad\qquad\square$

One can also give a closed form formula for α_i at every iteration of the above algorithm.

Lemma 6.3.14 *Given any* $\mathbf{z} \in \mathbb{R}^d$ *and* $\mathbf{s} \in \mathbb{R}^d$, *the minimizing* $\alpha \in \mathbb{R}$ *for* $\varphi(\alpha) = f(\mathbf{z} + \alpha\mathbf{s})$ *is given by* $\alpha = \frac{\mathbf{s}^T(\mathbf{c} - Q\mathbf{z})}{\langle \mathbf{s}, \mathbf{s} \rangle_Q}$.

Conjugate Gradient method: Improved.

1. Initialize an arbitrary $\mathbf{z}^0 \in \mathbb{R}^d$. Let $\mathbf{s}^0 = \mathbf{r}^0 = \mathbf{c} - Q\mathbf{z}^0 = -\nabla f(\mathbf{z}^0)$.
2. For $i = 0, 1, \ldots, d-1$:

   (a) Set $\alpha_i = \frac{(\mathbf{s}^i)^T \mathbf{r}^i}{\langle \mathbf{s}^i, \mathbf{s}^i \rangle_Q}$.

   (b) Set $\mathbf{z}^{i+1} := \mathbf{z}^i + \alpha_i \mathbf{s}^i$.

   (c) Set $\mathbf{r}^{i+1} := -\nabla f(\mathbf{z}^{i+1}) = \mathbf{c} - Q\mathbf{z}^{i+1}$.

   (d) Set $\mathbf{s}^{i+1} := \mathbf{r}^{i+1} - \frac{\langle \mathbf{r}^{i+1}, \mathbf{s}^i \rangle_Q}{\langle \mathbf{s}^i, \mathbf{s}^i \rangle_Q} \mathbf{s}^i$.

In each iteration, one needs two matrix-vector products: $Q\mathbf{z}^{i+1}$ and $Q\mathbf{s}^i$. One can reduce the number of matrix-vector products from two to one per iteration, involving just $Q\mathbf{s}^i$, as follows (the changes are Steps 2(a), 2(c), and 2(d)):

Conjugate Gradient method: Final.

1. Initialize an arbitrary $\mathbf{z}^0 \in \mathbb{R}^n$. Let $\mathbf{s}^0 = \mathbf{r}^0 = \mathbf{c} - Q\mathbf{z}^0 = -\nabla f(\mathbf{z}^0)$.
2. For $i = 0, 1, \ldots, d - 1$:

   (a) Set $\alpha_i = \frac{(\mathbf{r}^i)^T \mathbf{r}^i}{\langle \mathbf{s}^i, \mathbf{s}^i \rangle_Q}$.
   (b) Set $\mathbf{z}^{i+1} := \mathbf{z}^i + \alpha_i \mathbf{s}^i$.
   (c) Set $\mathbf{r}^{i+1} := \mathbf{r}^i - \alpha_i Q\mathbf{s}^i$.
   (d) Set $\mathbf{s}^{i+1} := \mathbf{r}^{i+1} + \frac{(\mathbf{r}^{i+1})^T \mathbf{r}^{i+1}}{(\mathbf{r}^i)^T \mathbf{r}^i} \mathbf{s}^i$.

Proof of correctness of Final version First note that in Step 2(c) of the Improved version,

$$\mathbf{r}^{i+1} = \mathbf{c} - Q\mathbf{z}^{i+1} = \mathbf{c} - Q(\mathbf{z}^i + \alpha_i \mathbf{s}^i) = \mathbf{r}^i - \alpha_i Q\mathbf{s}^i, \qquad (6.3.3)$$

where we used Step 2(b) (of either version). Next observe that $\alpha_0 = \frac{(\mathbf{s}^0)^T \mathbf{r}^0}{\langle \mathbf{s}^0, \mathbf{s}^0 \rangle_Q} = \frac{(\mathbf{r}^0)^T \mathbf{r}^0}{\langle \mathbf{s}^0, \mathbf{s}^0 \rangle_Q}$ since $\mathbf{s}^0 = \mathbf{r}^0$. We will show that in every iteration, the equality

$$\alpha_i = \frac{(\mathbf{s}^i)^T \mathbf{r}^i}{\langle \mathbf{s}^i, \mathbf{s}^i \rangle_Q} = \frac{(\mathbf{r}^i)^T \mathbf{r}^i}{\langle \mathbf{s}^i, \mathbf{s}^i \rangle_Q} \qquad (6.3.4)$$

is maintained. From (6.3.2), setting $j = k$, we obtain that

$$\langle \mathbf{r}^{k+1}, \mathbf{s}^k \rangle_Q = \frac{1}{\alpha_j} (\mathbf{r}^{k+1})^T (\mathbf{r}^k - \mathbf{r}^{k+1}) = -\frac{1}{\alpha_k} (\mathbf{r}^{k+1})^T \mathbf{r}^{k+1}.$$

Thus, iteratively, combining this with (6.3.4), we obtain that in Step 2(d) we have

$$\mathbf{s}^{i+1} := \mathbf{r}^{i+1} - \frac{\langle \mathbf{r}^{i+1}, \mathbf{s}^i \rangle_Q}{\langle \mathbf{s}^i, \mathbf{s}^i \rangle_Q} \mathbf{s}^i = \mathbf{r}^{i+1} + \frac{(\mathbf{r}^{i+1})^T \mathbf{r}^{i+1}}{(\mathbf{r}^i)^T \mathbf{r}^i} \mathbf{s}^i. \qquad (6.3.5)$$

Now take the standard inner product with $\mathbf{r}^{i+1}$ on both sides of (6.3.5) and recall that $(\mathbf{r}^{i+1})^T \mathbf{s}^i = 0$ because $\mathbf{r}^{i+1}$ is orthogonal to L_i with respect to the standard inner product. Therefore, $(\mathbf{s}^{i+1})^T \mathbf{r}^{i+1} = (\mathbf{r}^{i+1})^T \mathbf{r}^{i+1}$, showing that (6.3.4) is maintained iteratively. Making the changes from (6.3.3), (6.3.4), and (6.3.5) in Steps 2(c), 2(a), and 2(d) of the Improved version, respectively, we obtain the Final version of the Conjugate Gradient method. $\qquad \square$

6.3.1.3 The Pure Discrete Case: Closest Lattice Vector Problem (CVP)

We now turn to the case where $d = 0$ (no continuous variables) for minimizing $f(\mathbf{z}) = \frac{1}{2}\mathbf{z}^T Q\mathbf{z} - \mathbf{c}^T\mathbf{z}$. From the perspective of finding the closest (mixed)-integer point to $Q^{-1}\mathbf{c}$ (Exercise 2 from Section 6.3.3), the problem becomes a special case of the closest lattice vector problem (Definition 4.3.4): the lattice is $\mathbb{Z}^n$ and we want to find the closest point in $\mathbb{Z}^n$ to $\mathbf{t} = Q^{-1}\mathbf{c}$ in the norm N_Q induced by Q. It will be notationally easier to consider the following equivalent problem (Exercise 9 from Section 6.3.3):

> **CVP in ℓ_2.** Given a basis $\mathbf{b}^1, \ldots, \mathbf{b}^r$ for a lattice $\Lambda \subseteq \mathbb{R}^n$ of rank $r \leq n$ and a point $\mathbf{t} \in \mathbb{R}^n$, find the closest point to $\mathbf{t}$ in the standard Euclidean norm in the lattice Λ.

We will present an algorithm that solves the above problem with worst-case complexity bounded by $2^{O(r)}\,\text{poly}\left(\log\left(\frac{\|\mathbf{b}^{\max}\|_2}{\ell(\Lambda)}\right)\right)$, where $\mathbf{b}^{\max}$ is vector in the given basis with the largest Euclidean norm and $\ell(\Lambda)$ is the Euclidean length of a shortest lattice vector in Λ. As noted in the paragraph below Definition 4.3.4, without loss of generality, we may assume $\mathbf{t} \in \text{span}(\Lambda)$ since we can project $\mathbf{t}$ on to this subspace and then solve the problem. Below, we will use $B \in \mathbb{R}^{n \times r}$ to denote the matrix with columns $\mathbf{b}^1, \ldots, \mathbf{b}^r$.

Reduced bases. If one can construct an orthogonal basis $\tilde{\mathbf{b}}^1, \ldots, \tilde{\mathbf{b}}^r$ for the lattice Λ, then the problem becomes easy: the closest lattice point to $\mathbf{t}$ is $\lfloor\lambda_1\rceil\tilde{\mathbf{b}}^1 + \cdots + \lfloor\lambda_r\rceil\tilde{\mathbf{b}}^r$, where $\lambda_1, \ldots, \lambda_r \in \mathbb{R}$ are the coefficients that express $\mathbf{t}$ in the basis $\tilde{\mathbf{b}}^1, \ldots, \tilde{\mathbf{b}}^r$ (Exercise 10 from Section 6.3.3). However, not every lattice has an orthogonal basis (Exercise 12 from Section 6.3.3). Therefore, one tries to find a basis for Λ which is "as orthogonal as one can get." This technique is known as *basis reduction*. To quantify how close a basis is to being orthogonal, we introduce the following concept.

Definition 6.3.15   Given a basis B for a lattice Λ of $\mathbb{R}^n$, the *orthogonality defect* $\gamma(B)$ *of* B is defined as

$$\gamma(B) = \frac{\prod_{i=1}^r \|\mathbf{b}^i\|}{\sqrt{\det(B^T B)}} = \frac{\prod_{i=1}^r \|\mathbf{b}^i\|}{\det(\Lambda)}.$$

Let $\mathbf{b}^1_\star, \ldots, \mathbf{b}^r_\star$ denote the vectors one obtains by performing Gram–Schmidt Orthogonalization (Definition 6.3.8) on the vectors $\mathbf{b}^1, \ldots, \mathbf{b}^r$ with respect to the standard inner product in $\mathbb{R}^n$. Then the following relation holds:

$$B = B^\star R, \tag{6.3.6}$$

where B has $\mathbf{b}^i$'s as the columns, $B^\star$ has the $\mathbf{b}^i_\star$'s as the columns, and $R \in \mathbb{R}^{r \times r}$ is an upper triangular matrix with 1's on the diagonal. We will call this decomposition of B as the *Gram–Schmidt Factorization of B*. Thus, $\|\mathbf{b}^k\|^2 = \sum_{j=1}^{k-1} R_{jk}^2 \|\mathbf{b}^j_\star\|^2 + \|\mathbf{b}^k_\star\|^2$ since all the $\mathbf{b}^j_\star$'s are orthogonal for all $k \in \{1, \ldots, r\}$. Therefore, $\|\mathbf{b}^k\| \geq \|\mathbf{b}^k_\star\|$. Moreover, from (6.3.6), we have $\det(B^T B) = \det(R^T) \det((B^\star)^T B^\star) \det(R)$. Since $\det(R) = 1$ (R is upper triangular with 1's on the diagonal), we have that $\det(B^T B) = \det((B^\star)^T B^\star)$. Since $B^\star$ is an orthogonal matrix, $(B^\star)^T B^\star$ is an $r \times r$ diagonal matrix with $\|\mathbf{b}^i_\star\|^2$ as the diagonal entries. Therefore, $\sqrt{\det((B^\star)^T B^\star)} = \prod_{i=1}^{r} \|\mathbf{b}^i_\star\|$. Thus, we obtain

$$\sqrt{\det(B^T B)} = \sqrt{\det((B^\star)^T B^\star)} = \prod_{i=1}^{r} \|\mathbf{b}^i_\star\| \leq \prod_{i=1}^{r} \|\mathbf{b}^i\|. \tag{6.3.7}$$

This inequality immediately shows that the orthogonality defect of B is at least 1, and the orthogonality defect is 1 if and only if the basis vectors in B are orthogonal (Exercise 11 from Section 6.3.3). The inequality $\sqrt{\det(B^T B)} \leq \prod_{i=1}^{r} \|\mathbf{b}^i\|$ is also known as the *Hadamard inequality* and is often very useful for proving inequalities involving matrices and vector norms.

We now show how to use Gram–Schmidt Orthogonalization to obtain a basis with a provable bound on its orthogonality defect. We will use the following column operations on a matrix that guarantee that the new columns generate the same lattice as the original columns (see Exercise 17 from Section 4.1.1).

i. Exchanging two columns.
ii. Adding an integral multiple of a lower indexed column to a higher indexed column.

Observe that every operation above corresponds to multiplication of the basis matrix (on the right) by a unimodular matrix (Definition 4.1.9) obtained by performing that operation on the corresponding columns of the identity matrix. We now provide some of the intuition behind the basis reduction algorithm, before presenting the actual algorithm.

Normalization. Consider the basis $\mathbf{b}^1 = (1, 1)$, $\mathbf{b}^2 = (2, 3)$ for $\mathbb{Z}^2$. This looks like a pair of vectors which is far from being orthogonal. In fact, the orthogonality defect is 5.099. However, if we consider the basis $\mathbf{b}^1 = (1, 1), \mathbf{b}^2 - 2\mathbf{b}^1 = (0, 1)$, we immediately reduce the orthogonality defect down to 1.414. Notice that this corresponds to doing an elementary column operation of type ii. above on the basis matrix. Performing another appropriate elementary column operation we can obtain the basis $(1,0),(0,1)$ getting an orthogonal basis with defect equal to 1. The basic idea behind this is to perform elementary column operations so that we have small projections when projecting a basis vector onto the subspace spanned by the basis vectors preceding it. In the case of orthogonal bases, these projections are 0, and our goal is to get as close to 0 as possible. We make a useful observation at this juncture.

Lemma 6.3.16 *Let B be a basis for some lattice with Gram–Schmidt Factorization $B = B^{\star}R$. Consider an elementary operation of type ii. above. Then the matrix E corresponding to it is upper triangular with 1's on the diagonal. Moreover, the new basis $B' = BE$ has the same Gram–Schmidt vectors $B^{\star}$ and $B' = B^{\star}R'$ is the Gram–Schmidt Factorization for B' where $R' = RE$.*

Proof   Left as an exercise. $\square$

The idea behind getting small projections is simply the following: perform elementary column operations on R such that R remains an upper triangular matrix with 1's on the diagonal, with the further property that all the off-diagonal terms R_{ij} have absolute value at most $\frac{1}{2}$.

Lemma 6.3.17 *Let R be an upper triangular matrix R with 1's on the diagonal. Then there exists a sequence of operations of the type ii. above with corresponding matrices $E_1, E_2, \ldots, E_k$ such that $R' := RE_1E_2\ldots E_k$ is an upper triangular matrix with 1's on the diagonal and all off-diagonal entries have absolute value at most $\frac{1}{2}$.*

Proof   Left as an exercise. $\square$

Combining Lemmas 6.3.16 and 6.3.17 shows that starting with an arbitrary basis B with Gram–Schmidt Factorization $B = B^{\star}R$, we can obtain a new basis B' whose Gram–Schmidt Factorization is $B' = B^{\star}R'$, where the absolute value of every off diagonal entry in R' is at most $\frac{1}{2}$. This step will be called *normalization*. It is important to note that the normalization procedure does not change the orthogonal vectors in $B^{\star}$, i.e., $B^{\star}$ is still the Gram–Schmidt Orthogonalization vectors for B'.

Swapping. Unfortunately, normalizations are not enough to guarantee basis with small orthogonality defect. The problem is that the normalization step inherently depends on the ordering of the columns of the matrix. Consider the same example of the basis as above, except with the order switched $\mathbf{b}^1 = (2,3), \mathbf{b}^2 = (1,1)$. The corresponding Gram–Schmidt Factorization is:

$$\begin{bmatrix} 2 & 1 \\ 3 & 1 \end{bmatrix} = \begin{bmatrix} 2 & 0.2308 \\ 3 & -0.1538 \end{bmatrix}\begin{bmatrix} 1 & 0.3846 \\ 0 & 1 \end{bmatrix}.$$

The off-diagonal elements in R are already "normalized," i.e., have absolute value less than 0.5. But this is the same basis that we had above when discussing normalization, where we were able to normalize further. This shows that the order of the columns in the basis is also important when considering the orthogonality – in this particular order the second Gram–Schmidt vector is "too small" compared to the first one. When the norms of the Gram–Schmidt vectors drop very quickly in the given sequence of basis vectors, this can hide a deviation from orthogonality even though the vectors are normalized. This issue is resolved by a *swap* or exchange of two columns in B (operation i. above). This will lead to a change in the Gram–Schmidt Orthogonalization vectors associated with the new matrix, and the hope is that the norms of the new Gram–Schmidt Orthogonalization vectors are more similar to each other compared to before the exchange.

The Lenstra–Lenstra–Lovasz (LLL) algorithm. The Lenstra–Lenstra–Lovasz (LLL) basis reduction algorithm we describe below was designed by A. K. Lenstra, H. W. Lenstra, and L. Lovasz [164]. The algorithm repeatedly implements the above two steps of normalization and swapping until a stopping condition is reached. The stopping condition is dictated by the requirement that the norms of the Gram–Schmidt Orthogonalization vectors should not drop very fast as we move along the sequence of basis vectors.

Definition 6.3.18 (LLL algorithm)   Input: A basis $B \in \mathbb{R}^{n \times r}$ of a lattice of $\mathbb{R}^n$ with rank r.

1. Construct the Gram–Schmidt Factorization $B = B^\star R$.
2. Normalize B.
3. If there exists $i \in \{2, \ldots, r\}$ such that $\|\mathbf{b}_\star^i\|^2 < \frac{1}{2}\|\mathbf{b}_\star^{i-1}\|^2$, swap $\mathbf{b}^i$ and $\mathbf{b}^{i-1}$ and go back to Step 1. Else, STOP and output the current B as the reduced basis.

Theorem 6.3.19 *Let $B \in \mathbb{R}^{n \times r}$ be a basis of a lattice Λ of $\mathbb{R}^n$ that is given as input to the LLL algorithm. The algorithm terminates in a finite number of iterations bounded by $O\left(r^2 \log\left(\frac{\sqrt{r}\|\mathbf{b}^{\max}\|_2}{\ell(\Lambda)}\right)\right)$, where $\mathbf{b}^{\max}$ is the column of B with the largest Euclidean norm and $\ell(\Lambda)$ is the Euclidean length of a shortest nonzero lattice vector in Λ. The final output of the algorithm is a set of basis vectors for Λ with orthogonality defect at most $2^{r(r-1)/4}$.*

Proof   We first show that if the LLL algorithm terminates, we have an orthogonality defect of at most $2^{r(r-1)/4}$. Since the algorithm has stopped, for each $i \in \{2, \ldots, r\}$ we have $\|\mathbf{b}_\star^i\|^2 \geq \frac{1}{2}\|\mathbf{b}_\star^{i-1}\|^2$. This implies that for $1 \leq j \leq k \leq r$, $\|\mathbf{b}_\star^j\|^2 \leq 2^{k-j}\|\mathbf{b}_\star^k\|^2$. Then it follows that

$$\begin{aligned}
\|\mathbf{b}^k\|^2 &= \sum_{j=1}^{k-1} R_{jk}^2 \|\mathbf{b}_\star^j\|^2 + \|\mathbf{b}_\star^k\|^2 \\
&\leq \sum_{j=1}^{k-1} \tfrac{1}{4} \|\mathbf{b}_\star^j\|^2 + \|\mathbf{b}_\star^k\|^2 \\
&\leq \tfrac{1}{4}\Big(\sum_{j=1}^{k-1} 2^{k-j} \|\mathbf{b}_\star^k\|^2\Big) + \|\mathbf{b}_\star^k\|^2 \\
&\leq \big(\tfrac{1}{4}\sum_{j=1}^{k-1} 2^{k-j} + 1\big)\|\mathbf{b}_\star^k\|^2 \\
&\leq 2^{k-1}\|\mathbf{b}_\star^k\|^2.
\end{aligned}$$

This shows that $\prod_{k=1}^{r} \|\mathbf{b}^k\| \leq 2^{r(r-1)/4} \prod_{k=1}^{r} \|\mathbf{b}_\star^k\|$. Using the equalities from (6.3.7), this implies that $\prod_{k=1}^{r} \|\mathbf{b}^k\| \leq 2^{r(r-1)/4} \det(B)$ showing that the orthogonality defect is at most $2^{r(r-1)/4}$.

We now show that the LLL algorithm terminates. The intuition behind this is the following. The algorithm does not terminate only if it continues to make swaps. However, the Gram–Schmidt vectors keep getting "shorter" every time we make a swap and since we are dealing with basis vectors of a *lattice*, the Gram–Schmidt Orthogonalization vectors have a lower bound on their length. So the process has to stop at some point. This is formalized by considering the following potential function for any intermediate basis B.

$$\Phi(B) = \|\mathbf{b}_\star^1\|^{2r} \|\mathbf{b}_\star^2\|^{2(r-1)} \|\mathbf{b}_\star^3\|^{2(r-2)} \cdots \|\mathbf{b}_\star^r\|^2.$$

We now make an important observation about the swapping operation. When we do a swap of $\mathbf{b}^i$ and $\mathbf{b}^{i-1}$ for some $i \in \{2,\dots,r\}$, the Gram–Schmidt Factorization changes, but not by much. In fact, the only vectors that change in the Gram–Schmidt Orthogonalization are $\mathbf{b}_\star^i$ and $\mathbf{b}_\star^{i-1}$. More formally, if we had $B = B^\star R$ and $\widetilde{B}$ was obtained by swapping columns $\mathbf{b}^i$ and $\mathbf{b}^{i-1}$ in B, then $\widetilde{B} = \widetilde{B}^\star \widetilde{R}$ where $\widetilde{\mathbf{b}}_\star^k$ is different from $\mathbf{b}_\star^k$ only for $k = i-1, i$. One further observes that $\mathbf{b}^i = \sum_{j=1}^{i-2} R_{ji} \mathbf{b}_\star^j + R_{(i-1)i} \mathbf{b}_\star^{i-1} + \mathbf{b}_\star^i$ and $\widetilde{\mathbf{b}}^{i-1} = \mathbf{b}^i$ and so $\widetilde{\mathbf{b}}_\star^{i-1} = R_{(i-1)i} \mathbf{b}_\star^{i-1} + \mathbf{b}_\star^i$. This implies that $\|\widetilde{\mathbf{b}}_\star^{i-1}\|^2 = R_{(i-1)i}^2 \|\mathbf{b}_\star^{i-1}\|^2 + \|\mathbf{b}_\star^i\|^2$. Moreover, we made the swap in the algorithm because $\|\mathbf{b}_\star^i\|^2 < \tfrac{1}{2}\|\mathbf{b}_\star^{i-1}\|^2$. Combined with the fact that $R_{(i-1)i}^2 \leq \tfrac{1}{4}$ we get that $\|\widetilde{\mathbf{b}}_\star^{i-1}\|^2 \leq \tfrac{3}{4}\|\mathbf{b}_\star^{i-1}\|^2$. One further observes that a swap does not change the determinant of the basis/lattice $\det(B^T B) = \det((B^\star)^T B^\star) = \prod_{i=1}^{r} \|\mathbf{b}_\star^i\|^2$. Therefore, if $\|\mathbf{b}_\star^{i-1}\|^2$ is decreased by some factor $\alpha \leq \tfrac{3}{4}$, then $\|\mathbf{b}_\star^i\|^2$ has to *increase* by the same factor of α in the new Gram–Schmidt vectors, since all the other Gram–Schmidt Orthogonalization vectors remain unchanged. The potential function $\Phi(B)$ has one more term for $\|\mathbf{b}_\star^{i-1}\|^2$ as compared to $\|\mathbf{b}_\star^i\|^2$ and so every swap leads to a decrease by a factor of at least $\tfrac{3}{4}$.

We now bound the number of iterations of the algorithm. Let B_i be the first i columns of B (respectively define $B_i^\star$ as the submatrix of $B^\star$ consisting of the first i columns), and let R_i be the top $i \times i$ submatrix of R. Then we have $B_i = B_i^\star R_i$. So we have,

$$\begin{aligned}
\det(B_i^T B_i) &= \det((B_i^\star R_i)^T (B_i^\star R_i)) \\
&= \det(R_i^T B_i^{\star T} B_i^\star R_i) \\
&= \det(R_i^T) \det(B_i^{\star T} B_i^\star) \det(R_i) \\
&= \det(B_i^{\star T} B_i^\star).
\end{aligned}$$

Therefore,

$$\begin{aligned}
\Phi(B) &= (\|\mathbf{b}_\star^1\|^2) \cdot (\|\mathbf{b}_\star^1\|^2 \|\mathbf{b}_\star^2\|^2) \cdots \cdot (\|\mathbf{b}_\star^1\|^2 \|\mathbf{b}_\star^2\|^2 \cdots \|\mathbf{b}_\star^r\|^2) \\
&= \prod_{i=1}^r \det(B_i^{\star T} B_i^\star) \\
&= \prod_{i=1}^r \det(B_i^T B_i).
\end{aligned}$$

By the Cauchy–Binet formula (Theorem 1.2.6), for every $i = 1, \dots, r$, $\det(B_i^T B_i)$ equals the sum over all $i \times i$ submatrices of B_i of their squared determinants. By Hadamard's inequality (6.3.7), the determinant of these $i \times i$ submatrices is bounded by the products of the norms of the columns of these submatrices. Since these columns are subvectors of the columns of B, their norms are bounded by $\|\mathbf{b}^{\max}\|$, where $\mathbf{b}^{\max}$ is the maximum norm column of B. Therefore, $\varphi(B) \leq (2\|\mathbf{b}^{\max}\|)^{r(r+1)}$.

Note also that B_i is a basis for a sublattice of Λ. By Exercise 4 from Section 4.2.1, the determinant $\sqrt{\det(B_i^T B_i)}$ of this sublattice is bounded below by $\left(\frac{\ell(B_i)}{\sqrt{i}}\right)^i$, where $\ell(B_i)$ is the length of the shortest nonzero vector in this sublattice. Moreover, $\ell(\Lambda) \leq \ell(B_i)$ where $\ell(\Lambda)$ is norm of the shortest nonzero lattice vector in Λ. Therefore, $\varphi(B) \geq \frac{\ell(\Lambda)^{r(r+1)}}{r^{r(r+1)/2}}$.

In every iteration, $\varphi(B)$ decreases by a factor of at least $3/4$. Thus, there can be at most $r(r+1)\left(\log_{4/3}\left(\frac{2\sqrt{r}\|\mathbf{b}^{\max}\|}{\ell(\Lambda)}\right)\right)$ iterations. $\qquad\square$

Remark 6.3.20 Observe that if B is the output of the LLL algorithm, then the matrix B_i, formed by taking the first i columns of B, will not be modified by the LLL algorithm if B_i is given as input to it.

Computing CVP and Voronoi cells. We now have the tools in place to solve the closest lattice vector problem in the standard Euclidean norm. We will use the observation that given a lattice $\Lambda \subseteq \mathbb{R}^n$, and a target vector $\mathbf{t} \in \mathrm{span}(\Lambda)$, $\mathbf{z}$ is the closest lattice vector to $\mathbf{t}$ if and only if $\mathbf{t} - \mathbf{z}$ is in the Dirichlet–Voronoi cell $\mathcal{V}(\Lambda)$ of the lattice (see Section 4.3.2). Below we will design two algorithms: one that computes the Dirichlet–Voronoi cell and the second solves the CVP problem using the Dirichlet–Voronoi cell.

The nearest lattice plane algorithm. We begin with an algorithm that has worse complexity compared to our final algorithm but is much simpler to describe. It will also provide a crucial ingredient for the final algorithm.

Definition 6.3.21 (Nearest lattice plane algorithm)   Input: A basis $B \in \mathbb{R}^{n \times r}$ of a lattice of $\mathbb{R}^n$ with rank r, and $\mathbf{t} \in \text{span}(B)$ as the target vector.

1. If $r = 1$, then return $\mathbf{z} := \lfloor \lambda \rceil \mathbf{b}^1$, where $\mathbf{t} = \lambda \mathbf{b}^1$.
2. Else, apply the LLL algorithm on B to obtain $\widehat{B}$.
3. Construct the Gram–Schmidt Factorization $\widehat{B} = \widehat{B}^\star R$.
4. Express $\mathbf{t} = \lambda_1 \widehat{\mathbf{b}}^1 + \cdots \lambda_r \widehat{\mathbf{b}}^r$.
5. Set $\mathbf{t}' := \mathbf{t} + (\lceil \lambda_r \rceil - \lambda_r)\widehat{\mathbf{b}}_\star^r - \lfloor \lambda_r \rceil \widehat{\mathbf{b}}^r = \lambda_1 \widehat{\mathbf{b}}^1 + \cdots \lambda_{r-1} \widehat{\mathbf{b}}^{r-1} + (\lceil \lambda_r \rceil - \lambda_r)(\widehat{\mathbf{b}}_\star^r - \widehat{\mathbf{b}}^r)$.
6. Recursively solve the problem with the lattice spanned by the first $r - 1$ columns of $\widehat{B}$ and $\mathbf{t}'$ as the target vector, and get the output $\mathbf{z}'$. Return $\mathbf{z} := \mathbf{z}' + \lfloor \lambda_r \rceil \widehat{\mathbf{b}}^r$.

The base case $r = 1$ in the algorithm is clear. For rank greater than or equal to two, the algorithm projects $\mathbf{t}$ to the nearest affine lattice subspace parallel to $\text{span}(\widehat{\mathbf{b}}^1, \ldots, \widehat{\mathbf{b}}^{r-1})$ and then tries to find a lattice point close to this projection in that lattice hyperplane. This can be restated as the problem in Step 6 involving the lower rank lattice spanned by $\mathbf{b}^1, \ldots, \mathbf{b}^{r-1}$. Note that Remark 6.3.20 shows that the LLL algorithm needs to be run only once.

Theorem 6.3.22 *The output $\mathbf{z} \in \mathbb{R}^n$ of the nearest lattice plane algorithm (Definition 6.3.21) with input $B \in \mathbb{R}^{n \times r}$ and $\mathbf{t} \in \text{span}(B)$ belongs to the lattice $Z(B)$ and $\|\mathbf{z} - \mathbf{t}\| \le \frac{1}{2} \sum_{i=1}^r \|\widehat{\mathbf{b}}_\star^i\|$, where $\widehat{B}$ is the output of the LLL algorithm on B.*

Proof   The proof proceeds by induction on the rank r. When $r = 1$, $\|\lambda \mathbf{b}^1 - \lfloor \lambda \rceil \mathbf{b}^1\| \le \frac{1}{2}\|\mathbf{b}^1\|$ and we are done. For $r \ge 2$, the distance between $\mathbf{t}$ and the projection $\mathbf{t}''$ of $\mathbf{t}$ onto the nearest affine lattice subspace parallel to $\text{span}(\mathbf{b}^1, \ldots, \mathbf{b}^{r-1})$ is at most $\frac{1}{2}\|\widehat{\mathbf{b}}_\star^r\|$. By the induction hypothesis (and Remark 6.3.20), $\|\mathbf{z}' - \mathbf{t}'\| \le \frac{1}{2} \sum_{i=1}^{r-1} \|\widehat{\mathbf{b}}_\star^i\|$. We also observe that

$$
\begin{aligned}
\mathbf{z} - \mathbf{t}'' &= (\mathbf{z}' + \lfloor \lambda_r \rceil \widehat{\mathbf{b}}^r) - (\mathbf{t} + (\lceil \lambda_r \rceil - \lambda_r)\widehat{\mathbf{b}}_\star^r) \\
&= (\mathbf{z}' + \lfloor \lambda_r \rceil \widehat{\mathbf{b}}^r) - (\mathbf{t}' + \lfloor \lambda_r \rceil \widehat{\mathbf{b}}^r) \\
&= \mathbf{z}' - \mathbf{t}'.
\end{aligned}
$$

Therefore,

$$
\|\mathbf{z} - \mathbf{t}\| \le \|\mathbf{z} - \mathbf{t}''\| + \|\mathbf{t}'' - \mathbf{t}\| \le \frac{1}{2} \sum_{i=1}^r \|\widehat{\mathbf{b}}_\star^i\|,
$$

completing the proof. $\square$

Corollary 6.3.23 *Let $\widehat{B}$ be the output of the LLL algorithm on a basis of some lattice $\Lambda \subseteq \mathbb{R}^n$. Let $\mathbf{t} \in \text{span}(\Lambda)$ and let H be the closest affine lattice subpsace* *to* **t** *parallel to* $\mathrm{span}(\widehat{\boldsymbol{b}}^1, \ldots, \widehat{\boldsymbol{b}}^{r-1})$. *Then, the closest lattice vector* $CV(\Lambda, \mathbf{t})$ *must lie on one of the affine lattice subspaces* $H + \mu \widehat{\boldsymbol{b}}^r$, $\mu \in [-2^r, 2^r] \cap \mathbb{Z}$.

Proof Since $\widehat{B}$ is the output of the LLL algorithm, we have the relation $\|\mathbf{b}_\star^j\| \le 2^{(k-j)/2} \|\mathbf{b}_\star^k\|$ for all $j \le k$ (as was observed in the proof of Theorem 6.3.19). Consequently, by Theorem 6.3.22, the output of the nearest lattice plane algorithm returns a lattice point at distance at most $\frac{1}{2} \sum_{i=1}^r 2^{(r-i)/2} \|\mathbf{b}_\star^r\| \le 2^r \|\mathbf{b}_\star^r\|$. Consequently, the distance between **t** and $CV(\Lambda, \mathbf{t})$ is at most $2^r \|\mathbf{b}_\star^r\|$. Since the affine lattice subspaces parallel to $\mathrm{span}(\widehat{\mathbf{b}}^1, \ldots, \widehat{\mathbf{b}}^{r-1})$ are separated by a distance of $\|\mathbf{b}_\star^r\|$, the result follows. $\square$

Corollary 6.3.23 suggests the following modification of the nearest lattice plane algorithm to compute the closest lattice vector problem. Instead of recursing only on the nearest affine lattice subspace H parallel to $\mathrm{span}(\widehat{\mathbf{b}}^1, \ldots, \widehat{\mathbf{b}}^{r-1})$, we recurse on the $2^{O(r)}$ subspaces $H + \mu \widehat{\mathbf{b}}^r$, $\mu \in [-2^r, 2^r] \cap \mathbb{Z}$. This immediately gives us an algorithm to compute the closest lattice vector with worst-case complexity $2^{O(r^2)} \log\left(\frac{\|\mathbf{b}^{\max}\|_2}{\ell(\Lambda)}\right)$, where $\mathbf{b}^{\max}$ is the column of B with the largest Euclidean norm and $\ell(\Lambda)$ is the Euclidean length of a shortest nonzero lattice vector in Λ. To reduce the term $2^{O(r^2)}$ down to $2^{O(r)}$, we observe that lattice points in a affine lattice subspace parallel to H are simply translates of the lattice points in H. So one should be able to reuse some of the work that was done to compute the closest lattice point in H. The main idea is to compute the Dirichlet–Voronoi cell of the lower-dimensional lattice and reuse it for all the parallel affine lattice subspaces to solve the closest lattice vector problem.

Solving CVP given access to the Dirichlet–Voronoi cell. Suppose we are given the Voronoi relevant vectors $R(\Lambda)$ of a lattice $\Lambda \subseteq \mathbb{R}^n$ of rank r, i.e., we have an inequality description of the Dirichlet–Voronoi cell (see Section 4.3.2). We describe an algorithm that computes $CV(\Lambda, \mathbf{t})$, given $\mathbf{t} \in \mathrm{span}(\Lambda)$, with worst-case complexity $2^{O(r)} \log(N_{\mathcal{V}(\Lambda)}(\mathbf{t}))$, where $N_{\mathcal{V}(\Lambda)}$ is the norm associated with $\mathcal{V}(\Lambda)$ (see Theorem 3.3.14). The main insight is the following observation about Voronoi relevant vectors and Dirichlet–Voronoi cells (Figure 6.4).

Theorem 6.3.24 *Let* $\Lambda \subseteq \mathbb{R}^n$ *be a lattice and let* $\mathcal{V}(\Lambda)$ *be its Dirichlet–Voronoi cell. Let* $\mathbf{t} \in \mathrm{span}(\Lambda)$. *Let* $k \in \mathbb{Z}_+$ *be the smallest nonnegative integer such that* $\mathbf{t} \in 2^k \mathcal{V}(\Lambda)$.

Consider the graph $G = (V, E)$, *where* $V = (2^{k-1}\Lambda + \mathbf{t}) \cap (2^{k+1}\mathcal{V}(\Lambda))$ *and* $\mathbf{v}^1, \mathbf{v}^2 \in V$ *are connected by an edge if and only if* $\mathbf{v}^1 - \mathbf{v}^2$ *is a Voronoi relevant vector for the lattice* $2^{k-1}\Lambda$. *Then there exists* $\mathbf{v} \in V \cap 2^{k-1}\mathcal{V}(\Lambda)$ *such that* **t** *and* **v** *are connected in this graph by a simple path.*

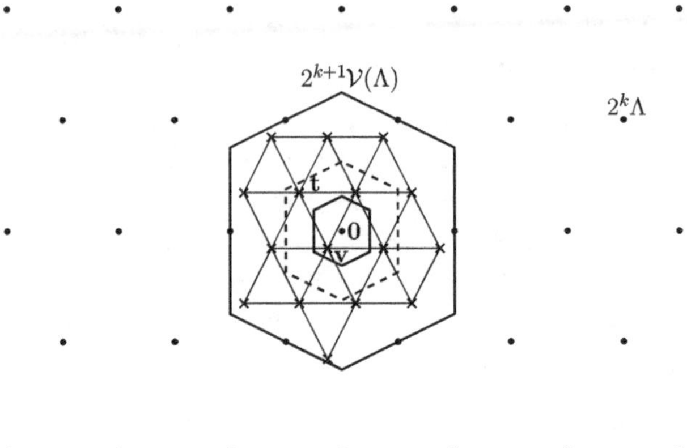

Figure 6.4 Illustration for Theorem 6.3.24. The lattice $2^k\Lambda$ is shown with disks. The Dirichlet–Voronoi cell for the lattice $2^k\Lambda$ is shown in dotted lines, and the Dirichlet–Voronoi cells for $2^{k-1}\Lambda$ and $2^{k+1}\Lambda$ are shown in solid lines. The graph in the statement of the theorem is shown with vertices $V = (2^{k-1}\Lambda + \mathbf{t}) \cap (2^{k+1}\mathcal{V}(\Lambda))$ marked by crosses and edges marked by solid thin lines.

Proof Exercise 17 from Section 4.3.3 shows that for any $\mu \geq 0$, $\mathcal{V}(\mu\Lambda) = \mu\mathcal{V}(\Lambda)$. This fact will be used without explicit mention in this proof.

Consider the line segment ℓ joining $\mathbf{0}$ and $\mathbf{t}$. By part 5. of Proposition 4.3.17 applied to the lattice $2^{k-1}\Lambda$, this line segment can be partitioned into finitely many smaller line segments $\ell \cap (2^{k-1}\mathcal{V}(\Lambda) + \mathbf{p})$, $\mathbf{p} \in 2^{k-1}\Lambda$. For any $\mathbf{p} \in 2^{k-1}\Lambda$ such that $2^{k-1}\mathcal{V}(\Lambda) + \mathbf{p}$ has a nonempty intersection with ℓ, we must have $\mathbf{p} \in 3 \cdot 2^{k-1}\mathcal{V}(\Lambda)$. Indeed, if $\mathbf{x} \in \ell \cap (2^{k-1}\mathcal{V}(\Lambda) + \mathbf{p})$, then $\mathbf{x} \in \ell \subseteq 2^k\mathcal{V}(\Lambda)$ since $\mathbf{t} \in 2^k\mathcal{V}(\Lambda)$ and $\mathcal{V}(\Lambda)$ is a $\mathbf{0}$-symmetric, convex set (part 2. of Proposition 4.3.17). Since $\mathbf{x} \in 2^{k-1}\mathcal{V}(\Lambda) + \mathbf{p}$, $\mathbf{p} - \mathbf{x} \in 2^{k-1}\mathcal{V}(\Lambda)$. Therefore, $\mathbf{p} = \mathbf{x} + (\mathbf{p} - \mathbf{x}) \in 2^k\mathcal{V}(\Lambda) + 2^{k-1}\mathcal{V}(\Lambda) = 3 \cdot 2^{k-1}\mathcal{V}(\Lambda)$ (see Exercise 3 from Section 2.1.1). Thus, the set $S := \{\mathbf{p} \in 2^{k-1}\Lambda : \ell \cap 2^{k-1}\mathcal{V}(\Lambda) + \mathbf{p} \neq \emptyset\} \subseteq 3 \cdot 2^{k-1}\mathcal{V}(\Lambda)$.

Exercises 22 and 23 from Section 4.3.3 imply the following: for any $\mathbf{p}^1, \mathbf{p}^2 \in S$ such that the corresponding line segments $\ell \cap 2^{k-1}\mathcal{V}(\Lambda) + \mathbf{p}^1$ and $\ell \cap 2^{k-1}\mathcal{V}(\Lambda) + \mathbf{p}^2$ have a common endpoint, a vertex of the graph in $2^{k-1}\mathcal{V}(\Lambda) + \mathbf{p}^1$ is connected to a vertex of the graph in $2^{k-1}\mathcal{V}(\Lambda) + \mathbf{p}^2$. Moreover, for any $\mathbf{p} \in S$, $2^{k-1}\mathcal{V}(\Lambda) + \mathbf{p} \subseteq 2^{k-1}\mathcal{V}(\Lambda) + 3 \cdot 2^{k-1}\mathcal{V}(\Lambda) = 2^{k+1}\mathcal{V}(\Lambda)$. Since $\mathbf{0} \in \ell \cap (2^{k-1}\mathcal{V}(\Lambda))$, this observation implies that there is a connected path between $\mathbf{t}$ and a vertex of the graph in $2^{k-1}\mathcal{V}(\Lambda)$. $\square$

Theorem 6.3.24 allows one to start from a point $\mathbf{t} \in 2^k\mathcal{V}(\Lambda)$ and arrive inside $2^{k-1}\mathcal{V}(\Lambda)$ in a finite number of steps by only moving in the directions of the Voronoi relevant vectors of $2^{k-1}\Lambda$ (which are just scalings of the Voronoi relevant vectors of Λ). The number of steps needed is finite because one never needs to explore any points outside the set $(2^{k-1}\Lambda + \mathbf{t}) \cap (2^{k+1}\mathcal{V}(\Lambda))$. This suggests the following algorithm to find the closest lattice point to $\mathbf{t}$: find the smallest $k \in \mathbb{Z}_+$ such that $\mathbf{t} \in 2^k\mathcal{V}(\Lambda)$, then make moves along (scaled) Voronoi relevant vectors to arrive inside $2^{k-1}\mathcal{V}(\Lambda)$, and repeat until we arrive at a point $\mathbf{v}$ in $\mathcal{V}(\Lambda)$. Then we are done since $\mathbf{t} - \mathbf{v}$ is the closest lattice vector to $\mathbf{t}$ by Proposition 4.3.16. We formalize this below.

Definition 6.3.25 (CVP algorithm given Voronoi relevant vectors)

Input: A set $R \subseteq \mathbb{R}^n$ that contains all Voronoi relevant vectors of a lattice $\Lambda \subseteq \mathbb{R}^n$, and $\mathbf{t} \in \mathrm{span}(\Lambda)$ as the target vector.

1. Set $\mathbf{z} := \mathbf{0}$.
2. Compute the smallest nonnegative integer $k \in \mathbb{Z}_+$ such that $2^k \geq N_{\mathcal{V}(\Lambda)}(\mathbf{t}) = \max_{\mathbf{r}\in R} \frac{2\langle \mathbf{r}, \mathbf{t}\rangle}{\|\mathbf{r}\|^2}$.
3. While $k \geq 1$, perform the following steps:

   (a) Construct a graph $G = (V, E)$ as follows.
   $V := (2^{k-1}\Lambda + \mathbf{t}) \cap (2^{k+1}\mathcal{V}(\Lambda))$ and $\mathbf{v}^1, \mathbf{v}^2 \in V$ are connected by a (directed) edge if and only if $\mathbf{v}^1 - \mathbf{v}^2 \in 2^{k-1}R$.
   (b) Perform a breadth-first search starting from $\mathbf{t}$ on G to obtain
   $\mathbf{v} \in (2^{k-1}\Lambda + \mathbf{t}) \cap 2^{k-1}\mathcal{V}(\Lambda)$.
   (c) Update $\mathbf{z} := \mathbf{z} + \mathbf{t} - \mathbf{v}$.
   (d) Update $\mathbf{t} := \mathbf{v}$.
   (e) Update $k := k - 1$.

4. Return $\mathbf{z}$.

Theorem 6.3.26 *Given a set $R \subseteq \Lambda$ containing all Voronoi relevant vectors of a lattice $\Lambda \subseteq \mathbb{R}^n$ with rank r and $\mathbf{t} \in \mathrm{span}(\Lambda)$ as input, the algorithm in Definition 6.3.25 computes $CV(\Lambda, \mathbf{t})$ with worst-case complexity $2^{O(r)} \cdot \#(R) \cdot \left\lceil \max\left\{0, \log\left(\frac{2\|\mathbf{t}\|_2}{\ell(R)}\right)\right\}\right\rceil$, where $\ell(R)$ is the length of the shortest vector in R.*

Proof By Proposition 4.3.16, $\mathbf{p} \in CV(\Lambda, \mathbf{t})$ if and only if $\mathbf{t} - \mathbf{p} \in \mathcal{V}(\Lambda)$. We now show that the algorithm computes a point from the coset $\Lambda + \mathbf{t}$ in the Dirichlet–Voronoi cell $\mathcal{V}(\Lambda)$. Let $k_{\mathrm{init}} \in \mathbb{Z}_+$ be the smallest nonnegative integer k such that $\mathbf{t} \in 2^k\mathcal{V}(\Lambda)$ (this number exists since $\mathcal{V}(\Lambda)$ is full dimensional by part 1 of Proposition 4.3.17). By Theorem 6.3.24, the breadth-first search in Step 3(b) is guaranteed to terminate with $\mathbf{v} \in (2^{k-1}\Lambda + \mathbf{t}) \cap 2^{k-1}\mathcal{V}(\Lambda)$. The While loop in Step 3 thus generates a sequence of points $\mathbf{t}^0 := \mathbf{t}, \mathbf{t}^1, \mathbf{t}^2, \ldots, \mathbf{t}^{k_{\mathrm{init}}}$ such that $\mathbf{t}^i \in 2^{k_{\mathrm{init}}-i}\mathcal{V}(\Lambda)$ and $\mathbf{t}^i \in 2^{k_{\mathrm{init}}-i}\Lambda + \mathbf{t} \subseteq \Lambda + \mathbf{t}$. Therefore,

$\mathbf{t}^{k_{\text{init}}} \in \mathcal{V}(\Lambda)$ and $\mathbf{t} - \mathbf{t}^{k_{\text{init}}} \in CV(\Lambda, \mathbf{t})$. We also have $\mathbf{z} := \mathbf{0} + (\mathbf{t}_0 - \mathbf{t}^1) + \cdots + (\mathbf{t}^{k_{\text{init}}-1} - \mathbf{t}^{k_{\text{init}}}) = \mathbf{t} - \mathbf{t}^{k_{\text{init}}}$, from Step 3(c).

We next bound the complexity of the algorithm. By the Cauchy–Schwarz inequality (1.1.6), $N_{\mathcal{V}(\Lambda)}(\mathbf{t}) = \max_{\mathbf{r} \in R} \frac{2\langle \mathbf{r}, \mathbf{t} \rangle}{\|\mathbf{r}\|^2} \leq 2\frac{\|\mathbf{t}\|}{\ell(R)}$; thus, $k_{\text{init}} = \left\lceil \max \left\{ 0, \log \left(\frac{2\|\mathbf{t}\|_2}{\ell(R)} \right) \right\} \right\rceil$. Hence the While loop in Step 3 of the algorithm loops at most $\left\lceil \log \left(2\frac{\|\mathbf{t}\|}{\ell(R)} \right) \right\rceil$ times. By Exercise 5 from Section 4.3.3, $\#(V) \leq 5^r$. Thus, by the handshaking lemma,[2] $\#(E) \leq \frac{5^r \cdot \#(R)}{2}$. Since breadth-first search runs in time $O(\#(E))$, we will find a point from the coset $2^{k-1}\Lambda + \mathbf{t}$ in $2^{k-1}\mathcal{V}(\Lambda)$ with worst-case complexity $2^{O(r)}\#(R)$. $\qquad\square$

Computing the Dirichlet–Voronoi cell. We now design a recursive algorithm that, given a basis for a lattice of rank r, computes a set of lattice vectors of size at most $2^{O(r)}$ that contains all the Voronoi relevant vectors.

Definition 6.3.27 (Dirichlet–Voronoi cell algorithm)

Input: A basis $B \in \mathbb{R}^{n \times r}$ of a lattice of $\mathbb{R}^n$ with rank r.

1. If $r = 1$, return $\left\{ -\mathbf{b}^1, \mathbf{b}^1 \right\}$ as the Voronoi relevant vectors.
2. Else, apply the LLL algorithm on B to obtain $\widehat{B}$.
3. Set $R := \emptyset$.
4. Define $\Lambda' := Z(2\widehat{\mathbf{b}}^1, \dots, 2\widehat{\mathbf{b}}^{r-1})$. With a recursive call on Λ', we compute a superset of the Voronoi relevant vectors for Λ'.
5. Construct the Gram–Schmidt Factorization $\widehat{B} = \widehat{B}^\star R$.
6. For every tuple $(\mu_1, \dots, \mu_r) \in \{0, 1\}^r \setminus \{\mathbf{0}\}$, repeat the following steps:

   (a) Define $\mathbf{t} := \mu_1 \mathbf{b}^1 + \cdots + \mu_r \mathbf{b}^r$ and $\lambda_i = \frac{1}{2}\mu_i$ for $i = 1, \dots, r$.
   (b) For every $\mu \in [-2^r, 2^r] \cap \mathbb{Z}$, define
   $$\mathbf{t}^\mu := \mathbf{t} + (\lceil \lambda_r \rceil + \mu - \lambda_r)2\widehat{\mathbf{b}}_\star^r - (\lceil \lambda_r \rceil + \mu)2\widehat{\mathbf{b}}^r =$$
   $$\lambda_1 2\widehat{\mathbf{b}}^1 + \cdots \lambda_{r-1} 2\widehat{\mathbf{b}}^{r-1} + (\lceil \lambda_r \rceil + \mu - \lambda_r)(2\widehat{\mathbf{b}}_\star^r - 2\widehat{\mathbf{b}}^r).$$
   Using the Voronoi relevant vectors computed for Λ', compute $\mathbf{z}^\mu \in CV(\Lambda', \mathbf{t}^\mu)$, as described above in Definition 6.3.25.
   (c) Set $\mathbf{z} := \mathbf{z}_\mu$ such that $\mu \in \operatorname{argmin}\{\|\mathbf{z}^\mu + (\lceil \lambda_r \rceil + \mu)2\widehat{\mathbf{b}}^r - \mathbf{t}\| : \mu \in [-2^r, 2^r] \cap \mathbb{Z}\}$.
   (d) Update $R := R \cup \{\pm(\mathbf{z} - \mathbf{t})\}$.

Theorem 6.3.28 *Given a basis $B \in \mathbb{R}^{n \times r}$ of a lattice $\Lambda \subseteq \mathbb{R}^n$ of rank r as input, the algorithm in Definition 6.3.27 computes a set $R \subseteq \Lambda$ of size at most $2(2^r - 1)$ that contains all the Voronoi relevant vectors $R(\Lambda)$ of Λ, with worst-case complexity bounded by $2^{O(r)} \log \left(\frac{\|\mathbf{b}^{\max}\|_2}{\ell(\Lambda)} \right)$, where $\mathbf{b}^{\max}$ is the column of B*

---

[2] The handshaking lemma is a simple counting argument in graph theory that says that the number of edges in an undirected graph is equal to half of the sum of the degrees of the vertices.

with the largest Euclidean norm and $\ell(\Lambda)$ is the Euclidean length of a shortest nonzero vector in Λ.

Proof Part 6. of Theorem 4.3.20 shows that $\mathbf{r} \in \Lambda \setminus \{0\}$ is Voronoi relevant vector if and only if $\mathbf{r}$ and $-\mathbf{r}$ are the shortest vectors in the coset of 2Λ that contains them. In Step 6 of the algorithm in Definition 6.3.27, we enumerate over all possible cosets of 2Λ and the shortest vectors in the coset are computed by solving the corresponding closest vector problem (Proposition 4.3.16), using Corollary 6.3.23; see also the discussion just below Corollary 6.3.23. Exercise 16 in Section 6.3.3 fills in the rest of the details. $\square$

6.3.1.4 The Mixed-Integer Case

We now put together the insights from the previous discussions in this section to solve the unconstrained convex quadratic minimization problem over $\mathbb{Z}^n \times \mathbb{R}^d$. As before, the function we are minimizing is given by $\mathbf{f}(\mathbf{z}) = \frac{1}{2}\mathbf{z}^T Q\mathbf{z} - \mathbf{c}^T\mathbf{z}$, or equivalently, we are minimizing $N_Q(\mathbf{z} - \mathbf{t})$, where $\mathbf{t} = Q^{-1}\mathbf{c}$, where $Q \in \mathbb{R}^{(n+d)\times(n+d)}$ is a positive definite matrix and $\mathbf{c} \in \mathbb{R}^n \times \mathbb{R}^d$.

Define $L := \{\mathbf{0}_n\} \times \mathbb{R}^d$. By Theorem 6.3.5, for every $\mathbf{z} \in \mathbb{R}^n \times \mathbb{R}^d$ there exist $\mathbf{p} \in L$ and $\mathbf{q} \in L_Q^{\perp}$ such that $\mathbf{z} = \mathbf{p} + \mathbf{q}$; let $\mathbf{p}' \in L$ and $\mathbf{q}' \in L_Q^{\perp}$ such that $\mathbf{t} = \mathbf{p}' + \mathbf{q}'$. Moreover, $f(\mathbf{z}) = f(\mathbf{p}) + f(\mathbf{q})$ and $N_Q(\mathbf{z} - \mathbf{t})^2 = N_Q(\mathbf{p} - \mathbf{p}')^2 + N_Q(\mathbf{q} - \mathbf{q}')^2$. In fact, combining Theorem 6.3.5 with Exercise 2 from Section 6.3.3, we know that $f(\mathbf{p}) = \frac{1}{2}N_Q(\mathbf{p} - \mathbf{p}')^2 + \frac{1}{2}N_Q(\mathbf{q}')^2 - \frac{1}{2}N_{Q^{-1}}(\mathbf{c})^2$ and $f(\mathbf{q}) = \frac{1}{2}N_Q(\mathbf{q} - \mathbf{q}')^2 + \frac{1}{2}N_Q(\mathbf{p}')^2 - \frac{1}{2}N_{Q^{-1}}(\mathbf{c})^2$. This implies that (see Exercise 17 from Section 6.3.3)

$$\underset{\mathbf{z} \in \mathbb{Z}^n \times \mathbb{R}^d}{\operatorname{argmin}} \ f(\mathbf{z}) = \underset{\mathbf{p} \in L}{\operatorname{argmin}} \ f(\mathbf{p}) + \underset{\mathbf{q} \in L_Q^{\perp} \cap (\mathbb{Z}^n \times \mathbb{R}^d)}{\operatorname{argmin}} \ N_Q(\mathbf{q} - \mathbf{q}'). \tag{6.3.8}$$

L is isomorphic to $\mathbb{R}^d$ and $L_Q^{\perp} \cap (\mathbb{Z}^n \times \mathbb{R}^d)$ can be verified to be a lattice of rank n (Lemma 6.3.29 below). Thus, we have decomposed the problem into the "continuous" part (the first term on the right-hand side of 6.3.8), and into the "lattice" part (the second term). We can use Conjugate Gradients or the analytic formulas to solve the first term (see Section 6.3.1.2). We can make some elementary transformations to convert the second term into a CVP problem in the ℓ_2 norm and then use the techniques from Section 6.3.1.3. Consider the basis $(\mathbf{0}_n, \mathbf{e}_d^i)$, $i = 1, \dots, d$ for L; appended with $(\mathbf{e}_n^i, \mathbf{0}_d)$ we have a basis B for $\mathbb{R}^n \times \mathbb{R}^d$; note that the matrix B is obtained from the identity matrix I_{n+d} by permuting the columns:

$$B = \begin{bmatrix} \mathbf{0}_{n \times d} & I_n \\ I_d & \mathbf{0}_{d \times n} \end{bmatrix}. \tag{6.3.9}$$

Consider the Gram–Schmidt Orthogonalization of B with respect to the inner product $\langle \cdot, \cdot \rangle_Q$ (Theorem 6.3.8):

$$B = B^\star R. \tag{6.3.10}$$

We denote by $B_n^\star$ the submatrix of $B^\star$ consisting of the last n columns; thus, $B_n^\star$ is a Q-conjugate basis for $L_Q^\perp$. We use $R_{n \times n}$ to denote the $n \times n$ submatrix of R obtained from the last n rows and columns.

Lemma 6.3.29 $L_Q^\perp \cap (\mathbb{Z}^n \times \mathbb{R}^d)$ *is a lattice of rank n with basis $B_n^\star R_{n \times n}$.*

Proof Left as an exercise. $\qquad\qquad\qquad\qquad\qquad\qquad\qquad\qquad\qquad\quad\square$

By Lemma 6.3.7, the projection $\mathbf{q}'$ of $\mathbf{t}$ on to $L_Q^\perp$ is given by

$$\mathbf{q}' = \sum_{i=1}^{n} \frac{\langle \mathbf{b}_\star^i, \mathbf{t} \rangle_Q}{N_Q(\mathbf{b}_\star^i)^2} \mathbf{b}_\star^i = \sum_{i=1}^{n} \frac{\langle \mathbf{b}_\star^i, \mathbf{c} \rangle}{N_Q(\mathbf{b}_\star^i)^2} \mathbf{b}_\star^i, \tag{6.3.11}$$

where $\mathbf{b}_\star^i$ is the ith column of $B_n^\star$ for $i = 1, \ldots, n$. Define

$$D := (B^\star)^T Q B^\star. \tag{6.3.12}$$

Since $B^\star$ has Q-conjugate columns, D is a diagonal matrix with strictly positive entries on the diagonal, given by squares of the norms of the columns of $B^\star$ in the norm induced by Q. Thus,[3]

$$Q = (B^\star)^{-T} D (B^\star)^{-1}. \tag{6.3.13}$$

Using (6.3.13), the second term on the right-hand side of (6.3.8) can be written as

$$\operatorname*{argmin}_{\mathbf{q} \in L_Q^\perp \cap (\mathbb{Z}^n \times \mathbb{R}^d)} N_Q(\mathbf{q} - \mathbf{q}') = \operatorname*{argmin}_{\mathbf{q} \in L_Q^\perp \cap (\mathbb{Z}^n \times \mathbb{R}^d)} \left\| \sqrt{D}(B^\star)^{-1}(\mathbf{q} - \mathbf{q}') \right\|_2,$$

where $\sqrt{D}$ denotes diagonal matrix with diagonal entries equal to the positive square root of the corresponding entry in D. We thus arrive at the following algorithm for mixed-integer unconstrained quadratic minimization (or finding the closest mixed-integer point in a norm induced by a positive definite matrix).

Definition 6.3.30 (Mixed-integer unconstrained quadratic minimization or closest mixed-integer point computation)

Input: A positive definite matrix $Q \in \mathbb{R}^{(n+d) \times (n+d)}$ and $\mathbf{c} \in \mathbb{R}^n \times \mathbb{R}^d$ (or $\mathbf{t} \in \mathbb{R}^n \times \mathbb{R}^d$), and we seek to find a point in $\operatorname{argmin}_{\mathbf{z} \in \mathbb{Z}^n \times \mathbb{R}^d} \frac{1}{2} \mathbf{z}^T Q \mathbf{z} - \mathbf{c}^T \mathbf{z}$ (or $\operatorname{argmin}_{\mathbf{z} \in \mathbb{Z}^n \times \mathbb{R}^d} N_Q(\mathbf{z} - \mathbf{t})$ respectively).

---

[3] This is related to the so-called *Cholesky decomposition* of Q, but we do not make use of this connection.

1. Define $L := \{\mathbf{0}_n\} \times \mathbb{R}^d$.
2. Compute the Gram–Schmidt Orthogonalization in (6.3.10).
3. Compute $\mathbf{p}^\star \in \operatorname{argmin}_{\mathbf{p} \in L} f(\mathbf{p})$ (or $\mathbf{p}^\star \in \operatorname{argmin}_{\mathbf{p} \in L} N_Q(\mathbf{p} - \mathbf{t})$) using the Conjugate Gradient method or the analytic solution.
4. Compute $\mathbf{q}'$ as in (6.3.11) and D as in (6.3.12).
5. Use the closest lattice vector (CVP) algorithm from Section 6.3.1.3 to compute $\mathbf{y}^\star \in \operatorname{argmin}_{\mathbf{y} \in \Lambda} \|\mathbf{y} - \sqrt{D}(B^\star)^{-1}\mathbf{q}'\|$, where $\Lambda \subseteq \mathbb{R}^n \times \mathbb{R}^d$ is the lattice generated by the columns of $\sqrt{D}(B^\star)^{-1} B_n^\star R_{n \times n} = \begin{bmatrix} \mathbf{0}_{d \times n} \\ \sqrt{D_{n \times n}} R_{n \times n} \end{bmatrix}$, where $D_{n \times n}$ is the $n \times n$ submatrix of D obtained from the last n rows and columns.
6. Compute $\mathbf{q}^\star = B^\star \sqrt{D^{-1}} \mathbf{y}^\star$, where $\sqrt{D^{-1}}$ denotes diagonal matrix with diagonal entries equal to the reciprocal of the positive square root of the corresponding entry in D.
7. Return $\mathbf{z}^\star = \mathbf{p}^\star + \mathbf{q}^\star$.

Theorem 6.3.31 *Given $Q \in \mathbb{R}^{(n+d) \times (n+d)}$ and $\mathbf{c} \in \mathbb{R}^n \times \mathbb{R}^d$ (or $\mathbf{t} \in \mathbb{R}^n \times \mathbb{R}^d$), the algorithm in Definition 6.3.30 returns a minimizer of $\frac{1}{2}\mathbf{z}^T Q \mathbf{z} - \mathbf{c}^T \mathbf{z}$ in $\mathbb{Z}^n \times \mathbb{R}^d$ (or the closest point in $\mathbb{Z}^n \times \mathbb{R}^d$ to $\mathbf{t}$), with worst-case complexity $O((n+d)^3) + 2^{O(n)} \log\left(\frac{\lambda_{\max}(Q)}{\lambda_{\min}(Q)} \max\left\{1, \frac{2\|\mathbf{c}\|_2}{\lambda_{\min}(Q)}\right\}\right)$ (or $O((n+d)^3) + 2^{O(n)} \log\left(\frac{\lambda_{\max}(Q)}{\lambda_{\min}(Q)} \max\{1, \|\mathbf{t}\|_2\}\right)$). If $n = 0$, then the complexity is $O(d^3)$.*

Proof The correctness of the algorithm should be clear from the discussions above. We will derive the upper bound on complexity. From the discussions in the pure continuous case, Steps 2, 3, and 4 take time $O((n+d)^3)$. If $n = 0$, the algorithm would stop here and we get the complexity of $O(d^3)$ (as discussed in the first paragraph of Section 6.3.1.2). In Step 5 we compute the closest lattice point to $\sqrt{D}(B^\star)^{-1}\mathbf{q}'$ in the lattice Λ generated by the basis $\begin{bmatrix} \mathbf{0}_{d \times n} \\ \sqrt{D_{n \times n}} R_{n \times n} \end{bmatrix}$. To bound the computational complexity of this step, we will use Theorems 6.3.28 and 6.3.26. We proceed to bound the required terms appearing in the complexity bounds in those theorems, i.e., the norm of $\sqrt{D}(B^\star)^{-1}\mathbf{q}'$, the norm of the shortest nonzero lattice vector in Λ and the norm of the largest basis vector in $\begin{bmatrix} \mathbf{0}_{d \times n} \\ \sqrt{D_{n \times n}} R_{n \times n} \end{bmatrix}$.

From (6.3.11), the vector $\sqrt{D}(B^\star)^{-1}\mathbf{q}' = (\mathbf{0}_d, v_1, \ldots, v_n)$, where $v_i = \frac{\langle \mathbf{b}_\star^i, \mathbf{t} \rangle_Q}{N_Q(\mathbf{b}_\star^i)}$ or $v_i = \frac{\langle \mathbf{b}_\star^i, \mathbf{c} \rangle}{N_Q(\mathbf{b}_\star^i)}$, depending on if we are working with $\mathbf{t}$ or $\mathbf{c}$, for $i = 1, \ldots, n$. Using the generalized Cauchy–Schwarz inequalities (Exercise 2 from Section 1.2.3), $|v_i| \leq \sqrt{\lambda_{\max}(Q)}\|\mathbf{t}\|_2$ or $|v_i| \leq \frac{\|\mathbf{c}\|_2}{\sqrt{\lambda_{\min}(Q)}}$.

We next lower bound the Euclidean norm of the shortest nonzero vector in Λ. Λ is the image of the lattice $L_Q^\perp \cap (\mathbb{Z}^n \times \mathbb{R}^d)$ under the invertible transformation $\sqrt{D}(B^\star)^{-1}$. Since $L \cap L_Q^\perp = \{0\}$, any nonzero vector in the lattice $L_Q^\perp \cap (\mathbb{Z}^n \times \mathbb{R}^d)$ has at least one integral coordinate and therefore has Euclidean norm at least 1. From (6.3.13), we have $Q = (B^\star)^{-T} D (B^\star)^{-1} = (\sqrt{D}(B^\star)^{-1})^T (\sqrt{D}(B^\star)^{-1})$ and so, the smallest singular value of $\sqrt{D}(B^\star)^{-1}$ is $\sqrt{\lambda_{\min}(Q)}$ (Proposition 1.2.18). Therefore, the smallest norm of a vector in Λ is at least $\sqrt{\lambda_{\min}(Q)}$ (see Exercise 4 from Section 1.2.3).

Finally, the maximum squared Euclidean norm of a vector in the basis
$\begin{bmatrix} \mathbf{0}_{d \times n} \\ \sqrt{D_{n \times n}} R_{n \times n} \end{bmatrix}$ is upper bounded by $\lambda_{\max}(Q)$ (Exercise 20 from Section 6.3.3).

Putting everything into the complexity bounds from Theorems 6.3.28 and 6.3.26, we obtain the stated complexity bound. $\qquad\square$

Remark 6.3.32 The quantity $\frac{\lambda_{\max}(Q)}{\lambda_{\min}(Q)}$ that appears in Theorem 6.3.31 is known as the *condition number of the matrix* Q.

6.3.1.5 Shortest Lattice Vector Problem (SVP) from CVP

It turns out that an algorithm for solving the closest lattice vector problem (CVP) can be used to solve the shortest nonzero lattice vector problem (SVP) as well (Definition 4.3.1). This follows from the following theorem.

Theorem 6.3.33 *Let* $\Lambda \subseteq \mathbb{R}^n$ *be a lattice with basis* $\mathbf{b}^1, \ldots, \mathbf{b}^k$*. Let N be any norm on* $\mathbb{R}^n$*. Then*

$$N(SV(N, \Lambda)) = \min_{i=1,\ldots,k} N(CV(\mathbf{b}^i, \Lambda_i) - \mathbf{b}^i),$$

where Λ_i *is the lattice spanned by* $\mathbf{b}^1, \ldots, 2\mathbf{b}^i, \ldots, \mathbf{b}^k$ *for* $i = 1, \ldots, k$*. [The closest lattice vector* $CV(\mathbf{b}^i, \Lambda_i)$ *is also defined in the norm N.]*

Proof We first establish that $N(SV(N, \Lambda)) \leq N(CV(\mathbf{b}^i, \Lambda_i) - \mathbf{b}^i)$ for all $i = 1, \ldots, k$, and therefore, $N(SV(N, \Lambda)) \leq \min_{i=1,\ldots,k} N(CV(\mathbf{b}^i, \Lambda_i) - \mathbf{b}^i)$. For any $i = 1, \ldots, k$, consider $\mathbf{u} \in CV(\mathbf{b}^i, \Lambda_i)$ and express $\mathbf{u} = \mu_1 \mathbf{b}^1 + \cdots, \mu_i(2\mathbf{b}^i) + \mu_k \mathbf{b}^k$ for some integers $\mu_1, \ldots, \mu_k$. Then, $\mathbf{u} - \mathbf{b}^i = \mu_1 \mathbf{b}^1 + \cdots, (2\mu_i - 1)\mathbf{b}^i + \mu_k \mathbf{b}^k$. Since $2\mu_i - 1$ is odd and therefore nonzero, $\mathbf{u} - \mathbf{b}^i$ is a nonzero vector in the original lattice Λ. Thus, $N(SV(N, \Lambda)) \leq N(\mathbf{u} - \mathbf{b}^i) = N(CV(\mathbf{b}^i, \Lambda_i) - \mathbf{b}^i)$.

Next, we show that there exists $i \in \{1, \ldots, k\}$ such that $N(SV(N, \Lambda)) \geq N(CV(\mathbf{b}^i, \Lambda_i) - \mathbf{b}^i)$. This will complete the proof. Let $\mathbf{v} \in SV(N, \Lambda)$. We first observe that if we express $\mathbf{v} = \mu_1 \mathbf{b}^1 + \cdots + \mu_k \mathbf{b}^k$ where $\mu_1, \ldots, \mu_k \in \mathbb{Z}$, then at least one of $\mu_1, \ldots, \mu_k$ is odd. This is because if every μ_i is even for

$i = 1, \ldots, k$, then $\frac{\mathbf{v}}{2} \in \Lambda \setminus \{\mathbf{0}\}$ and has strictly smaller norm than $\mathbf{v}$. Let $i^\star \in \{1, \ldots, k\}$ be such that $\mu_{i^\star}$ is odd. Define

$$\mathbf{u} := \mathbf{v} + \mathbf{b}^{i^\star} = \mu_1 \mathbf{b}^1 + \cdots + (\mu_{i^\star} + 1)\mathbf{b}^{i^\star} + \cdots + \mu_k \mathbf{b}^k$$

$$= \mu_1 \mathbf{b}^1 + \cdots + \left(\frac{\mu_{i^\star} + 1}{2}\right)(2\mathbf{b}^{i^\star}) + \cdots + \mu_k \mathbf{b}^k.$$

In other words, $\mathbf{u} \in \Lambda_{i^\star}$ and therefore, $N(CV(\mathbf{b}^{i^\star}, \Lambda_{i^\star}) - \mathbf{b}^{i^\star}) \leq N(\mathbf{u} - \mathbf{b}^{i^\star}) = N(\mathbf{v}) = N(SV(N, \Lambda))$. $\qquad\square$

Corollary 6.3.34 *One can solve the SVP for a lattice of rank r by solving at most r closest lattice vector problems.*

6.3.2 General Mixed-Integer Convex Optimization

We will now use the unconstrained quadratic case as a subroutine within a recursive algorithm to solve the general mixed-integer convex optimization problem (6.1.1). The algorithm will use recursion on the number n of integer variables. Before we present the formal algorithm and its analysis, let us discuss the overall idea. At every stage of our algorithm we will maintain an ellipsoid $E(A, \mathbf{c})$ (see Definition 2.7.2) which is guaranteed to contain an optimal solution of the instance $(f, I) \in \mathcal{I}_{n,d,R,\rho,M}$; our initial ellipsoid will simply be the Euclidean ball of radius $R\sqrt{n+d}$ centered at $\mathbf{0}_n \times \mathbf{0}_d$. We first compute the closest mixed-integer point $\mathbf{z}^\star \in \mathbb{Z}^n \times \mathbb{R}^d$ to the center $\mathbf{c}$ in the norm induced by A^{-1} using the techniques discussed in Section 6.3.1. If $N_{A^{-1}}(\mathbf{z}^\star - \mathbf{c}) > \frac{1}{n+d+1}$, then the ellipsoid $\mathbf{c} + \frac{1}{n+d+1}(E(A, \mathbf{c}) - \mathbf{c})$ contains no point from $\mathbb{Z}^n \times \mathbb{R}^d$ and consequently, the projection $\widetilde{E}$ of this ellipsoid on to $\mathbb{R}^n \times \{\mathbf{0}_d\}$ contains no point from $\Lambda := \mathbb{Z}^n \times \{\mathbf{0}_d\}$. By Theorem 4.4.6, the Λ-width of $\widetilde{E}$ is at most $n^{3/2} + 1$. This implies that the projection of $E(A, \mathbf{c})$ has Λ-width at most $(n^{3/2} + 1)(n + d + 1)$. We can therefore recurse on these $O(n^{3/2}(n + d))$ hyperplanes, where we can reformulate the problem with one less integer variable. If $N_{A^{-1}}(\mathbf{z}^\star - \mathbf{c}) \leq \frac{1}{n+d+1}$, then we query the separation oracle for C at $\mathbf{z}^\star$. If $\mathbf{z}^\star \in C$, we make a first-order functional query to obtain the function value and a subgradient $\mathbf{s} \in \partial f(\mathbf{z}^\star)$, and define $H := \left\{\mathbf{z} \in \mathbb{R}^n \times \mathbb{R}^d : \langle \mathbf{s}, \mathbf{z} - \mathbf{z}^\star \rangle \leq 0\right\}$. Otherwise, if $\mathbf{z}^\star \notin C$ let $H \subseteq \mathbb{R}^n \times \mathbb{R}^d$ be a separating halfspace containing C. Update $A, \mathbf{c}$ according to Theorem 2.7.6 with $\beta = \frac{1}{n+d+1}$ to obtain a new ellipsoid whose volume is at most $e^{-\frac{1}{2(n+d+1)^3}} \mathrm{vol}(E(A, \mathbf{c}))$ and we continue the iterations. The stopping condition for the iteration loop looks at the volume of the current ellipsoid. If this volume is less than $\left(\frac{\epsilon \min\{\rho, 1\}}{4MR}\right)^{n+d}$, then the iterations stop. The algorithm then checks if the volume of the projection $\widetilde{E}$ of the current ellipsoid on to $\mathbb{R}^n \times \{\mathbf{0}_d\}$ is greater than or equal to 1. If yes, then the algorithm returns the feasible point seen so far with the smallest function value as an ϵ-approximate solution. If the volume of $\widetilde{E}$ is strictly less than 1, then by Exercise 1 from Section 4.2.1, there is a translate of $\widetilde{E}$ that does not contain any point from $\Lambda = \mathbb{Z}^n \times \{\mathbf{0}_d\}$. As argued above, by Theorem 4.4.6, the Λ-width of $\widetilde{E}$ is at most $n^{3/2} + 1$ and we can recurse on these $O(n^{3/2})$ hyperplanes, where we can reformulate the problem with one less integer variable.

Definition 6.3.35 (Lenstra style algorithm for mixed-integer convex optimization) Input: Access to first-order separation and functional oracle for $\mathcal{I}_{n,d,R,\rho,M}$.

1. Initialize $E := E(A, \mathbf{c})$ with $A = \frac{1}{R\sqrt{n+d}} I_{n+d}$ and $\mathbf{c} = \mathbf{0}$.

2. While $\mathrm{vol}(E) \geq \left(\frac{\epsilon \min\{\rho, 1\}}{4MR} \right)^{n+d}$:

   (a) Compute the closest mixed-integer point $\mathbf{z}^\star \in \mathbb{Z}^n \times \mathbb{R}^d$ to $\mathbf{c}$ in the norm $N_{A^{-1}}$, using the algorithm from Definition 6.3.30.

   (b) If $N_{A^{-1}}(\mathbf{z}^\star - \mathbf{c}) > \frac{1}{n+d+1}$, do the following steps.

      i. Define A' to be the $n \times n$ submatrix of A corresponding to the integer constrained variables, and $\mathbf{c}'$ to be the projection of $\mathbf{c}$ on to the space of the integer variables.

      ii. Compute

$$\mathbf{w}^\star \in \arg\min_{\mathbf{w} \in \mathbb{Z}^n \setminus \{\mathbf{0}\}} \| \mathbf{w} \|_{\widetilde{A}},$$

      where $\widetilde{A} := \frac{1}{4} A'^{-1}$, using Corollary 6.3.34.

      iii. Consider the instances of (6.1.1) where the feasible region is a section of C given by $C \cap \{(\mathbf{x}, \mathbf{y}) \in \mathbb{R}^n \times \mathbb{R}^d : \langle \mathbf{w}^\star, \mathbf{x} \rangle = m\}$ for $m = \lceil \langle \mathbf{w}^\star, \mathbf{c}' \rangle - n^{3/2}(n + d + 1) \rceil, \lceil \langle \mathbf{w}^\star, \mathbf{c}' \rangle - n^{3/2}(n + d + 1) \rceil + 1, \ldots, \lfloor \langle \mathbf{w}^\star, \mathbf{c}' \rangle + n^{3/2}(n + d + 1) \rfloor$, and the objective function is a restriction of f to these hyperplanes. By a change of coordinates in the integer constrained variables, these problems involve $n - 1$ integer variables and we recurse on these $O(n^{3/2}(n + d))$ subproblems.

      iv. Report the feasible solution with the smallest objective value seen so far in the algorithm, or report infeasibility if no feasible points have been encountered. STOP.

   (c) Else, $N_{A^{-1}}(\mathbf{z}^\star - \mathbf{c}) \leq \frac{1}{n+d+1}$. Do the following steps:

   i. Query the separation oracle for $C \cap \mathrm{dom}(f)$ at $\mathbf{z}^\star$. If $\mathbf{z}^\star \in C \cap \mathrm{dom}(f)$, we make a first-order functional query to obtain the function value and a subgradient $\mathbf{s} \in \partial f(\mathbf{z}^\star)$, and define $H := \left\{ \mathbf{z} \in \mathbb{R}^n \times \mathbb{R}^d : \left\langle \frac{\mathbf{s}}{\|\mathbf{s}\|_2}, \mathbf{z} - \mathbf{z}^\star \right\rangle \leq 0 \right\}$. Otherwise, if $\mathbf{z}^\star \notin C \cap \mathrm{dom}(f)$ let $\mathbf{s} \in \mathbb{R}^d$ be a unit norm vector and $\delta \in \mathbb{R}$ be such that $H^{\leq}(\mathbf{s}, \delta)$ is a separating halfspace containing $C \cap \mathrm{dom}(f)$.

   ii. Update $A, \mathbf{c}$ according to the formulas in Exercise 14 from Section 2.7.2 with $\beta = \frac{1}{n+d+1}$ to obtain a new ellipsoid whose volume is at most $e^{-\frac{1}{2(n+d+1)^3}} \mathrm{vol}(E(A, \mathbf{c}))$.

3. If there are no integer-constrained variables, then if no functional oracle call has been made, report that the instance is infeasible, i.e., $C \cap \mathrm{dom}(f) = \emptyset$. Else, report the feasible point with the smallest objective value seen so far. STOP.

4. Else, compute $\mathrm{vol}(E(A', \mathbf{c}'))$, where A' is the $n \times n$ submatrix of A corresponding to the integer constrained variables, and $\mathbf{c}'$ to be the projection of $\mathbf{c}$ on to the space of the integer variables.

5. If $\mathrm{vol}(E(A', \mathbf{c}')) > 1$, then if no functional oracle call has been made, report that the instance is infeasible, i.e., $C \cap \mathrm{dom}(f) \cap (\mathbb{Z}^n \times \mathbb{R}^d) = \emptyset$. Else, report the feasible point with the smallest objective value seen so far in the algorithm. STOP.

6. Else, $\mathrm{vol}(E(A', \mathbf{c}')) \leq 1$. Do the following steps.

   (a) Compute

$$\mathbf{w}^\star \in \arg\min_{\mathbf{w} \in \mathbb{Z}^n \setminus \{\mathbf{0}\}} \|\mathbf{w}\|_{\widetilde{A}},$$

   where $\widetilde{A} := \frac{1}{4} A'^{-1}$, using Corollary 6.3.34.

   (b) Consider the instances of (6.1.1) where the feasible region is a section of C given by $C \cap \{(\mathbf{x}, \mathbf{y}) \in \mathbb{R}^n \times \mathbb{R}^d : \langle \mathbf{w}^\star, \mathbf{x} \rangle = m\}$ for $m = \lceil \langle \mathbf{w}^\star, \mathbf{c}' \rangle - n^{3/2} \rceil, \lceil \langle \mathbf{w}^\star, \mathbf{c}' \rangle - n^{3/2} \rceil + 1, \ldots, \lfloor \langle \mathbf{w}^\star, \mathbf{c}' \rangle + n^{3/2} \rfloor$, and the objective function is a restriction of f to these hyperplanes. By a change of coordinates in the integer constrained variables, these problems involve $n - 1$ integer variables and we recurse on these $O(n^{3/2})$ subproblems.

   (c) Report the feasible solution with the smallest objective value seen so far in the algorithm, or report infeasibility if no feasible points have been encountered. STOP.

Note that the volume computations in the above algorithm correspond to determinant computations, by Theorem 2.7.3.

Theorem 6.3.36 *Let $\epsilon \leq 2MR$. The algorithm in Definition 6.3.35 is an ϵ-approximation algorithm for the class $\mathcal{I}_{n,d,R,\rho,M}$ with worst-case complexity*

$$2^{O(n\log(n+d))} \, \mathrm{poly}\left(\log\left(\frac{MR}{\min\{\rho,1\}\epsilon}\right)\right),$$

where $\mathrm{poly}(\cdot)$ is some fixed polynomial function.

Proof We first verify the correctness of the algorithm, assuming it stops. We proceed by induction on the number of integer variables.

If there are no integer variables in the problem, then there is no recursion in Step 2 and Steps 4, 5, and 6 are not executed. In other words, the algorithm simply runs the While loop in Step 2, executing Step 2c, and ends with Step 3. Lemma 6.1.5 says that if there exist feasible solutions then the feasible region contains an $\|\cdot\|_\infty$ ball X of radius $\frac{\epsilon\rho}{2MR}$ of ϵ-approximate solutions. Since the While loop stops with an ellipsoid whose volume is strictly less than $\left(\frac{\epsilon\min\{\rho,1\}}{4MR}\right)^{n+d} \leq \left(\frac{\epsilon\rho}{2MR}\right)^d$, either the instance is infeasible or a functional oracle call must have been made which intersects this ball X. In the first case, the algorithm correctly reports infeasibility in Step 3. In the second case, let the functional call be made at the query point $\mathbf{z}^\star$ with reported subgradient $\mathbf{s}$, yielding the corresponding halfspace $H := \{\mathbf{z} : \langle \mathbf{s}, \mathbf{z} - \mathbf{z}^\star\rangle \leq 0\}$ that does not fully contain the ball X. In other words, there must exist some ϵ-approximate solution $\hat{\mathbf{z}} \in X$ such that $\langle \mathbf{s}, \hat{\mathbf{z}} - \mathbf{z}^\star\rangle > 0$. Combined with the subgradient inequality $f(\hat{\mathbf{z}}) \geq f(\mathbf{z}^\star) + \langle \mathbf{s}, \hat{\mathbf{z}} - \mathbf{z}^\star\rangle > f(\mathbf{z}^\star)$, this implies that $\mathbf{z}^\star$ is also an ϵ-approximate solution. Since the algorithm reports the feasible query point with the minimum objective value, the reported point must also be an ϵ-approximate solution.

If there are integer variables and if the algorithm enters the recursions in Steps 2b or 6, then by Theorem 4.4.6 and the induction hypothesis, we are guaranteed to get an ϵ-approximate solution, or correctly report infeasibility. Thus, we may assume that there are no recursions, i.e., we always enter Step 2c in the While loop of Step 2 and the algorithm ends with Step 5. The While loop of Step 2 ends with an ellipsoid in $\mathbb{R}^n \times \mathbb{R}^d$ of volume strictly less than $\left(\frac{\epsilon\min\{\rho,1\}}{4MR}\right)^{n+d}$. Since we end with Step 5, the projection of the ellipsoid on to the space of the integer variables, i.e., $\mathbb{R}^n$, is greater than 1. By the Rogers–Shephard inequality (Theorem 3.5.9), this implies that the optimal (in fact, any) integer fiber intersected with the final ellipsoid has d-dimensional volume strictly less than $\left(\frac{\epsilon\min\{\rho,1\}}{2MR}\right)^{n+d} \leq \left(\frac{\epsilon\rho}{MR}\right)^d$. By the same argument as in the case with no integer variables above, either the algorithm correctly reports infeasibility (if no feasible point has been encountered), or one of the functional queries must be at an ϵ-approximate solution and so must be the reported point.

We now analyze the worst-case complexity. Since there are at most n integer variables, the recursion in the algorithm can be of depth at most n. Since at each level of recursion at most $O(n^{3/2}(n+d))$ recursive calls are made (see Steps 2b and 6). Thus, we have at most $O\left((n^{3/2}(n+d))^n\right) = 2^{O(n\log(n+d))}$ recursive calls. Within each recursive call, let us bound the number of iterations the While loop in Step 2 can go for. The initial ellipsoid from Step 1 has volume at most $2^{n+d}R^{(n+d)}(n+d)^{(n+d)/2}$ (we bound the volume of a sphere by the $\|\cdot\|_\infty$ ball of radius $R\sqrt{n+d}$). We stop at the first iteration where the volume falls below $\left(\frac{\epsilon\min\{\rho,1\}}{2MR}\right)^{(n+d)}$. At every iteration, if we do not go into a recursive call, we update the ellipsoid in Step 2c. By Theorem 2.7.6, the new ellipsoid has volume reduced by at least $e^{-\frac{1}{2(n+d+1)^3}}$ compared to the previous ellipsoid. Thus, we can have at most $O\left((n+d)^4\log\left(\frac{MR(n+d)}{\min\{\rho,1\}\epsilon}\right)\right)$ iterations of the While loop.

The final thing to verify is that the closest mixed-integer point and shortest lattice vector calculations also have worst-case complexity bounded by $O((n+d)^3) + 2^{O(n)}\log\left(\frac{MR}{\min\{\rho,1\}\epsilon}\right)$. This will follow from Theorem 6.3.31 and Corollary 6.3.34 as long as we can bound the condition number (ratio of largest to smallest eigenvalues) of the positive definite matrices involved and the norms of the centers. Note that the starting condition number is 1 and the starting norm is 0. By Exercise 14 from Section 2.7.2, in every update of the ellipsoid the condition number increases by at most $\frac{\sigma'}{\sigma} = \frac{(n+d)(n+d+1)}{\sqrt{(n+d-1)(n+d)(n+d+1)(n+d+2)}}$, where σ,σ' are defined in (2.7.2), where the dimension d must be changed to $n+d$ and $\beta = \frac{1}{n+d+1}$. By the same exercise, the norm of the center increases by at most $\frac{\|As\|}{\sqrt{s^T As}}$ where s is the unit norm vector that defines the halfspace for the ellipsoidal update. Since $\frac{\|As\|}{\sqrt{s^T As}} \le \frac{\lambda_{\max}(A)}{\lambda_{\min}(A)}$ and the number of iterations is bounded by $O\left((n+d)^4\log\left(\frac{MR(n+d)}{\min\{\rho,1\}\epsilon}\right)\right)$, we obtain the desired result. $\square$

Remark 6.3.37 (Pure continuous case)   Stronger results can be obtained in the pure continuous case, i.e., $n = 0$. First, in Step 2 of the algorithm, we can use $\beta = 0$ instead of $\beta = \frac{1}{n+d+1}$, reducing the volume of the ellipsoid by a factor $e^{-\frac{1}{2(n+d+1)}} = e^{-\frac{1}{2(d+1)}}$ every time. Thus we make a factor of $(n+d+1)^2$ less number of iterations in Step 2 of the algorithm. Moreover, no CVP/SVP computations are needed and no recursion is needed, i.e., Steps 2a, 2b, 4, 5, and 6 are not executed. Thus, the algorithm's information complexity is

$O\left(d^2 \log\left(\frac{MRd}{\rho\epsilon}\right)\right)$ with an additional computational overhead of $O(d^2)$ per iteration for computing the new ellipsoids. This is the classical *ellipsoid algorithm* for convex optimization. Thus, one obtains an ϵ-approximation algorithm for the optimization problem that differs only by a factor of the dimension d from the ϵ-information complexity bound given in Theorem 6.2.1.[4] Vaidya [226] designed an algorithm whose information complexity matches Theorem 6.2.1's ϵ-information complexity bound of $O\left(d \log\left(\frac{MR}{\rho\epsilon}\right)\right)$, with the same overall complexity as the ellipsoid algorithm. See [11, 148, 163] for improvements in the overall complexity of Vaidya's algorithm. Lemma 2.7.1 with $\beta > 0$ is also used in continuous convex optimization under the name of the *shallow cut ellipsoid method*; see [123] for details.

Remark 6.3.38 (Pure integer case)   For the pure integer case, i.e., $d = 0$ one can strengthen the complexity bound in Theorem 6.3.36 and also report exact optimal solutions, as opposed to ϵ-approximate solutions (recall that $\rho = +\infty$ for the pure integer case as per Definition 6.1.4, and the bound stated in Theorem 6.3.36 becomes $2^{O(n \log(n))} \operatorname{poly}\left(\log\left(\frac{MR}{\epsilon}\right)\right)$). One observes that in Step 2, the stopping condition for the While loop can be modified to the condition that the volume of the ellipsoid falls below 1. Then the algorithm directly goes to Step 6 and recurses on dimension. Thus, there is no dependence on the parameters ϵ and M and we obtain an exact optimization algorithm with information complexity $2^{O(n \log(n))} \log(R)$ and overall complexity $2^{O(n \log(n))} \operatorname{poly}(\log(nR))$.

6.3.3 Exercises

1. Let $Q \in \mathbb{R}^{k \times k}$ be a positive definite matrix and let $L \subseteq \mathbb{R}^k$ be a linear subspace. Show that $L_Q^\perp = \{\mathbf{y} \in \mathbb{R}^k : \langle \mathbf{x}, \mathbf{y}\rangle_Q = 0 \quad \forall \mathbf{x} \in L\}$ is a linear subspace.

2. Let $Q \in \mathbb{R}^{k \times k}$ be a positive definite matrix and let $\mathbf{c} \in \mathbb{R}^k$. Show that

$$\frac{1}{2}\mathbf{z}^T Q\mathbf{z} - \mathbf{c}^T\mathbf{z} = \frac{1}{2}N_Q(\mathbf{z} - \mathbf{t})^2 - \frac{1}{2}N_{Q^{-1}}(\mathbf{c})^2$$

   for all $\mathbf{z} \in \mathbb{R}^k$, where $\mathbf{t} := Q^{-1}\mathbf{c}$.

3. Let $f(\mathbf{z}) = \frac{1}{2}\mathbf{z}^T Q\mathbf{z} - \mathbf{b}^T\mathbf{z}$, where $Q \in \mathbb{R}^{k \times k}$ is a positive definite matrix and $\mathbf{b} \in \mathbb{R}^k$. Show that if one has access to an algebraic oracle that reports

---

[4] There is a slight discrepancy because of the use of the $\|\cdot\|_\infty$-norm for the information complexity bound (see Theorem 6.2.1), and the use of $\|\cdot\|_2$-norm here. This adds a $\log(d)$ factor to the complexity of the ellipsoid algorithm, compared to the information complexity bound.

any entry of Q or $\mathbf{b}$, then one can implement a first-order functional query oracle for f by making $O(k^2)$ calls to the algebraic oracle.

4. Prove Theorem 6.3.5 and Lemma 6.3.7.

5. Show that the set of vectors $\mathbf{s}^1, \ldots, \mathbf{s}^p$ defined in the proof of Theorem 6.3.8 are Q-conjugate and for all $i = 1, \ldots, p$, $\mathrm{span}(\mathbf{s}^1, \ldots, \mathbf{s}^i) = \mathrm{span}(\mathbf{z}^1, \ldots, \mathbf{z}^i)$.

6. Let $Q \in \mathbb{R}^{k \times k}$ be a positive definite matrix and $\mathbf{b} \in \mathbb{R}^k$. Define $f(\mathbf{z}) = \frac{1}{2}\mathbf{z}^T Q\mathbf{z} - \mathbf{b}^T\mathbf{z}$. Let L be an arbitrary linear subspace of $\mathbb{R}^k$. Consider all possible translates of this linear subspace, i.e., $L_\mathbf{x} := \mathbf{x} + L$ for $\mathbf{x} \in \mathbb{R}^k$. For any such translate $L_\mathbf{x}$, consider the minimizer of f restricted to $L_\mathbf{x}$. Then the set of all such minimizers forms a translate of $L_Q^\perp$.

7. Prove Lemmas 6.3.10, 6.3.11, and 6.3.14.

8. Verify that the Final version of the Conjugate Gradient method has total complexity $O(d^3)$ in the arithmetic model of computation.

9. Let $Q \in \mathbb{R}^{n \times n}$ be a positive definite matrix and $\mathbf{t} \in \mathbb{R}^n$. Show that $\mathbf{z}^\star \in \mathrm{argmin}\{N_Q(\mathbf{z} - \mathbf{t}) : \mathbf{z} \in \mathbb{Z}^n\}$ if and only if $B\mathbf{z}^\star \in \mathrm{argmin}\{\|\mathbf{y} - B\mathbf{t}\|_2 : \mathbf{y} \in Z(B)\}$, where B is any square root of Q (see Definition 1.2.17).

10. Let $\mathbf{v}^1, \ldots, \mathbf{v}^k$ be an orthogonal basis for a lattice Λ. Let $\mathbf{t} \in \mathrm{span}(\Lambda)$. Show that the closest lattice point in Λ to $\mathbf{t}$ is given by $\lfloor\lambda_1\rceil\mathbf{v}^1 + \cdots + \lfloor\lambda_k\rceil\mathbf{v}^k$, where $\lambda_1, \ldots, \lambda_k \in \mathbb{R}$ are the coefficients that express $\mathbf{t}$ in the basis $\mathbf{v}^1, \ldots, \mathbf{v}^k$.

11. Show that the orthogonality defect of a basis of a lattice is 1 if and only if the basis vectors are orthogonal.

12. Construct a lattice in $\mathbb{R}^2$ that does not have an orthogonal basis.

13. Prove Lemmas 6.3.16 and 6.3.17.

14.* Show that if B is an $n \times r$ matrix with rational entries ($r \leq n$), then the LLL algorithm (Definition 6.3.18) has worst-case complexity $\mathrm{poly}(n, r, L)$, where L is the sum of the binary encoding sizes of the entries of B.

15. Use Theorem 6.3.22 to show that if $\mathbf{z}$ is the output of the nearest lattice plane algorithm (Definition 6.3.21) with input $B \in \mathbb{R}^{n \times r}$ and $\mathbf{t} \in \mathbb{R}^n$, then $\|\mathbf{z} - \mathbf{t}\| \leq r 2^{r/2}\|CV(Z(B), \mathbf{t}) - \mathbf{t}\|$, i.e., the nearest lattice plane algorithm returns a $r2^{r/2}$-approximation to the closest lattice vector.

16. Complete the proof of Theorem 6.3.28.

17.* Verify (6.3.8).

18.* Prove Lemma 6.3.29.

19.* Show that if Q is an $(n+d) \times (n+d)$ positive definite matrix with rational entries and $\mathbf{c} \in \mathbb{R}^n \times \mathbb{R}^d$ (or $\mathbf{t} \in \mathbb{R}^n \times \mathbb{R}^d$) is a vector with rational coordinates, then the algorithm from Definition 6.3.30 for minimizes $\frac{1}{2}\mathbf{z}^T Q\mathbf{z} - \mathbf{c}^T\mathbf{z}$ over $\mathbb{Z}^n \times \mathbb{R}^d$ (or computing the closest point in $\mathbb{Z}^n \times \mathbb{R}^d$ to

$\mathbf{t}$ in the norm N_Q) has worst-case complexity $2^{O(n)}\,\mathrm{poly}(d, L)$, where L is the total binary encoding sizes of the entries of Q and $\mathbf{c}$ (or $\mathbf{t}$).

20.* Let Q be an $(n + d) \times (n + d)$ positive definite matrix. Let B be the matrix defined in (6.3.9) and define $B^\star$ and D as in (6.3.10) and (6.3.12), respectively. Show that the maximum squared Euclidean norm of a column of $\begin{bmatrix} \mathbf{0}_{d \times n} \\ \sqrt{D_{n \times n}} R_{n \times n} \end{bmatrix}$ is upper bounded by $\lambda_{\max}(Q)$, where $R_{n \times n}$ is the $n \times n$ submatrix of R obtained from the last n rows and columns and $D_{n \times n}$ is the $n \times n$ submatrix of D obtained from the last n rows and columns.

6.4 The Branch-and-Cut Method

The algorithm in Definition 6.3.35, which we shall call a "Lenstra style algorithm" below, is a special case of a general framework that is termed as *branch-and-cut*. Branch-and-cut algorithms for mixed-integer convex optimization are based on two main ideas. The first one, called *branching*, is a way to systematically explore different parts of the feasible region. This was done in the Lenstra style algorithm in the recursions in Steps 2b and 6. The second aspect, that of *cutting planes*, is useful when one is working with a superset of the feasible region and uses separating hyperplanes to remove parts of this superset that do not contain feasible or optimal points. Such a superset of the feasible points is called a *relaxation of the feasible region*.

Definition 6.4.1   A *disjunction* for $\mathbb{Z}^n \times \mathbb{R}^d$ is a union of closed convex sets $D = Q_1 \cup \cdots \cup Q_k$ such that $\mathbb{Z}^n \times \mathbb{R}^d \subseteq D$. A family $\mathcal{D}$ of disjunctions is called a *branching scheme*.

Example 6.4.2   The most widely studied example of disjunctions for $\mathbb{Z}^n \times \mathbb{R}^d$ is the family of *split disjunctions* that are of the form $\{\mathbf{z} \in \mathbb{R}^n \times \mathbb{R}^d : \langle \mathbf{p}, \mathbf{z} \rangle \le K\} \cup \{\mathbf{z} \in \mathbb{R}^n \times \mathbb{R}^d : \langle \mathbf{p}, \mathbf{z} \rangle \ge K + 1\}$, where $\mathbf{p} \in \mathbb{Z}^n \times \{0\}^d$ and $K \in \mathbb{Z}$. When the first n coordinates of $\mathbf{p}$ correspond to a standard unit vector, we get *variable disjunctions*, i.e., disjunctions of the form $\{\mathbf{z} : \mathbf{z}_i \le K\} \cup \{\mathbf{z} : \mathbf{z}_i \ge K + 1\}$, for $i = 1, \ldots, n$. Several researchers in this area have also considered the intersection of t different split disjunctions to get a disjunction [86, 88, 166]; these are known as *t-branch split disjunctions*.

Definition 6.4.3   For any set X in some Euclidean space, a *cutting plane* for X is a halfspace H such that $X \subseteq H$; i.e., a cutting plane is simply another term for valid inequality/halfspace (Definition 2.4.1) in the context of optimization. If X is of the form $C \cap (\mathbb{Z}^n \times \mathbb{R}^d)$, then the cutting plane is *trivial* if $C \subseteq H$, while it is said to be *nontrivial* otherwise. A *cutting-plane paradigm* is a map $\mathcal{CP}$ that takes as input any closed, convex set C and $\mathcal{CP}(C)$ is a family of cutting planes for $C \cap (\mathbb{Z}^n \times \mathbb{R}^d)$. $\mathcal{CP}(C)$ may contain trivial and/or nontrivial cutting planes, and it may even be empty for certain inputs C.

The words "trivial" and "nontrivial" are used here in a purely technical sense. For a complicated convex set C, we may have a simple polyhedral relaxation $R \supseteq C$ such as those used in the proofs of upper bounds in Theorem 6.2.1, and the separation oracle for C can return *trivial* cutting planes that shave off parts of R. Similarly, in Lenstra style algorithms an ellipsoidal relaxation was maintained at every recursive level and the separation oracle was used to obtain trivial cutting planes that reduced the volume of the ellipsoidal relaxations. However, if the separation oracle is difficult to implement, there may be nothing trivial about obtaining such a cutting plane. Our terminology comes from settings where C has a simple description and separating from C is not a big deal; rather, the interesting work is in removing parts of C that do not contain any point from $X = C \cap (\mathbb{Z}^n \times \mathbb{R}^d)$.

Example 6.4.4   1. As mentioned above, for any convex set C contained in a polyhedron P, a separation oracle for C can return trivial cutting planes if a point from $P \setminus C$ is queried.

2. *Chvátal–Gomory cutting-plane paradigm:* given any convex set $C \subseteq \mathbb{R}^n \times \mathbb{R}^d$, define

$$\mathcal{CP}(C) := \{H' : H' = \operatorname{conv}(H \cap (\mathbb{Z}^n \times \mathbb{R}^d)),$$
$$H \text{ rational halfspace with } H \supseteq C\}.$$

A standard way of obtaining such cutting planes is by starting with halfspaces H of the form $\{(\mathbf{x}, \mathbf{y}) \in \mathbb{Z}^n \times \mathbb{R}^d : \langle \mathbf{a}, \mathbf{x} \rangle \leq \delta\}$ for some $\mathbf{a} \in \mathbb{Z}^n$ with relatively prime coordinates and $\delta \notin \mathbb{Z}$, in which case $H' = \{(\mathbf{x}, \mathbf{y}) \in \mathbb{Z}^n \times \mathbb{R}^d : \langle \mathbf{a}, \mathbf{x} \rangle \leq \lfloor \delta \rfloor\}$.

3. *Disjunctive cuts:* given any family of disjunctions (branching scheme) $\mathcal{D}$, the *disjunctive cutting-plane paradigm based on $\mathcal{D}$* is defined as

$$\mathcal{CP}(C) := \{H' \text{ halfspace} : H' \supseteq C \cap D, \ D \in \mathcal{D}\}.$$

The collection of halfspaces H' valid for $C \cap D$ are said to be the *cutting planes derived from the disjunction D*. These are cutting planes since $\mathbb{Z}^n \times \mathbb{R}^d \subseteq D$ by definition of a disjunction, and therefore $C \cap (\mathbb{Z}^n \times \mathbb{R}^d) \subseteq C \cap D \subseteq H'$. A disjunction D produces nontrivial cutting planes for a compact, convex set C if and only if at least one extreme point of C is not contained in D, due to the Krein–Milman theorem (Theorem 2.4.16).

A branch-and-cut algorithm is based on a family of disjunctions (branching scheme) and a cutting-plane paradigm. The general framework is as follows.

Definition 6.4.5 (Branch-and-cut framework based on a branching scheme $\mathcal{D}$ and cutting-plane paradigm $\mathcal{CP}$)

Input: A closed, convex set $C \subseteq \mathbb{R}^n \times \mathbb{R}^d$, a convex function $f : \mathbb{R}^n \times \mathbb{R}^d \to \mathbb{R} \cup \{+\infty\}$, error guarantee $\epsilon > 0$, and a relaxation $X \subseteq \mathbb{R}^n \times \mathbb{R}^d$ which is closed, convex and contains $C \cap \mathrm{dom}(f)$.

Output: A point $\mathbf{z}^\star \in C \cap \mathrm{dom}(f) \cap (\mathbb{Z}^n \times \mathbb{R}^d)$ such that $f(\mathbf{z}^\star) \leq f(\mathbf{z}) + \epsilon$ for all $\mathbf{z} \in C \cap \mathrm{dom}(f) \cap (\mathbb{Z}^n \times \mathbb{R}^d)$.

1. Initialize a set $L = \{X\}$. Initialize $UB = +\infty$.
2. While $L \neq \emptyset$ do:

   (a) [Node selection] Select an element $N \in L$.
   (b) [Pruning] If it can be verified that $\inf\{f(\mathbf{z}) : \mathbf{z} \in N\} \geq UB - \epsilon$, then update $L := L \setminus \{N\}$ and continue the While loop.
   Else, select a test point $\hat{\mathbf{z}}$ in N.
   (c) If $\hat{\mathbf{z}} \in C \cap \mathrm{dom}(f) \cap (\mathbb{Z}^n \times \mathbb{R}^d)$, obtain a subgradient $\mathbf{s} \in \partial f(\hat{\mathbf{z}})$ and add the subgradient halfspace
   $H = \{\mathbf{z} : \langle \mathbf{s}, \mathbf{z} - \hat{\mathbf{z}} \rangle \leq 0\}$ to all the elements in L; i.e., update
   $N' \supseteq N \cap H$ for all $N \in L$.
   Additionally, if $f(\hat{\mathbf{z}}) < UB$, then update $UB = f(\hat{\mathbf{z}})$ and set $\mathbf{z}^\star = \hat{\mathbf{z}}$.
   (d) If $\hat{\mathbf{z}} \notin C \cap \mathrm{dom}(f) \cap (\mathbb{Z}^n \times \mathbb{R}^d)$, decide whether to BRANCH or CUT.
   If BRANCH, then choose a disjunction $D = Q_1 \cup \cdots \cup Q_k$ in $\mathcal{D}$ such that $\hat{\mathbf{z}} \notin D$.
   Select sets (relaxations) $N_1, \ldots, N_k$ such that $N \cap Q_i \subseteq N_i$.
   Update $L := L \setminus \{N\} \cup \{N_1, \ldots, N_k\}$.
   If CUT, then choose a cutting plane $H \in \mathcal{CP}(C \cap \mathrm{dom}(f) \cap N)$ such that $\hat{\mathbf{z}} \notin H$.
   Select a set (relaxation) N' such that $N \cap H \subseteq N'$. Update
   $L := L \setminus \{N\} \cup \{N'\}$.

3. If $UB = +\infty$ (i.e., no test point was in $C \cap \mathrm{dom}(f) \cap (\mathbb{Z}^n \times \mathbb{R}^d)$), return "INFEASIBLE." Else return $\mathbf{z}^\star$.

We obtain a specific branch-and-cut procedure once we specify the following things in the framework above.

1. In Step 2a, we must decide on a strategy to select an element from L. In the case of the algorithm from Definition 6.3.35, this would be the choice of a lower-dimensional section of C to recurse on.

2. In Step 2b, we must decide on a strategy to verify the condition $\inf\{f(\mathbf{z}) : \mathbf{z} \in N\} \geq UB - \epsilon$. In the case of the algorithm from Definition 6.3.35, this is determined by a volume condition on N (which is an ellipsoid). Another common strategy is used in linear integer optimization, where N is a polyhedron and linear optimization methods like the simplex method or an interior-point algorithm is used to determine $\inf\{f(\mathbf{z}) : \mathbf{z} \in N\}$. More generally, one could have a continuous convex optimization subroutine suitable for the class of problems under study.

3. In Step 2b, if one has $\inf\{f(\mathbf{z}) : \mathbf{z} \in N\} < UB - \epsilon$, then a test point $\hat{\mathbf{z}} \in N$ must be selected and one must have a procedure/subroutine for this. In the algorithm from Definition 6.3.35, this was chosen using a quadratic minimization subroutine to select the closest mixed-integer point to the center of the ellipsoid N. In most mixed-integer optimization solvers, this test point is taken as an optimal or ϵ-approximate solution to the convex optimization problem $\inf\{f(\mathbf{z}) : \mathbf{z} \in C \cap N\}$ or $\inf\{f(\mathbf{z}) : \mathbf{z} \in N\}$.

4. In Step 2d, one must have a strategy for deciding whether to branch or to cut and, in either case, have a strategy for selecting a disjunction or a cutting plane. The decision to branch might fail because there is no disjunction D in the chosen branching scheme $\mathcal{D}$ that does not contain $\hat{\mathbf{z}}$. In such a case, we may choose to add a cutting plane instead or simply continue the While loop. In the algorithm from Definition 6.3.35, the branching scheme used was the family of split disjunctions defined in Example 6.4.2: the sections of C can be seen as branching on the disjunctions $\{(\mathbf{x}, \mathbf{y}) : \langle \mathbf{w}^\star, \mathbf{x} \rangle \leq j\} \cup \{(\mathbf{x}, \mathbf{y}) : \langle \mathbf{w}^\star, \mathbf{x} \rangle \geq j + 1\}$ for different values of $j \in \mathbb{Z}$. The algorithm from Definition 6.3.35 differs slightly from the branch-and-cut framework in Definition 6.4.5 because the branching may happen even if the test point is in $C \cap \mathrm{dom}(f) \cap (\mathbb{Z}^n \times \mathbb{R}^d)$, since the decision is made based on the distance of the closest mixed-integer point to the center of the current ellipsoid (in the norm induced by the ellipsoid).

   On the other hand, if the decision is to add a cutting plane in Step 2d, one may add a trivial cutting plane valid for $C \cap N$, as was done in the algorithm from Definition 6.3.35. One may also fail to find a cutting plane that removes $\hat{\mathbf{z}}$, either because the cutting-plane paradigm can produce no such cutting plane, i.e., $\mathcal{CP}(C \cap N) = \emptyset$, or because the strategy chosen fails to find such a cutting plane in $\mathcal{CP}(C \cap N)$ even though one exists. In such a case, we may decide to branch instead or simply continue the While loop.

   Finally, in Steps 2c and 2d we must have a strategy to select the relaxations $N_1, \ldots, N_k$ of $N \cap Q_1, \ldots, N \cap Q_k$ respectively, if the decision is to branch, or we must have a strategy to select a relaxation N' of $N \cap H$ if the decision is to cut. In the algorithm from Definition 6.3.35, these relaxations were taken as ellipsoids. In most solvers, the relaxations N_i are simply taken to be $N \cap Q_i$ in a decision to branch, and the relaxation N' is simply taken to be $N \cap H$ in a decision to cut.

5. If an algorithm based on the branch-and-cut framework above decides never to branch in Step 2d, it is called a *pure cutting-plane* algorithm. If an algorithm decides never to use cutting planes in Step 2d, it is called a *pure branch-and-bound* algorithm.

6.4.1 The Relationship between Branching and Cutting Planes

In practice, pruning and nontrivial cutting planes make a huge difference [47, 48, 167]. Turning these off will make most of the state-of-the-art mixed-integer optimization solvers take an inordinate amount of time for even small-scale problems. Nevertheless, from a theoretical perspective, researchers have not been able to improve upon the complexity of the algorithm from Definition 6.3.35 by utilizing pruning and nontrivial cutting planes for the general problem; see [30, 99, 179] for examples of positive results in restricted settings. Another empirical fact is that if branching is completely turned off and only cutting planes are used, then again the solvers' performance degrades massively. Recently, some results have been obtained that provide some theoretical basis to these empirical observations that the combination of branching and cutting planes performs significantly better than branching alone or using cutting planes alone. We present some of these results now.

The next definition is inspired by the following simple intuition. It has been established that certain branching schemes can be simulated by certain cutting-plane paradigms in the sense that for the problem class under consideration, if we have a pure branch-and-bound algorithm based on the branching scheme, then there exists a pure cutting-plane algorithm for the same class that has complexity at most a polynomial factor worse than the branch-and-bound algorithm. Similarly, there are results that establish the reverse. See [31, 32, 39, 83, 84, 117] for results of this type. In such situations, combining branching and cutting planes into branch-and-cut is likely to give no substantial improvement since one method can always do the job of the other, up to polynomial factors.

Definition 6.4.6   Let $\mathcal{I}$ be a family of mixed-integer convex optimization problems of the form (6.1.1) that has a representation in the arithmetic model of computation[5] (see Definition 1.4.1). To make the following discussion

---

[5] The entire discussion works in the Turing machine model of computation as well.

easier, we assume that the objective function is linear. This is without loss of generality since we can introduce an auxiliary variable v, introduce the epigraph constraint $f(z) \leq v$, and use the linear objective "inf v."

A cutting-plane paradigm $\mathcal{CP}$ and a branching scheme $\mathcal{D}$ are *complementary for $\mathcal{I}$* if there is a family of instances $\mathcal{I}_{\mathcal{CP}<\mathcal{D}} \subseteq \mathcal{I}$ such that there is a pure cutting-plane algorithm based on $\mathcal{CP}$ that has polynomial (in the encoding size of the instances) complexity and any branch-and-bound algorithm based on $\mathcal{D}$ is exponential (in the encoding size of the instances), and there is another family of instances $\mathcal{I}_{\mathcal{CP}>\mathcal{D}} \subseteq \mathcal{I}$ where $\mathcal{D}$ gives a polynomial complexity pure branch-and-bound algorithm while any pure cutting-plane algorithm based on $\mathcal{CP}$ is exponential.

We wish to formalize the intuition that branch-and-cut is expected to be exponentially better than branch-and-bound or cutting planes alone for complementary pairs of branching schemes and cutting-plane paradigms. For this we need some mild assumptions about the branching schemes and cutting-plane paradigms. All known branching schemes and cutting-plane methods from the literature satisfy the following conditions.

Definition 6.4.7 A branching scheme is said to be *regular* if no disjunction involves a continuous variable, i.e., each closed, convex set in the disjunction is of the form $C \times \mathbb{R}^d$ for some closed, convex set $C \subseteq \mathbb{R}^n$.

A branching scheme $\mathcal{D}$ is said to be *embedding closed* if disjunctions from higher dimensions can be applied to lower dimensions. More formally, let n_1, $n_2, d_1, d_2 \in \mathbb{N}$. If $D \in \mathcal{D}$ is a disjunction in $\mathbb{R}^{n_1} \times \mathbb{R}^{d_1} \times \mathbb{R}^{n_2} \times \mathbb{R}^{d_2}$ with respect to $\mathbb{Z}^{n_1} \times \mathbb{R}^{d_1} \times \mathbb{Z}^{n_2} \times \mathbb{R}^{d_2}$, then the disjunction $D \cap (\mathbb{R}^{n_1} \times \mathbb{R}^{d_1} \times \{0\}^{n_2} \times \{0\}^{d_2})$, interpreted as a set in $\mathbb{R}^{n_1} \times \mathbb{R}^{d_1}$, is also in $\mathcal{D}$ for the space $\mathbb{R}^{n_1} \times \mathbb{R}^{d_1}$ with respect to $\mathbb{Z}^{n_1} \times \mathbb{R}^{d_1}$ (note that $D \cap (\mathbb{R}^{n_1} \times \mathbb{R}^{d_1} \times \{0\}^{n_2} \times \{0\}^{d_2})$, interpreted as a set in $\mathbb{R}^{n_1} \times \mathbb{R}^{d_1}$, is certainly a disjunction with respect to $\mathbb{Z}^{n_1} \times \mathbb{R}^{d_1}$; we want $\mathcal{D}$ to be closed with respect to such restrictions).

A cutting-plane paradigm $\mathcal{CP}$ is said to be *regular* if it has the following property, which says that adding "dummy variables" to the formulation of the instance should not change the power of the paradigm. Formally, let $C \subseteq \mathbb{R}^n \times \mathbb{R}^d$ be any closed, convex set and let $C' = \{(\mathbf{x},t) \in \mathbb{R}^n \times \mathbb{R}^d \times \mathbb{R} : \mathbf{x} \in C,\ t = \langle \mathbf{c},\mathbf{x} \rangle\}$ for some $\mathbf{c} \in \mathbb{R}^n$. Then if a cutting plane $\langle \mathbf{a},\mathbf{x} \rangle \leq b$ for $C \cap (\mathbb{Z}^n \times \mathbb{R}^d)$ is derived by $\mathcal{CP}$ applied to C, i.e., this inequality is in $\mathcal{CP}(C)$, then it should also be in $\mathcal{CP}(C')$ as a cutting plane for $C' \cap (\mathbb{Z}^n \times \mathbb{R}^d \times \mathbb{R})$, and conversely, if $\langle \mathbf{a},\mathbf{x} \rangle + \mu t \leq b$ is in $\mathcal{CP}(C')$ as a cutting plane for $C' \cap (\mathbb{Z}^n \times \mathbb{R}^d \times \mathbb{R})$, then the equivalent inequality $\langle \mathbf{a} + \mu\mathbf{c},\mathbf{x} \rangle \leq b$ should be in $\mathcal{CP}(C)$ as a cutting plane for $C \cap (\mathbb{Z}^n \times \mathbb{R}^d)$.

A cutting-plane paradigm $\mathcal{CP}$ is said to be *embedding closed* if cutting planes from higher dimensions can be applied to lower dimensions. More formally, let $n_1, n_2, d_1, d_2 \in \mathbb{N}$. Let $C \subseteq \mathbb{R}^{n_1} \times \mathbb{R}^{d_1}$ be any closed, convex set. If the inequality $\langle \mathbf{c}_1, \mathbf{x}_1 \rangle + \langle \mathbf{a}_1, \mathbf{y}_1 \rangle + \langle \mathbf{c}_2, \mathbf{x}_2 \rangle + \langle \mathbf{a}_2, \mathbf{y}_2 \rangle \leq \gamma$ is a cutting plane for $C \times \{0\}^{n_2} \times \{0\}^{d_2}$ with respect to $\mathbb{Z}^{n_1} \times \mathbb{R}^{d_1} \times \mathbb{Z}^{n_2} \times \mathbb{R}^{d_2}$ that can be derived by applying $\mathcal{CP}$ to $C \times \{0\}^{n_2} \times \{0\}^{d_2}$, then the cutting plane $\langle \mathbf{c}_1, \mathbf{x}_1 \rangle + \langle \mathbf{a}_1, \mathbf{y}_1 \rangle \leq \gamma$ that is valid for $C \cap (\mathbb{Z}^{n_1} \times \mathbb{R}^{d_1})$ should also belong to $\mathcal{CP}(C)$.

A cutting-plane paradigm $\mathcal{CP}$ is said to be *inclusion closed* if for any two closed convex sets $C \subseteq C'$, we have $\mathcal{CP}(C') \subseteq \mathcal{CP}(C)$. In other words, any cutting plane derived for C' can also be derived for a subset C.

Theorem 6.4.8 (31, Theorem 1.12) *Let $\mathcal{D}$ be a regular, embedding closed branching scheme and let $\mathcal{CP}$ be a regular, embedding closed, and inclusion closed cutting-plane paradigm such that $\mathcal{D}$ includes all variable disjunctions and $\mathcal{CP}$ and $\mathcal{D}$ form a complementary pair for a mixed-integer convex optimization problem class $\mathcal{I}$. Then there is a family of instances in $\mathcal{I}$ such that there exists a polynomial complexity branch-and-cut algorithm, whereas any branch-and-bound algorithm based on $\mathcal{D}$ and any cutting-plane algorithm based on $\mathcal{CP}$ are of exponential complexity.*

The rough idea of the proof of Theorem 6.4.8 is to embed pairs of instances from $\mathcal{I}_{\mathcal{CP}>\mathcal{D}}$ and $\mathcal{I}_{\mathcal{CP}<\mathcal{D}}$ as faces of a convex set such that a single variable disjunction results in these instances as the subproblems in a branch-and-cut algorithm. On one subproblem, one uses cutting planes and on the other subproblem one uses branching. However, since $\mathcal{CP}$ and $\mathcal{D}$ are complementary, a pure cutting plane or pure branch-and-bound algorithm takes exponential time in processing one or the other of the faces. The details get technical and the reader is referred to [31].

Example 6.4.9   We now present a concrete example of a complementary pair that satisfies the conditions of Theorem 6.4.8. Let $\mathcal{I}$ be the family of mixed-integer linear optimization problems described in point 2. of Example 5.1.2, with standard "binary encoding" oracles described in point 2. of Example 5.2.2 and the parameterization based on encoding size (point 1. of Example 5.4.2). Let $\mathcal{CP}$ be the Chvátal–Gomory paradigm described in point 2. in Example 6.4.4, and $\mathcal{D}$ to be the family of variable disjunctions described in Example 6.4.2. They are both regular and embedding closed; the Chvátal–Gomory paradigm is also inclusion closed – see Exercises 1 and 2 from Section 6.4.2.

Consider the so-called *Jeroslow instances*: for every odd natural number $n \in \mathbb{N}$, $\max\{\sum_{i=1}^{n} \mathbf{x}_i : \sum_{i=1}^{n} \mathbf{x}_i \leq \frac{n}{2}, \quad \mathbf{x} \in [0,1]^n, \quad \mathbf{x} \in \mathbb{Z}^n\}$. The single

Chvátal–Gomory cut $\sum_{i=1}^{n} \mathbf{x}_i \leq \lfloor \frac{n}{2} \rfloor$ proves optimality, whereas any branch-and-bound algorithm based on variable disjunctions has complexity at least $2^{\lfloor \frac{n}{2} \rfloor}$ [147]. On the other hand, consider the *Schrijver triangle* $T_h \in \mathbb{R}^2$, where $T_h = \text{conv}\{(0,0),(1,0),(\frac{1}{2},h)\}$ and consider the family of problems for $h \in \mathbb{N}$: $\max\{\mathbf{x}_2 : \mathbf{x} \in T_h, \ \mathbf{x} \in \mathbb{Z}^2\}$. Any Chvátal–Gomory paradigm–based algorithm has complexity at least $\Omega(h)$ [212, Section 23.3], which is exponential in the size of the input $O(\log(h))$. On the other hand, a single disjunction on the variable $\mathbf{x}_1$ solves the problem.

Example 6.4.9 shows that the classical Chvátal–Gomory cuts and variable branching are complementary and thus Theorem 6.4.8 implies that they give rise to a superior branch-and-cut routine when combined, compared to their stand-alone use. The Chvátal–Gomory cutting-plane paradigm and variable disjunctions are the most widely used pairs in state-of-the-art branch-and-cut solvers. We thus have some theoretical basis for explaining the success of this particular combination. The literature on complementary pairs of branching schemes and cutting-plane paradigms is discussed in Section 6.6.

6.4.2 Exercises

1. Show that the branching schemes discussed in Example 6.4.2 are regular and embedding closed (Definition 6.4.7).
2. Show that the cutting-plane paradigms discussed in Example 6.4.4 are regular, embedding closed and inclusion closed (Definition 6.4.7).
3. Show that if a particular realization of the branch-and-cut framework described in Definition 6.4.5 terminates on an instance, then $\mathbf{z}^\star$ is an ϵ-approximate solution for that instance.
4. Show that for the Jeroslow instances described in Example 6.4.9, any branch-and-bound algorithm based on variable disjunctions (Example 6.4.2) must have complexity at least $2^{\lfloor \frac{n}{2} \rfloor}$.
5. Show that for the Schrijver triangle T_h described in Example 6.4.9, one needs to generate at least $\Omega(h)$ Chvátal–Gomory cutting planes (point 2. in Example 6.4.4) such that resulting relaxation has optimal value equal to 0 when maximizing $\mathbf{x}_2$ (which is the correct optimal value for the [pure] integer optimization problem).

6.5 The Special Case of Continuous Convex Optimization

All of the discussion so far has focused on the general mixed-integer case. In this section, we will discuss the purely continuous convex optimization problem with $n = 0$, i.e., we have no integer-constrained variables. In particular, we will discuss the important class of Subgradient descent algorithms which have been shown to work only in this special case.

6.5.1 Cutting-Plane Methods

Let us first revisit the branch-and-cut type algorithms we have already discussed, specialized to this setting. Recall that for establishing the information complexity upper bound, as well as the algorithmic upper bound, we had pure cutting-plane methods (see point 5. in the discussion below Definition 6.4.5 and Remark 6.3.37). Let us rewrite the branch-and-cut framework (Definition 6.4.5) without the extra frills of branching and nontrivial cutting planes for the mixed-integer points.

Definition 6.5.1 (General cutting-plane method for continuous convex optimization)

Input: Access to first-order separation and functional oracle for $\mathcal{I}_{0,d,R,\rho,M}$.

Output: For any instance $(f, C) \in \mathcal{I}_{0,d,R,\rho,M}$, compute a point $\mathbf{z}^\star \in C \cap \mathrm{dom}(f)$ such that $f(\mathbf{z}^\star) \le f(\mathbf{z}) + \epsilon$ for all $\mathbf{z} \in C \cap \mathrm{dom}(f)$, or report infeasibility: $C \cap \mathrm{dom}(f) = \emptyset$.

1. Initialize a closed, convex set $X \subseteq \mathbb{R}^d$ such that $[-R, R]^d \subseteq X$. Initialize $UB = +\infty$.

2. While $\mathrm{vol}(X) > \left(\frac{\epsilon \min\{\rho, 1\}}{MR} \right)^d$:

   (a) Select a test point $\hat{\mathbf{z}}$ in X.

   (b) If $\hat{\mathbf{z}} \in C \cap \mathrm{dom}(f)$, obtain a subgradient $\mathbf{s} \in \partial f(\hat{\mathbf{z}})$ and find a relaxation of the intersection of X with the subgradient halfspace $H = \{\mathbf{z} : \langle \mathbf{s}, \mathbf{z} - \hat{\mathbf{z}} \rangle \le 0\}$, i.e., let E be a suitable closed convex set containing $X \cap H$. Additionally, if $f(\hat{\mathbf{z}}) < UB$, then update $UB = f(\hat{\mathbf{z}})$ and set $\mathbf{z}^\star = \hat{\mathbf{z}}$.

   (c) If $\hat{\mathbf{z}} \notin C \cap \mathrm{dom}(f)$, obtain a separating hyperplane H and let E be a suitable closed convex set containing $X \cap H$.

   (d) Update $X := E$.

3. If $UB = +\infty$ (i.e., no test point was in $C \cap \mathrm{dom}(f)$), return "INFEASIBLE." Else return $\mathbf{z}^\star$.

Information complexity. For the information complexity upper bound of $O\left(d \log \left(\frac{MR}{\epsilon \min\{\rho, 1\}} \right) \right)$, we made the following choices:

- $X = [-R, R]^d$ in the initialization Step 1.
- $\hat{\mathbf{z}}$ was the centroid of X in Step 2a.
- $E = X \cap H$ in Steps 2b and 2c.

The bound then followed from the following argument. We assume $\epsilon \leq 2MR$. Grünbaum's theorem (Theorem 6.2.5 with $n = 0$) implies that at every iteration of the While loop, the volume of X must decrease by at least $(1 - \frac{1}{e})$, where e is Euler's constant. Since $\text{vol}(X) = (2R)^d$ initially, the While loop can go for at most $O\left(d \log\left(\frac{MR}{\epsilon \min\{\rho, 1\}}\right)\right)$ iterations. At the end, X has volume less than $\left(\frac{\epsilon \min\{\rho, 1\}}{MR}\right)^d$. If no test point was in $C \cap \text{dom}(f)$, the algorithm always executed Step 2c. This implies that $C \cap \text{dom}(f) \subseteq X$ at the end and the volume of $C \cap \text{dom}(f)$ is less than $\left(\frac{\epsilon \min\{\rho, 1\}}{MR}\right)^d \leq \rho^d$. From the definition of $\mathcal{I}_{0,d,R,\rho,M}$, this actually implies that $C \cap \text{dom}(f)$ is empty and we have correctly reported the problem to be infeasible. Else, at least one test point was feasible and in $C \cap \text{dom}(f)$. Lemma 6.1.5 implies that C contains an ℓ_∞ ball of radius $\frac{\epsilon \rho}{2MR}$ consisting of ϵ-approximate solutions; this ball must necessarily be in $C \cap \text{dom}(f)$. The volume of this ball is $\left(\frac{\epsilon \rho}{MR}\right)^d \geq \left(\frac{\epsilon \min\{\rho, 1\}}{MR}\right)^d$. Therefore, at least one subgradient halfspace from a test point $\hat{\mathbf{z}}$ must not be valid for this ball and cut off an ϵ-approximate solution $\bar{\mathbf{z}}$, i.e., $\langle \mathbf{s}, \bar{\mathbf{z}} - \hat{\mathbf{z}} \rangle > 0$, where $\mathbf{s}$ is the subgradient returned at $\hat{\mathbf{z}}$. This implies $f(\bar{\mathbf{z}}) \geq f(\hat{\mathbf{z}}) + \langle \mathbf{s}, \bar{\mathbf{z}} - \hat{\mathbf{z}} \rangle > f(\hat{\mathbf{z}})$. So, $\hat{\mathbf{z}}$ must be an ϵ-approximate solution and so must the feasible test point $\mathbf{z}^\star$ with the minimum value that is returned by the algorithm.

Algorithmic complexity. For the algorithmic complexity upper bound of $O\left(d^2 \log\left(\frac{MRd}{\epsilon \min\{\rho, 1\}}\right)\right)$, we made the following choices (Remark 6.3.37):

- $X = B_2(0, R)$ in the initialization Step 1.
- $\hat{\mathbf{z}}$ was the centroid/center of X in Step 2a.
- E is the smallest volume ellipsoid containing $X \cap H$ in Steps 2b and 2c, using the formulas in Exercise 14 from Section 2.7.2 with $\beta = 0$.

The bound then followed from the following argument. Theorem 2.7.6 implies that at every iteration of the While loop, the volume of X must decrease by at least $e^{-\frac{1}{2(d+1)}}$, where e is Euler's constant. Since $\text{vol}(X) = (2R\sqrt{d})^d$ initially, the While loop can go for at most $O\left(d^2 \log\left(\frac{MRd}{\epsilon \min\{\rho, 1\}}\right)\right)$ iterations. At the end, X has volume less than $\left(\frac{\epsilon \min\{\rho, 1\}}{MR}\right)^d$. The argument is now exactly the same as made above for information complexity.

6.5.2 Subgradient Descent

There is a class of methods for continuous convex optimization that are of a very different nature compared to the cutting-plane methods discussed above. For the purpose of this section, we will restrict attention to instances where the objective f is finite valued everywhere and thus subgradients always exist. The methods we will discuss require access to subgradients of f as we have been using, but a *stronger* oracle for C which, given any $\mathbf{x} \in \mathbb{R}^d$, can report the closest point $\mathrm{Proj}_C(\mathbf{x})$ (Section 2.3) in C to $\mathbf{x}$ (in standard Euclidean norm), or reports that C is empty.

Note that an oracle that reports $\mathrm{Proj}_C(\mathbf{x})$ for any $\mathbf{x} \in \mathbb{R}^d$ is stronger than a separation oracle for C, because $\mathrm{Proj}_C(\mathbf{x}) = \mathbf{x}$ if and only if $\mathbf{x} \in C$, and when $\mathrm{Proj}_C(\mathbf{x}) \neq \mathbf{x}$, then one can use $\mathbf{a} = \mathbf{x} - \mathrm{Proj}_C(\mathbf{x})$ and $\delta = \langle \mathbf{a}, \mathrm{Proj}_C(\mathbf{x}) \rangle$ as a separating hyperplane; see the proof of Theorem 2.4.2. Even so, for "simple" sets C, computing $\mathrm{Proj}_C(\mathbf{x})$ is not a difficult task. For example, when $C = \mathbb{R}^d_+$, then $\mathrm{Proj}_C(\mathbf{x}) = \mathbf{y}$, where $\mathbf{y}_i = \max\{0, \mathbf{x}_i\}$ for all $i = 1, \ldots, d$. Note that, in particular, when we have no constraints, i.e., $C = \mathbb{R}^d$, then $\mathrm{Proj}_C(\mathbf{x}) = \mathbf{x}$ for all $\mathbf{x} \in \mathbb{R}^d$. Therefore, this algorithm can be used for *unconstrained optimization of general convex functions* with only a first-order oracle for f.

Definition 6.5.2 (Subgradient descent algorithm)

Input: Access to first-order functional oracle for the objective function f and projection oracle for the feasible region C.

1. Choose any sequence $\alpha_0, \alpha_1, \ldots,$ of strictly positive numbers. Let $\mathbf{x}^0 \in \mathbb{R}^d$.
2. For $i = 0, 1, 2, \ldots,$ do:

   (a) Use the first-order oracle for f to get some $\mathbf{s}^i \in \partial f(\mathbf{x}^i)$.
   (b) Set $\mathbf{x}^{i+1} = \mathrm{Proj}_C\left(\mathbf{x}^i - \alpha_i \frac{\mathbf{s}^i}{\|\mathbf{s}^i\|}\right)$.

The points $\mathbf{x}^0, \mathbf{x}^1, \ldots$ are called the *iterates* of the Subgradient descent algorithm and the numbers $\alpha_0, \alpha_1, \ldots$ are called the *step sizes* used in the algorithm.

Theorem 6.5.3 *Let* $f : \mathbb{R}^d \to \mathbb{R}$ *be a convex function,* $C \subseteq \mathbb{R}^d$ *be a closed, convex set and let* $\mathbf{x}^\star \in \arg\min_{\mathbf{x} \in C} f(\mathbf{x})$. *Suppose* $\mathbf{x}_0 \in B(\mathbf{x}^\star, R)$ *for some real number* $R \geq 0$. *Let* M *be a Lipschitz constant for* f, *guaranteed to exist by Theorem 3.2.3, i.e.,* $|f(\mathbf{x}) - f(\mathbf{y})| \leq M\|\mathbf{x} - \mathbf{y}\|$ *for all* $\mathbf{x}, \mathbf{y} \in B(\mathbf{x}^\star, R)$. *Let* $\mathbf{x}^0, \mathbf{x}^1, \ldots$ *be the sequence of iterates obtained by the Subgradient descent algorithm from Definition 6.5.2. Then, for all* $k \geq 0$,

$$\min_{i=0,\ldots,k} f(\mathbf{x}^i) \leq f(\mathbf{x}^\star) + M\left(\frac{R^2 + \sum_{i=0}^{k} \alpha_i^2}{2\sum_{i=0}^{k} \alpha_i}\right).$$

Proof For $i \geq 0$, define $d_i = \frac{\langle s^i, x^i - x^\star \rangle}{\|s^i\|}$. By Problem 11 from Section 2.1.1, d_i as the Euclidean distance of $x^\star$ from the hyperplane passing through x^i with normal s^i. We next observe that

$$\begin{aligned}
\|x^{i+1} - x^\star\|^2 &= \left\|\text{Proj}_C\left(x^i - \alpha_i \frac{s^i}{\|s^i\|}\right) - x^\star\right\|^2 \\
&\leq \left\|x^i - \alpha_i \frac{s^i}{\|s^i\|} - x^\star\right\|^2 \\
&= \|x^i - x^\star\|^2 + \alpha_i^2 - 2\alpha_i d_i,
\end{aligned}$$

where the inequality follows from Proposition 2.3.3 with $x = x^i - \alpha_i \frac{s^i}{\|s^i\|}$ and $y = x^\star$. Adding these inequalities for $i = 0, 1, \ldots, k$, we obtain that

$$\|x^{k+1} - x^\star\|^2 \leq \|x^0 - x^\star\|^2 + \sum_{i=0}^{k} \alpha_i^2 - 2\sum_{i=0}^{k} \alpha_i d_i. \tag{6.5.1}$$

Let $d_{\min} = \min_{i=0,\ldots,k} d_i$ and let $i^{\min}$ be such that $d_{\min} = d_{i^{\min}}$. Using the fact that $\|x^0 - x^\star\|^2 \leq R^2$, and that $\|x^{k+1} - x^\star\|^2 \geq 0$, we obtain from (6.5.1) that

$$d_{\min}\left(2\sum_{i=0}^{k} \alpha_i\right) \leq 2\sum_{i=0}^{k} \alpha_i d_i \leq R^2 + \sum_{i=0}^{k} \alpha_i^2.$$

Consequently,

$$d_{\min} \leq \frac{R^2 + \sum_{i=0}^{k} \alpha_i^2}{2\sum_{i=0}^{k} \alpha_i}. \tag{6.5.2}$$

Consider the hyperplane $H := H(s^{i^{\min}}, \langle s^{i^{\min}}, x^{i^{\min}}\rangle)$ passing through $x^{i^{\min}}$, orthogonal to $s^{i^{\min}}$. Let $\bar{x}$ be the point on H closest to $x^\star$ and so $d_{\min} = \|\bar{x} - x^\star\|$; see Figure 6.5. Moreover, $d_{\min} \leq d_0 \leq \|x^0 - x^\star\| \leq R$. Therefore, $\bar{x} \in B(x^\star, R)$. Using the Lipschitz constant M, we obtain that $f(\bar{x}) \leq f(x^\star) + M d_{\min}$. Finally, since $s^{i^{\min}} \in \partial f(x^{i^{\min}})$, we must have that $f(\bar{x}) \geq f(x^{i^{\min}}) + \langle s^{i^{\min}}, \bar{x} - x^{i^{\min}}\rangle = f(x^{i^{\min}})$, since $\bar{x} \in H$. Therefore, we obtain

$$\min_{i=0,\ldots,k} f(x^i) \leq f(x^{i^{\min}}) \leq f(\bar{x}) \leq f(x^\star) + M d_{\min}$$

$$\leq f(x^\star) + M\left(\frac{R^2 + \sum_{i=0}^{k} \alpha_i^2}{2\sum_{i=0}^{k} \alpha_i}\right),$$

where the last inequality follows from (6.5.2). $\qquad\square$

If we fix the number of steps of the algorithm to be $N \in \mathbb{N}$, then the choice of step sizes $\alpha_0, \ldots, \alpha_N$ that minimizes $\frac{R^2 + \sum_{i=0}^{k} \alpha_i^2}{2\sum_{i=0}^{k} \alpha_i}$ is where $\alpha_i = \frac{R}{\sqrt{N+1}}$ for all $i = 0, \ldots, N$, which yields the following corollary.

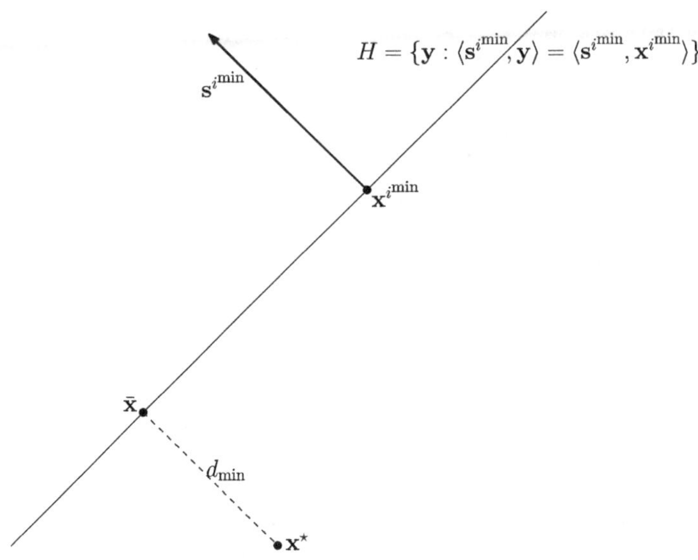

Figure 6.5 Using $d_{\min}$ to bound the function value.

Corollary 6.5.4 *Let $f : \mathbb{R}^d \to \mathbb{R}$ be a convex function, and let $\mathbf{x}^\star \in \arg\min_{\mathbf{x}\in\mathbb{R}^d} f(\mathbf{x})$. Suppose $\mathbf{x}_0 \in B(\mathbf{x}^\star, R)$ for some real number $R \geq 0$. Let $M := M(B(\mathbf{x}^\star, R))$ be a Lipschitz constant for f. Let $N \in \mathbb{N}$ be any natural number, and set $\alpha_i = \frac{R}{\sqrt{N+1}}$ for all $i = 0, \ldots, N$. Then the iterates of the Subgradient descent algorithm in Definition 6.5.2, with this choice of α_i, satisfy*

$$\min_{i=0,\ldots,N} f(\mathbf{x}^i) \leq f(\mathbf{x}^\star) + \frac{MR}{\sqrt{N+1}}.$$

Turning this around, if we want to be within ϵ of the optimal value $f(\mathbf{x}^\star)$ for some $\epsilon > 0$, we should run the Subgradient algorithm for $\frac{M^2R^2}{\epsilon^2}$ iterates, with step size $\alpha_i = \frac{\epsilon}{M}$.

If we theoretically let the algorithm run for infinitely many steps, we would hope to make the difference between $\min_i f(\mathbf{x}^i)$ and $f(\mathbf{x}^\star)$ go to 0 in the limit. This, of course, depends on the choice of the step sizes $\alpha_0, \alpha_1, \ldots$ so that the expression $\frac{R^2 + \sum_{i=0}^{k} \alpha_i^2}{2\sum_{i=0}^{k} \alpha_i} \to 0$ as $k \to \infty$. There is a general sufficient condition that guarantees this. Let $\{\alpha_i\}_{i=0}^\infty$ be a sequence of strictly positive real numbers such that $\lim_{i\to\infty} \alpha_i = 0$ and $\sum_{i=1}^\infty \alpha_i = \infty$. Then, for any real number R, $\lim_{k\to\infty} \frac{R^2 + \sum_{i=0}^{k} \alpha_i^2}{2\sum_{i=0}^{k} \alpha_i} = 0$.

Remark 6.5.5 Corollary 6.5.4 shows that the information complexity of the subgradient algorithm is $O(\frac{M^2R^2}{\epsilon^2})$, which is *independent* of the dimension for *fixed M* and *R* (see Exercise 1 from Section 6.5.3). This is one reason why these methods are very popular in practice.

6.5.2.1 Smooth Gradients

It turns out that the Subgradient descent algorithm can be shown to have even faster convergence if we assume that the function f is differentiable and the gradient map ∇f is Lipschitz continuous. This additional structure sharpens some of the properties we know about differentiable convex functions from Section 3.2.2. We first begin with an observation that is independent of convexity.

Theorem 6.5.6 *Let* $g\colon \mathbb{R}^d \to \mathbb{R} \cup \{+\infty\}$ *be differentiable over some open subset* $X \subseteq \mathrm{dom}(f)$. *Suppose* $\nabla g\colon \mathbb{R}^d \to \mathbb{R}^d$ *is Lipschitz continuous over X, i.e.,* $\|\nabla g(\mathbf{x}) - \nabla g(\mathbf{y})\| \le L\|\mathbf{x} - \mathbf{y}\|$ *for all* $\mathbf{x}, \mathbf{y} \in X$. *Then*

$$g(\mathbf{x})+\langle\nabla g(\mathbf{x}),\mathbf{y}-\mathbf{x}\rangle-\frac{1}{2}L\|\mathbf{y}-\mathbf{x}\|^2 \le g(\mathbf{y}) \le g(\mathbf{x})+\langle\nabla g(\mathbf{x}),\mathbf{y}-\mathbf{x}\rangle+\frac{1}{2}L\|\mathbf{y}-\mathbf{x}\|^2$$

$$(6.5.3)$$

$\forall \mathbf{x}, \mathbf{y} \in X.$

Proof Left as exercise. $\qquad\square$

Definition 6.5.7 A $g\colon \mathbb{R}^d \to \mathbb{R} \cup \{+\infty\}$ is said to be *L-smooth* over an open subset $X \subseteq \mathrm{dom}(g)$ if it is differentiable over X and ∇g is Lipschitz continuous with constant L.

Below, we will consider convex functions with domain equal to all of $\mathbb{R}^d$ and that are *L*-smooth over all of $\mathbb{R}^d$. However, this is just for ease of presentation. Everything can be adapted for classes of convex functions defined over a convex subset C of some open subset X of $\mathbb{R}^d$.

Recall the following properties that are equivalent to convexity of a differentiable function $f\colon \mathbb{R}^d \to \mathbb{R}$ (Theorem 3.2.8).

$$\begin{aligned}
\forall \mathbf{x}, \mathbf{y} \in \mathbb{R}^d : \quad & f(\mathbf{x}) + \langle\nabla f(\mathbf{x}),\mathbf{y} - \mathbf{x}\rangle \le f(\mathbf{y}) \\
\Leftrightarrow \forall \mathbf{x}, \mathbf{y} \in \mathbb{R}^d : \quad & \langle\nabla f(\mathbf{y}) - \nabla f(\mathbf{x}),\mathbf{y} - \mathbf{x}\rangle \ge 0.
\end{aligned}$$

$$(6.5.4)$$

If one combines the property of being convex with *L*-smoothness, then one can sharpen these inequalities.

Theorem 6.5.8 *Let* $f\colon \mathbb{R}^d \to \mathbb{R}$ *be a convex, L-smooth function. Then we have*

$$\forall \mathbf{x}, \mathbf{y} \in \mathbb{R}^d : \quad f(\mathbf{x}) + \langle\nabla f(\mathbf{x}),\mathbf{y} - \mathbf{x}\rangle + \frac{1}{2L}\|\nabla f(\mathbf{y}) - \nabla f(\mathbf{x})\|^2 \le f(\mathbf{y})$$

$$(6.5.5)$$

and

$$\forall \mathbf{x}, \mathbf{y} \in \mathbb{R}^d : \quad \langle \nabla f(\mathbf{y}) - \nabla f(\mathbf{x}), \mathbf{y} - \mathbf{x} \rangle \geq \frac{1}{L} \|\nabla f(\mathbf{y}) - \nabla f(\mathbf{x})\|^2. \quad (6.5.6)$$

Proof It suffices to prove (6.5.5) because (6.5.6) follows from (6.5.5) by considering the inequality with $\mathbf{x}$ and $\mathbf{y}$ switched and adding to the original inequality.

Consider an arbitrary $\mathbf{x}^0 \in \mathbb{R}^d$ and the corresponding function $h(\mathbf{y}) = f(\mathbf{y}) - \langle \nabla f(\mathbf{x}^0), \mathbf{y} \rangle$. Note that $\nabla h(\mathbf{y}) = \nabla f(\mathbf{y}) - \nabla f(\mathbf{x}^0)$. In particular, h is also L-smooth and $\nabla h(\mathbf{x}^0) = \mathbf{0}$. Using the first inequality in (6.5.4) applied to h, $h(\mathbf{x}^0) \leq h(\mathbf{y})$ for all $\mathbf{y} \in \mathbb{R}^d$. Using the second inequality from (6.5.3), we also have that $h\left(\mathbf{y} - \frac{1}{L}\nabla h(\mathbf{y})\right) \leq h(\mathbf{y}) - \frac{1}{2L}\|\nabla h(\mathbf{y})\|^2$ for all $\mathbf{y} \in \mathbb{R}^d$. Thus, we must have

$$h(\mathbf{x}^0) \leq h(\mathbf{y}) - \frac{1}{2L}\|\nabla h(\mathbf{y})\|^2$$

for all $\mathbf{y} \in \mathbb{R}^d$. Substituting $h(\mathbf{x}^0) = f(\mathbf{x}^0) - \langle \nabla f(\mathbf{x}^0), \mathbf{x}^0 \rangle$ and $h(\mathbf{y}) = f(\mathbf{y}) - \langle \nabla f(\mathbf{x}^0), \mathbf{y} \rangle$ and rearranging gives us (6.5.5) with $\mathbf{x} = \mathbf{x}^0$. Since the choice of $\mathbf{x}_0$ was arbitrary, the inequality holds for all $\mathbf{x}, \mathbf{y} \in \mathbb{R}^d$. $\square$

We now show that convexity and L-smoothness together improve the convergence rate of (sub)gradient descent to $O\left(\frac{1}{\epsilon}\right)$.

Theorem 6.5.9 *Let $f : \mathbb{R}^d \to \mathbb{R}$ be a convex, L-smooth function. Assume that f has a minimizer $\mathbf{x}^\star$ with value $f^\star = f(\mathbf{x}^\star)$. If we start with an arbitrary $\mathbf{x}^0 \in \mathbb{R}^d$ and define iterates $\mathbf{x}^{k+1} := \mathbf{x}^k - \frac{1}{L}\nabla f(\mathbf{x}^k)$ for $k = 0, 1, 2, \ldots$, then for any $T \geq 1$, $f(\mathbf{x}^T) \leq f^\star + \frac{2L\|\mathbf{x}^0 - \mathbf{x}^\star\|^2}{T+4}$.*

Proof First note that $\nabla f(\mathbf{x}^\star) = \mathbf{0}$; otherwise, using the second inequality from (6.5.3) we will have $f(\mathbf{x}^\star - \frac{1}{L}\nabla f(\mathbf{x}^\star)) \leq f(\mathbf{x}^\star) - \frac{1}{2L}\|\nabla f(\mathbf{x}^\star)\|^2 < f(\mathbf{x}^\star)$, contradicting the fact that $\mathbf{x}^\star$ is the minimizer.

We first show that the iterates keep getting closer to $\mathbf{x}^\star$.

$$\begin{aligned}
\|\mathbf{x}^{k+1} - \mathbf{x}^\star\|^2 &= \|\mathbf{x}^k - \tfrac{1}{L}\nabla f(\mathbf{x}^k) - \mathbf{x}^\star\|^2 \\
&= \|\mathbf{x}^k - \mathbf{x}^\star\|^2 - \tfrac{2}{L}\langle \nabla f(\mathbf{x}^k), \mathbf{x}^k - \mathbf{x}^\star \rangle + \tfrac{1}{L^2}\|\nabla f(\mathbf{x}^k)\|^2 \\
&\leq \|\mathbf{x}^k - \mathbf{x}^\star\|^2 - \tfrac{2}{L^2}\|\nabla f(\mathbf{x}^k)\|^2 + \tfrac{1}{L^2}\|\nabla f(\mathbf{x}^k)\|^2 \\
&= \|\mathbf{x}^k - \mathbf{x}^\star\|^2 - \tfrac{1}{L^2}\|\nabla f(\mathbf{x}^k)\|^2,
\end{aligned}$$

where the inequality follows from (6.5.6) with $\mathbf{y} = \mathbf{x}_k$ and $\mathbf{x} = \mathbf{x}^\star$ and the fact that $\nabla f(\mathbf{x}^\star) = 0$. Thus, for any $k \geq 0$, we must have $\|\mathbf{x}^k - \mathbf{x}^\star\| \leq \|\mathbf{x}^0 - \mathbf{x}^\star\|$. Further,

$$\delta_k := f(\mathbf{x}^k) - f^\star \le \langle \nabla f(\mathbf{x}^k), \mathbf{x}^k - \mathbf{x}^\star \rangle \le \|\nabla f(\mathbf{x}^k)\| \|\mathbf{x}^k - \mathbf{x}^\star\|$$
$$\le \|\nabla f(\mathbf{x}^k)\| \|\mathbf{x}^0 - \mathbf{x}^\star\|, \tag{6.5.7}$$

where the first inequality follows from the top inequality in (6.5.4) with $\mathbf{y} = \mathbf{x}^\star$ and $\mathbf{x} = \mathbf{x}^k$, the second inequality follows from Cauchy–Schwarz (Theorem 1.1.6), and the last inequality uses the fact that the iterates keep getting closer to the optimum $\mathbf{x}^\star$. Using the second inequality from (6.5.3) with $\mathbf{x} = \mathbf{x}^k$ and $\mathbf{y} = \mathbf{x}^{k+1}$, we also have

$$\delta_{k+1} = f(\mathbf{x}^{k+1}) - f^\star \le f(\mathbf{x}^k) - f^\star - \frac{1}{2L}\|\nabla f(\mathbf{x}^k)\|^2 = \delta_k - \frac{1}{2L}\|\nabla f(\mathbf{x}^k)\|^2.$$

Using (6.5.7), we obtain

$$\begin{aligned}
\delta_{k+1} &\le \delta_k - \tfrac{1}{2L}\|\nabla f(\mathbf{x}^k)\|^2 \\
&\le \delta_k - \tfrac{1}{2L\|\mathbf{x}^0-\mathbf{x}^\star\|^2}\delta_k^2
\end{aligned} \tag{6.5.8}$$

Dividing through by $\delta_k \delta_{k+1}$ and rearranging we obtain

$$\begin{aligned}
\frac{1}{\delta_{k+1}} &\ge \frac{1}{\delta_k} + \frac{1}{2L\|\mathbf{x}^0-\mathbf{x}^\star\|^2}\frac{\delta_k}{\delta_{k+1}} \\
&\ge \frac{1}{\delta_k} + \frac{1}{2L\|\mathbf{x}^0-\mathbf{x}^\star\|^2},
\end{aligned}$$

where the second inequality follows from the fact that $\delta_{k+1} \le \delta_k$ by (6.5.8). Unrolling the recursion above, we obtain that

$$\frac{1}{\delta_T} \ge \frac{1}{\delta_0} + \frac{T}{2L\|\mathbf{x}^0 - \mathbf{x}^\star\|^2}. \tag{6.5.9}$$

Using the second inequality from (6.5.3) with $\mathbf{y} = \mathbf{x}^0$ and $\mathbf{x} = \mathbf{x}^\star$, and the fact that $\nabla f(\mathbf{x}^\star) = \mathbf{0}$, we obtain $f(\mathbf{x}^0) \le f^\star + \frac{L}{2}\|\mathbf{x}^0 - \mathbf{x}^\star\|^2$. Thus, $\delta_0 \le \frac{L}{2}\|\mathbf{x}^0 - \mathbf{x}^\star\|^2$. Plugging this into (6.5.9), we obtain $\frac{1}{\delta_T} \ge \frac{2}{L\|\mathbf{x}^0-\mathbf{x}^\star\|^2} + \frac{T}{2L\|\mathbf{x}^0-\mathbf{x}^\star\|^2}$. Rearranging, we obtain that $f(\mathbf{x}_T) - f^\star = \delta_T \le \frac{2L\|\mathbf{x}^0-\mathbf{x}^\star\|^2}{T+4}$. $\qquad\square$

6.5.2.2 Strong Convexity

We can get an exponential speedup if we assume the stronger property of strong convexity (Definition 3.1.3), i.e., in $O\left(\log\left(\frac{1}{\epsilon}\right)\right)$ steps, we can get to an ϵ-approximate solution.

Theorem 6.5.10 *Let $f: \mathbb{R}^d \to \mathbb{R}$ be strongly convex with modulus $c > 0$ and L-smooth. Assume that f has a minimizer $\mathbf{x}^\star$ with value $f^\star = f(\mathbf{x}^\star)$. If we start with an arbitrary $\mathbf{x}^0 \in \mathbb{R}^d$ and define iterates $\mathbf{x}^{k+1} := \mathbf{x}^k - \frac{1}{L}\nabla f(\mathbf{x}^k)$ for $k = 0, 1, 2, \ldots$, then for any $T \ge 1$,*

$$f(\mathbf{x}^T) \le f^\star + \left(1 - \frac{c}{L}\right)^T (f(\mathbf{x}^0) - f^\star) \le f^\star + \left(1 - \frac{c}{L}\right)^T \frac{L\|\mathbf{x}^0 - \mathbf{x}^\star\|^2}{2}.$$

Proof For any $k \geq 1$, the characterization of strongly convex functions in Theorem 3.2.8 gives us

$$\forall \mathbf{y} \in \mathbb{R}^d : \quad f(\mathbf{x}^k) + \langle \nabla f(\mathbf{x}^k), \mathbf{y} - \mathbf{x}^k \rangle + \frac{c}{2} \|\mathbf{y} - \mathbf{x}^k\|^2 \leq f(\mathbf{y}).$$

If we think of the left-hand side as a quadratic function $h(\mathbf{y})$, then we have $h(\mathbf{y}) \leq f(\mathbf{y})$ for all $\mathbf{y} \in \mathbb{R}^d$. Thus, the minimum value of $h(\mathbf{y})$ is less than or equal to the minimum value of f which is $f^\star$. Since h is a convex quadratic, we have a closed form for its minimizer $\mathbf{y}^\star = \mathbf{x}^k - \frac{1}{c} \nabla f(\mathbf{x}^k)$, and the minimum value $h(\mathbf{y}^\star) = f(\mathbf{x}^k) - \frac{1}{2c} \|\nabla f(\mathbf{x}^k)\|^2$ (Exercise 5 from Section 6.5.3). Thus we obtain $f^\star \geq f(\mathbf{x}^k) - \frac{1}{2c} \|\nabla f(\mathbf{x}^k)\|^2$. Setting $\delta_k := f(\mathbf{x}^k) - f^\star$, we obtain that $-2c\delta_k \geq -\|\nabla f(\mathbf{x}^k)\|^2$ for every iteration $k \geq 1$. Plugging this into (6.5.8), we obtain

$$\delta_{k+1} \leq \left(1 - \frac{c}{L}\right) \delta_k.$$

Unrolling the recursion, we obtain the desired inequality $f(\mathbf{x}^T) - f^\star \leq \left(1 - \frac{c}{L}\right)^T (f(\mathbf{x}^0) - f^\star)$ for all $T \geq 1$.

Note that $\nabla f(\mathbf{x}^\star) = \mathbf{0}$, otherwise, using the second inequality from (6.5.3) we will have $f(\mathbf{x}^\star - \frac{1}{L} \nabla f(\mathbf{x}^\star)) \leq f(\mathbf{x}^\star) - \frac{1}{2L} \|\nabla f(\mathbf{x}^\star)\|^2 < f(\mathbf{x}^\star)$, contradicting the fact that $\mathbf{x}^\star$ is the minimizer. Using the second inequality from (6.5.3) again with $\mathbf{y} = \mathbf{x}^0$ and $\mathbf{x} = \mathbf{x}^\star$, we obtain $f(\mathbf{x}^0) \leq f^\star + \frac{L}{2} \|\mathbf{x}^0 - \mathbf{x}^\star\|^2$, which yields the second inequality in the statement. $\qquad \square$

6.5.3 Exercises

1. Consider the family of instances $\mathcal{I}_{0,d,R,\rho,M}$ from Definition 6.1.4. From Corollary 6.5.4, we have an upper bound of $O\left(\frac{1}{\epsilon^2}\right)$ on the information complexity that is seemingly dimension independent (see Remark 6.5.5). However, we established a lower bound of $\Omega\left(d \log\left(\frac{MR}{\epsilon}\right)\right)$ in Theorem 6.2.1; see also Section 6.5.1. So when the dimension d is much larger than $\frac{1}{\epsilon^2}$, we seem to have an upper bound that is smaller than the lower bound. What gives?

2. Prove Theorem 6.5.6.

3. Let $f : \mathbb{R}^d \to \mathbb{R} \cup \{+\infty\}$ be twice differentiable over an open subset $X \subseteq \mathrm{dom}(f)$. Show that f is L-smooth function over X if and only if $\sigma_{\max}(\nabla^2 f(\mathbf{x})) \leq L$ for all $\mathbf{x} \in X$, where $\sigma_{\max}(\nabla^2 f(\mathbf{x}))$ is the largest singular value of the Hessian $\nabla^2 f(\mathbf{x})$.

4. Show that if $f : \mathbb{R}^d \to \mathbb{R}$ is strongly convex with modulus $c > 0$ and also L-smooth, then $c \leq L$. When does equality hold?

5. Let $f \colon \mathbb{R}^d \to \mathbb{R}$ be strongly convex with modulus $c > 0$. Let $\mathbf{x} \in \mathbb{R}^d$. Show that $\mathbf{y}^\star = \mathbf{x} - \frac{1}{c}\nabla f(\mathbf{x})$ is the unique minimizer of $h(\mathbf{y}) := f(\mathbf{x}) + \langle \nabla f(\mathbf{x}), \mathbf{y} - \mathbf{x} \rangle + \frac{c}{2}\|\mathbf{y} - \mathbf{x}\|^2$ and the minimum value is $h(\mathbf{y}^\star) = f(\mathbf{x}) - \frac{1}{2c}\|\nabla f(\mathbf{x})\|^2$.

6.6 Notes and Related Literature

The information complexity of mixed-integer convex optimization has a long history. The seminal results were established in the pure continuous case by Nemirovskii and Yudin in the 1970s which gave matching upper and lower bounds, up to a constant (see Theorem 6.2.1 and the discussion below it). These were summarized in their influential monograph [182]. The information complexity for problems involving integer-constrained variables was first considered in Timm Oertel's Ph.D. thesis [186], where the upper bound in Theorem 6.2.1 was obtained using the idea of centerpoints as presented in Section 6.2.3. The first lower bounds in the presence of integer variables were established in [38], but only for a restricted class of algorithms. The unconditional lower bounds on information complexity for general mixed-integer convex optimization presented in Theorem 6.2.1 were established in [36]. A partial unconditional result was proved earlier in [25] using the ideas from [38]. A more general result was obtained in [37], where it was shown that if ℓ is a lower bound on the information complexity of the pure continuous problem, then the mixed-integer problem has a lower bound of $2^n \ell$, where the information complexity is measured with respect to a broad class of oracles that generalize the first-order oracles studied in this book. Combined with the classical results of Nemirovskii and Yudin [182], this immediately gives the mixed-integer lower bounds established in Theorem 6.2.1.

The concept of a centerpoint from Definition 6.2.4 has been used as a formalization of the "most symmetric" point in convex geometry [64, 126, 127, 221], as a generalization of the median for higher-dimensional probability distributions – so-called *Tukey medians* – in the statistics literature [107, 156, 178, 203], and has also appeared in the computational geometry literature [61, 68, 110, 203]. As mentioned above, its importance for mixed-integer optimization was first observed in Timm Oertel's thesis [186] and subsequently elaborated upon in [38].

On the algorithmic side, progress on mixed-integer convex optimization happened on multiple fronts over several decades. The special case of the closest vector problem (convex quadratic minimization over the integers) has been known to be NP-hard to solve exactly, as well as approximately within small factors (constant or subpolynomial in the dimension) [6, 12, 105, 227]. Under standard computational complexity assumptions such as $P \neq NP$ and the *Exponential Time Hypothesis (ETH)*, we do not expect an algorithm with dependence better than $2^{O(n)}$ asymptotically. Until 2010, the best known algorithm for CVP was due to Blömer [51] with complexity $O(n!)$, which is asymptotically the same complexity as the previous best known algorithm due to Kannan [150], which was a general (pure) integer optimization algorithm with complexity $2^{O(n\log(n))}$. In 2010, Micciancio and Voulgaris achieved a breakthrough with a $2^{O(n)}$ complexity algorithm [174, 175]. Section 6.3.1.3 is an exposition of their algorithm. Subsequently, other algorithms with 2^n complexity have been obtained [2–4], improving on Micciancio and Voulgaris' $2^{O(n)}$ dependence. These algorithms are based on completely different and powerful new techniques – *Discrete Gaussian Sampling* – and the applications of these new tools are still being actively explored. Prior to Micciancio and Voulgaris' work, *randomized* algorithms had been discovered with $2^{O(n)}$ time for solving the related shortest vector problem (SVP) exactly (see Section 6.3.1.5), and the CVP up to a $(1 + \epsilon)$ factor for any $\epsilon > 0$, in two seminal papers by Ajtai, Kumar, and Sivakumar [7, 8]. In these papers, the authors introduced an influential *sieving technique* for algorithmic geometry of numbers that spawned a flurry of research activity in the next two decades. In particular, this technique was used to solve CVP approximately, and to solve the related SVP problem exactly (with randomization); see, e.g., [52] and the references therein. No constant factor approximation algorithm for CVP with complexity better than $2^{O(n)}$ was known until [113]. The method of using nearest lattice planes, which is an important ingredient in the Micciancio–Voulgaris algorithm (see Section 6.3.1.3), was known to yield a $2^{O(n)}$ approximation algorithm with polynomial in n running time [18] (see Exercise 15 in Section 6.3.3), and all known polynomial time approximation algorithms give exponential approximation factors. The CVP and SVP problems in norms different from ellipsoidal norms have recently been investigated; see [113, 202] and the references therein.

In parallel to the special case of the convex quadratic minimization (CVP) problem, algorithms for the general mixed-integer convex optimization problem with provable complexity bounds were being developed, starting from a seminal paper by Lenstra [165]. This paper investigated the mixed-integer *linear* case, i.e., when C is a rational polytope and f is a rational linear function, and the complexity of the algorithm was $2^{O(n^3)}\,\mathrm{poly}(d, T)$, where T is the binary encoding size of C and f. This paper provided the key insight of recursing on lower-dimensional problems by considering sections of C by intersecting with parallel hyperplanes and controlling the number of sections required in the recursion (see Definition 6.3.35). Lenstra's insights were soon extended to handle the general nonlinear case in [123] for *pure* integer optimization problems with the same $2^{O(n^3)}$ dependence on dimension. Kannan [150] achieved a breakthrough in the complexity bounds – improving from $2^{O(n^3)}$ dependence on the number of integer variables to $2^{O(n \log n)}$ – by modifying the algorithm to recurse over lower-dimensional affine spaces, as opposed to hyperplanes as discussed above. We refer to [80, 111, 134, 138, 152] for a representative sample of important papers giving algorithms, all with $2^{O(n \log(n))}$ complexity. The sharpest constant in the exponent of $2^{O(n \log n)}$ was derived in Daniel Dadush's Ph.D. thesis [80]; in particular, Dadush's algorithm has complexity $2^{O(n)} n^n$. The $2^{O(n \log(n))}$ complexity presented in Section 6.3.2 (see Remark 6.3.38) is the best known *deterministic* complexity for the pure integer optimization problem. Reis and Rothvoss obtained a structural result in the geometry of numbers that can be combined with ideas from [80] to give a *randomized* algorithm with complexity $(\log(2n))^{O(n)}$. As mentioned in Section 4.5, this structural result also gives the best bound of $O(n \log^3(2n))$ for the lattice width of a lattice-free convex body. On a related note, the general mixed-integer convex optimization algorithm presented in Section 6.3.2 uses Theorem 4.4.6 to bound the lattice width of a lattice-free ellipsoid by $n^{3/2}$. Banaszczyk [21] showed that this bound can be improved to n; however, this improvement from $n^{3/2}$ to n does not make much of a difference in the overall complexity.

The general mixed-integer convex optimization algorithm presented in Section 6.3.2 is an adaptation of the pure integer optimization ideas discussed in the previous paragraph to the mixed-integer case. To the best of our knowledge, no computational complexity results have been formally stated in the literature for the mixed-integer case, beyond the setting of mixed-integer *linear* optimization in the original paper [165] of Lenstra mentioned above. It is possible that the complexity bounds derived in Theorem 6.3.36 can be improved, even though they reduce to the best known (deterministic) complexity bounds in the pure integer case (Remark 6.3.38).

The general branch-and-cut framework of Section 6.4 is a confluence of ideas originating in the seminal paper of Land and Doig [161] and a long line of research on *polyhedral combinatorics*. We recommend [74, 180, 212] as excellent textbooks that explore the varied aspects of this method not discussed in this book.

Obtaining concrete lower bounds for branch-and-cut algorithms has a long history within the optimization, discrete mathematics, and computer science communities. In particular, there is a rich interplay of ideas between optimization and proof complexity arising from the fact that branch-and-cut can be used to certify emptiness of sets of the form $P \cap \{0,1\}^n$, where P is a polyhedron. This question is of fundamental importance in computer science because the *satisfiability* question in logic can be modeled in this way. We provide here a necessarily incomplete but representative sample of references for the interested reader [39, 54, 55, 63, 69–72, 76–78, 81, 83–85, 112, 120–122, 146, 157, 192, 193, 196, 201]. The complexity of branch-and-cut has seen a resurgence in research focus with many new fundamental results in [81, 87, 99–102, 117].

In [31, 32], the authors explore whether certain widely used cutting-plane paradigms and branching schemes form complementary pairs (Definition 6.4.6). We summarize their results here in informal terms and refer to the two papers cited for precise statements and proofs.

1. *Lift-and-project* cutting planes (disjunctive cutting planes based on variable disjunctions – see point 2. in Example 6.4.4) and variable disjunctions are *not* a complementary pair for 0/1 pure integer convex optimization problems, i.e., $C \subseteq [0,1]^n$. Any branch-and-bound algorithm based on variable disjunctions can be simulated by a cutting-plane algorithm with the same complexity.[6] Moreover, there are instances of graph stable set problems where there is a lift-and-project cutting-plane algorithm with polynomial complexity, but any branch-and-bound algorithm based on variable disjunctions has exponential complexity. See Theorems 2.1 and 2.2 in [32] and results in [83, 84].

2. Lift-and-project cutting planes and variable disjunctions *do form* a complementary pair for *general* mixed-integer convex optimization problems, i.e., C is not restricted to be in the 0/1 hypercube.[7] See Theorems 2.2 and 2.9 in [32].

3. Split cutting planes (disjunctive cutting planes based on split disjunctions – see point 2. in Example 6.4.4) and split disjunctions are *not* a complementary pair for general pure integer convex problems. Any cutting-plane algorithm based on split cuts can be simulated by a branch-and-bound algorithm based on split disjunctions with the same complexity (up to constant factors).[8] See Theorem 1.8 in [31].

---

[6] There is a technical point that arises here concerning the notions of algorithm and *proof*. We have omitted all discussions of cutting-plane and branch-and-bound proofs here, which are powerful tools to prove unconditional lower bounds on these algorithms. The precise statement is that any branch-and-bound proof based on variable disjunctions can be replaced by a lift-and-project cutting-plane proof of the same size. See [32] for details.

[7] "Lift-and-project cuts" here mean disjunctive cutting planes based on the variable disjunctions (typically the phrase "lift-and-project" is reserved for 0/1 problems).

[8] The same caveat as in point 1. regarding algorithms versus proofs applies.

The material in Section 6.5.2 can be found in several convex optimization textbooks and is included here for the sake of completing the story on classical convex optimization and presenting its main insights. Our presentation here is heavily based on Yurii Nesterov's excellent monograph [184]; in fact, the analysis in Theorem 6.5.3 and Corollary 6.5.4, as well as all of the material in Section 6.5.2.1, is taken with no modifications from Sections 3.2.3 and 2.1.5 of [184], respectively. The derivation in the proof of Theorem 6.5.10 is from Section 3.3 of Stephen Wright's survey on coordinate descent methods [231].

Arithmetic versus Turing machine models of computation. As mentioned at the beginning of this chapter, since the query points and responses for the first-order oracle, and the range of the (continuous) variables are allowed to be arbitrary real numbers, these cannot be handled in the Turing machine model of computation. One could restrict the variables to take rational values only and insist on first-order oracles whose queries can be made only at points with rational coordinates and whose responses are always rational numbers and vectors with rational coordinates. To the best of our knowledge, a thorough study of information complexity of mixed-integer convex optimization has not been undertaken in this setting. Some results in this direction involving inexact first-order oracles have been discussed [89, 97, 139, 154, 160, 188, 209, 216], but tight upper and lower bounds on information complexity are not known in this setting. Some partial progress has been made recently in [37, 172].

Let us discuss the different algorithms we present in this book for the algorithmic upper bound. The Conjugate Gradient method from Section 6.3.1.2 uses only the arithmetic operations of addition, subtraction, multiplication, and division. Therefore, if the matrix Q and the vector $\mathbf{c}$ defining the quadratic function are rational, all intermediate calculations are rational and everything works in the Turing machine model as well. It is also possible to allow real inputs in the Turing machine model enhanced with real number oracles (Definition 1.4.2). Given a desired (rational) accuracy for the final solution, the Conjugate Gradient method can be implemented on such an oracle Turing machine model with overhead proportional to the binary encoding size of the desired accuracy. In the same way, the Lenstra–Lenstra–Lovasz (LLL) algorithm from Section 6.3.1.3 will compute with only rational numbers, when the basis B is a rational matrix and therefore can be implemented in the Turing machine model for rational input; see Exercise 14 from Section 6.3.3. The same comments hold for the CVP algorithm (Definition 6.3.25) and the algorithm for computing the Voronoi relevant vectors (Definition 6.3.27). To the best of our knowledge, the LLL algorithm, the CVP algorithm and the Voronoi relevant vectors algorithm have not been analyzed in the Turing machine model equipped with real number oracles to allow for nonrational inputs. The algorithms for unconstrained mixed-integer quadratic minimization (Definition 6.3.30) and general mixed-integer convex optimization (Definition 6.3.35) are of a different nature. For the quadratic minimization case, even if the input matrix Q defining the quadratic function and the vector $\mathbf{c}$ (or $\mathbf{t}$) are rational, the algorithm requires the computation of square roots. Notice the difference from the CVP algorithm where the input is the basis of the lattice, while in quadratic minimization, the input is the matrix defining the quadratic and extracting the basis of the relevant lattice (to reduce to a CVP problem) inevitably requires square root operations. To the best of our knowledge, this problem has not been analyzed in the Turing machine model (with or without real number oracles). For general mixed-integer convex optimization, the issue is even more confounded because apart from having to solve the mixed-integer quadratic minimization problems as subroutines, one also has to compute the intermediate ellipsoids using the formulas in Exercise 14 from Section 2.7.2. These also require the use of square roots. As far as we know, versions of the algorithm presented in Definition 6.3.35 with the same complexity have not been designed in the oracle Turing machine model. However, an algorithm for *pure integer* convex optimization with a weaker complexity guarantee that works in the oracle Turing machine model was fully worked out in [123]. The complexity guarantee is $2^{O(n^3)} \operatorname{poly}(\log(nR))$ as opposed to the state-of-the-art $2^{O(n \log n)} \operatorname{poly}(\log(nR))$ (deterministic) algorithmic complexity presented in this book (Remark 6.3.38). For the subgradient descent-type algorithms presented in Section 6.5.2 for continuous convex optimization, implementations in the Turing machine model can be done using the work on inexact oracles mentioned above.

7

Certificates and Duality

The focus of Chapter 6 was on the information and algorithmic complexity of convex optimization. We turn now to an important aspect of optimization problems beyond an analysis of lower and upper bounds on their complexity. In particular, let us ask the following question: when an algorithm returns a solution to an optimization problem, how do we know if it is an ϵ-approximate solution (whatever be the notion of approximation; see Section 5.1)? One response can be to go through the analysis of the algorithm's correctness and convince ourselves that the correct execution of the algorithm is the "proof" that the output is indeed an ϵ-approximate solution. This response to the question is unsatisfactory on two fronts: (1) It seems awkward to have to go through a complicated correctness proof of an algorithm to verify the (approximate) optimality of a solution; and (2) even if we believe the proof is correct, an actual execution of the algorithm on a physical machine may go wrong for several reasons (e.g., incorrect implementation, hardware glitches, or some hard-to-trace execution time error on the physical machine). How will we know that the theoretically proven algorithm actually was implemented and executed correctly just by looking at the final output?

For this reason, the notion of a *certificate* of optimality or ϵ-optimality (i.e., certifying that a solution is an ϵ-approximate solution) is very useful in optimization. We first establish results for the classical continuous case and then apply these to the study of certificates in the mixed-integer setting.

7.1 Continuous Optimization: The Case with No Integer Variables

Recall that our optimization problem (6.1.1) with $n = 0$ becomes

$$\inf_{\mathbf{x} \in C} f(\mathbf{x}), \tag{7.1.1}$$

where $f \colon \mathbb{R}^d \to \mathbb{R}$ is a closed, convex function, and $C \subseteq \mathbb{R}^d$ is a closed, convex set. In this chapter we restrict ourselves to objectives functions that are real valued; the statements of results take clean and elegant forms if we disallow $+\infty$ values. The central idea behind producing an "easily verifiable" certificate is the subgradient inequality: given any convex function $f \colon \mathbb{R}^d \to \mathbb{R}$ and any $\hat{\mathbf{x}} \in \mathbb{R}^d$, we must have

$$f(\mathbf{y}) \geq f(\hat{\mathbf{x}}) + \langle \mathbf{s}, \mathbf{y} - \hat{\mathbf{x}} \rangle \tag{7.1.2}$$

for all $\mathbf{y} \in \mathbb{R}^d$, where $\mathbf{s} \in \partial f(\hat{\mathbf{x}})$ is any subgradient at $\hat{\mathbf{x}}$ (recall Definition 3.1.15). For instance, in the simplest case of unconstrained optimization, i.e., $C = \mathbb{R}^d$, this immediately gives the following sufficient condition for exact optimality: $\mathbf{x}^\star$ is an optimal solution ($\epsilon = 0$) if $\mathbf{0} \in \partial f(\mathbf{x}^\star)$. Below we will see that this condition is also necessary (Corollary 7.1.6). We will first study certificates for *exact optimality* where complete characterizations are often available and then apply these ideas to certificates for ϵ-optimality.

7.1.1 Exact Optimality Certificates

We will derive a characterization of exact minimizers of (7.1.1) in terms of the local geometry of C and the first-order properties of f, i.e., its subdifferential ∂f. We first need some concepts related to the local geometry of a convex set.

Definition 7.1.1 Let $C \subseteq \mathbb{R}^d$ be a convex set, and let $\mathbf{x} \in C$. Define the *cone of feasible directions* as

$$F_C(\mathbf{x}) = \{\mathbf{r} \in \mathbb{R}^d : \exists \epsilon > 0 \text{ such that } \mathbf{x} + \epsilon \mathbf{r} \in C\}.$$

$F_C(\mathbf{x})$ may not be a closed cone – consider C as the unit circle in $\mathbb{R}^2$ and $\mathbf{x} = (-1, 0)$; then $F_C(\mathbf{x}) = \{\mathbf{r} \in \mathbb{R}^2 : \mathbf{r}_1 > 0\} \cup \{\mathbf{0}\}$. It is much nicer to work with its closure.

Definition 7.1.2 Let $C \subseteq \mathbb{R}^d$ be a convex set, and let $\mathbf{x} \in C$. The *tangent cone of C at* $\mathbf{x}$ is $T_C(\mathbf{x}) := \mathrm{cl}(F_C(\mathbf{x}))$.

The other important concept related to the local geometry of closed, convex sets is the *normal cone*.

Definition 7.1.3 Let $C \subseteq \mathbb{R}^d$ be a convex set, and let $\mathbf{x} \in C$. The *normal cone of C at* $\mathbf{x}$ is $N_C(\mathbf{x}) := \{\mathbf{r} \in \mathbb{R}^d : \langle \mathbf{r}, \mathbf{x} \rangle \geq \langle \mathbf{r}, \mathbf{y} \rangle \ \forall \mathbf{y} \in C\}$.

The normal cone $N_C(\mathbf{x})$ is the set of vectors $\mathbf{r} \in \mathbb{R}^d$ such that $\mathbf{x}$ is the maximizer over C for the corresponding linear functional $\langle \mathbf{r}, \cdot \rangle$, i.e., $\langle \mathbf{r}, \mathbf{x} \rangle = \sup_{\mathbf{y} \in C} \langle \mathbf{r}, \mathbf{y} \rangle$. Moreover, since $N_C(\mathbf{x}) = \{\mathbf{r} \in \mathbb{R}^d : \langle \mathbf{r}, \mathbf{y} - \mathbf{x} \rangle \leq 0 \ \forall \mathbf{y} \in C\}$

which is an intersection of halfspaces with the origin on the boundary, it is immediate that N_C is a closed, convex cone. Note that any nonzero vector $\mathbf{r} \in N_C(\mathbf{x})$ defines a supporting hyperplane $H^=(\mathbf{r}, \langle \mathbf{r}, \mathbf{x} \rangle)$ for C at $\mathbf{x}$.

Proposition 7.1.4 *Let* $C \subseteq \mathbb{R}^d$ *be a convex set, and let* $\mathbf{x} \in C$. *Then* $F_C(\mathbf{x}), T_C(\mathbf{x})$, *and* $N_C(\mathbf{x})$ *are all convex cones, with* $T_C(\mathbf{x}), N_C(\mathbf{x})$ *being closed, convex cones. Moreover,* $N_C(\mathbf{x}) = T_C(\mathbf{x})^\circ$, *i.e., the tangent cone and the normal cone are polars of each other.*

Proof See Exercise 2 from Section 7.1.3. $\square$

We are now ready to state the characterization of a global minimizer of (7.1.1), in terms of the local geometry of C and the first-order information of f. We will utilize the notion of directional derivative from Definition 1.3.16.

Theorem 7.1.5 *Let* $f \colon \mathbb{R}^d \to \mathbb{R}$ *be a closed, convex function, and* C *be a closed, convex set. Then the following are all equivalent.*

1. $\mathbf{x}^\star$ *is a global minimizer of* (7.1.1).
2. $f'(\mathbf{x}^\star; \mathbf{y} - \mathbf{x}^\star) \geq 0$ *for all* $\mathbf{y} \in C$.
3. $f'(\mathbf{x}^\star; \mathbf{r}) \geq 0$ *for all* $\mathbf{r} \in T_C(\mathbf{x}^\star)$.
4. $\mathbf{0} \in \partial f(\mathbf{x}^\star) + N_C(\mathbf{x}^\star)$.

Proof (1. $\implies$ 2.) Since $f(\mathbf{z}) \geq f(\mathbf{x}^\star)$ for all $\mathbf{z} \in C$, in particular this holds for $\mathbf{z} = \mathbf{x}^\star + t(\mathbf{y} - \mathbf{x}^\star)$ for all $0 \leq t \leq 1$. Therefore, $\frac{f(\mathbf{x}^\star + t(\mathbf{y} - \mathbf{x}^\star)) - f(\mathbf{x}^\star)}{t} \geq 0$ for all $t \in (0, 1)$. Taking the limit as $t \to 0$, we obtain that $f'(\mathbf{x}^\star; \mathbf{y} - \mathbf{x}^\star) \geq 0$.

(2. $\implies$ 3.) We first show that $f'(\mathbf{x}^\star; \mathbf{r}) \geq 0$ for all $\mathbf{r} \in F_C(\mathbf{x})$. Let $\epsilon > 0$ such that $\mathbf{y} = \mathbf{x}^\star + \epsilon\mathbf{r} \in C$. By assumption, $0 \leq f'(\mathbf{x}^\star; \mathbf{y} - \mathbf{x}^\star) = f'(\mathbf{x}^\star; \epsilon\mathbf{r}) = \epsilon f'(\mathbf{x}; \mathbf{r})$, using the positive homogeneity of $f'(\mathbf{x}^\star; \cdot)$, since $f'(\mathbf{x}^\star; \cdot)$ is sublinear by Proposition 3.4.2. Diving by ϵ, we obtain that $f'(\mathbf{x}^\star; \mathbf{r}) \geq 0$ for all $\mathbf{r} \in F_C(\mathbf{x}^\star)$. Since $f'(\mathbf{x}^\star; \cdot)$ is sublinear and finite valued everywhere by Proposition 3.4.2, it is convex by Proposition 3.3.2, and thus, it is continuous by Theorem 3.2.3. Consequently, it must be nonnegative on $T_C(\mathbf{x}) = \mathrm{cl}(F_C(\mathbf{x}))$, because it is nonnegative on $F_C(\mathbf{x})$.

(3. $\implies$ 4.) Suppose to the contrary that $\mathbf{0} \notin \partial f(\mathbf{x}^\star) + N_C(\mathbf{x}^\star)$. Since f is assumed to be finite valued everywhere, $\mathrm{dom}(f) = \mathbb{R}^d$. Thus, by Exercise 16 from Section 3.1.1, $\partial f(\mathbf{x}^\star)$ is a compact, convex set. Moreover, $N_C(\mathbf{x}^\star)$ is a closed, convex cone by Proposition 7.1.4. Therefore, by Exercise 1 from Section 1.3.1, $\partial f(\mathbf{x}^\star) + N_C(\mathbf{x}^\star)$ is a closed, convex set. By the separating hyperplane theorem (Theorem 2.4.2), there exist $\mathbf{a} \in \mathbb{R}^d \setminus \{\mathbf{0}\}, \delta \in \mathbb{R}$ such that $0 = \langle \mathbf{a}, \mathbf{0} \rangle > \delta \geq \langle \mathbf{a}, \mathbf{v} \rangle$ for all $\mathbf{v} \in \partial f(\mathbf{x}^\star) + N_C(\mathbf{x}^\star)$.

First, we claim that $\langle \mathbf{a}, \mathbf{n} \rangle \leq 0$ for all $\mathbf{n} \in N_C(\mathbf{x}^\star)$. Otherwise, consider $\bar{\mathbf{n}} \in N_C(\mathbf{x}^\star)$ such that $\langle \mathbf{a}, \bar{\mathbf{n}} \rangle > 0$. Since $N_C(\mathbf{x}^\star)$ is a convex cone, $\lambda \bar{\mathbf{n}} \in N_C(\mathbf{x}^\star)$ for all $\lambda \geq 0$. But then consider any $\mathbf{s} \in \partial f(\mathbf{x})$ (which is nonempty by Exercise 16 from Section 3.1.1) and the set of points $\mathbf{s} + \lambda \bar{\mathbf{n}}$. Since $\langle \mathbf{a}, \bar{\mathbf{n}} \rangle > 0$, we can find $\lambda \geq 0$ large enough such that $\langle \mathbf{a}, \mathbf{s} + \lambda \bar{\mathbf{n}} \rangle > \delta$, contradicting that $\delta \geq \langle \mathbf{a}, \mathbf{v} \rangle$ for all $\mathbf{v} \in \partial f(\mathbf{x}^\star) + N_C(\mathbf{x}^\star)$.

Since $\langle \mathbf{a}, \mathbf{n} \rangle \leq 0$ for all $\mathbf{n} \in N_C(\mathbf{x}^\star)$, we obtain that $\mathbf{a} \in N_C(\mathbf{x}^\star)^\circ = T_C(\mathbf{x}^\star)$, by Proposition 7.1.4. Now we use the fact that $\partial f(\mathbf{x}^\star) \subseteq \partial f(\mathbf{x}^\star) + N_C(\mathbf{x}^\star)$, since $\mathbf{0} \in N_C(\mathbf{x}^\star)$. This implies that $\langle \mathbf{a}, \mathbf{s} \rangle \leq \delta < 0$ for all $\mathbf{s} \in \partial f(\mathbf{x}^\star)$. Since $\partial f(\mathbf{x}^\star)$ is a compact, convex set, this implies that $\sup_{\mathbf{s} \in \partial f(\mathbf{x}^\star)} \langle \mathbf{a}, \mathbf{s} \rangle < 0$. From Theorem 3.4.3, $f'(\mathbf{x}^\star; \mathbf{a}) = \sigma_{\partial f(\mathbf{x}^\star)}(\mathbf{a}) = \sup_{\mathbf{s} \in \partial f(\mathbf{x}^\star)} \langle \mathbf{a}, \mathbf{s} \rangle < 0$. This contradicts the assumption of 3., because we showed above that $\mathbf{a} \in T_C(\mathbf{x}^\star)$.

$(4. \implies 1.)$ Consider any $\mathbf{y} \in C$. Since $\mathbf{0} \in \partial f(\mathbf{x}^\star) + N_C(\mathbf{x}^\star)$, there exist $\mathbf{s} \in \partial f(\mathbf{x}^\star)$ and $\mathbf{n} \in N_C(\mathbf{x}^\star)$ such that $\mathbf{0} = \mathbf{s} + \mathbf{n}$. Now, $\mathbf{y} - \mathbf{x}^\star \in T_C(\mathbf{x}^\star)$ and so $\langle \mathbf{y} - \mathbf{x}^\star, \mathbf{n} \rangle \leq 0$ by Proposition 7.1.4. Since we have

$$0 = \langle \mathbf{y} - \mathbf{x}^\star, \mathbf{0} \rangle = \langle \mathbf{y} - \mathbf{x}^\star, \mathbf{s} \rangle + \langle \mathbf{y} - \mathbf{x}^\star, \mathbf{n} \rangle,$$

this implies that $\langle \mathbf{y} - \mathbf{x}^\star, \mathbf{s} \rangle \geq 0$. By definition of subgradient, $f(\mathbf{y}) \geq f(\mathbf{x}^\star) + \langle \mathbf{s}, \mathbf{y} - \mathbf{x}^\star \rangle \geq f(\mathbf{x}^\star)$. Since the choice of $\mathbf{y} \in C$ was arbitrary, this shows that $\mathbf{x}^\star$ is a global minimizer. $\qquad \square$

Corollary 7.1.6 *Let $\mathbf{x}^\star$ be a minimizer for (7.1.1). If $\mathbf{x}^\star \in \mathrm{int}(C)$, then $\mathbf{0} \in \partial f(\mathbf{x}^\star)$. In particular, if $C = \mathbb{R}^d$, then a minimizer of f must contain $\mathbf{0}$ in its subdifferential. Consequently, if f is differentiable everywhere, the gradient at the minimizer must be $\mathbf{0}$.*

Proof This follows from the fact that for any convex set C and any $\mathbf{y} \in \mathrm{int}(C)$, $N_C(\mathbf{y}) = \{\mathbf{0}\}$. $\qquad \square$

The main point of Theorem 7.1.5 is that one can certify the optimality of a solution $\mathbf{x}^\star$ by exhibiting $\mathbf{s} \in \partial f(\mathbf{x}^\star)$ such that $-\mathbf{s} \in N_C(\mathbf{x}^\star)$; moreover, one can *always* produce such a certificate. The big drawback with this general statement is how to verify that $-\mathbf{s} \in N_C(\mathbf{x}^\star)$ (technically, one also has to verify $\mathbf{s} \in \partial f(\mathbf{x}^\star)$, but if f is differentiable then this can be done by computing partial derivatives). This can be done when C is given in a more explicit manner. For instance, if C is a polyhedron then the normal and tangent cones at a point can be described in terms of the defining linear inequalities; see Exercise 5 from Section 7.1.3. In other words, when C is a polyhedron given by $A\mathbf{x} \leq \mathbf{b}$, a readily checkable certificate of optimality is the following: $\mathbf{x}^\star$, along with $\mathbf{s} \in \mathbb{R}^d$ and nonnegative scalars λ_i, $i \in \mathrm{tight}(\mathbf{x}^\star)$ such that $\mathbf{s} \in \partial f(\mathbf{x}^\star)$ and $-\mathbf{s} = \sum_{i \in \mathrm{tight}(\mathbf{x}^\star)} \lambda_i \mathbf{a}^i$ (here we use the notation $\mathrm{tight}(\mathbf{x})$, $A_{\mathrm{tight}(\mathbf{x})}$ and $\mathbf{b}_{\mathrm{tight}(\mathbf{x})}$

for any $\mathbf{x} \in P$ from Definition 2.5.1 to denote the tight inequalities at $\mathbf{x}$). More generally, when C is given as the intersection of sublevel sets of more general convex inequalities (as opposed to linear inequalities), a description of the normal cone can often be obtained by considering all nonnegative combinations of different subgradients of the tight constraints (i.e., constraints satisfied at equality) at $\mathbf{x}^\star$. This is not always possible and one needs so-called *constraint qualification conditions* for such a description to hold; see Section 7.3 for references on this topic. We will revisit this from a different perspective in our discussion of Lagrangian duality below.

7.1.2 Approximate Optimality Certificates

Unlike the case of exact optimality certificates where we can guarantee the existence of a certificate, for ϵ-approximate solutions there is no known notion of a certificate that is always known to exist. Thus, we have to rely on certificates that, if shown to exist, will indeed verify the ϵ-optimality of the solution but these certificates will not necessarily exist for all instances. We will also strive to come up with easily checkable conditions under which these certificates can be guaranteed to exist.

7.1.2.1 Volume-Based Certificates

The first type of certificate we present is based on ideas we have seen in Chapter 6.

Theorem 7.1.7 *Let* $f : \mathbb{R}^d \to \mathbb{R}$ *be a convex function that is Lipschitz continuous with constant M, and $C \subseteq \mathbb{R}^d$ be a closed, convex set contains $B_\infty(\mathbf{z}, \rho) \subseteq C \subseteq B_\infty(\mathbf{0}, R)$ for some $\mathbf{z} \in C$ and $\rho, R > 0$. Let $\epsilon \le 2MR$. Let $\hat{\mathbf{x}} \in C$ such that* $\mathrm{vol}(C \cap H) < \left(\frac{\epsilon\rho}{MR}\right)^d$, *where $H = \{\mathbf{y} : \langle \mathbf{s}, \mathbf{y} - \hat{\mathbf{x}} \rangle < 0\}$ for some $\mathbf{s} \in \partial f(\hat{\mathbf{x}})$. Then $\hat{\mathbf{x}}$ is an ϵ-approximate solution.*

Proof By Lemma 6.1.5, there exists $\mathbf{v} \in C$ such that all points in $B_\infty\left(\mathbf{v}, \frac{\epsilon\rho}{2MR}\right)$ are ϵ-approximate solutions. Since $\mathrm{vol}(C \cap H) < \left(\frac{\epsilon\rho}{MR}\right)^d$, there exists $\hat{\mathbf{y}} \in B_\infty\left(\mathbf{v}, \frac{\epsilon\rho}{2MR}\right)$ such that $\langle \mathbf{s}, \hat{\mathbf{y}} - \hat{\mathbf{x}} \rangle \ge 0$. Thus, $f(\hat{\mathbf{y}}) \ge f(\hat{\mathbf{x}}) + \langle \mathbf{s}, \hat{\mathbf{y}} - \hat{\mathbf{x}} \rangle \ge f(\hat{\mathbf{x}})$. Thus, $\hat{\mathbf{x}}$ must also be an ϵ-approximate solution. $\square$

Thus, one can certify that $\hat{\mathbf{x}}$ is an ϵ-approximate solution by providing $\mathbf{s} \in \partial f(\hat{\mathbf{x}})$ and checking that $\mathrm{vol}(C \cap H) < \left(\frac{\epsilon\rho}{MR}\right)^d$, where $H = \{\mathbf{y} : \langle \mathbf{s}, \mathbf{y} - \hat{\mathbf{x}} \rangle < 0\}$. A special case of this is when $\mathbf{s} \in -N_C(\hat{\mathbf{x}})$ because then by the definition of $N_C(\hat{\mathbf{x}})$, $C \cap H = \emptyset$, and we have a certificate of exact optimality (deriving the implication 4. $\Longrightarrow$ 1. in Theorem 7.1.5). Computing volumes of convex sets is a challenging computation problem and this becomes a drawback of such volume-based certificates.

7.1.2.2 Lagrangian Duality with Generalized Constraints

We now come to the topic of *Lagrangian duality* which is another way to certify ϵ-optimality of a solution. We begin by reviewing the notion of a partially ordered set and adapting it to the setting of Euclidean space.

Definition 7.1.8 Let X be any set. A *partial order* on X is a binary relation on X, i.e., a subset $\mathcal{R} \subseteq X \times X$ that satisfies certain conditions. We will denote $x \preccurlyeq y$ for $x, y \in X$ if $(x, y) \in \mathcal{R}$. The conditions are as follows:

1. $x \preccurlyeq x$ for all $x \in X$.
2. $x \preccurlyeq y$ and $y \preccurlyeq z$ implies $x \preccurlyeq z$.
3. $x \preccurlyeq y$ and $y \preccurlyeq x$ if and only if $x = y$.

We now define partial orders on $\mathbb{R}^m$ for any $m \geq 1$ that respect the vector space structure of $\mathbb{R}^m$.

Definition 7.1.9 We will say that a binary relation on $\mathbb{R}^m$ is a *generalized inequality* if it satisfies the following conditions.

1. $\mathbf{x} \preccurlyeq \mathbf{x}$ for all $\mathbf{x} \in \mathbb{R}^m$.
2. $\mathbf{x} \preccurlyeq \mathbf{y}$ and $\mathbf{y} \preccurlyeq \mathbf{z}$ implies $\mathbf{x} \preccurlyeq \mathbf{z}$.
3. $\mathbf{x} \preccurlyeq \mathbf{y}$ and $\mathbf{y} \preccurlyeq \mathbf{x}$ if and only if $\mathbf{x} = \mathbf{y}$.
4. $\mathbf{x} \preccurlyeq \mathbf{y}$ implies $\mathbf{x} + \mathbf{z} \preccurlyeq \mathbf{y} + \mathbf{z}$ for all $\mathbf{z} \in \mathbb{R}^m$.
5. $\mathbf{x} \preccurlyeq \mathbf{y}$ implies $\lambda\mathbf{x} \preccurlyeq \lambda\mathbf{y}$ for all $\lambda \geq 0$.

Generalized inequalities have an elegant geometric characterization.

Proposition 7.1.10 *Let $K \subseteq \mathbb{R}^m$ be a convex, pointed cone. Then, the relation on $\mathbb{R}^m$ defined by $\mathbf{x} \preccurlyeq_K \mathbf{y}$ if and only if $\mathbf{y} - \mathbf{x} \in K$ is a generalized inequality. In this case, we say that $\preccurlyeq_K$ is the generalized inequality induced by K.*

Conversely, any generalized inequality $\preccurlyeq$ is induced by a unique convex, pointed cone given by $K_{\preccurlyeq} = \{\mathbf{x} \in \mathbb{R}^d : \mathbf{0} \preccurlyeq \mathbf{x}\}$. In other words, $\preccurlyeq$ is the same relation as $\preccurlyeq_{K_{\preccurlyeq}}$.

Proof Left as an exercise. $\qquad\square$

Example 7.1.11 Here are some examples of generalized inequalities.

1. $K = \mathbb{R}^m_+$ induces the generalized inequality $\mathbf{x} \preccurlyeq_K \mathbf{y}$ if and only if $\mathbf{x}_i \leq \mathbf{y}_i$ for all $i = 1, \ldots, m$. This is often abbreviated to $\mathbf{x} \leq \mathbf{y}$ and is sometimes called the "canonical" generalized inequality on $\mathbb{R}^m$.

2. $K = \{\mathbf{x} \in \mathbb{R}^d : \sqrt{\mathbf{x}_1^2 + \cdots + \mathbf{x}_{d-1}^2} \leq \mathbf{x}_d\}$. This cone is called the *Lorentz cone*, and the corresponding generalized inequality is called a *second-order cone constraint (SOCC)*.

3. Let $m = n^2$ for some $n \in \mathbb{N}$, i.e., consider the space $\mathbb{R}^{n^2}$. Identifying $\mathbb{R}^{n^2} = \mathbb{R}^{n \times n}$ using some (arbitrary) ordering of the coordinates, we think of $\mathbb{R}^{n^2}$ as the space of all $n \times n$ matrices. Let K be the cone of all symmetric matrices that are positive semidefinite; see Definition 1.2.17. The corresponding generalized inequality on $\mathbb{R}^{n^2}$ is called the *positive semidefinite cone constraint*.

We can extend the notion of convex functions to vector valued maps using the notion of generalized inequalities.

Definition 7.1.12   Let $\preccurlyeq_K$ be a generalized inequality on $\mathbb{R}^m$ induced by the cone K. We say that $G\colon \mathbb{R}^d \to \mathbb{R}^m$ is a *K-convex mapping* if

$$G(\lambda \mathbf{x} + (1 - \lambda)\mathbf{y}) \preccurlyeq_K \lambda G(\mathbf{x}) + (1 - \lambda)G(\mathbf{y})$$

for all $\mathbf{x}, \mathbf{y} \in \mathbb{R}^d$ and $\lambda \in (0, 1)$.

Example 7.1.13   Here are some examples of K-convex mappings.

1. Let $K \subseteq \mathbb{R}^m$ be any closed, convex, pointed cone. If $G\colon \mathbb{R}^d \to \mathbb{R}^m$ is an affine map, i.e., there exist a matrix $A \in \mathbb{R}^{m \times d}$ and a vector $\mathbf{b} \in \mathbb{R}^m$ such that $G(\mathbf{x}) = A\mathbf{x} + \mathbf{b}$, then G is a K-convex mapping.

2. Let $m = n^2$ for some $n \in \mathbb{N}$, i.e., consider the space $\mathbb{R}^{n^2}$ and let $\preccurlyeq$ be the *positive semidefinite cone constraint* from part 3. of Example 7.1.11, i.e., induced by the cone K of positive semidefinite matrices. Let $A_0, A_1, \ldots, A_d$ be fixed $p \times n$ matrices for some $p \in \mathbb{N}$ (not necessarily equal to n). Define $G\colon \mathbb{R}^d \times \mathbb{R} \to \mathbb{R}^{n^2}$ to be the mapping

$$G(\mathbf{x}, s) = (A_0 + \mathbf{x}_1 A_1 + \cdots + \mathbf{x}_d A_d)^T (A_0 + \mathbf{x}_1 A_1 + \cdots + \mathbf{x}_d A_d) - sI - B,$$

where I is the $n \times n$ identity matrix and B is an arbitrary $n \times n$ symmetric matrix. Then G is a K-convex mapping.

3. Let $K = \mathbb{R}^m_+$, and let $g_1, \ldots, g_m \colon \mathbb{R}^d \to \mathbb{R}$ be convex functions. Let $G\colon \mathbb{R}^d \to \mathbb{R}^m$ be defined as $G(\mathbf{x}) = (g_1(\mathbf{x}), \ldots, g_m(\mathbf{x}))$, then G is a K-convex mapping.

We can now define a very general framework for (continuous) convex optimization problems, which is more concrete than the abstraction level of black-box first-order oracles, but it is still flexible enough to incorporate the majority of convex optimization problems that show up in practice. This will enable us to come up with a notion of an optimality certificate that can be used for ϵ-approximate solutions.

Definition 7.1.14   Let $f \colon \mathbb{R}^d \to \mathbb{R}$ be a convex function, let $K \subseteq \mathbb{R}^m$ be a closed, convex, pointed cone, and let $G \colon \mathbb{R}^d \to \mathbb{R}^m$ be a K-convex mapping. Then f, K, G define a *convex optimization problem with generalized constraints* given as follows:

$$\inf\{f(\mathbf{x}) \ : \ G(\mathbf{x}) \preccurlyeq_K \mathbf{0}\}. \tag{7.1.3}$$

Exercise 10 from Section 7.1.3 shows that the set $C = \{\mathbf{x} \in \mathbb{R}^d : G(\mathbf{x}) \preccurlyeq_K \mathbf{0}\}$ is a convex set, when G is a K-convex mapping. Thus, (7.1.3) is a special case of (7.1.1).

Example 7.1.15   Let us look at some concrete examples of (7.1.3).

1. **Linear/quadratic programming.** Let $f(\mathbf{x}) = \langle \mathbf{c}, \mathbf{x} \rangle$ for some $\mathbf{c} \in \mathbb{R}^d$, let $K = \mathbb{R}^m_+$, and let $G \colon \mathbb{R}^d \to \mathbb{R}^m$ be an affine map, i.e., $G(\mathbf{x}) = A\mathbf{x} - \mathbf{b}$ for some matrix $A \in \mathbb{R}^{m \times d}$ and a vector $\mathbf{b} \in \mathbb{R}^m$. Then (7.1.3) becomes

$$\inf\{\langle \mathbf{c}, \mathbf{x} \rangle \ : \ A\mathbf{x} \leq \mathbf{b}\},$$

   which is the problem of minimizing a linear function over a polyhedron. This is more commonly known as a *linear program*, in accordance with the fact that the objective and the constraints are all linear.

   If $f(\mathbf{x}) = \mathbf{x}^T Q \mathbf{x} + \langle \mathbf{c}, \mathbf{x} \rangle$ where Q is a given $d \times d$ positive semidefinite matrix, and $\mathbf{c} \in \mathbb{R}^d$, then f is a convex function (see Exercise 11 from Section 3.2.4). With K and G as above, (7.1.3) is called a *convex quadratic program*.

2. **Semidefinite programming.** Let $m = n^2$ for some $n \in \mathbb{N}$ and consider the space $\mathbb{R}^{n^2}$. Let $f(\mathbf{x}) = \langle \mathbf{c}, \mathbf{x} \rangle$ for some $\mathbf{c} \in \mathbb{R}^d$, let $K \subseteq \mathbb{R}^{n^2}$ be the positive semidefinite cone, inducing the positive semidefinite cone constraint from part 3. of Example 7.1.11, and let $G \colon \mathbb{R}^d \to \mathbb{R}^{n^2}$ be an affine map, i.e., there exist $n \times n$ matrices $F_0, F_1, \ldots, F_d$ such that $G(\mathbf{x}) = F_0 + \mathbf{x}_1 F_1 + \cdots + \mathbf{x}_d F_d$. Then (7.1.3) becomes

$$\inf\{\langle \mathbf{c}, \mathbf{x} \rangle \ : \ -F_0 - \mathbf{x}_1 F_1 - \cdots - \mathbf{x}_d F_d \text{ is a PSD matrix}\}.$$

   This is known as a *semidefinite program*.

3. **Convex optimization with explicit constraints.** Let $f, g_1, \ldots, g_m \colon \mathbb{R}^d \to \mathbb{R}$ be convex functions. Define $K = \mathbb{R}^m_+$ and define $G \colon \mathbb{R}^d \to \mathbb{R}^m$ as $G(\mathbf{x}) = (g_1(\mathbf{x}), \ldots, g_m(\mathbf{x}))$, which is the K-convex mapping from part 3. in Example 7.1.13. Then (7.1.3) becomes

$$\inf\{f(\mathbf{x}) \ : \ g_1(\mathbf{x}) \leq 0, \ldots, g_m(\mathbf{x}) \leq 0\}.$$

Given that one can obtain certificates of optimality for unconstrained optimization by just exhibiting that the subdifferential contains $\mathbf{0}$ (Corollary 7.1.6), we try to convert constrained optimization problems into unconstrained ones. This is the main idea behind Lagrangian duality.

Note that problem (7.1.3) is equivalent to the problem

$$\inf_{\mathbf{x} \in \mathbb{R}^d} f(\mathbf{x}) + I_{-K}(G(\mathbf{x})), \tag{7.1.4}$$

where I_{-K} is the indicator function for the cone $-K$ (see Example 3.1.11, part 1.). It can be shown that the function $I_{-K} \circ G$ is a convex function – see Exercise 11 from Section 7.1.3. Thus, problem (7.1.4) is an unconstrained convex optimization problem. However, indicator functions are nasty to deal with because they are not finite valued, and thus, obtaining subgradients at all points becomes impossible. Thus, we try to replace I_{-K} with a "nicer" penalty function $p: \mathbb{R}^m \to \mathbb{R}$, which is not that wildly discontinuous and is finite valued everywhere. So we would be looking at the problem

$$\inf_{\mathbf{x} \in \mathbb{R}^d} f(\mathbf{x}) + p(G(\mathbf{x})). \tag{7.1.5}$$

What properties should we require from our penalty function? First we would like problem (7.1.5) to be a convex problem; thus, we impose that

$$p \circ G: \mathbb{R}^d \to \mathbb{R} \text{ is a convex function.} \tag{7.1.6}$$

Next, from an optimization perspective, we would like to have a guaranteed relationship between the function $f(\mathbf{x}) + I_{-K}(G(\mathbf{x}))$ and the function $f(\mathbf{x}) + p(G(\mathbf{x}))$. It turns out that a nice property to have is the guarantee that $f(\mathbf{x}) + p(G(\mathbf{x})) \leq f(\mathbf{x}) + I_{-K}(G(\mathbf{x}))$ for all $\mathbf{x} \in \mathbb{R}^d$. This can be achieved by imposing that

$$p \text{ is an under-estimator of } I_{-K}, \text{ i.e., } p \leq I_{-K}. \tag{7.1.7}$$

Lagrangian duality theory is the study of penalty functions p that are *linear* on $\mathbb{R}^m$ and satisfy the two conditions highlighted above. The following proposition characterizes linear functions that satisfy the two conditions above.

Proposition 7.1.16 *Let $p: \mathbb{R}^m \to \mathbb{R}$ be a linear function given by $p(\mathbf{z}) = \langle \mathbf{c}, \mathbf{z} \rangle$ for some $\mathbf{c} \in \mathbb{R}^m$. Then the following are equivalent:*

1. *p satisfies condition (7.1.7).*
2. *$\mathbf{c} \in -K^\circ$, i.e., $-\mathbf{c}$ is in the polar of K.*
3. *p satisfies conditions (7.1.6) and (7.1.7).*

Proof (1. $\implies$ 2.) Condition (7.1.7) is equivalent to saying that $p(\mathbf{z}) \leq 0$ for all $\mathbf{z} \in -K$, i.e.,

$$
\begin{aligned}
\langle \mathbf{c}, \mathbf{z} \rangle &\leq 0 && \text{for all } \mathbf{z} \in -K \\
\Leftrightarrow \langle \mathbf{c}, -\mathbf{z} \rangle &\leq 0 && \text{for all } \mathbf{z} \in K \\
\Leftrightarrow \langle -\mathbf{c}, \mathbf{z} \rangle &\leq 0 && \text{for all } \mathbf{z} \in K \\
\Leftrightarrow -\mathbf{c} &\in K^\circ \\
\Leftrightarrow \mathbf{c} &\in -K^\circ.
\end{aligned}
$$

(2. $\implies$ 3.) We showed above that assuming $\mathbf{c} \in -K^\circ$ is equivalent to condition (7.1.7). We now check that $\mathbf{c} \in -K^\circ$ implies (7.1.6). Since G is a K-convex mapping, we have that for all $\mathbf{x}, \mathbf{y} \in \mathbb{R}^d$ and $\lambda \in (0,1)$,

$$
\begin{aligned}
& \langle \mathbf{c}, \lambda G(\mathbf{x}) + (1-\lambda)G(\mathbf{y}) - G(\lambda \mathbf{x} + (1-\lambda)\mathbf{y}) \rangle \geq 0 \\
\implies & \qquad \langle \mathbf{c}, \lambda G(\mathbf{x}) \rangle + \langle \mathbf{c}, (1-\lambda)G(\mathbf{y}) \rangle \geq \langle \mathbf{c}, G(\lambda \mathbf{x} + (1-\lambda)\mathbf{y}) \rangle \\
\implies & \qquad \lambda \langle \mathbf{c}, G(\mathbf{x}) \rangle + (1-\lambda)\langle \mathbf{c}, G(\mathbf{y}) \rangle \geq \langle \mathbf{c}, G(\lambda \mathbf{x} + (1-\lambda)\mathbf{y}) \rangle \\
\implies & \qquad \lambda p(G(\mathbf{x})) + (1-\lambda)p(G(\mathbf{y})) \geq p(G(\lambda \mathbf{x} + (1-\lambda)\mathbf{y})).
\end{aligned}
$$

Hence, condition (7.1.6) is satisfied.

(3. $\implies$ 1.) Trivial. $\qquad\square$

Definition 7.1.17 The set $-K^\circ$ is important in Lagrangian duality, and a separate notation and name have been invented: $-K^\circ$ is called the *dual cone of K* and is denoted by $K^\star$.

The above discussions show that for any $\mathbf{y} \in K^\star$, the optimal value of (7.1.5), with p given by $p(\mathbf{z}) = \langle \mathbf{y}, \mathbf{z} \rangle$, is a lower bound on the optimal value of (7.1.3). This motivates the definition of the so-called *dual function* $\mathcal{L} \colon \mathbb{R}^m \to \mathbb{R}$ associated with (7.1.3) as follows:

$$
\mathcal{L}(\mathbf{y}) := \inf_{\mathbf{x} \in \mathbb{R}^d} f(\mathbf{x}) + \langle \mathbf{y}, G(\mathbf{x}) \rangle. \tag{7.1.8}
$$

We state the lower bound property formally.

Proposition 7.1.18 (Weak Duality) *Let $f \colon \mathbb{R}^d \to \mathbb{R}$ be convex; let $K \subseteq \mathbb{R}^m$ be a closed, convex, pointed cone; and let $G \colon \mathbb{R}^d \to \mathbb{R}^m$ be a K-convex mapping. Let $\mathcal{L} \colon \mathbb{R}^m \to \mathbb{R}$ be as defined in (7.1.8). Then, for all $\bar{\mathbf{x}} \in \mathbb{R}^d$ such that $G(\bar{\mathbf{x}}) \preccurlyeq_K \mathbf{0}$ and all $\bar{\mathbf{y}} \in K^\star$, we must have $\mathcal{L}(\bar{\mathbf{y}}) \leq f(\bar{\mathbf{x}})$. Consequently, $\mathcal{L}(\bar{\mathbf{y}}) \leq \inf\{f(\mathbf{x}) : G(\mathbf{x}) \preccurlyeq_K \mathbf{0}\}$.*

Proof We simply follow the inequalities

$$
\begin{aligned}
\mathcal{L}(\bar{\mathbf{y}}) &= \inf_{\mathbf{x} \in \mathbb{R}^d} f(\mathbf{x}) + \langle \bar{\mathbf{y}}, G(\mathbf{x}) \rangle \\
&\leq f(\bar{\mathbf{x}}) + \langle \bar{\mathbf{y}}, G(\bar{\mathbf{x}}) \rangle \\
&\leq f(\bar{\mathbf{x}}),
\end{aligned}
$$

where the last inequality holds because $G(\mathbf{x}) \preccurlyeq_K \mathbf{0}$ and $\bar{\mathbf{y}} \in K^\star$, and so $\langle \bar{\mathbf{y}}, G(\bar{\mathbf{x}}) \rangle \leq 0$. $\qquad\qquad\qquad\qquad\qquad\qquad\qquad\qquad\qquad\qquad\qquad\square$

Proposition 7.1.18 shows that any $\mathbf{y} \in K^\star$ provides the lower bound $\mathcal{L}(\mathbf{y})$ on the optimal value of the optimization problem (7.1.3). The *Lagrangian dual optimization problem* is the problem of finding the $\mathbf{y} \in K^\star$ that provides the *best/largest* lower bound. In other words, the Lagrangian dual problem is defined as

$$\sup_{\mathbf{y} \in K^\star} \mathcal{L}(\mathbf{y}), \tag{7.1.9}$$

and Proposition 7.1.18 can be restated as

$$\sup\{\mathcal{L}(\mathbf{y}) : \mathbf{y} \in K^\star\} \;\leq\; \inf\{f(\mathbf{x}) \;:\; G(\mathbf{x}) \preccurlyeq_K \mathbf{0}\}. \tag{7.1.10}$$

The original problem $\inf\{f(\mathbf{x}) : G(\mathbf{x}) \preccurlyeq_K \mathbf{0}\}$ is called the *primal optimization problem*.

Explicit examples of the Lagrangian dual. We will now explore some special settings of convex optimization problems with generalized inequalities and see that the Lagrangian dual has a particularly nice form.

1. **Conic optimization.** Let $K \subseteq \mathbb{R}^m$ be a closed, convex, pointed cone. Let $G \colon \mathbb{R}^d \to \mathbb{R}^m$ be an affine map given by $G(\mathbf{x}) = A\mathbf{x} - \mathbf{b}$, where $A \in \mathbb{R}^{m \times d}$ and $\mathbf{b} \in \mathbb{R}^m$. Let $f \colon \mathbb{R}^d \to \mathbb{R}$ be a linear function given by $f(\mathbf{x}) = \langle \mathbf{c}, \mathbf{x} \rangle$ for some $\mathbf{c} \in \mathbb{R}^d$. Then Problem (7.1.3) becomes

$$\inf\{\langle \mathbf{c}, \mathbf{x} \rangle \;:\; A\mathbf{x} \preccurlyeq_K \mathbf{b}\}. \tag{7.1.11}$$

   For a fixed cone K, problems of the form (7.1.11) are called *conic optimization problems over the cone K*. As we pick different data $A, \mathbf{b}, \mathbf{c}$, we get different instances of a conic optimization problem over the cone K. A special case is when $K = \mathbb{R}^m_+$, which is known as *linear programming or linear optimization* – see part 1. in Example 7.1.15 – which is the problem of optimizing a linear function over a polyhedron.

   Let us investigate the dual function of (7.1.11). Recall that $\mathcal{L}(\mathbf{y}) = \inf_{\mathbf{x} \in \mathbb{R}^d} f(\mathbf{x}) + \langle \mathbf{y}, G(\mathbf{x}) \rangle$, which in this case becomes

$$\begin{aligned}
\inf_{\mathbf{x} \in \mathbb{R}^d} \langle \mathbf{c}, \mathbf{x} \rangle + \langle \mathbf{y}, A\mathbf{x} - \mathbf{b} \rangle &= \inf_{\mathbf{x} \in \mathbb{R}^d} \langle \mathbf{c}, \mathbf{x} \rangle + \langle \mathbf{y}, A\mathbf{x} \rangle - \langle \mathbf{y}, \mathbf{b} \rangle \\
&= \inf_{\mathbf{x} \in \mathbb{R}^d} \langle \mathbf{c}, \mathbf{x} \rangle + \langle A^T \mathbf{y}, \mathbf{x} \rangle - \langle \mathbf{y}, \mathbf{b} \rangle \\
&= \inf_{\mathbf{x} \in \mathbb{R}^d} \langle \mathbf{c} + A^T \mathbf{y}, \mathbf{x} \rangle - \langle \mathbf{y}, \mathbf{b} \rangle.
\end{aligned}$$

Now, if $\mathbf{c} + A^T \mathbf{y} \neq \mathbf{0}$, then the infimum above is clearly $-\infty$. And if $\mathbf{c} + A^T \mathbf{y} = \mathbf{0}$, then the infimum is $-\langle \mathbf{b}, \mathbf{y} \rangle$. Therefore, for (7.1.11), the dual function is given by

$$\mathcal{L}(\mathbf{y}) = \begin{cases} -\infty & \text{if } \mathbf{c} + A^T\mathbf{y} \neq \mathbf{0}, \\ -\langle \mathbf{b}, \mathbf{y} \rangle & \text{if } \mathbf{c} + A^T\mathbf{y} = \mathbf{0}. \end{cases} \qquad (7.1.12)$$

Therefore,

$$\sup_{\mathbf{y} \in K^\star} \mathcal{L}(\mathbf{y}) = \sup\{-\langle \mathbf{b}, \mathbf{y} \rangle : A^T\mathbf{y} = -\mathbf{c}, \ \mathbf{y} \in K^\star\}$$

$$= -\inf\{\langle \mathbf{b}, \mathbf{y} \rangle : A^T\mathbf{y} = -\mathbf{c}, \ \mathbf{y} \in K^\star\}.$$

To remove the slightly annoying minus sign in front of $\mathbf{c}$ above, it is more standard to write (7.1.11) as $-\sup\{\langle -\mathbf{c}, \mathbf{x} \rangle : A\mathbf{x} \preccurlyeq_K \mathbf{b}\}$, and then replace $-\mathbf{c}$ with $\mathbf{c}$ throughout the above derivation. Thus, the standard primal dual pairs for conic optimization problems are

$$\sup\{\langle \mathbf{c}, \mathbf{x} \rangle : A\mathbf{x} \preccurlyeq_K \mathbf{b}\} \leq \inf\{\langle \mathbf{b}, \mathbf{y} \rangle : A^T\mathbf{y} = \mathbf{c}, \ \mathbf{y} \in K^\star\}. \quad (7.1.13)$$

Linear Programming/Optimization. Specializing to the linear programming case with $K = \mathbb{R}^m_+$ and observing that $K^\star = K = \mathbb{R}^m_+$ (see Problem 8 from Section 2.4.4), we obtain the primal dual pair

$$\sup\{\langle \mathbf{c}, \mathbf{x} \rangle : A\mathbf{x} \leq \mathbf{b}\} \leq \inf\{\langle \mathbf{b}, \mathbf{y} \rangle : A^T\mathbf{y} = \mathbf{c}, \ \mathbf{y} \geq \mathbf{0}\}. \quad (7.1.14)$$

Semidefinite Programming/Optimization. Another special case is that of semidefinite optimization (part 2. in Example 7.1.15). This is the situation when $m = n^2$ and K is the cone of positive semidefinite matrices. $G \colon \mathbb{R}^d \to \mathbb{R}^{n^2}$ is an affine map from $\mathbb{R}^d$ to the space of $n \times n$ matrices. To avoid dealing with asymmetric matrices, G is always assumed to be of the form $G(\mathbf{x}) = \mathbf{x}_1 F_1 + \cdots + \mathbf{x}_d F_d - F_0$, where $F_0, F_1, \ldots, F_d$ are $n \times n$ symmetric matrices.[1] If one works through the algebra in this case and uses the fact that the positive semidefinite cone is *self-dual*, i.e., $K = K^\star$ (Exercise 14 from Section 7.1.3), (7.1.13) becomes

$$\sup\{\langle \mathbf{c}, \mathbf{x} \rangle : \mathbf{x}_1 F_1 + \cdots + \mathbf{x}_d F_d - F_0 \text{ is a PSD matrix}\}$$

$$\leq \inf\{\langle F_0, Y \rangle : \langle F_i, Y \rangle = \mathbf{c}_i, \ Y \text{ is a PSD matrix}\},$$

where $\langle X, Z \rangle = \sum_{i,j} X_{ij} Z_{ij}$ for any pair X, Z of $n \times n$ symmetric matrices.

2. **Convex optimization with explicit constraints and objective.** Recall part 3. of Example 7.1.15, where $K = \mathbb{R}^m_+$, $f, g_1, \ldots, g_m \colon \mathbb{R}^d \to \mathbb{R}$ are convex functions, and $G \colon \mathbb{R}^d \to \mathbb{R}^m$ was defined as $G(\mathbf{x}) = (g_1(\mathbf{x}), \ldots, g_m(\mathbf{x}))$, giving the explicit problem

$$\inf\{f(\mathbf{x}) : g_1(\mathbf{x}) \leq 0, \ldots, g_m(\mathbf{x}) \leq 0\}.$$

---

[1] Dealing with asymmetric matrices is not hard, but it involves little details that can be overlooked for this exposition and do not provide any great insight.

In this case, since $K^\star = K = \mathbb{R}^m_+$ (see Problem 8 from Section 2.4.4), the dual problem is

$$\sup_{\mathbf{y} \in K^\star} \mathcal{L}(\mathbf{y}) = \sup_{\mathbf{y} \geq \mathbf{0}} \inf_{\mathbf{x} \in \mathbb{R}^d} \{ f(\mathbf{x}) + \mathbf{y}_1 g_1(\mathbf{x}) + \cdots, \mathbf{y}_m g_m(\mathbf{x}) \}.$$

ϵ-optimality certificates from the Lagrangian dual. We can now see how the Lagrangian dual problem can provide ϵ-optimality certificates. One produces $\mathbf{x}^\star \in \mathbb{R}^d$, $\mathbf{y}^\star \in \mathbb{R}^m$ and $\hat{\mathbf{x}} \in \mathbb{R}^d$ such that $G(\mathbf{x}^\star) \preccurlyeq_K \mathbf{0}$, $\mathbf{y}^\star \in K^\star$, $\hat{\mathbf{x}}$ is an optimal solution to the problem $\mathcal{L}(\mathbf{y}^\star)$, i.e., $\hat{\mathbf{x}}$ is a minimizer of $\inf_{\mathbf{x} \in \mathbb{R}^d} f(\mathbf{x}) + \langle \mathbf{y}^\star, G(\mathbf{x}) \rangle$, and check that $\mathcal{L}(\mathbf{y}^\star) = f(\hat{\mathbf{x}}) + \langle \mathbf{y}^\star, G(\hat{\mathbf{x}}) \rangle \geq f(\mathbf{x}^\star) - \epsilon$ (note that $\hat{\mathbf{x}}$ need not be a feasible solution to the primal problem, i.e., $-G(\hat{\mathbf{x}})$ may not be in K). Since $\inf_{\mathbf{x} \in \mathbb{R}^d} f(\mathbf{x}) + \langle \mathbf{y}^\star, G(\mathbf{x}) \rangle$ is an unconstrained optimization problem, $\hat{\mathbf{x}}$ is an exact optimal solution if and only if $\mathbf{0}$ is in the subgradient of the convex function $f(\mathbf{x}) + \langle \mathbf{y}^\star, G(\mathbf{x}) \rangle$ at $\hat{\mathbf{x}}$ (Corollary 7.1.6). In the special case of explicit convex constraints (part 3. of Example 7.1.15), this subgradient condition can be expressed very concretely: we want $\mathbf{0}$ to be in the subgradient of the function $f(\mathbf{x}) + \mathbf{y}_1^\star g_i(\mathbf{x}) + \cdots + \mathbf{y}_m^\star g_m(\mathbf{x})$ at $\hat{\mathbf{x}}$. By Theorem 3.4.5, this means $\mathbf{0} \in \partial f(\hat{\mathbf{x}}) + \mathbf{y}_1^\star \partial g_1(\hat{\mathbf{x}}) + \cdots + \mathbf{y}_m^\star \partial g_m(\hat{\mathbf{x}})$, and if $f, g_1, \ldots, g_m$ are all differentiable, then this becomes the condition $\nabla f(\hat{\mathbf{x}}) + \mathbf{y}_1^\star \nabla g_1(\hat{\mathbf{x}}) + \cdots + \mathbf{y}_m^\star \nabla g_m(\hat{\mathbf{x}}) = \mathbf{0}$. In this way, $\mathbf{y}^\star$ and $\hat{\mathbf{x}}$ become a certificate of ϵ-optimality of $\mathbf{x}^\star$. See also below under the paragraph titled "Strong duality, complementary slackness and the KKT conditions."

If we have equality in (7.1.10), then we can produce an exact optimality solution in the manner above. Moreover, to solve (7.1.3), one can instead solve (7.1.9). This merits a definition.

Definition 7.1.19 (Strong Duality) We say that we have a *zero duality gap* if equality holds in (7.1.10). In addition, if the supremum in (7.1.9) is attained for some $\mathbf{y} \in K^\star$, then we say that *strong duality* holds.

Strong duality, complementary slackness, and the KKT conditions. For general conic optimization problems, or a convex optimization problem with generalized inequalities, one may not have zero duality gap. But when we do have strong duality, there is an important characterization of the solutions for the primal and dual problems.

Theorem 7.1.20 *Let* $f : \mathbb{R}^d \to \mathbb{R}$ *be convex, let* $K \subseteq \mathbb{R}^m$ *be a closed, convex, pointed cone, and let* $G : \mathbb{R}^d \to \mathbb{R}^m$ *be a K-convex mapping. Let* $\mathcal{L} : \mathbb{R}^m \to \mathbb{R}$ *be as defined in (7.1.8). Let* $\mathbf{x}^\star$ *be such that* $G(\mathbf{x}^\star) \preccurlyeq_K \mathbf{0}$ *and* $\mathbf{y}^\star \in K^\star$. *Then,* $f(\mathbf{x}^\star) = \mathcal{L}(\mathbf{y}^\star)$ *if and only if* $\langle \mathbf{y}^\star, G(\mathbf{x}^\star) \rangle = 0$ *and* $\mathbf{x}^\star$ *is a minimizer of* $\mathcal{L}(\mathbf{y}^\star)$, *i.e.,* $\mathbf{x}^\star \in \arg\inf_{\mathbf{x} \in \mathbb{R}^d} f(\mathbf{x}) + \langle \mathbf{y}^\star, G(\mathbf{x}) \rangle$.

Proof Since $G(\mathbf{x}^\star) \preccurlyeq_K \mathbf{0}$ and $\mathbf{y}^\star \in K^\star$, we must have $\langle \mathbf{y}^\star, G(\mathbf{x}^\star) \rangle \leq 0$. Therefore,

$$f(\mathbf{x}^\star) \geq f(\mathbf{x}^\star) + \langle \mathbf{y}^\star, G(\mathbf{x}^\star) \rangle \geq \inf_{\mathbf{x} \in \mathbb{R}^d} f(\mathbf{x}) + \langle \mathbf{y}^\star, G(\mathbf{x}) \rangle = \mathcal{L}(\mathbf{y}^\star).$$

Thus, $f(\mathbf{x}^\star) = \mathcal{L}(\mathbf{y}^\star)$ if and only if the two inequalities are equalities. The first inequality is an equality if and only if $\langle \mathbf{y}^\star, G(\mathbf{x}^\star) \rangle = 0$, and the second inequality is an equality if and only if $\mathbf{x}^\star$ is a minimizer of $\mathcal{L}(\mathbf{y}^\star)$. $\square$

The condition $\langle \mathbf{y}^\star, G(\mathbf{x}^\star) \rangle = 0$ is known as *complementary slackness*. For the case of convex optimization with explicit constraints and objective (part 3. of Example 7.1.15), the above characterization translates to the following, in the light of Theorems 7.1.5 and 3.4.5: $\mathbf{x}^\star$ and $\mathbf{y}^\star$ are optimal solutions to the primal and Lagrangian dual problems with zero duality gap if and only if all the following conditions hold:

(i) (Primal feasibility) $g_i(\mathbf{x}^\star) \leq 0$ for all $i = 1, \ldots, m$.
(ii) (Dual feasibility) $\mathbf{y}_i^\star \geq 0$ for all $i = 1, \ldots, m$.
(iii) (Complementary slackness) $\sum_{i=1}^{m} \mathbf{y}_i^\star g_i(\mathbf{x}^\star) = 0$; equivalently, $\mathbf{y}_i^\star g_i(\mathbf{x}^\star) = 0$ for all $i = 1, \ldots, m$.
(iv) (Lagrangian optimality) $0 \in \partial f(\mathbf{x}^\star) + \mathbf{y}_1^\star \partial g_1(\mathbf{x}^\star) + \cdots + \mathbf{y}_m^\star \partial g_m(\mathbf{x}^\star)$. This becomes $\nabla f(\mathbf{x}^\star) + \mathbf{y}_1^\star \nabla g_1(\mathbf{x}^\star) + \cdots + \mathbf{y}_m^\star \nabla g_m(\mathbf{x}^\star) = 0$, when $f, g_1, \ldots, g_m$ are all differentiable.

The above necessary and sufficient conditions are known as the *Karush–Kuhn–Tucker (KKT) conditions* for convex optimization with explicit constraints.

The above characterization is useful when one has a pair $\mathbf{x}^\star, \mathbf{y}^\star$ of feasible primal and dual solutions and wishes to certify optimality of both: one checks $\langle \mathbf{y}^\star, G(\mathbf{x}^\star) \rangle = 0$ and $\mathbf{x}^\star \in \arg\inf_{\mathbf{x} \in \mathbb{R}^d} f(\mathbf{x}) + \langle \mathbf{y}^\star, G(\mathbf{x}) \rangle$. While the second condition has a nice characterization in the case of explicit constraints (KKT conditions above), it is not so easy to certify in general. Moreover, strong duality is a property of the problem but in the above characterization, one has to check the properties of particular solutions $\mathbf{x}^\star, \mathbf{y}^\star$. We now supply two different sufficient conditions that can be checked without having a primal-dual pair $\mathbf{x}^\star, \mathbf{y}^\star$ under which strong duality is obtained. These two conditions will be properties of the problem rather than of a pair of solutions $\mathbf{x}^\star, \mathbf{y}^\star$. Linear programming strong duality will be a special case of the second sufficient condition.

Slater's condition for strong duality. The following is perhaps the most well-known sufficient condition in convex optimization that guarantees strong duality.

Theorem 7.1.21 (Slater's condition) *Let $f \colon \mathbb{R}^d \to \mathbb{R}$ be convex, let $K \subseteq \mathbb{R}^m$ be a closed, convex, pointed cone, and let $G \colon \mathbb{R}^d \to \mathbb{R}^m$ be a K-convex mapping. Let $\mathcal{L} \colon \mathbb{R}^m \to \mathbb{R}$ be as defined in (7.1.8). If there exists $\bar{\mathbf{x}}$ such that $-G(\bar{\mathbf{x}}) \in \mathrm{int}(K)$ and $\inf\{ f(\mathbf{x}) : G(\mathbf{x}) \preccurlyeq_K \mathbf{0}\}$ is a finite value, then there exists $\mathbf{y}^\star \in K^\star$ such that $\mathcal{L}(\mathbf{y}^\star) = \sup_{\mathbf{y} \in K^\star} \mathcal{L}(\mathbf{y}) = \inf\{ f(\mathbf{x}) : G(\mathbf{x}) \preccurlyeq_K \mathbf{0}\}$, i.e., strong duality holds.*

Proof   Let $OPT := \inf\{ f(\mathbf{x}) : G(\mathbf{x}) \preccurlyeq_K \mathbf{0}\} \in \mathbb{R}$. Define the sets

$$A := \{(\mathbf{z}, r) \in \mathbb{R}^m \times \mathbb{R} : \exists \mathbf{x} \in \mathbb{R}^d \text{ such that } f(\mathbf{x}) \le r,\ G(\mathbf{x}) \preccurlyeq_K \mathbf{z}\},$$
$$B := \{(\mathbf{z}, r) \in \mathbb{R}^m \times \mathbb{R} : r < OPT,\ \mathbf{z} \preccurlyeq_K \mathbf{0}\}.$$

It is not hard to verify that A, B are convex. Moreover, it is also not hard to verify that $A \cap B = \emptyset$. By Exercise 5 from Section 2.4.4, there exists $\mathbf{a} \in \mathbb{R}^m \setminus \{\mathbf{0}\}, \gamma \in \mathbb{R}$ such that

$$\langle \mathbf{a}, \mathbf{z}_1 \rangle + \gamma r_1 \ge \langle \mathbf{a}, \mathbf{z}_2 \rangle + \gamma r_2 \tag{7.1.15}$$

for all $(\mathbf{z}_1, r_1) \in A$ and $(\mathbf{z}_2, r_2) \in B$.

Claim 3 $\mathbf{a} \in K^\star$ and $\gamma \ge 0$.

Proof of Claim   Suppose to the contrary that $\mathbf{a} \notin K^\star$. Then $\mathbf{a} \notin -K^\circ = (-K)^\circ$. Thus, there exists $\bar{\mathbf{z}} \in -K$, i.e., $\bar{\mathbf{z}} \preccurlyeq_K \mathbf{0}$, such that $\langle \mathbf{a}, \bar{\mathbf{z}} \rangle > 0$. Now (7.1.15) holds with $\mathbf{z}_1 = G(\bar{\mathbf{x}})$ ($\bar{\mathbf{x}}$ is the point in the hypothesis of the theorem), $r_1 = f(\bar{\mathbf{x}})$, $r_2 = OPT - 1$ and $\mathbf{z}_2 = \lambda \bar{\mathbf{z}}$ for all $\lambda \ge 0$. But since $\langle \mathbf{a}, \bar{\mathbf{z}} \rangle > 0$, the inequality (7.1.15) would be violated for large enough λ. Thus, we must have $\mathbf{a} \in K^\star$.

Similarly, (7.1.15) holds with $\mathbf{z}_1 = G(\bar{\mathbf{x}})$, $r_1 = f(\bar{\mathbf{x}})$, $\mathbf{z}_2 = \mathbf{0}$, and *all* $r_2 < OPT$. If $\gamma < 0$, then letting $r_2 \to -\infty$ would violate (7.1.15). $\square$

We now show that, in fact, $\gamma > 0$ because of the existence of $\bar{\mathbf{x}}$ assumed in the hypothesis of the theorem. Substitute $\mathbf{z}_1 = G(\bar{\mathbf{x}})$, $r_1 = f(\bar{\mathbf{x}})$, $r_2 = OPT - 1$, and $\mathbf{z}_2 = \mathbf{0}$ in (7.1.15). If $\gamma = 0$, then this relation becomes

$$\langle \mathbf{a}, G(\bar{\mathbf{x}}) \rangle \ge 0.$$

However, $-G(\bar{\mathbf{x}}) \in \mathrm{int}(K)$ and $\mathbf{a} \in K^\star \setminus \{\mathbf{0}\}$, and therefore, $\langle \mathbf{a}, G(\bar{\mathbf{x}}) \rangle < 0$ (see Problem 11 from Section 2.2.3).

Let $\mathbf{y}^\star := \frac{\mathbf{a}}{\gamma}$; by Claim 3, $\mathbf{y}^\star \in K^\star$. We will now show that for every $\epsilon > 0$, $\mathcal{L}(\mathbf{y}^\star) \ge OPT - \epsilon$. This will establish the result because this means $\mathcal{L}(\mathbf{y}^\star) \ge OPT$ and since $\mathcal{L}(\mathbf{y}) \le OPT$ for all $\mathbf{y} \in K^\star$ by Proposition 7.1.18, we must have $\mathcal{L}(\mathbf{y}^\star) = \sup_{\mathbf{y} \in K^\star} \mathcal{L}(\mathbf{y}) = OPT$. Consider any $\mathbf{x} \in \mathbb{R}^d$; then $\mathbf{z}_1 = G(\mathbf{x})$ and $r_1 = f(\mathbf{x})$ gives a point in A. Substituting into (7.1.15) with $\mathbf{z}_2 = \mathbf{0}$ and $r_2 = OPT - \epsilon$, we obtain that $\langle \mathbf{a}, G(\mathbf{x}) \rangle + \gamma f(\mathbf{x}) \ge \gamma(OPT - \epsilon)$. Dividing through by γ, we obtain

$$\langle \mathbf{y}^\star, G(\mathbf{x}) \rangle + f(\mathbf{x}) \geq OPT - \epsilon$$

for all $\mathbf{x} \in \mathbb{R}^d$. This implies that $\mathcal{L}(\mathbf{y}^\star) = \inf_{\mathbf{x} \in \mathbb{R}^d} \langle \mathbf{y}^\star, G(\mathbf{x}) \rangle + f(\mathbf{x}) \geq OPT - \epsilon.$ $\qquad\square$

Closed cone condition for strong duality in conic optimization. Slater's condition applied to conic optimization problems translates into requiring that there is some $\bar{\mathbf{x}}$ such that $\mathbf{b} - A\bar{\mathbf{x}} \in \text{int}(K)$. Another very useful strong duality condition uses topological properties of the dual cone $K^\star$.

Theorem 7.1.22 (Closed cone condition) *Consider the conic optimization primal dual pair* (7.1.13). *Suppose the set* $\{(A^T\mathbf{y}, \langle \mathbf{b}, \mathbf{y} \rangle) \in \mathbb{R}^d \times \mathbb{R} : \mathbf{y} \in K^*\}$ *is closed and the dual is feasible, i.e., there exists* $\mathbf{y} \in K^\star$ *such that* $A^T\mathbf{y} = \mathbf{c}$. *Then we have zero duality gap. If the optimal dual value is finite, then strong duality holds in* (7.1.13).

Proof Since the dual is feasible, its optimal value is either $-\infty$ or finite. By weak duality (Proposition 7.1.18), in the first case we must have zero duality gap and the primal is infeasible. So we consider the case when the optimal value of the dual is finite, say, $OPT \in \mathbb{R}$. Let us label the set $S := \{(A^T\mathbf{y}, \langle \mathbf{b}, \mathbf{y} \rangle) : \mathbf{y} \in K^*\} \subseteq \mathbb{R}^d \times \mathbb{R}$. Notice that the optimal value of the dual is $OPT = \inf\{r \in \mathbb{R} : (\mathbf{c}, r) \in S\}$. Since S is assumed to be closed, the set $\{r \in \mathbb{R} : (\mathbf{c}, r) \in S\}$ is closed because it is topologically the same as $S \cap (\mathbf{c} \times \mathbb{R})$. Therefore the infimum in $\inf\{r \in \mathbb{R} : (\mathbf{c}, r) \in S\}$ is over a closed subset of the real line. Hence, $(\mathbf{c}, OPT) \in S$ and so there exists $\mathbf{y}^\star \in K^\star$ such that $A^T\mathbf{y}^\star = \mathbf{c}$ and $\langle \mathbf{b}, \mathbf{y}^\star \rangle = OPT$.

Since $OPT = \inf\{r \in \mathbb{R} : (\mathbf{c}, r) \in S\}$, for every $\epsilon > 0$, $(\mathbf{c}, OPT - \epsilon) \notin S$. Therefore, there exists a separating hyperplane $(\mathbf{a}, \gamma) \in \mathbb{R}^d \times \mathbb{R}$ and $\delta \in \mathbb{R}$ such that $\langle \mathbf{a}, A^T\mathbf{y} \rangle + \gamma \cdot \langle \mathbf{b}, \mathbf{y} \rangle \leq \delta$ for all $\mathbf{y} \in K^\star$, and $\langle \mathbf{a}, \mathbf{c} \rangle + \gamma(OPT - \epsilon) > \delta$. Note that S is a cone since it is the linear transformation of a cone, and by Exercise 43 from Section 2.4.4, we may assume $\delta = 0$. Therefore, we have

$$\langle \mathbf{a}, A^T\mathbf{y} \rangle + \gamma \cdot \langle \mathbf{b}, \mathbf{y} \rangle \leq 0 \quad \text{for all } \mathbf{y} \in K^\star, \tag{7.1.16}$$

$$\langle \mathbf{a}, \mathbf{c} \rangle + \gamma(OPT - \epsilon) > 0. \tag{7.1.17}$$

Substituting $\mathbf{y}^\star$ in (7.1.16), we obtain that $\langle \mathbf{a}, \mathbf{c} \rangle + \gamma OPT \leq 0$, and (7.1.17) tells us that $\langle \mathbf{a}, \mathbf{c} \rangle + \gamma OPT > \gamma\epsilon$. This implies that $\gamma < 0$ since $\epsilon > 0$. Now (7.1.16) can be rewritten as $\langle A\mathbf{a} + \gamma\mathbf{b}, \mathbf{y} \rangle \leq 0$ for all $\mathbf{y} \in K^\star$ and (7.1.17) can be rewritten as $\langle \mathbf{a}, \mathbf{c} \rangle > -\gamma(OPT - \epsilon)$. Dividing through both these relations by $-\gamma > 0$, and setting $\mathbf{x} = \frac{\mathbf{a}}{-\gamma}$, we obtain that $\langle A\mathbf{x} - \mathbf{b}, \mathbf{y} \rangle \leq 0$ for all $\mathbf{y} \in K^\star$ implying that $A\mathbf{x} \preccurlyeq_K \mathbf{b}$ (by part 2. of Theorem 2.4.10), and $\langle \mathbf{x}, \mathbf{c} \rangle > OPT - \epsilon$. Thus, we have a feasible solution $\mathbf{x}$ for the primal with

Table 7.1 *Different possibilities for primal and dual pairs of linear programs*

Primal \ Dual	Infeasible	Finite	Unbounded
Infeasible	Possible	Impossible	Possible
Finite	Impossible	Possible, strong duality	Impossible
Unbounded	Possible	Impossible	Impossible

value of at least $OPT - \epsilon$. Since $\epsilon > 0$ was chosen arbitrarily, this shows that for every $\epsilon > 0$, the primal has optimal value better than $OPT - \epsilon$. Therefore, the primal value must be OPT and we have zero duality gap. The existence of $\mathbf{y}^\star$ shows that we have strong duality. $\qquad\square$

Linear programming strong duality. The closed cone condition for strong duality implies that linear programs always enjoy strong duality when either the primal or the dual (or both) are feasible. This is because the cone $K = \mathbb{R}^m_+$ is a polyhedral cone and also self-dual, i.e., $K^\star = K = \mathbb{R}^m_+$. Since linear transformations of polyhedral cones are polyhedral (see part 8e. of Problem 8 from Section 2.5.6), and hence closed, linear programs always satisfy the condition in Theorem 7.1.22. One therefore has Table 7.1 for the possible outcomes in the primal dual linear programming pair.

An alternate proof of zero duality gap for linear programming follows from our results on polyhedral theory. We outline it here to illustrate that linear programming duality can be approached in different ways (although ultimately both proofs go back to the separating hyperplane theorem – Theorem 2.4.2). We consider two cases:

1. Primal is infeasible. In this case, we will show that if the dual is feasible, then the dual must be unbounded. Since the primal is infeasible, the polyhedron $A\mathbf{x} \leq \mathbf{b}$ is empty. By Theorem 2.5.22, there exists $\hat{\mathbf{y}} \geq \mathbf{0}$ such that $A^T\hat{\mathbf{y}} = \mathbf{0}$ and $\langle \mathbf{b}, \hat{\mathbf{y}} \rangle = -1$. Since the dual is feasible, consider any $\bar{\mathbf{y}} \geq \mathbf{0}$ such that $A^T\bar{\mathbf{y}} = \mathbf{c}$. Now, all points of the form $\bar{\mathbf{y}} + \lambda\hat{\mathbf{y}}$ with $\lambda \geq 0$ are also feasible to the dual, and the corresponding value $\langle \mathbf{b}, \bar{\mathbf{y}} + \lambda\hat{\mathbf{y}} \rangle$ can be made to go to $-\infty$ because $\langle \mathbf{b}, \hat{\mathbf{y}} \rangle = -1$.

2. Primal is feasible. If the primal is unbounded, then by weak duality, the dual must be infeasible. So let us consider the case that the primal has a finite value OPT. This means that the inequality $\langle \mathbf{c}, \mathbf{x} \rangle \leq OPT$ is a valid inequality for the polyhedron $A\mathbf{x} \leq \mathbf{b}$. By Theorem 2.5.19, there exists $\hat{\mathbf{y}} \geq \mathbf{0}$ such that $A^T\hat{\mathbf{y}} = \mathbf{c}$ and $\langle \mathbf{b}, \hat{\mathbf{y}} \rangle \leq OPT$. Therefore the dual has a

solution $\hat{\mathbf{y}}$ whose objective value is equal to the primal value OPT. This guarantees strong duality.

Solving the Lagrangian dual problem. We observed that if one has zero duality gap or strong duality, in order to solve (7.1.3), one can instead solve (7.1.9). Let us try to see how one could use the subgradient descent algorithm (Definition 6.5.2) to solve (7.1.9).

Proposition 7.1.23 $\mathcal{L}(\mathbf{y})$ *is a concave function (Definition 3.1.1) of* $\mathbf{y}$.

Proof We have to show that $-\mathcal{L}(\mathbf{y})$ is a convex function of $\mathbf{y}$. This follows from the fact that

$$
\begin{aligned}
-\mathcal{L}(\mathbf{y}) &= -\inf_{\mathbf{x}\in\mathbb{R}^d} f(\mathbf{x}) + \langle \mathbf{y}, G(\mathbf{x})\rangle \\
&= \sup_{\mathbf{x}\in\mathbb{R}^d} -f(\mathbf{x}) + \langle \mathbf{y}, -G(\mathbf{x})\rangle,
\end{aligned}
$$

that is, $-\mathcal{L}(\mathbf{y})$ is the supremum of affine linear functions of $\mathbf{y}$ of the form $-f(\mathbf{x}) + \langle \mathbf{y}, -G(\mathbf{x})\rangle$. By part 2. of Theorem 3.1.13, $-\mathcal{L}(\mathbf{y})$ is convex in $\mathbf{y}$. $\square$

We could now use the subgradient algorithm to solve (7.1.9) if we had a first-order oracle for $-\mathcal{L}(\mathbf{y})$ and an algorithm to project to $K^\star$. We show that a subgradient for $-\mathcal{L}(\mathbf{y})$ can be found by solving an unconstrained convex optimization problem.

Proposition 7.1.24 *Let* $\bar{\mathbf{y}} \in \mathbb{R}^m$ *and let* $\bar{\mathbf{x}} \in \arg\inf_{\mathbf{x}\in\mathbb{R}^d} f(\mathbf{x}) + \langle\bar{\mathbf{y}}, G(\mathbf{x})\rangle$. *Then* $-G(\bar{\mathbf{x}}) \in \partial(-\mathcal{L})(\bar{\mathbf{y}})$.

Proof We express $-\mathcal{L}(\mathbf{y}) = \sup_{\mathbf{x}\in\mathbb{R}^d} -f(\mathbf{x}) + \langle \mathbf{y}, -G(\mathbf{x})\rangle$ as the supremum of affine linear functions and use part 3. of Theorem 3.4.5, and the fact that the subdifferential of the affine linear function $-f(\bar{\mathbf{x}}) + \langle \mathbf{y}, -G(\bar{\mathbf{x}})\rangle$, at $\bar{\mathbf{y}}$ is simply $\{-G(\bar{\mathbf{x}})\}$. $\square$

Now if we have an algorithm that can compute $\mathrm{Proj}_{K^\star}(\mathbf{y})$ for all $\mathbf{y} \in \mathbb{R}^m$, then using Propositions 7.1.23 and 7.1.24, one can solve the Lagrangian dual problem (7.1.9), where in each iteration of the algorithm, one solves the unconstrained problem $\inf_{\mathbf{x}\in\mathbb{R}^d} f(\mathbf{x}) + \langle\bar{\mathbf{y}}, G(\mathbf{x})\rangle$ for a given $\bar{\mathbf{y}} \in K^\star$. This can, in turn, be solved by the subgradient algorithm if one has the appropriate first-order oracles for $f(\mathbf{x})$ and $\langle\bar{\mathbf{y}}, G(\mathbf{x})\rangle$.

Saddle point interpretation of the Lagrangian dual. Let us go back to the original problem (7.1.3) and revisit the dual function $\mathcal{L}(\mathbf{y})$. Define the function

$$
\hat{\mathcal{L}}(\mathbf{x}, \mathbf{y}) := f(\mathbf{x}) + \langle \mathbf{y}, G(\mathbf{x})\rangle, \tag{7.1.18}
$$

which is often called the *Lagrangian function* associated with (7.1.3). A characterization of a pair of optimal solutions to (7.1.3) and (7.1.9) can be obtained using saddle points of the Lagrangian function.

Theorem 7.1.25 *Let $f: \mathbb{R}^d \to \mathbb{R}$ be convex, let $K \subseteq \mathbb{R}^m$ be a closed, convex, pointed cone, and let $G: \mathbb{R}^d \to \mathbb{R}^m$ be a K-convex mapping. Let $\mathcal{L}: \mathbb{R}^m \to \mathbb{R}$ be as defined in (7.1.8) and $\hat{\mathcal{L}}: \mathbb{R}^d \times \mathbb{R}^m \to \mathbb{R}$ be as defined in (7.1.18). Let $\mathbf{x}^\star$ be such that $G(\mathbf{x}^\star) \preccurlyeq_K \mathbf{0}$ and $\mathbf{y}^\star \in K^\star$. Then the following are equivalent.*

1. $\mathcal{L}(\mathbf{y}^\star) = f(\mathbf{x}^\star)$.
2. $\hat{\mathcal{L}}(\mathbf{x}^\star, \hat{\mathbf{y}}) \leq \hat{\mathcal{L}}(\mathbf{x}^\star, \mathbf{y}^\star) \leq \hat{\mathcal{L}}(\hat{\mathbf{x}}, \mathbf{y}^\star)$ for all $\hat{\mathbf{x}} \in \mathbb{R}^d$ and $\hat{\mathbf{y}} \in K^\star$.

Proof $(1. \implies 2.)$ Consider any $\hat{\mathbf{x}} \in \mathbb{R}^d$ and $\hat{\mathbf{y}} \in K^\star$. We now derive the following chain of inequalities:

$$
\begin{aligned}
\hat{\mathcal{L}}(\mathbf{x}^\star, \hat{\mathbf{y}}) &= f(\mathbf{x}^\star) + \langle \hat{\mathbf{y}}, G(\mathbf{x}^\star) \rangle \\
&\leq f(\mathbf{x}^\star) && \text{since } \langle \hat{\mathbf{y}}, G(\mathbf{x}^\star) \rangle \leq 0 \text{ because } \hat{\mathbf{y}} \in K^\star, G(\mathbf{x}^\star) \preccurlyeq_K \mathbf{0} \\
&= f(\mathbf{x}^\star) + \langle \mathbf{y}^\star, G(\mathbf{x}^\star) \rangle && = \hat{\mathcal{L}}(\mathbf{x}^\star, \mathbf{y}^\star) \text{ since } \langle \mathbf{y}^\star, G(\mathbf{x}^\star) \rangle = 0 \text{ by Theorem 7.1.20} \\
&= \mathcal{L}(\mathbf{y}^\star) && \text{since } \mathcal{L}(\mathbf{y}^\star) = f(\mathbf{x}^\star) \\
&= \inf_{\mathbf{x} \in \mathbb{R}^d} f(\mathbf{x}) + \langle \mathbf{y}^\star, G(\mathbf{x}) \rangle \\
&\leq f(\hat{\mathbf{x}}) + \langle \mathbf{y}^\star, G(\hat{\mathbf{x}}) \rangle \\
&= \hat{\mathcal{L}}(\hat{\mathbf{x}}, \mathbf{y}^\star).
\end{aligned}
$$

$(2. \implies 1.)$ Since $\hat{\mathcal{L}}(\mathbf{x}^\star, \hat{\mathbf{y}}) \leq \hat{\mathcal{L}}(\mathbf{x}^\star, \mathbf{y}^\star)$ for all $\hat{\mathbf{y}} \in K^\star$, we have that

$$
\hat{\mathcal{L}}(\mathbf{x}^\star, \mathbf{y}^\star) \geq \sup_{\mathbf{y} \in K^\star} \hat{\mathcal{L}}(\mathbf{x}^\star, \hat{\mathbf{y}}) = \sup_{\mathbf{y} \in K^\star} f(\mathbf{x}^\star) + \langle \mathbf{y}, G(\mathbf{x}^\star) \rangle = f(\mathbf{x}^\star),
$$

where the last equality follows from the fact that $\langle \mathbf{y}, G(\mathbf{x}^\star) \rangle \leq 0$ for all $\mathbf{y} \in K^\star$. So the supremum is achieved for $\mathbf{y} = \mathbf{0}$. On the other hand, since $\hat{\mathcal{L}}(\mathbf{x}^\star, \mathbf{y}^\star) \leq \hat{\mathcal{L}}(\hat{\mathbf{x}}, \mathbf{y}^\star)$ for all $\hat{\mathbf{x}} \in \mathbb{R}^d$, we have that

$$
\hat{\mathcal{L}}(\mathbf{x}^\star, \mathbf{y}^\star) \leq \inf_{\mathbf{x} \in \mathbb{R}^d} \hat{\mathcal{L}}(\mathbf{x}, \mathbf{y}^\star) = \inf_{\mathbf{x} \in \mathbb{R}^d} f(\mathbf{x}) + \langle \mathbf{y}^\star, G(\mathbf{x}) \rangle = \mathcal{L}(\mathbf{y}^\star).
$$

Thus, combined with weak duality (Proposition 7.1.18), we get $\mathcal{L}(\mathbf{y}^\star) \leq f(\mathbf{x}^\star) \leq \hat{\mathcal{L}}(\mathbf{x}^\star, \mathbf{y}^\star) \leq \mathcal{L}(\mathbf{y}^\star)$. Therefore, we obtain that $f(\mathbf{x}^\star) = \hat{\mathcal{L}}(\mathbf{x}^\star, \mathbf{y}^\star) = \mathcal{L}(\mathbf{y}^\star)$. $\square$

Theorem 7.1.25 says that $\mathbf{x}^\star$ and $\mathbf{y}^\star$ are solutions for the primal problem (7.1.3) and dual problem (7.1.9), respectively, with zero duality gap, if and only if $(\mathbf{x}^\star, \mathbf{y}^\star)$ is a saddle point for the function $\hat{\mathcal{L}}(\mathbf{x}, \mathbf{y})$ of the type that $\mathbf{x}^\star$ is the minimizer when $\mathbf{y}$ is fixed at $\mathbf{y}^\star$ and $\mathbf{y}^\star$ is the maximizer when $\mathbf{x}$ is fixed at $\mathbf{x}^\star$. This can be used to directly solve (7.1.3) and (7.1.9) simultaneously by searching for such saddle points of the function $\hat{\mathcal{L}}(\mathbf{x}, \mathbf{y})$. This approach can be useful if one has analytical forms for f and G (with sufficient differentiable properties) so that finding saddle points is a reasonable option.

7.1.3 Exercises

1. Let $g \colon \mathbb{R}^d \to \mathbb{R}$ be any function (not necessarily convex) and let $X \subseteq \mathbb{R}^d$ be any set (not necessarily convex). Then $\mathbf{x}^\star \in X$ is said to be a *local minimizer* for the problem $\inf_{\mathbf{x} \in X} g(\mathbf{x})$ if there exists $\epsilon > 0$ such that $g(\mathbf{y}) \geq g(\mathbf{x}^\star)$ for all $\mathbf{y} \in B(\mathbf{x}^\star, \epsilon) \cap X$. $\mathbf{x}^\star \in X$ is said to be a *global minimizer* if $g(\mathbf{y}) \geq g(\mathbf{x}^\star)$ for all $\mathbf{y} \in X$.

   Show that any local minimizer for (7.1.1) is a global minimizer.

2. Show that the normal cone and the tangent cones are polars of each other, i.e., for any closed, convex set $C \subseteq \mathbb{R}^d$ and $\mathbf{x} \in C$, $T_C(\mathbf{x})^\circ = N_C(\mathbf{x})$.

3. Let $C \subseteq \mathbb{R}^d$ be a closed, convex set. Show that $\cap_{\mathbf{x} \in C} T_C(\mathbf{x}) = \operatorname{rec}(C)$. Conclude that $\operatorname{cl}(\operatorname{conv}(\cup_{\mathbf{x} \in C} N_C(\mathbf{x}))) = \operatorname{rec}(C)^\circ$.

4. Let $C \subseteq \mathbb{R}^d$ be a closed, convex set and let $\mathbf{x} \in C$. Show that $\mathbf{r} \in \operatorname{int}(N_C(\mathbf{x}))$ implies that $\mathbf{x}$ is the unique solution (maximizer) to the problem $\sup_{\mathbf{y} \in C} \langle \mathbf{r}, \mathbf{y} \rangle$. Is the converse true? Conclude that if $\mathbf{x}, \mathbf{y} \in C$ are distinct points (i.e., $\mathbf{x} \neq \mathbf{y}$), then $\operatorname{int}(N_C(\mathbf{x})) \cap \operatorname{int}(N_C(\mathbf{y})) = \emptyset$.

5. Let $P = \{\mathbf{x} \in \mathbb{R}^d : A\mathbf{x} \leq \mathbf{b}\}$ be a polyhedron. Recall the definition of $\operatorname{tight}(\mathbf{x})$, $A_{\operatorname{tight}(\mathbf{x})}$, and $\mathbf{b}_{\operatorname{tight}(\mathbf{x})}$ for any $\mathbf{x} \in P$ from Definition 2.5.1 to denote the tight inequalities at $\mathbf{x}$. Show that $T_C(\mathbf{x}) = \{\mathbf{r} \in \mathbb{R}^d : A_{\operatorname{tight}(\mathbf{x})}\mathbf{r} \leq \mathbf{0}\}$, and $N_C(\mathbf{x})$ is given by the cone formed by the rows of $A_{\operatorname{tight}(\mathbf{x})}$.

6. Let $C = \{\mathbf{x} \in \mathbb{R}^d : g_1(\mathbf{x}) \leq 0, \ldots, g_m(\mathbf{x}) \leq 0\}$ be a closed, convex set given by explicit closed, convex constraints $g_1, \ldots, g_m$ (Example 7.1.15, part 3.). Let $\mathbf{x}^\star \in C$. Let $J = \{j \in \{1, \ldots, m\} : g_j(\mathbf{x}^\star) = 0\}$, i.e., the set of constraints that are satisfied at equality at $\mathbf{x}^\star$. Show that

$$\left\{ \sum_{j \in J} \lambda_j \mathbf{s}^j : \mathbf{s}^j \in \partial g_j(\mathbf{x}^\star), \lambda_j \geq 0 \ \forall j \in J \right\} \subseteq N_C(\mathbf{x}^\star).$$

7. Prove Proposition 7.1.10 from the lecture notes.

8. Let $m = n^2$ and consider the subspace of $\mathbb{R}^m$ corresponding to symmetric matrices and let $\preceq$ be the *positive semidefinite cone constraint* from part 3. of Example 7.1.11, i.e., induced by the cone K of positive semidefinite matrices. Let $G \colon \mathbb{R}^d \to \mathbb{R}^m$ be some map. Show that G is a K-convex mapping if and only if $\mathbf{z}^T G(\cdot)\mathbf{z}$ is a convex function, viewed as a map from $\mathbb{R}^d$ to $\mathbb{R}$, for every $\mathbf{z} \in \mathbb{R}^m$.

9. Establish parts 1., 2., and 3. from Example 7.1.13.

10. Let $K \subseteq \mathbb{R}^m$ be a closed, convex, pointed cone, and let $G \colon \mathbb{R}^d \to \mathbb{R}^m$ be a K-convex mapping. Show that the set $C = \{\mathbf{x} \in \mathbb{R}^d : G(\mathbf{x}) \preceq_K \mathbf{0}\}$ is a convex set.

11. Let $K \subseteq \mathbb{R}^m$ be a closed, convex, pointed cone, and let $G : \mathbb{R}^d \to \mathbb{R}^m$ be a K-convex mapping. Show that the function $I_{-K} \circ G : \mathbb{R}^d \to \mathbb{R} \cup \{+\infty\}$ is a convex function, where I_{-K} is the indicator function for the cone $-K$ (see Example 3.1.11, part 1. for the definition of indicator function).

12. Let $K \subseteq \mathbb{R}^m$ be a closed, convex, pointed cone, and let $G : \mathbb{R}^d \to \mathbb{R}^m$ be a K-convex mapping. Let $p : \mathbb{R}^m \to \mathbb{R} \cup \{+\infty\}$ be a sublinear function such that $p(\mathbf{x}) \leq 0$ for all $\mathbf{x} \in -K$. Show that $p \circ G : \mathbb{R}^d \to \mathbb{R} \cup \{+\infty\}$ is a convex function.

13. Let $f : \mathbb{R}^d \to \mathbb{R}$ be a convex function. Let $K \subseteq \mathbb{R}^m$ be a closed, convex, pointed cone, and let $G : \mathbb{R}^d \to \mathbb{R}^m$ be a K-convex mapping. Recall the Lagrangian function $\hat{\mathcal{L}}(\mathbf{x}, \mathbf{y}) := f(\mathbf{x}) + \langle \mathbf{y}, G(\mathbf{x}) \rangle$ associated with the convex optimization problem $\inf\{f(\mathbf{x}) : G(\mathbf{x}) \preceq_K \mathbf{0}\}$. The Lagrangian dual problem was defined as $\sup_{\mathbf{y} \in K^\star} \inf_{\mathbf{x} \in \mathbb{R}^d} \hat{\mathcal{L}}(\mathbf{x}, \mathbf{y})$. Show that

$$\inf_{\mathbf{x} \in \mathbb{R}^d} \sup_{\mathbf{y} \in K^\star} \hat{\mathcal{L}}(\mathbf{x}, \mathbf{y}) = \inf\{f(\mathbf{x}) : G(\mathbf{x}) \preceq_K \mathbf{0}\}.$$

14. Show that, in the space of symmetric matrices, the cone of positive semidefinite matrices is its own dual.

7.2 The General Mixed-Integer Case

We next consider the notion of a certificate for the mixed-integer convex optimization problem (6.1.1). The idea is to use certificates for continuous convex optimization explored in Section 7.1 and the idea of disjunctions from Definition 6.4.1 as building blocks.

Proposition 7.2.1 *Let* $f : \mathbb{R}^n \times \mathbb{R}^d \to \mathbb{R}$ *be a convex function and* $C \subseteq \mathbb{R}^n \times \mathbb{R}^d$. *Let* $\hat{\mathbf{z}} \in C \cap (\mathbb{Z}^n \times \mathbb{R}^d)$ *and let* D *be a disjunction (Definition 6.4.1); i.e.,* $D = Q_1 \cup \cdots \cup Q_k$ *is the union of closed, convex sets with* $\mathbb{Z}^n \times \mathbb{R}^d \subseteq D$. *Then* $f(\hat{\mathbf{z}}) \geq \min_{i=1,\dots,k} \inf_{\mathbf{y} \in C \cap Q_i} f(\mathbf{y})$.

Proof This follows from the fact that $C \cap (\mathbb{Z}^n \times \mathbb{R}^d) \subseteq C \cap D = \bigcup_{i=1}^k C \cap Q_i$. Thus, $\hat{\mathbf{z}} \in C \cap Q_{\hat{i}}$ for some $\hat{i} = 1, \dots, k$ and so $f(\hat{\mathbf{z}}) \geq \inf_{\mathbf{y} \in C \cap Q_{\hat{i}}} f(\mathbf{y})$. $\square$

Theorem 7.2.2 *Let* $f : \mathbb{R}^n \times \mathbb{R}^d \to \mathbb{R}$ *be a convex function and* $C \subseteq \mathbb{R}^n \times \mathbb{R}^d$. *Let* $\epsilon > 0$. *Let* $\mathbf{z}^\star \in C \cap (\mathbb{Z}^n \times \mathbb{R}^d)$ *and let* D *be a disjunction (Definition 6.4.1), i.e.,* $D = Q_1 \cup \cdots \cup Q_k$ *is the union of closed, convex sets with* $\mathbb{Z}^n \times \mathbb{R}^d \subseteq D$. *Let* $\mathbf{y}^i$ *be* $\frac{\epsilon}{2}$-*approximate solutions for the problems* $\inf_{\mathbf{y} \in C \cap Q_i} f(\mathbf{y})$ *such that* $f(\mathbf{z}^\star) \leq f(\mathbf{y}^i) + \frac{\epsilon}{2}$ *for all* $i = 1, \dots, k$. *Then* $\mathbf{z}^\star$ *is an* ϵ-*approximate solution for* (6.1.1).

Proof By Proposition 7.2.1,

$$\inf\{f(\mathbf{z}) : \mathbf{z} \in C \cap (\mathbb{Z}^n \times \mathbb{R}^d)\} \geq \min_{i=1,\ldots,k} \inf\{f(\mathbf{y}) : \mathbf{y} \in C \cap Q_i\}$$
$$\geq \min_{i=1,\ldots,k} f(\mathbf{y}^i) - \tfrac{\epsilon}{2}$$
$$\geq f(\mathbf{z}^\star) - \tfrac{\epsilon}{2} - \tfrac{\epsilon}{2}$$
$$\geq f(\mathbf{z}^\star) - \epsilon,$$

where the second inequality follows from the fact that $\mathbf{y}^i$ is an $\tfrac{\epsilon}{2}$-approximate solution to $\inf_{\mathbf{y} \in C \cap Q_i} f(\mathbf{y})$, and the third inequality follows from the hypothesis that $f(\mathbf{z}^\star) \leq f(\mathbf{y}^i) + \tfrac{\epsilon}{2}$ for all $i = 1,\ldots,k$. $\square$

Theorem 7.2.2 suggests that an ϵ-optimality certificate of a mixed-integer solution $\mathbf{z}^\star$ is given by a disjunction and k certificates of $\tfrac{\epsilon}{2}$-optimality of continuous convex optimization problems where k is the number of terms in the disjunction. Notice that the branch-and-cut method from Section 6.4 produces such a certificate of optimality at the end where the disjunction is given by the feasible regions of the continuous convex optimization problems where the pruning step (Step 2b in Definition 6.4.5) continues the while loop.

Typically, for a disjunction-based certificate of optimality, one also needs to provide a proof or certificate that $D = Q_1 \cup \cdots \cup Q_k$ indeed covers $\mathbb{Z}^n \times \mathbb{R}^d$. This is often done by taking each $Q_i := \{\mathbf{z} : \langle \mathbf{a}^i, \mathbf{z} \rangle \leq \delta_i\}$ to be a halfspace and showing that the polyhedron $\{z : \langle \mathbf{a}^i, \mathbf{z} \rangle \geq \delta_i\}$ has no point from $\mathbb{Z}^n \times \mathbb{R}^d$ in its interior; equivalently, the projection of the polyhedron on to $\mathbb{R}^n \times \{0\}$ has no point from $\mathbb{Z}^n$ in its interior (assuming both polyhedra are full-dimensional). This connects with the theory of lattice-free convex sets (Section 4.4), and under certain conditions, it can be shown that such certificates always exist. Thus, one has sufficient conditions for strong duality in the mixed-integer case. See Section 7.3 for references to this line of work.

7.3 Notes and Related Literature

As noted in the discussion below Theorem 7.1.5, to be able to use the necessary and sufficient condition $\mathbf{0} \in \partial f(\mathbf{x}^\star) + N_C(\mathbf{x}^\star)$ for $\mathbf{x}^\star$ to be an optimal solution to $\inf\{f(\mathbf{x}) : \mathbf{x} \in C\}$, one usually needs an explicit description of $N_C(\mathbf{x}^\star)$. When C is a polyhedron, this is easy enough (Exercise 5 from Section 7.1.3). One can sometimes generalize this to the nonlinear situation where C is given by explicit convex constraints $g_1(\mathbf{x}) \leq 0, \ldots, g_m(\mathbf{x}) \leq 0$: one considers the constraints g_j, $j \in J \subseteq \{1,\ldots,m\}$ that are satisfied at equality at $\mathbf{x}^\star$ and by Exercise 6 from Section 7.1.3, conical combinations of subgradients of g_j at $\mathbf{x}^\star$ are always contained in the normal cone $N_C(\mathbf{x}^\star)$.

Therefore, a set of subgradients $\mathbf{s}^j \in \partial g_j(\mathbf{x}^\star)$ and scalars $\lambda_j \geq 0$ for $j \in J$ such that $-\sum_{j \in J} \lambda_j \mathbf{s}^j \in \partial f(\mathbf{x}^\star)$ gives a certificate of optimality of $\mathbf{x}^\star$. Conditions on C under which the normal cone $N_C(\mathbf{x}^\star)$ is precisely the set of all conical combinations of subgradients of g_j at $\mathbf{x}^\star$, i.e., equality holds in the subset relation in Exercise 6 from Section 7.1.3, are called *constraint qualification conditions*. As noted above, one such condition is that C is a polyhedron. Another condition is when strong duality holds; this was discussed as the Karush–Kuhn–Tucker conditions under Theorem 7.1.20. Thus, Slater's condition (Theorem 7.1.21) and the closed cone condition (Theorem 7.1.22) are both considered constraint qualification conditions. More on constraint qualification can be found in the monographs [140] and [185].

Our presentation of convex optimization with generalized inequalities, Lagrangian duality and Slater's condition, including most of the proofs, very closely follows the exposition in [170]. The closed cone condition (Theorem 7.1.22) for strong duality and its proof is taken from [24].

In our presentation of Lagrangian duality for convex optimization with generalized inequalities, the dual function $\mathcal{L}(\mathbf{y}) = \inf_{\mathbf{x} \in \mathbb{R}^d} f(\mathbf{x}) + \langle \mathbf{y}, G(\mathbf{x}) \rangle$ is presented as an unconstrained problem. There is a more general way to formulate this. Instead of only allowing constraints of the form $G(\mathbf{x}) \preccurlyeq_K \mathbf{0}$, one considers the problem $\inf\{f(\mathbf{x}) : G(\mathbf{x}) \preccurlyeq_K \mathbf{0}, \mathbf{x} \in S\}$, where $S \subseteq \mathbb{R}^d$ is some additional set of constraints, possibly nonconvex. For example, one can impose integrality constraints on some variables (thus, obtain a mixed-integer convex optimization problem), or some other kind of constraints. The corresponding dual function is then defined as

$$\mathcal{L}(\mathbf{y}) = \inf_{\mathbf{x} \in S} f(\mathbf{x}) + \langle \mathbf{y}, G(\mathbf{x}) \rangle,$$

where we now have a *constrained* optimization problem in this dual. The motivation is that sometimes one can "decompose" the feasible region of the optimization problem into "easy" constraints modeled by S and "hard" constraints modeled by $G(\mathbf{x}) \preccurlyeq_K \mathbf{0}$, and the constrained problem $\inf_{\mathbf{x} \in S} f(\mathbf{x}) + \langle \mathbf{y}, G(\mathbf{x}) \rangle$ is still tractable since S is "easy" to handle. One can develop the entire theory in a similar way as done in this book, and conditions for strong duality/zero duality gap now involve S in addition to G and K. We recommend [170] where the setting with convex S is considered and refer to [74, 180, 212] when S is considered to be the set of all mixed-integer points in a polyhedron.

The use of disjunctions to provide optimality certificates in the general mixed-integer convex case originates in the branch-and-cut framework, as discussed in the paragraph below Theorem 7.2.2, although most of this work is focused on the setting with linear constraints and objectives. The idea of using the complements of (maximal) lattice-free polyhedra as the disjunctions for the general nonlinear mixed-integer convex problem first appeared in [19], with some follow-up work in [26], where it was shown that strong duality holds under some mild conditions on f and C, i.e., there always exists an optimality certificate given by disjunctions coming from the complements of maximal lattice-free polyhedra. Moreover, one can restrict attention to specially structured polyhedra whose defining halfspaces come from subgradient inequalities of the objective function evaluated at mixed-integer points on the boundary of the lattice-free polyhedra.

A completely different, algebraic approach to duality and optimality certificates for mixed-integer convex optimization was taken in [177]. Their approach is based on the theory of subadditive functions applied to mixed-integer optimization. This fascinating branch of mixed-integer optimization lies outside the scope of this book; the reader is referred to the surveys [29, 34, 35, 198] and the references therein for an entry into the topic.

Hints to Selected Exercises

Section 2.3.1

Exercise 1. Using the inner product inequality in Proposition 2.3.1, first establish that

$$\langle \mathbf{x} - \mathbf{y}, \mathrm{Proj}_C(\mathbf{x}) - \mathrm{Proj}_C(\mathbf{y}) \rangle \geq \| \mathrm{Proj}_C(\mathbf{x}) - \mathrm{Proj}_C(\mathbf{y}) \|^2$$

for all $\mathbf{x}, \mathbf{y} \in \mathbb{R}^d$.

Section 2.4.4

Exercise 1. Consider the set $C = Y - X = Y + (-1)X$.

Exercise 28. Let $\mathbf{x}$ be an extreme point of a closed, convex set C. Use Exercise 27 from Section 2.4.4 to show that for any $\epsilon > 0$, there exists a halfspace $H^{\leq}(\mathbf{a}, \delta)$ given by $\mathbf{a} \in \mathbb{R}^d \setminus \{\mathbf{0}\}$ and $\delta \in \mathbb{R}$ such that $\langle \mathbf{a}, \mathbf{x} \rangle < \delta$ and $C \cap H^{\leq}(\mathbf{a}, \delta) \subseteq C \cap B(\mathbf{x}, \epsilon)$. Next consider $\mathbf{y}$ of the form $\mathbf{y} = \mathbf{x} + \lambda \mathbf{a}$ with $\lambda > 0$ large enough such that the point $\mathbf{p}$ in C that is farthest in Euclidean distance to $\mathbf{y}$ is in $C \cap H^{\leq}(\mathbf{a}, \delta)$. Show that $\mathbf{p}$ is an exposed point of C (consider the ball $B(\mathbf{y}, \|\mathbf{p} - \mathbf{y}\|)$).

Section 2.6.1

Exercise 5. For each C_i, define a set $\tilde{C}_i$ as the set of all $\mathbf{u}$ such that $C_i \cap (X + \mathbf{u}) \neq \emptyset$.

Exercise 7. For each $\mathbf{y} \in T$, define the set $A_{\mathbf{y}} := \{(\mu_1, \ldots, \mu_k) \in \mathbb{R}^k : |g(\mathbf{y}) - \sum_{i=1}^{k} \mu_i f_i(\mathbf{y})| \leq \epsilon\}$.

Section 2.7.2

Exercise 8. Use the change of variables formula for integrals from multivariable calculus and the fact that a matrix with orthonormal columns has determinant 1 to establish part 3. Then derive parts 1. and 2. using the reasoning from the proof of Theorem 2.7.1.

Exercise 11. Consider the function that takes as input $\mathbf{v}, \mathbf{p}^1, \ldots, \mathbf{p}^d, \sigma_1, \ldots, \sigma_d,$ $\mathbf{x} \in \mathbb{R}^d$; computes the unique $\lambda_1, \ldots, \lambda_d \in \mathbb{R}$ such that $\mathbf{x} = \mathbf{v} + \sum_{i=1}^d \lambda_i \mathbf{p}^i$; and outputs $\frac{\lambda_1^2}{\sigma_1^2} + \cdots + \frac{\lambda_d^2}{\sigma_d^2}$.

Exercise 12. Use polars.

Section 3.2.4

Exercise 9. It suffices to prove the inequality for $\mathbf{x}, \mathbf{y} \in \mathbb{R}^d$ such that $\|\mathbf{x}\|_p = \|\mathbf{y}\|_q = 1$, which can be derived from Young's inequality (Exercise 8 from Section 3.2.4).

Section 3.4.1

Exercise 1. Use Exercise 15 from Section 3.3.5 and Theorem 3.4.3.

Section 3.5.4

Exercise 6. Instead of considering $\mathbf{y}^\star + \frac{1}{2}\left(\text{Proj}_{L^\perp}(C) - \mathbf{y}^\star\right)$, use $\mathbf{y}^\star + \alpha\left(\text{Proj}_{L^\perp}(C) - \mathbf{y}^\star\right)$ for every $\alpha \in [0,1]$ and the fact that $\int_0^1 (1 - t)^{d-k} k t^{k-1} dt = \binom{d}{k}^{-1}$ (using the so-called *beta functions*).

Section 4.1.1

Exercise 18. Sufficiency of the stated condition is clear. For the other direction, by considering arbitrary bases of Λ and Λ', first reduce to the case where $\Lambda = \mathbb{Z}^k$ and Λ' is a rank k sublattice of $\mathbb{Z}^k$. Let B be any $k \times k$ matrix with integer entries whose columns form a basis for Λ'. Show that it suffices to construct two unimodular matrices $U, V \in \mathbb{Z}^{k \times k}$ such that $BU = VD$ where $D \in \mathbb{Z}^{k \times k}$ is a diagonal matrix with diagonal entries $u_1, \ldots, u_k \geq 1$. Equivalently, it suffices to show the existence of unimodular matrices U and V such that $V^{-1}BU$ is an integer diagonal matrix with nonzero entries on the diagonal.

Consider the elementary column operations defined in Exercise 17 from Section 4.1.1. Define row operations of the same form. Show that one can perform these elementary column and row operations on any integer matrix B to convert it into a diagonal matrix.

Exercise 19. The Cauchy–Binet formula (Theorem 1.2.6) for determinants could be useful.

Section 4.2.1

Exercise 7. Consider the vector $\mathbf{v} = (\alpha_1, \ldots, \alpha_d)$ and use the type of reasoning from the proof of Lemma 4.2.7.

Exercise 8. Consider the subspace $\bigcap_{\epsilon > 0} \mathrm{span}(B(\mathbf{0}, \epsilon) \cap \mathrm{Proj}_L(\Lambda))$.

Section 4.3.3

Exercise 1. $\mathbf{z} \in \Lambda$ if and only if $\mathbf{z} = B\mathbf{y}$ for some $\mathbf{y} \in \mathbb{Z}^r$.

Exercise 5. $C + \mathbf{z} \subseteq (\lambda + 1)C$ for all $\mathbf{z} \in \lambda C \cap (\Lambda + \mathbf{t})$.

Exercise 6. Let $\mathbf{b}^1, \ldots, \mathbf{b}^k \in \mathbb{R}^d$ form a basis for Λ. Show that there exist linearly independent vectors $\mathbf{a}^1, \ldots, \mathbf{a}^k \in \mathrm{span}(\Lambda)$ such that $\langle \mathbf{a}^i, \mathbf{b}^j \rangle = \delta_{ij}$ for all $i, j \in \{1, \ldots, k\}$. Then show that $\Lambda^* = Z(\mathbf{a}^1, \ldots, \mathbf{a}^k)$.

Section 4.4.3

Exercise 2. There are multiple ways to prove this. Perhaps the shortest proof comes by employing Zorn's lemma (see, e.g., [145, Section 7]). But there are proofs that avoid the axiom of choice.

Exercise 5. Consider the set of lattice points in the relative interiors of facets (see Theorem 4.4.4) and revisit the argument made in the proof of Theorem 2.6.13.

Section 6.2.4

Exercise 10. Use the concavity of $\frac{f_{C,\mathbf{a}}(t)}{V(d-1)^{1/(d-1)}}$ and the fact that the function $p(t)$ from the proof of Theorem 6.2.5 is an affine function.

Section 6.3.3

Exercise 14. Use Exercise 1 from Section 4.3.3.

Exercise 17. Exercise 6 from Section 6.3.3 can be useful.

Exercise 18. Any vector $\lambda = (\lambda_n, \lambda_d) \in \mathbb{R}^n \times \mathbb{R}^d$ is equal to $B\tilde{\lambda} = B^{\star} R\tilde{\lambda}$, where $\tilde{\lambda} = \begin{bmatrix} \lambda_d \\ \lambda_n \end{bmatrix}$. If we require $\lambda \in \mathbb{Z}^n \times \mathbb{R}^d$, then $\lambda_n \in \mathbb{Z}^n$. If $\lambda \in L_Q^{\perp}$, then the first d coordinates of $R\tilde{\lambda}$ should be zero. Thus, $B^{\star} R\tilde{\lambda}$ is given by $B_n^{\star}(R\tilde{\lambda})_n$, where $(R\tilde{\lambda})_n$ is the vector in $\mathbb{R}^n$ formed by the last n coordinates of $R\tilde{\lambda}$. Finally, observe that $(R\tilde{\lambda})_n = R_{n \times n}\lambda_n$ because R is upper triangular.

Exercise 19. Gershgorin discs [219] can be useful which bound the largest eigenvalue by the largest sum of absolute values of entries in a row: apply to Q and Q^{-1} to bound $\lambda_{\max}(Q)$ and $\lambda_{\min}(Q)$.

Exercise 20. Use the fact $B^T Q B = R^T D R = (\sqrt{D}R)^T(\sqrt{D}R)$ using (6.3.10) and (6.3.12). Observe that, for $i = 1, \ldots, n$, the norm of the i-column of $\begin{bmatrix} \mathbf{0}_{d \times n} \\ \sqrt{D_{n \times n}}R_{n \times n} \end{bmatrix}$ is less than or equal to the norm of the $(i+d)$th column of $\sqrt{D}R$. $B^T Q B = (\sqrt{D}R)^T(\sqrt{D}R)$ implies that the $(i+d)$th column of $\sqrt{D}R$ has norm $N_Q(B_{i+d})$ which is at most $\sqrt{\lambda_{\max}(Q)}$, where B_{i+d} is the $(i+d)$th column of B.

References

[1] Fred G. Abramson. Effective computation over the real numbers. In *12th Annual Symposium on Switching and Automata Theory (SWAT 1971)*, pages 33–37. IEEE Computer Society, 1971.

[2] Divesh Aggarwal, Daniel Dadush, Oded Regev, and Noah Stephens-Davidowitz. Solving the shortest vector problem in 2^n time using discrete Gaussian sampling. In *Proceedings of the Forty-Seventh Annual ACM Symposium on Theory of Computing (STOC)*, pages 733–742. Association for Computing Machinery, 2015.

[3] Divesh Aggarwal, Daniel Dadush, and Noah Stephens-Davidowitz. Solving the closest vector problem in 2^n time – the discrete Gaussian strikes again! In *2015 IEEE 56th Annual Symposium on Foundations of Computer Science (FOCS)*, pages 563–582. IEEE, 2015.

[4] Divesh Aggarwal and Noah Stephens-Davidowitz. Just take the average! An embarrassingly simple 2^n-time algorithm for SVP (and CVP). *arXiv preprint arXiv:1709.01535*, 2017.

[5] Alfred V. Aho, John E. Hopcroft, and Jeffrey D. Ullman. *The Design and Analysis of Computer Algorithms*. Computer Science and Information Processing. Addison-Wesley Publishing Company, 1974.

[6] Miklós Ajtai. The shortest vector problem in L_2 is NP-hard for randomized reductions. In *STOC '98: Proceedings of the Thirtieth Annual ACM Symposium on Theory of Computing*, pages 10–19. Association for Computing Machinery, 1998.

[7] Miklós Ajtai, Ravi Kumar, and Dandapani Sivakumar. A sieve algorithm for the shortest lattice vector problem. In *Proceedings of the Thirty-Third Annual ACM Symposium on Theory of Computing (STOC)*, pages 601–610. Association for Computing Machinery, 2001.

[8] Miklos Ajtai, Ravi Kumar, and Dandapani Sivakumar. Sampling short lattice vectors and the closest lattice vector problem. In *CCC '02: Proceedings of the 17th IEEE Annual Conference on Computational Complexity*. IEEE Computer Society, 2002.

[9] Iskander Aliev, Robert Bassett, Jesús A. De Loera, and Quentin Louveaux. A quantitative Doignon-Bell-Scarf theorem. *Combinatorica*, 37(3):313–332, 2017.

[10] Nina Amenta, Jesús A. De Loera, and Pablo Soberón. Helly's theorem: new variations and applications. *arXiv preprint arXiv:1508.07606*, 2015.

[11] Kurt M. Anstreicher. On Vaidya's volumetric cutting plane method for convex programming. *Mathematics of Operations Research*, 22(1):63–89, 1997.

[12] Sanjeev Arora, László Babai, Jacques Stern, and Z. Sweedyk. The hardness of approximate optima in lattices, codes, and systems of linear equations. *Journal of Computer and System Sciences*, 54(2):317–331, 1997.

[13] Sanjeev Arora and Boaz Barak. *Computational Complexity: A Modern Approach*. Cambridge University Press, 2009.

[14] Gennadiy Averkov. On maximal S-free sets and the Helly number for the family of S-convex sets. *SIAM Journal on Discrete Mathematics*, 27(3):1610–1624, 2013.

[15] Gennadiy Averkov. A proof of Lovász's theorem on maximal lattice-free sets. *Beiträge zur Algebra und Geometrie*, 54(1):105–109, 2013.

[16] Gennadiy Averkov, Bernardo González Merino, Ingo Paschke, Matthias Schymura, and Stefan Weltge. Tight bounds on discrete quantitative Helly numbers. *Advances in Applied Mathematics*, 89:76–101, 2017.

[17] Gennadiy Averkov and Robert Weismantel. Transversal numbers over subsets of linear spaces. *Advances in Geometry*, 12(1):19–28, 2012.

[18] László Babai. On Lovász lattice reduction and the nearest lattice point problem. *Combinatorica*, 6(1):1–13, 1986.

[19] Michel Baes, Timm Oertel, and Robert Weismantel. Duality for mixed-integer convex minimization. *Mathematical Programming*, 158(1):547–564, 2016.

[20] Keith Ball. An elementary introduction to modern convex geometry. *Flavors of Geometry*, 31:1–58, 1997.

[21] Wojciech Banaszczyk. Inequalities for convex bodies and polar reciprocal lattices in R^n II: Application of K-convexity. *Discrete & Computational Geometry*, 16(3):305–311, 1996.

[22] Wojciech Banaszczyk, Alexander E. Litvak, Alain Pajor, and Stanislaw J. Szarek. The flatness theorem for nonsymmetric convex bodies via the local theory of Banach spaces. *Mathematics of Operations Research*, 24(3):728–750, 1999.

[23] Imre Bárány. *Combinatorial Convexity*. American Mathematical Society, 2021.

[24] Alexander Barvinok. *A Course in Convexity*. American Mathematical Society, 2002.

[25] Amitabh Basu. Complexity of optimizing over the integers. *Mathematical Programming, Series B*, 200:739–780, 2023.

[26] Amitabh Basu, Michele Conforti, Gérard Cornuéjols, Robert Weismantel, and Stefan Weltge. Optimality certificates for convex minimization and Helly numbers. *Operations Research Letters*, 45(6):671–674, 2017.

[27] Amitabh Basu, Michele Conforti, Gérard Cornuéjols, and Giacomo Zambelli. Maximal lattice-free convex sets in linear subspaces. *Mathematics of Operations Research*, 35:704–720, 2010.

[28] Amitabh Basu, Michele Conforti, Gérard Cornuéjols, and Giacomo Zambelli. Minimal inequalities for an infinite relaxation of integer programs. *SIAM Journal on Discrete Mathematics*, 24:158–168, February 2010.

[29] Amitabh Basu, Michele Conforti, and Marco Di Summa. A geometric approach to cut-generating functions. *Mathematical Programming*, 151(1):153–189, 2015.

[30] Amitabh Basu, Michele Conforti, Marco Di Summa, and Hongyi Jiang. Split cuts in the plane. *SIAM Journal on Optimization*, 31(1):331–347, 2021.

[31] Amitabh Basu, Michele Conforti, Marco Di Summa, and Hongyi Jiang. Complexity of branch-and-bound and cutting planes for mixed-integer optimization–II. *Combinatorica*, 42:971–996, 2022.

[32] Amitabh Basu, Michele Conforti, Marco Di Summa, and Hongyi Jiang. Complexity of branch-and-bound and cutting planes for mixed-integer optimization. *Mathematical Programming*, 198(1):787–810, 2023.

[33] Amitabh Basu, Gérard Cornuéjols, and Giacomo Zambelli. Convex sets and minimal sublinear functions. *Journal of Convex Analysis*, 18:427–432, 2011.

[34] Amitabh Basu, Robert Hildebrand, and Matthias Köppe. Light on the infinite group relaxation I: Foundations and taxonomy. *4OR*, 14(1):1–40, 2016.

[35] Amitabh Basu, Robert Hildebrand, and Matthias Köppe. Light on the infinite group relaxation II: Sufficient conditions for extremality, sequences, and algorithms. *4OR*, 14(2):1–25, 2016.

[36] Amitabh Basu, Hongyi Jiang, Phillip Kerger, and Marco Molinaro. Information complexity of mixed-integer convex optimization. In *International Conference on Integer Programming and Combinatorial Optimization*, pages 1–13. Springer, 2023.

[37] Amitabh Basu, Hongyi Jiang, Phillip Kerger, and Marco Molinaro. Information complexity of mixed-integer convex optimization. *https://arxiv.org/abs/2308.11153*, 2023.

[38] Amitabh Basu and Timm Oertel. Centerpoints: A link between optimization and convex geometry. *SIAM Journal on Optimization*, 27(2):866–889, 2017.

[39] Paul Beame, Noah Fleming, Russell Impagliazzo, Antonina Kolokolova, Denis Pankratov, Toniann Pitassi, and Robert Robere. Stabbing Planes. In Anna R. Karlin, editor, *9th Innovations in Theoretical Computer Science Conference (ITCS 2018)*, Leibniz International Proceedings in Informatics (LIPIcs), pages 10:1–10:20. Schloss Dagstuhl–Leibniz-Zentrum fuer Informatik, 2018.

[40] Michael J. Beeson. *Foundations of Constructive Mathematics: Metamathematical Studies*. Springer Science & Business Media, 2012.

[41] David E. Bell. A theorem concerning the integer lattice. *Studies in Applied Mathematics*, 56(2):187–188, 1977.

[42] Aharon Ben-Tal and Arkadii Nemirovski. *Lectures on Modern Convex Optimization*. MPS/SIAM Series on Optimization. Society for Industrial and Applied Mathematics (SIAM), 2001.

[43] Alberto Bertoni, Giancarlo Mauri, and Nicoletta Sabadini. Simulations among classes of random access machines and equivalence among numbers succinctly represented. In *North-Holland Mathematics Studies*, pages 65–89. Elsevier, 1985.

[44] Dimitri P. Bertsekas. *Convex Optimization Theory*. Athena Scientific, 2009.

[45] Dimitri P. Bertsekas. *Convex Optimization Algorithms*. Athena Scientific, 2015.

[46] Errett Bishop. *Foundations of Constructive Analysis*. McGraw-Hill, 1967.

[47] Robert E. Bixby. A brief history of linear and mixed-integer programming computation. *Documenta Mathematica*, 107–121, 2012.

[48] Robert E. Bixby, Mary Fenelon, Zonghao Gu, Ed Rothberg, and Roland Wunderling. Mixed integer programming: A progress report. In *The Sharpest Cut*, pages 309–325. MPS-SIAM Series on Optimization. Society for Industrial and Applied Mathematics, 2004.

[49] Robert G. Bland, Donald Goldfarb, and Michael J. Todd. The ellipsoid method: A survey. *Operations Research*, 29(6):1039–1091, 1981.

[50] Wilhelm Blaschke. Über affine Geometrie vii: Neue Extremeigenschaften von Ellipse und Ellipsoid. *Ber. Verh. Sächs. Akad. Wiss. Leipzig, Math.-Phys. Kl*, 69:306–318, 1917.

[51] Johannes Blömer. Closest vectors, successive minima, and dual HKZ-bases of lattices. In *Proceedings of 27th ICALP*, pages 248–259. Lecture Notes in Computer Science. Springer, 2002.

[52] Johannes Blömer and Stefanie Naewe. Sampling methods for shortest vectors, closest vectors and successive minima. *Theoretical Computer Science*, 410(18):1648–1665, 2009.

[53] Lenore Blum, Mike Shub, and Steve Smale. On a theory of computation and complexity over the real numbers: W-completeness, recursive functions and universal machines. *Bulletin of the American Mathematical Society*, 21(1):1–46, 1989.

[54] Alexander Bockmayr, Friedrich Eisenbrand, Mark Hartmann, and Andreas S. Schulz. On the Chvátal rank of polytopes in the 0/1 cube. *Discrete Applied Mathematics*, 98(1–2):21–27, 1999.

[55] Maria Bonet, Toniann Pitassi, and Ran Raz. Lower bounds for cutting planes proofs with small coefficients. *The Journal of Symbolic Logic*, 62(3):708–728, 1997.

[56] Allan Borodin and Ian Munro. *The Computational Complexity of Algebraic and Numeric Problems*. Elsevier, 1975.

[57] Jonathan Borwein and Adrian S. Lewis. *Convex Analysis and Nonlinear Optimization: Theory and Examples*. Springer Science & Business Media, 2010.

[58] Jean Bourgain and Vitali D. Milman. Sections Euclidiennes et volume des corps symétriques convexes dans $\mathbb{R}^n$. *CR Acad. Sci. Paris*, 300:435–437, 1985.

[59] Stephen Boyd and Lieven Vandenberghe. *Convex Optimization*. Cambridge University Press, 2004.

[60] Gábor Braun, Cristóbal Guzmán, and Sebastian Pokutta. Lower bounds on the oracle complexity of nonsmooth convex optimization via information theory. *IEEE Transactions on Information Theory*, 63(7):4709–4724, 2017.

[61] David Bremner, Dan Chen, John Iacono, Stefan Langerman, and Pat Morin. Output-sensitive algorithms for Tukey depth and related problems. *Statistics and Computing*, 18(3):259–266, 2008.

[62] Herbert Busemann. Volume in terms of concurrent cross-sections. *Pacific Journal of Mathematics*, 3:1–12, 1953.

[63] Samuel R. Buss and Peter Clote. Cutting planes, connectivity, and threshold logic. *Archive for Mathematical Logic*, 35(1):33–62, 1996.

[64] Andrew Caplin and Barry Nalebuff. On 64%-majority rule. *Econometrica: Journal of the Econometric Society*, 787–814, 1988.

[65] Constantin Carathéodory. Über den Variabilitätsbereich der Koeffizienten von Potenzreihen, die gegebene Werte nicht annehmen. *Mathematische Annalen*, 64(1):95–115, 1907.

[66] Yair Carmon. *The Complexity of Optimization Beyond Convexity*. Ph.D. thesis, Stanford University, August 2020.

[67] John William Scott Cassels. *An Introduction to the Geometry of Numbers*. Springer Science & Business Media, 1997.

[68] Timothy M Chan. An optimal randomized algorithm for maximum Tukey depth. In *Proceedings of the Fifteenth Annual ACM-SIAM Symposium on Discrete Algorithms*, pages 430–436. Society for Industrial and Applied Mathematics, 2004.

[69] Vašek Chvátal. Hard knapsack problems. *Operations Research*, 28(6):1402–1411, 1980.

[70] Vašek Chvátal. *Cutting-Plane Proofs and the Stability Number of a Graph, Report Number 84326-OR*. Institut für Ökonometrie und Operations Research, Universität Bonn, Bonn, 1984.

[71] Vašek Chvátal, William J. Cook, and Mark Hartmann. On cutting-plane proofs in combinatorial optimization. *Linear Algebra and Its Applications*, 114:455–499, 1989.

[72] Peter Clote. Cutting planes and constant depth Frege proofs. In *Proceedings of the Seventh Annual IEEE Symposium on Logic in Computer Science*, pages 296–307. IEEE, 1992.

[73] Michele Conforti, Gérard Cornuéjols, Aris Daniilidis, Claude Lemaréchal, and Jérôme Malick. Cut-generating functions and S-free sets. *Mathematics of Operations Research*, 40(2): 276–301, 2015.

[74] Michele Conforti, Gérard Cornuéjols, and Giacomo Zambelli. *Integer Programming*, volume 271. Berlin, Springer, 2014.

[75] Michele Conforti and Marco Di Summa. Maximal S-free convex sets and the Helly number. *SIAM Journal on Discrete Mathematics*, 30(4):2206–2216, 2016.

[76] William J. Cook, Collette R. Coullard, and Gy Turán. On the complexity of cutting-plane proofs. *Discrete Applied Mathematics*, 18(1):25–38, 1987.

[77] William J. Cook and Sanjeeb Dash. On the matrix-cut rank of polyhedra. *Mathematics of Operations Research*, 26(1):19–30, 2001.

[78] William J. Cook and Mark Hartmann. On the complexity of branch and cut methods for the traveling salesman problem. *Polyhedral Combinatorics*, 1:75–82, 1990.

[79] Thomas H. Cormen, Charles E. Leiserson, Ronald L. Rivest, and Clifford Stein. *Introduction to Algorithms*. MIT Press, 2022.

[80] Daniel Dadush. *Integer Programming, Lattice Algorithms, and Deterministic Volume Estimation*. Ph.D. thesis, Georgia Institute of Technology, August 2012.

[81] Daniel Dadush and Samarth Tiwari. On the complexity of branching proofs. In *Proceedings of the 35th Computational Complexity Conference*. Association for Computing Machinery, 2020.

[82] Ludwig Danzer, Branko Grünbaum, and Victor Klee. Helly's theorem and its relatives. In *Proceedings of Symposia in Pure Mathematics VII*, pages 101–180. American Mathematical Society, 1963.

[83] Sanjeeb Dash. An exponential lower bound on the length of some classes of branch-and-cut proofs. In *International Conference on Integer Programming and Combinatorial Optimization (IPCO)*, pages 145–160. Springer, 2002.

[84] Sanjeeb Dash. Exponential lower bounds on the lengths of some classes of branch-and-cut proofs. *Mathematics of Operations Research*, 30(3):678–700, 2005.

[85] Sanjeeb Dash. On the complexity of cutting-plane proofs using split cuts. *Operations Research Letters*, 38(2):109–114, 2010.

[86] Sanjeeb Dash, Neil B. Dobbs, Oktay Günlük, Tomasz J. Nowicki, and Grzegorz M. Świrszcz. Lattice-free sets, multi-branch split disjunctions, and mixed-integer programming. *Mathematical Programming*, 145(1–2):483–508, 2014.

[87] Sanjeeb Dash and Yatharth Dubey. On polytopes with linear rank with respect to generalizations of the split closure. *arXiv preprint arXiv:2110.04344*, 2021.

[88] Sanjeeb Dash and Oktay Günlük. On t-branch split cuts for mixed-integer programs. *Mathematical Programming*, 141(1–2):591–599, 2013.

[89] Alexandre d'Aspremont. Smooth optimization with approximate gradient. *SIAM Journal on Optimization*, 19(3):1171–1183, 2008.

[90] Mark De Berg, Marc van Kreveld, Mark Overmars, and Otfried Schwarzkopf. *Computational Geometry: Algorithms and Applications*. Springer Science & Business Media, 2000.

[91] Jesús A. De Loera, Xavier Goaoc, Frédéric Meunier, and Nabil Mustafa. The discrete yet ubiquitous theorems of Carathéodory, Helly, Sperner, Tucker, and Tverberg. *Bulletin of the American Mathematical Society*, 56(3):415–511, 2019.

[92] Jesús A. De Loera and Thomas Hogan. Stochastic Tverberg theorems with applications in multiclass logistic regression, separability, and centerpoints of data. *SIAM Journal on Mathematics of Data Science*, 2(4):1151–1166, 2020.

[93] Jesús A. De Loera, Reuben N. La Haye, Déborah Oliveros, and Edgardo Roldán-Pensado. Helly numbers of algebraic subsets of $\mathbb{R}^d$ and an extension of Doignon's theorem. *Advances in Geometry*, 17(4):473–482, 2017.

[94] Jesús A. De Loera, Reuben N. La Haye, David Rolnick, and Pablo Soberón. Quantitative Tverberg theorems over lattices and other discrete sets. *Discrete & Computational Geometry*, 58:435–448, 2017.

[95] Jesús A. De Loera, Sonja Petrović, and Despina Stasi. Random sampling in computational algebra: Helly numbers and violator spaces. *Journal of Symbolic Computation*, 77:1–15, 2016.

[96] Satyan L. Devadoss and Joseph O'Rourke. *Discrete and Computational Geometry*. Princeton University Press, 2011.

[97] Olivier Devolder, François Glineur, and Yurii Nesterov. First-order methods of smooth convex optimization with inexact oracle. *Mathematical Programming*, 146:37–75, 2014.

[98] Luc Devroye, László Györfi, and Gábor Lugosi. *A Probabilistic Theory of Pattern Recognition*. Springer Science & Business Media, 2013.

[99] Santanu S. Dey, Yatharth Dubey, and Marco Molinaro. Branch-and-bound solves random binary packing IPs in polytime. Mathematical Programming, 200:569–587, 2023.

[100] Santanu S Dey, Yatharth Dubey, and Marco Molinaro. Lower bounds on the size of general branch-and-bound trees. *Mathematical Programming*, 198(1):539–559, 2023.

[101] Santanu S. Dey, Yatharth Dubey, Marco Molinaro, and Prachi Shah. A theoretical and computational analysis of full strong-branching. *Mathematical Programming*, 1–34, 2023.

[102] Santanu S. Dey and Prachi Shah. Lower bound on size of branch-and-bound trees for solving lot-sizing problem. *Operations Research Letters*, 50(5):430–433, 2022.

[103] Santanu S. Dey and Laurence A. Wolsey. Constrained infinite group relaxations of MIPs. *SIAM Journal on Optimization*, 20(6):2890–2912, 2010.

[104] Marco Di Summa, 2019. Personal communication.

[105] Irit Dinur, Guy Kindler, Ran Raz, and Shmuel Safra. Approximating CVP to within almost-polynomial factors is NP-Hard. *Combinatorica*, 23:205–243, April 2003.

[106] J.-P. Doignon. Convexity in cristallographical lattices. *Journal of Geometry*, 3:71–85, 1973.

[107] David L. Donoho and Miriam Gasko. Breakdown properties of location estimates based on halfspace depth and projected outlyingness. *The Annals of Statistics*, 1803–1827, 1992.

[108] Rodney G. Downey and Michael R. Fellows. *Parameterized Complexity*. Springer Science & Business Media, 2012.

[109] Rodney G. Downey and Michael R. Fellows. *Fundamentals of Parameterized Complexity*. Springer, 2013.

[110] R. Dyckerhoff and Pavlo Mozharovskyi. Exact computation of halfspace depth. *arXiv preprint arXiv:1411.6927v2*, 2015.

[111] Friedrich Eisenbrand. Integer programming and algorithmic geometry of numbers. In M. Jünger, T. Liebling, D. Naddef, W. Pulleyblank, G. Reinelt, G. Rinaldi, and L. Wolsey, editors, *50 Years of Integer Programming 1958–2008*. Springer-Verlag, 2010.

[112] Friedrich Eisenbrand and Andreas S. Schulz. Bounds on the Chvátal rank of polytopes in the 0/1-cube. *Combinatorica*, 23(2):245–261, 2003.

[113] Friedrich Eisenbrand and Moritz Venzin. Approximate CVP_p in time $2^{0.802n}$. *Journal of Computer and System Sciences*, 124:129–139, 2022.

[114] Ivar Ekeland and Roger Temam. *Convex Analysis and Variational Problems*. Society for Industrial and Applied Mathematics, 1999.

[115] Jeff Erickson, Ivor Van Der Hoog, and Tillmann Miltzow. Smoothing the gap between NP and ER. *SIAM Journal on Computing*, 2022.

[116] Günter Ewald. *Combinatorial Convexity and Algebraic Geometry*. Springer Science & Business Media, 2012.

[117] Noah Fleming, Mika Göös, Russell Impagliazzo, Toniann Pitassi, Robert Robere, Li-Yang Tan, and Avi Wigderson. On the power and limitations of branch and cut. *arXiv preprint arXiv:2102.05019*, 2021.

[118] Harvey Friedman. Algorithmic procedures, generalized Turing algorithms, and elementary recursion theory. In *Logic Colloquium '69*, pages 361–389. Elsevier, 1971.

[119] Harry Furstenberg and Isaac Tzkoni. Spherical functions and integral geometry. *Israel Journal of Mathematics*, 10:327–338, 1971.

[120] Andreas Goerdt. Cutting plane versus Frege proof systems. In *International Workshop on Computer Science Logic*, pages 174–194. Springer, 1990.

[121] Andreas Goerdt. The cutting plane proof system with bounded degree of falsity. In *International Workshop on Computer Science Logic*, pages 119–133. Springer, 1991.

[122] Dima Grigoriev, Edward A. Hirsch, and Dmitrii V. Pasechnik. Complexity of semi-algebraic proofs. In *Annual Symposium on Theoretical Aspects of Computer Science (STACS)*, pages 419–430. Springer, 2002.

[123] Martin Grötschel, László Lovász, and Alexander Schrijver. *Geometric Algorithms and Combinatorial Optimization*, Algorithms and Combinatorics: Study and Research Texts. Springer-Verlag, 1988.

[124] Peter M. Gruber. *Convex and Discrete Geometry*, Grundlehren der Mathematischen Wissenschaften [Fundamental Principles of Mathematical Sciences]. Springer, 2007.

[125] Peter M. Gruber and Cornelis G. Lekkerkerker. *Geometry of Numbers*, Second edition, North-Holland Mathematical Library. North-Holland Publishing Co., Amsterdam, 1987.

[126] Branko Grünbaum. Partitions of mass-distributions and of convex bodies by hyperplanes. *Pacific Journal of Mathematics*, 10:1257–1261, 1960.

[127] Branko Grünbaum. Measures of symmetry for convex sets. In *Convexity: Proceedings of the Seventh Symposium in Pure Mathematics of the American Mathematical Society*, page 233. American Mathematical Society, 1963.

[128] Branko Grünbaum. *Convex Polytopes*. Springer Science & Business Media, 2013.

[129] Andrzej Grzegorczyk. Computable functionals. *Fundamenta Mathematicae*, 42:168–202, 1955.

[130] Heinrich W. Guggenheimer. A formula of Furstenberg-Tzkoni type. *Israel Journal of Mathematics*, 14(3):281–282, 1973.

[131] Osman Güler. *Foundations of Optimization*. Springer Science & Business Media, 2010.

[132] Paul R. Halmos. *Finite-Dimensional Vector Spaces*. Courier Dover Publications, 2017.

[133] Juris Hartmanis and Janos Simon. On the power of multiplication in random access machines. In *15th Annual Symposium on Switching and Automata Theory (swat 1974)*, pages 13–23. IEEE, 1974.

[134] Sebastian Heinz. Complexity of integer quasiconvex polynomial optimization. *Journal of Complexity*, 21(4):543–556, 2005.

[135] Eduard Helly. Über Mengen konvexer Körper mit gemeinschaftlichen Punkte. *Jahresbericht der Deutschen Mathematiker-Vereinigung*, 32:175–176, 1923.

[136] Eduard Helly. Uber Systeme abgeschlossenen Mengen mit gemeinschaftlichen Punkten. *Monatshefte für Mathematik*, 37:281–302, 1930.

[137] Martin Henk. Löwner-John ellipsoids. *Documenta Mathematica*, Extra volume ISMP:95–106, 2012.

[138] Robert Hildebrand and Matthias Köppe. A new Lenstra-type algorithm for quasiconvex polynomial integer minimization with complexity $2^{O(n \log n)}$. *Discrete Optimization*, 10(1):69–84, 2013.

[139] Michael Hintermüller. A proximal bundle method based on approximate subgradients. *Computational Optimization and Applications*, 20:245–266, 2001.

[140] Jean-Baptiste Hiriart-Urruty, and Claude Lemaréchal. *Convex Analysis and Minimization Algorithms I*, Grundlehren der mathematischen Wissenschaften. Springer-Verlag, 1993.

[141] Jean-Baptiste Hiriart-Urruty and Claude Lemaréchal. *Convex Analysis and Minimization Algorithms II*, Grundlehren der Mathematischen Wissenschaften. Springer-Verlag, 1993.

[142] Alan J. Hoffman. Binding constraints and Helly numbers. *Annals of the New York Academy of Sciences*, 319:284–288, 1979.

[143] Richard B. Holmes. *Geometric Functional Analysis and Its Applications*. Springer-Verlag, 1975.

[144] Rodney R. Howell. On asymptotic notation with multiple variables. *Tech. Rep.*, 2008.

[145] Thomas W. Hungerford. *Algebra*. Springer Science & Business Media, 2012.

[146] Russell Impagliazzo, Toniann Pitassi, and Alasdair Urquhart. Upper and lower bounds for tree-like cutting planes proofs. In *Proceedings Ninth Annual IEEE Symposium on Logic in Computer Science*, pages 220–228. IEEE, 1994.

[147] Robert G. Jeroslow. Trivial integer programs unsolvable by branch-and-bound. *Mathematical Programming*, 6(1):105–109, December 1974.

[148] Haotian Jiang, Yin Tat Lee, Zhao Song, and Sam Chiu-wai Wong. An improved cutting plane method for convex optimization, convex-concave games, and its applications. In *Proceedings of the 52nd Annual ACM SIGACT Symposium on Theory of Computing*, pages 944–953. Association for Computing Machinery, 2020.

[149] Fritz John. Extremum problems with inequalities as subsidiary conditions. *Studies and Essays*, Courant Anniversary Volume:187–204, 1948.

[150] Ravindran Kannan. Minkowski's convex body theorem and integer programming. *Mathematics of Operations Research*, 12(3):415–440, 1987.

[151] Ravindran Kannan and László Lovász. Covering minima and lattice-point-free convex bodies. *Annals of Mathematics*, 577–602, 1988.

[152] Leonid Khachiyan and Lorant Porkolab. Integer optimization on convex semialgebraic sets. *Discrete & Computational Geometry*, 23(2):207–224, 2000.

[153] David Kirkpatrick and Stefan Reisch. Upper bounds for sorting integers on random access machines. *Theoretical Computer Science*, 28(3):263–276, 1983.

[154] Krzysztof C. Kiwiel. A proximal bundle method with approximate subgradient linearizations. *SIAM Journal on Optimization*, 16(4):1007–1023, 2006.

[155] Ker-I Ko. *Complexity Theory of Real Functions*. Birkhauser Boston Inc., 1991.

[156] Gleb A. Koshevoy. The Tukey depth characterizes the atomic measure. *Journal of Multivariate Analysis*, 83(2):360–364, 2002.

[157] Jan Krajíček. Discretely ordered modules as a first-order extension of the cutting planes proof system. *The Journal of Symbolic Logic*, 63(4):1582–1596, 1998.

[158] Daniel Lacombe. Extension de la notion de fonction récursive aux fonctions d'une ou plusieurs variables réelles. *Comptes Rendus Académie des Sciences Paris*, 241:151–153, 1955.

[159] Jeffrey C. Lagarias, Hendrik W. Lenstra, Jr., and Claus-Peter Schnorr. Korkin-Zolotarev bases and successive minima of a lattice and its reciprocal lattice. *Combinatorica*, 10(4):333–348, 1990.

[160] Guanghui Lan. *Convex Optimization Under Inexact First-Order Information*. Georgia Institute of Technology, 2009.

[161] Ailsa H. Land and Alison G. Doig. An automatic method of solving discrete programming problems. *Econometrica*, 28(3):497–520, 1960.

[162] Lap Chi Lau. Convexity and optimization. *https://cs.uwaterloo.ca/~lapchi/cs798/notes.html*, 2017.

[163] Yin Tat Lee, Aaron Sidford, and Sam Chiu-wai Wong. A faster cutting plane method and its implications for combinatorial and convex optimization. In *2015 IEEE 56th Annual Symposium on Foundations of Computer Science*, pages 1049–1065. IEEE, 2015.

[164] Arjen K. Lenstra, Hendrik W. Lenstra, Jr., and László Lovász. Factoring polynomials with rational coefficients. *Mathematische Annalen*, 261(4):515–534, 1982.

[165] Hendrik W. Lenstra, Jr. Integer programming with a fixed number of variables. *Mathematics of Operations Research*, 8(4):538–548, 1983.

[166] Yanjun Li and Jean-Philippe P. Richard. Cook, Kannan and Schrijver's example revisited. *Discrete Optimization*, 5(4):724–734, 2008.

[167] Andrea Lodi. Mixed integer programming computation. In *50 Years of Integer Programming 1958–2008*, pages 619–645. Springer, 2010.

[168] László Lovász. *An Algorithmic Theory of Numbers, Graphs, and Convexity*, volume 50. Society for Industrial and Applied Mathematics, 1986.

[169] László Lovász. Geometry of numbers and integer programming. In M. Iri and K. Tanabe, editors, *Mathematical Programming: State of the Art*, pages 177–201. Mathematical Programming Society, 1989.

[170] David G. Luenberger. *Optimization by Vector Space Methods*. Wiley-Interscience, 1996.

[171] Erwin Lutwak. On some ellipsoid formulas of Busemann, Furstenberg and Tzkoni, Guggenheimer, and Petty. *Journal of Mathematical Analysis and Applications*, 159(1):18–26, 1991.

[172] Annie Marsden, Vatsal Sharan, Aaron Sidford, and Gregory Valiant. Efficient convex optimization requires superlinear memory. In *Conference on Learning Theory*, pages 2390–2430. Proceedings of Machine Learning Research., 2022.

[173] Jiri Matousek. *Lectures on Discrete Geometry*. Springer Science & Business Media, 2002.

[174] Daniele Micciancio and Panagiotis Voulgaris. A deterministic single exponential time algorithm for most lattice problems based on Voronoi cell computations. In *Proceedings of the 42nd ACM Symposium on Theory of Computing*, pages 351–358. Association for Computing Machinery, 2010.

[175] Daniele Micciancio and Panagiotis Voulgaris. A deterministic single exponential time algorithm for most lattice problems based on Voronoi cell computations. *SIAM Journal on Computing*, 42(3):1364–1391, 2013.

[176] Diego A. Morán R. and Santanu S. Dey. On maximal S-free convex sets. *SIAM Journal on Discrete Mathematics*, 25(1):379, 2011.

[177] Diego A. Morán R., Santanu S. Dey, and Juan Pablo Vielma. A strong dual for conic mixed-integer programs. *SIAM Journal on Optimization*, 22(3):1136–1150, 2012.

[178] Karl Mosler. Depth statistics. In *Robustness and Complex Data Structures*, pages 17–34. Springer, 2013.

[179] Mohammad Javad Naderi, Austin Buchanan, and Jose L. Walteros. Worst-case analysis of clique MIPs. *Mathematical Programming*, 195(1):517–551, 2022.

[180] George L. Nemhauser and Laurence A. Wolsey. *Integer and Combinatorial Optimization*. Wiley, 1988.

[181] Arkadii Nemirovski. Efficient methods in convex programming. Lecture Notes. www2.isye.gatech.edu/~nemirovs/Lect_EMCO.pdf, 1994.

[182] Arkadii S. Nemirovski and David B. Yudin. *Problem Complexity and Method Efficiency in Optimization*. John Wiley, 1983.

[183] Yuri Nesterov. Rounding of convex sets and efficient gradient methods for linear programming problems. *Optimization Methods Software*, 23(1):109–128, 2008.

[184] Yurii E. Nesterov. *Introductory Lectures on Convex Optimization*, Applied Optimization. Kluwer Academic Publishers, 2004.

[185] Jorge Nocedal and Stephen Wright. *Numerical Optimization*. Springer Science & Business Media, 2006.

[186] Timm Oertel. *Integer Convex Minimization in Low Dimensions*. Ph.D. thesis, Diss., Eidgenössische Technische Hochschule ETH Zürich, Nr. 22288, 2014.

[187] Gilles Pisier. *The Volume of Convex Bodies and Banach Space Geometry*. Cambridge University Press, 1999.

[188] Boris T. Polyak. *Introduction to Optimization*. Translations Series in Mathematics and Engineering. Optimization Software Inc. Publications Division, 1987.

[189] Marian Boykan Pour-El and Ian Richards. Computability and noncomputability in classical analysis. *Transactions of the American Mathematical Society*, 275(2):539–560, 1983.

[190] Vaughan R. Pratt, Michael O. Rabin, and Larry J. Stockmeyer. A characterization of the power of vector machines. In *Proceedings of the Sixth Annual ACM Symposium on Theory of Computing*, pages 122–134. Association for Computing Machinery, 1974.

[191] Franco P. Preparata and Michael I. Shamos. *Computational Geometry: An Introduction*. Springer Science & Business Media, 2012.

[192] Pavel Pudlák. Lower bounds for resolution and cutting plane proofs and monotone computations. *The Journal of Symbolic Logic*, 62(3):981–998, 1997.

[193] Pavel Pudlák. On the complexity of the propositional calculus. In *On the Complexity of the Propositional Calculus*, pages 197–218, London Mathematical Society Lecture Note Series. Cambridge University Press, 1999.

[194] Johann Radon. Uber eine Erweiterung des Begriffs der konvexen Functionen mit einer Anwendung auf die Theorei der Konvexen korper. *Sitzungsberichte der Wiener Akademie*, 125:241–258, 1916.

[195] Johann Radon. Mengen konvexer Körper, die einen gemeinsamen Punkt enthalten. *Mathematische Annalen*, 83(1–2):113–115, 1921.

[196] Alexander A. Razborov. On the width of semialgebraic proofs and algorithms. *Mathematics of Operations Research*, 42(4):1106–1134, 2017.

[197] Victor Reis and Thomas Rothvoss. The subspace flatness conjecture and faster integer programming. *arXiv preprint arXiv:2303.14605*, 2023.

[198] Jean-Philippe P. Richard and Santanu S. Dey. The group-theoretic approach in mixed integer programming. In Michael Jünger, Thomas M. Liebling, Denis Naddef, George L. Nemhauser, William R. Pulleyblank, Gerhard Reinelt, Giovanni Rinaldi, and Laurence A. Wolsey, editors, *50 Years of Integer Programming 1958–2008*, pages 727–801. Springer, 2010.

[199] R. Tyrell Rockafellar. *Convex Analysis*. Princeton Mathematical Series. Princeton University Press, 1970.

[200] Thomas Rothvoss. Asymptotic convex geometry. *https://sites.math.washington .edu/~rothvoss/lecturenotes/AsymptoticConvexGeometry.pdf*, 2021.

[201] Thomas Rothvoß and Laura Sanità. 0/1 polytopes with quadratic Chvátal rank. In *International Conference on Integer Programming and Combinatorial Optimization (IPCO)*, pages 349–361. Springer, 2013.

[202] Thomas Rothvoss and Moritz Venzin. Approximate CVP in time $2^{0.802n}$ – now in any norm! In *International Conference on Integer Programming and Combinatorial Optimization*, pages 440–453. Springer, 2022.

[203] Peter J. Rousseeuw and Ida Ruts. The depth function of a population distribution. *Metrika*, 49(3):213–244, 1999.

[204] Halsey L. Royden and Patrick M. Fitzpatrick. *Real Analysis*, Fourth edition. Prentice Hall, 2010.

[205] Mark Rudelson. Distances between non-symmetric convex bodies and the MM^*-estimate. *Positivity*, 4(2):161–178, 2000.

[206] Walter Rudin. *Principles of Mathematical Analysis*, Third edition. McGraw-Hill, 1976.

[207] Luis A. Santaló. Un invariante afin para los cuerpos convexos del espacio de n dimensiones. *Portugaliae Mathematica*, 8:155–161, 1949.

[208] Herbert E. Scarf. An observation on the structure of production sets with indivisibilities. *Proceedings of the National Academy of Sciences*, 74(9):3637–3641, 1977.

[209] Mark Schmidt, Nicolas Roux, and Francis Bach. Convergence rates of inexact proximal-gradient methods for convex optimization. *Advances in Neural Information Processing Systems*, 24, 2011.

[210] Rolf Schneider. *Convex Bodies: The Brunn-Minkowski Theory*, Encyclopedia of Mathematics and Its Applications. Cambridge University Press, 1993.

[211] Arnold Schönhage. On the power of random access machines. In *International Colloquium on Automata, Languages, and Programming*, pages 520–529. Springer, 1979.

[212] Alexander Schrijver. *Theory of Linear and Integer Programming*. John Wiley & Sons, 1986.

[213] Naum Z. Shor. Convergence rate of the gradient descent method with dilatation of the space. *Cybernetics*, 6(2):102–108, 1970.

[214] Naum Z. Shor. Utilization of the operation of space dilatation in the minimization of convex functions. *Cybernetics*, 6(1):7–15, 1972.

[215] Naum Z. Shor. Cut-off method with space extension in convex programming problems. *Cybernetics*, 13(1):94–96, 1977.

[216] Naum Z. Shor. *Minimization Methods for Non-Differentiable Functions*. Springer Series in Computational Mathematics. Springer, 1985.

[217] Carl Ludwig Siegel. *Lectures on the Geometry of Numbers*. Springer Science & Business Media, 1989.

[218] Josef Stoer and Christoph Witzgall. *Convexity and Optimization in Finite Dimensions I*. Springer Science & Business Media, 1970.

[219] Gilbert Strang. *Linear Algebra and Its Applications*. Thomson, Brooks/Cole, 2006.

[220] Jan van Tiel. *Convex Analysis*. John Wiley, 1984.

[221] Gabor Toth. *Measures of Symmetry for Convex Sets and Stability*. Springer, 2015.

[222] Joseph Frederick Traub, Grzegorz Włodzimierz Wasilkowski, and Henryk Woźniakowski. *Information, Uncertainty, Complexity*. Addison-Wesley, 1983.

[223] Joseph Frederick Traub, Grzegorz Włodzimierz Wasilkowski, and Henryk Woźniakowski. *Information-Based Complexity*. Academic Press, 1988.

[224] Alan Mathison Turing. On computable numbers, with an application to the "Entscheidungsproblem." *Proceedings of the London Mathematical Society*, 42(2):230–265, 1936.

[225] Alan Mathison Turing. On computable numbers, with an application to the "Entscheidungsproblem." a correction. *Proceedings of the London Mathematical Society*, 43(2):544–546, 1937.

[226] Pravin M. Vaidya. A new algorithm for minimizing convex functions over convex sets. *Mathematical Programming*, 73(3):291–341, 1996.

[227] Peter van Emde Boas. Another NP-complete problem and the complexity of computing short vectors in a lattice. Technical Report 81-04, Mathematisch Instituut, University of Amsterdam, 1981.

[228] Vladimir N. Vapnik and Alexey Y. Chervonenkis. On the uniform convergence of relative frequencies of events to their probabilities. *Theory of Probability & Its Applications*, 16(2):264–280, 1971.

[229] Cédric Villani. *Optimal Transport: Old and New*. Springer, 2009.

[230] Klaus Weihrauch. *Computable Analysis: An Introduction*. Springer Science & Business Media, 2000.

[231] Stephen J. Wright. Coordinate descent algorithms. *Mathematical Programming*, 151(1):3–34, 2015.

[232] David B. Yudin and Arkadii S. Nemirovskii. Evaluation of the informational complexity of mathematical programming problems. *Matekon*, 13(2):3–25, 1976.

[233] David B. Yudin and Arkadii S. Nemirovskii. Informational complexity and efficient methods for the solution of convex extremal problems. *Matekon*, 13(3):25–45, 1977.

[234] Günter M. Ziegler. *Lectures on Polytopes*. Springer, 1995.

Index